Methods in Molecular Biology

Series Editor
John M. Walker
School of Life and Medical Sciences
University of Hertfordshire
Hatfield, Hertfordshire, UK

For further volumes:
http://www.springer.com/series/7651

For over 35 years, biological scientists have come to rely on the research protocols and methodologies in the critically acclaimed *Methods in Molecular Biology* series. The series was the first to introduce the step-by-step protocols approach that has become the standard in all biomedical protocol publishing. Each protocol is provided in readily-reproducible step-by-step fashion, opening with an introductory overview, a list of the materials and reagents needed to complete the experiment, and followed by a detailed procedure that is supported with a helpful notes section offering tips and tricks of the trade as well as troubleshooting advice. These hallmark features were introduced by series editor Dr. John Walker and constitute the key ingredient in each and every volume of the *Methods in Molecular Biology* series. Tested and trusted, comprehensive and reliable, all protocols from the series are indexed in PubMed.

RNA Interference and CRISPR Technologies

Technical Advances and New Therapeutic Opportunities

Edited by

Mouldy Sioud

Department of Cancer Immunology, Institute for Cancer Research, Oslo University Hospital-Radiumhospitalet, Oslo, Norway

Humana Press

Editor
Mouldy Sioud
Department of Cancer Immunology
Institute for Cancer Research
Oslo University Hospital-Radiumhospitalet
Oslo, Norway

ISSN 1064-3745 ISSN 1940-6029 (electronic)
Methods in Molecular Biology
ISBN 978-1-0716-0289-8 ISBN 978-1-0716-0290-4 (eBook)
https://doi.org/10.1007/978-1-0716-0290-4

This Humana imprint is published by the registered company Springer Science+Business Media, LLC, part of Springer Nature.
The registered company address is: 1 New York Plaza, New York, NY 10004, U.S.A.

Preface

RNA interference (RNAi) is a natural cellular mechanism which silences gene expression in a sequence-specific manner. Additionally, it constitutes the primary natural plant immune system against viruses. This endogenous silencing pathway can be triggered by small interfering RNAs (siRNAs), short hairpin RNAs (shRNAs), microRNAs, as well as other non-coding RNAs. Although the technology has rapidly become one of the key methods used to experimentally regulate gene expression, RNAi-based therapy is challenging because of issues related to RNA stability, delivery, endosomal escape, and off-target effects. Therefore, new delivery approaches and chemical modifications are needed to overcome these challenges.

By analogy to RNAi-mediated plant protection against pathogen attack, the clustered regularly interspaced short palindromic repeats (CRISPR) loci and CRISPR-associated (Cas) nucleases provide acquired immunity against viruses and other mobile genetic elements in bacteria and archaea. These repeats were initially discovered in the 1980s in *E. coli* and remained mostly as mystery until 2007, when Barrangou and colleagues demonstrated that the *Streptococcus thermophilus* strains that survived initial phage challenge were resistant to subsequent phage challenge due to the incorporation of sequences from the phage genome into their CRISPR loci. The bacteria use the genetic information stored in these CRISPR loci to recognize and chop up phage DNA using endogenous nucleases such as Cas9. By feeding the Cas9 with an artificial RNA, Doudna and Charpentier ingeniously adapted the CRISPR-Cas9 system to edit any genome at any place they wanted. Since then, the technology has attracted attention for its promise in basic research and gene editing-based therapeutics, and it is continually being refined to increase effectiveness and reduce off-target effects. This book addresses important technical, translational research, and therapeutic aspects of RNAi and CRISPR technologies.

Chapter 1 reviews the discovery of these two interference tools and how they work. It also highlights several parameters such as design, delivery, and off-target effects that are important for successful experiments and therapeutic applications. Among the available methods for siRNA production, chemical synthesis remains the predominant and most reliable method for generating siRNAs because it provides the fastest production capability, the highest purity, and the easiest scalability for high-throughput strategies and clinical use. Advances in RNA delivery and chemical modification should also benefit researchers who are developing gene editing therapies based on CRISPR-Cas9 systems. In Chapters 2 and 3, experts describe a variety of chemical modifications that can be applied to siRNAs and CRISPR-guide RNAs (gRNAs). Chemically modified self-delivered siRNAs provide an alternative RNAi strategy because they enable siRNAs to enter cells without the need for a transfection reagent. Additionally, chemical modifications at the 2′-position of the sugar ring, such as 2′-O-Me are used to increase siRNA and gRNA stability, efficiency, and reduce off-target effects.

The efficacy and safety of siRNA drugs heavily depends on their delivery and regulation. Selectively delivering siRNAs to diseased cells or tissues increases their accumulation at the site of interest, thus increasing the silencing potency and limiting toxicity to normal tissues. A block of six chapters (Chapters 4–9) covers a variety of delivery and regulation strategies that are expected to facilitate the therapeutic translation of RNA therapeutics.

Shortly after its discovery, RNAi was recognized as a potential alternative to traditional antiviral therapy, which limits viral infection via direct silencing of viral genes or host genes involved in virus recognition and/or replication. Chapter 10 describes the in vivo expression of anti-hepatitis B virus (HBV) microRNA inhibitors using an ancestral adeno-associated viral vector. Chronic hepatitis B, a liver disease resulting from persisting HBV infection, remains a global health challenge despite availability of an effective vaccine. Similarly, a protocol for the design, generation, and functional assessment of DNA cassettes encoding single and multimeric pri-miRNA mimics targeting HBV is described in Chapter 11. In Chapter 12, the authors describe an effective strategy to express the intronic microR-302 cluster in human-induced pluripotent stem cells (iPSCs), resulting in the regulation of their tumorigenic potential. In addition to gene regulation, a significant amount of miRNAs were detected in all biological fluids including blood plasma, urine, and cerebrospinal fluids. Chapter 13 describes the recent technical aspects of urine microRNA detection and their usage as clinical biomarkers in various human diseases.

Cancer is one of the most common leading causes of death in men and women worldwide. Radiation therapy or chemotherapy is the major treatment for most of the cancers. However, due to severe side effects, most of the cancer patients stumble to death. Immunotherapy is an alternative innovative therapy for cancer, which uses cancer patient's own body cells to stimulate their own immune system to kill cancer cells. In this respect, research has been trying to develop DC vaccines to fight cancer for decades, but this has proven to be harder than was first thought because of the immunosuppressive factors expressed by DCs and/or tumor cells. Targeting these factors using siRNA blockers in combination with DC vaccines or adoptive T cell therapies (ACT) are now being used to treat cancer patients. One of the most well-established siRNA delivery methods for DCs is electroporation (*see* Chapter 7). While Chapter 14 describes the generation of siRNA-programmable DC vaccines for clinical use, Chapter 15 reviews the recent advances in incorporating checkpoint siRNA blockers in DC vaccines and adoptive T cell therapy to enhance cancer immunotherapy. This strategy is relatively mature, with multiple clinical trials completed or ongoing.

DC cancer vaccines are often given along other substances (called adjuvants) that help boost the body's immune response, which might help the vaccines work better. Chapter 16 describes a new lipid-immunostimulatory siRNA formulation that activates DCs and simultaneously blocks the expression of immunosuppressive factors such as interleukin-10, resulting in a better response to the vaccine. Although abnormalities of cancer genes are essential contributors to cancer, tumor microenvironment also plays an important role in the initiation and progression of solid tumors. Several cell types of myeloid origin such as macrophages have been implicated in various human disorders including cancer. Hence, they represent an attractive therapeutic target. Chapter 17 reviews the current targeting strategies, including siRNAs, to modulate or kill tumor-associated macrophages.

The genetic retargeting of T cells to cancer can be achieved either by transduction with a chimeric antigen receptor (CAR) or a T-cell receptor (TCR) specific for an antigen of choice. The use of a transgenic TCR in primary patient-autologous T cells is hampered by the presence of endogenous TCRs in these cells, leading to the formation of the so-called mixed TCR dimers. Mixed TCR dimers can exhibit autoreactivity. In Chapter 18, the authors describe a new strategy for improving the expression and reactivity of transduced tumor-specific TCRs in human lymphocytes by specific gene silencing of endogenous TCRs.

The discovery and characterization of CRISPR and Cas genes provided a genetic tool that was previously unavailable. A critical step in doing gene editing experiments using the

CRISPR-Cas9 system is the design of guide RNA (gRNA). Chapter 19 describes the state-of-the-art design rules for maximizing on-target Cas9 activity while minimizing off-target activity. Chapter 20 details strategies for gene editing using donor vector made by Gibson assembly, a well-known molecular technique for allowing easy assembly of multiple DNA fragments with region homology at their ends, which are typically created by PCR. The authors used Cas9D10A to minimize off-target effects. It should be noted that efforts are continually being made to make CRISPR-Cas9 genome editing more precise and efficient. In this respect, multiple design or selection strategies have been applied to develop several Cas9 version proteins with reduced off-target effects. In Chapter 21, the authors describe the development of new high-fidelity Cas9 expression vectors that will enable scientists to perform more accurate genome editing.

With respect to adoptive T cell therapy, the majority of investigators have used autologous cells, which impose significant manufacturing, logistical, and cost issues. Moreover, many patients may not be eligible for treatment because of low T cell numbers and poor T cell quality. Hence, the use of allogeneic modified T cells derived from healthy donors could potentially address these limitations. However, the expression of endogenous TCRs by allogeneic modified T cells is a major problem because such T cells have the potential to induce graft versus host disease. The solution to this challenge is to eliminate endogenous TCR and/or HLA molecules via gene editing to generate safe “universal” donor T cells. In Chapters 22 and 23, the authors review CRISPR gene editing in T cells to improve their reactivity or generate off-the-shelf products for universal use. Similarly, Chapter 24 describes methods for gene editing and genome engineering of blood B cells. Interestingly, high editing frequencies of up to 90% were achieved. Reprogramming the antibody specificity of peripheral blood B cells could have clinical importance. Chapter 25 also describes methods for gene editing using CRIPR-Cas9 in lymphoma cell lines.

Human-induced pluripotent stem cells (iPSCs) are reprogrammed from somatic cells and are self-renewal cells that can produce different types of cells. This iPSC technology has created new therapeutic opportunities for a wide range of human diseases. The introduction of specific mutations in human iPSCs or hematopoietic stem cells is significant because these cells carry the potential to be differentiated into any lineage. In Chapter 26, the authors describe an optimized CRISPR-Cas9 RNP nucleofection protocol for gene knockout in hematopoietic stem and progenitor cells (HSPCs). A protocol for gene editing in patient-derived iPSC is described in Chapter 27. Both protocols walk through the step-by-step process of knocking out or correcting a gene of interest in cell culture to produce the desired cell pool that can be further differentiated for therapeutic use.

In conclusion, this book provides a helpful compilation of protocols and reviewers that will be of use not only to those with experience in RNAi and CRISPR technologies, but also to those wishing to use these two technologies for the first time. Additionally, this collection of chapters will be a great asset and reference book for young scientists, who are beginning their research on these novel technologies and who are interested in translational research and medicine.

Finally, I would like to thank the authors for their contributions, the series editors, John Walkers, and all those involved in the production of the book at Springer Nature.

Oslo, Norway *Mouldy Sioud*

Contents

Contributors

DIETMAR ABRAHAM • *Division of Cell and Developmental Biology, Center for Anatomy and Cell Biology, Medical University of Vienna, Vienna, Austria*

PATRICK ARBUTHNOT • *Wits/SAMRC Antiviral Gene Therapy Research Unit, School of Pathology, Health Sciences Faculty, University of the Witwatersrand, Johannesburg, South Africa*

BAOYAN BAI • *Department of Cancer Immunology, Institute for Cancer Research, Oslo University Hospital, Oslo, Norway; KG. Jebsen Centre for B cell Malignancies, University of Oslo, Oslo, Norway*

MARK A. BEHLKE • *Integrated DNA Technologies, Inc., Coralville, IA, USA*

JASON N. BERMAN • *CHEO Research Institute, Ottawa, ON, Canada*

LUCIA CACERES • *AGADA Biosciences, Halifax, NS, Canada*

NICHOLAS PAUL CASEY • *Department of Cancer Immunology, Institute for Cancer Research, Oslo University Hospital-Radiumhospitalet, Oslo, Norway*

CHANGMAI CHEN • *State Key Laboratory of Natural and Biomimetic Drug, Center of Noncoding RNA Medicine, The School of Pharmaceutical Sciences, Beijing, China*

JACK S. CHEN • *WJWU & LYNN Institute for Stem Cell Research, Santa Fe Springs, CA, USA*

YUHAO CHEN • *Department of Radiation Oncology, Washington University School of Medicine, St. Louis, MO, USA*

CHEN CHENG • *State Key Laboratory of Stem Cell and Reproductive Biology, Institute of Zoology, Chinese Academy of Sciences, Beijing, China; Institute for Stem Cell and Regeneration, Chinese Academy of Sciences, Beijing, China; School of Life Sciences, University of Science and Technology of China, Hefei, China*

IVAN V. CHERNIKOV • *Institute of Chemical Biology and Fundamental Medicine SB RAS, Novosibirsk, Russia*

ELENA L. CHERNOLOVSKAYA • *Institute of Chemical Biology and Fundamental Medicine SB RAS, Novosibirsk, Russia*

ANNA CORDEIRO-SANTANACH • *Department of Pediatrics, Dalhousie University, Halifax, NS, Canada*

VICTORIA CUELLO • *Department of Biology, University of Texas—Rio Grande Valley, Edinburg, TX, USA*

BENJAMIN DANNENMANN • *Department of Oncology, Hematology, Immunology, and Rheumatology, University Hospital Tuebingen, Tuebingen, Germany*

JAFAR EZZATI NAZHAD DOLATABADI • *Drug Applied Research Center, Tabriz University of Medical Sciences, Tabriz, Iran*

ABDULLAH ELY • *Wits-SAMRC Antiviral Gene Therapy Research Unit, Department of Molecular Medicine & Haematology, Faculty of Health Sciences, University of the Witwatersrand, Johannesburg, South Africa*

HIROSHI FUJIWARA • *Department of Personalized Cancer Immunotherapy, Mie University Graduate School of Medicine, Tsu, Mie, Japan*

MARYAM GHAFFARI • *Immunology Research Center, Tabriz University of Medical Sciences, Tabriz, Iran*

MICHAEL R. HAMBLIN • *Wellman Center for Photomedicine, Massachusetts General Hospital, Boston, MA, USA; Department of Dermatology, Harvard Medical School, Boston, MA, USA; Laser Research Centre, Faculty of Health Science, University of Johannesburg, Doornfontein, South Africa*

KASIDET HIRANNIRAMOL • *Department of Radiation Oncology, Washington University School of Medicine, St. Louis, MO, USA*

PER OLE IVERSEN • *Department of Hematology, Oslo University Hospital-Rikshospitalet, Oslo, Norway*

YANG JIN • *Division of Pulmonary and Critical Care Medicine, Department of Medicine, Boston University, Boston, MA, USA*

NANNAN JING • *State Key Laboratory of Natural and Biomimetic Drug, Center of Noncoding RNA Medicine, The School of Pharmaceutical Sciences, Beijing, China*

MATTHEW J. JOHNSON • *Department of Pediatrics, Cardiovascular Research, Masonic Cancer Center, Center for Genome Engineering, University of Minnesota, Minneapolis, MN, USA*

JAROSLAV JURACEK • *Central European Institute of Technology, Masaryk University, Brno, Czech Republic*

MEGAN KENIRY • *Department of Biology, University of Texas—Rio Grande Valley, Edinburg, TX, USA*

SHIVA KHEIRIABAD • *Immunology Research Center, Tabriz University of Medical Sciences, Tabriz, Iran*

MAKSIM KLIMIANKOU • *Translational Oncology Division, Internal Medicine II, University Hospital Tuebingen, Tuebingen, Germany*

JON AMUND KYTE • *Department of Cancer Immunology, Institute for Cancer Research, Oslo University Hospital-Radiumhospitalet, Oslo, Norway*

KANUT LAOHARAWEE • *Department of Pediatrics, Cardiovascular Research, Masonic Cancer Center, Center for Genome Engineering, University of Minnesota, Minneapolis, MN, USA*

BENEDICT LAW • *Department of Radiology, Weill Cornell Medicine, Molecular Imaging Innovations Institute, New York, NY, USA*

HEEDOO LEE • *Division of Pulmonary and Critical Care Medicine, Department of Medicine, Boston University, Boston, MA, USA*

SEUNG KOO LEE • *Department of Radiology, Weill Cornell Medicine, Molecular Imaging Innovations Institute, New York, NY, USA*

KIM A. LENNOX • *Integrated DNA Technologies, Inc., Coralville, IA, USA*

DUANWEI LIANG • *State Key Laboratory of Natural and Biomimetic Drug, Center of Noncoding RNA Medicine, The School of Pharmaceutical Sciences, Beijing, China*

SHI-LUNG LIN • *WJWU & LYNN Institute for Stem Cell Research, Santa Fe Springs, CA, USA*

CARL LITIF • *Department of Biology, University of Texas—Rio Grande Valley, Edinburg, TX, USA*

MOHUBE BETTY MAEPA • *Wits/SAMRC Antiviral Gene Therapy Research Unit, School of Pathology, Health Sciences Faculty, University of the Witwatersrand, Johannesburg, South Africa*

MARIYA I. MESCHANINOVA • *Institute of Chemical Biology and Fundamental Medicine SB RAS, Novosibirsk, Russia*

PERIHAN MIR • *Translational Oncology Division, Internal Medicine II, University Hospital Tuebingen, Tuebingen, Germany*

NJABULO MNYANDU • *Wits/SAMRC Antiviral Gene Therapy Research Unit, School of Pathology, Health Sciences Faculty, University of the Witwatersrand, Johannesburg, South Africa*
ANNE MOBERGSLIEN • *Department of Immunology, Institute for Cancer Research, Oslo University Hospital-Radiumhospitalet, Oslo, Norway*
BRANDEN S. MORIARITY • *Department of Pediatrics, Cardiovascular Research, Masonic Cancer Center, Center for Genome Engineering, University of Minnesota, Minneapolis, MN, USA*
JULIE A. MUSTARD • *Department of Biology, University of Texas—Rio Grande Valley, Edinburg, TX, USA*
JUNE HELEN MYKLEBUST • *Department of Cancer Immunology, Institute for Cancer Research, Oslo University Hospital, Oslo, Norway; KG. Jebsen Centre for B cell Malignancies, University of Oslo, Oslo, Norway*
MASOUD NASRI • *Department of Oncology, Hematology, Immunology, and Rheumatology, University Hospital Tuebingen, Tuebingen, Germany*
SERGEY V. PRYKHOZHIJ • *CHEO Research Institute, Ottawa, ON, Canada*
MALTE RITTER • *Translational Oncology Division, Internal Medicine II, University Hospital Tuebingen, Tuebingen, Germany*
STEIN SÆBØE-LARSSEN • *Department of Cellular Therapy, Institute for Cancer Research, Oslo University Hospital-Radium, Ullernchausseen, Norway*
NIRAKAR SAHOO • *Department of Biology, University of Texas—Rio Grande Valley, Edinburg, TX, USA*
MOULDY SIOUD • *Department of Cancer Immunology, Institute for Cancer Research, Oslo University Hospital-Radiumhospitalet, Oslo, Norway*
JULIA SKOKOWA • *Translational Oncology Division, Internal Medicine II, University Hospital Tuebingen, Tuebingen, Germany; Department of Oncology, Hematology, Immunology, and Rheumatology, University Hospital Tuebingen, Tuebingen, Germany*
ONDREJ SLABY • *Central European Institute of Technology, Masaryk University, Brno, Czech Republic*
WEN SUN • *State Key Laboratory of Stem Cell and Reproductive Biology, Institute of Zoology, Chinese Academy of Sciences, Beijing, China; Institute for Stem Cell and Regeneration, Chinese Academy of Sciences, Beijing, China*
XINJING TANG • *State Key Laboratory of Natural and Biomimetic Drug, Center of Noncoding RNA Medicine, The School of Pharmaceutical Sciences, Beijing, China*
CHING-HSUAN TUNG • *Department of Radiology, Weill Cornell Medicine, Molecular Imaging Innovations Institute, New York, NY, USA*
SHREYA UDAWANT • *Department of Biology, University of Texas—Rio Grande Valley, Edinburg, TX, USA*
FIONA T. VAN DEN BERG • *Wits-SAMRC Antiviral Gene Therapy Research Unit, Department of Molecular Medicine & Haematology, Faculty of Health Sciences, University of the Witwatersrand, Johannesburg, South Africa*
SÉBASTIEN WÄLCHLI • *Department of Cellular Therapy, Department of Oncology, Oslo University Hospital-Radiumhospitalet, Oslo, Norway*
HAOYI WANG • *State Key Laboratory of Stem Cell and Reproductive Biology, Institute of Zoology, Chinese Academy of Sciences, Beijing, China; University of Chinese Academy of Sciences, Beijing, China; Institute for Stem Cell and Regeneration, Chinese Academy of Sciences, Beijing, China*

XIAOWEI WANG • *Department of Radiation Oncology, Washington University School of Medicine, St. Louis, MO, USA*
KARL WELTE • *Translational Oncology Division, Internal Medicine II, University Hospital Tuebingen, Tuebingen, Germany; Department of General Paediatrics, Oncology/Haematology, The University Children's Hospital Tuebingen, Tuebingen, Germany; Department of Oncology, Hematology, Immunology, and Rheumatology, University Hospital Tuebingen, Tuebingen, Germany*
SHAO-YAO YING • *Department of Integrative Anatomical Sciences, Keck School of Medicine, University of Southern California, Los Angeles, CA, USA*
LIJIA YU • *National Center for Occupational Safety and Health, National Health Commission of the People's Republic of China, Beijing, China*
DUO ZHANG • *Division of Pulmonary and Critical Care Medicine, Department of Medicine, Boston University, Boston, MA, USA*
XINGYING ZHANG • *State Key Laboratory of Stem Cell and Reproductive Biology, Institute of Zoology, Chinese Academy of Sciences, Beijing, China; University of Chinese Academy of Sciences, Beijing, China; Institute for Stem Cell and Regeneration, Chinese Academy of Sciences, Beijing, China*
KARIN ZINS • *Division of Cell and Developmental Biology, Center for Anatomy and Cell Biology, Medical University of Vienna, Vienna, Austria*

Chapter 1

RNA and CRISPR Interferences: Past, Present, and Future Perspectives

Mouldy Sioud

Abstract

RNA interference (RNAi), a natural gene silencing process, is a widely used technique in basic research, preclinical studies, and drug development strategies. Although the technique has great potential to generate new human therapies and treat undruggable diseases, the clinical application of RNAi is still challenging primarily because of the delivery problem and potential off-target effects. Over the past two decades, great efforts have been undertaken to develop delivery agents and chemical modifications to overcome these challenges. Such advances in RNA delivery and chemical modifications have benefited researchers who are developing gene-editing therapies based on CRISPR-Cas9, an RNA-guided endonuclease, which is already having a major impact on biology and medicine. Here, I review the discovery of these two interference tools, identify the technical challenges yet to be overcome and provide some perspectives on how these two RNA-based technologies can be harnessed to treat human diseases.

Key words RNA interference, CRISPR-Cas9, Innate immunity, Gene regulation, Off-target effects, Chemical modifications, Gene therapy

1 Introduction

RNA interference (RNAi) is a posttranscriptional gene regulation mechanism by which small interfering RNAs (siRNA) induce the sequence-specific degradation of homologous messenger RNA (mRNA). With respect to discovery, Napoli and Jorgensen were the first to report an RNAi type of phenomenon in plants in 1990 [1]. RNA silencing was first documented in animals by Guo and Kemphues, who observed that the introduction of sense or antisense RNA to par-1 mRNA resulted in degradation of par-1 message in *Caenorhabditis elegans* [2]. At that time, antisense was one of the most attractive strategies of inhibiting gene expression. Surprisingly, when Guo and colleagues performed control experiments using only the sense part-1 RNA, which would not hybridize with the endogenous part-1 transcript, the part-1 message was still targeted for degradation.

Mouldy Sioud (ed.), *RNA Interference and CRISPR Technologies: Technical Advances and New Therapeutic Opportunities*, Methods in Molecular Biology, vol. 2115, https://doi.org/10.1007/978-1-0716-0290-4_1,

In 1998, Fire and Mello showed in the worm *Caenorhabditis elegans* that double-stranded RNA (dsRNA), but not single-stranded RNA (ssRNA), was involved in gene silencing [3]. Their initial idea was that the inconsistent finding of Guo and Kemphues showing that introduction of sense RNA leads to gene silencing could have been due to the contamination of preparation of ssRNA by dsRNA resulting from the activity of bacteriophage RNAs polymerases. Therefore, the authors extensively purified the sense and antisense RNA preparations and then directly compared their effects on the unc-22 gene. While their work established a new concept for the effects of RNA on gene silencing by highlighting a role for dsRNA, a number of questions remained regarding the mechanism by which dsRNAs trigger the degradation of endogenous mRNAs [3].

It was later demonstrated that dsRNA was converted to shorter intermediates, small interfering RNAs, capable of binding to their homologous target mRNAs, resulting in the cleavage of the transcript. Using *Drosophila* cell extracts, Tuschl and colleagues showed that chemically synthesized 21–23-nt dsRNAs can mediate the cleavage of the target mRNAs [4, 5]. These fragments were termed small interfering (si) RNAs and proved to be the main effectors of RNAi induction in mammalian cells [6]. Up to this point, the use of RNAi was limited to flies, worms, and plants, as the introduction of long dsRNA into mammalian cells induces an interferon response that causes a general inhibition of translation abrogating the specificity of RNAi. However, the finding by Tuschl and colleagues that short dsRNAs can bypass the interferon pathway facilitated the use of RNAi in mammalian cells.

Analysis of RNAi in various species revealed that RNAi is a widespread natural phenomenon that is conserved across fungi, plants, and animals. RNAi is induced not only by exogenous siRNAs but also by several endogenous small RNA species such as microRNAs (miRNAs) and piwi-interacting RNAs (piRNAs) [7]. RNAi is involved in several processes such as cell growth, tissue differentiation, heterochromatin formation, and cell proliferation. Accordingly, RNAi dysfunction is linked to cardiovascular diseases, neurological disorders, and cancers [8, 9]. Although the use of synthetic siRNAs shows great therapeutic promise, challenges with delivery as well as harmful off-target effects need to be resolved. Notably, the recent approval by the US Food and Drug Administration on August 10, 2018, for the first RNAi drug (patisiran) is encouraging. Patisiran works by silencing the gene that underlies a rare disease called hereditary transthyretin amyloidosis where mutated forms of the protein transthyretin accumulate in the body, sometimes impairing heart and nerve function.

2 Mechanisms of RNAi

The sequence-specific silencing of cognate genes by RNAi can be triggered by small interfering RNAs (siRNAs), short hairpin RNAs (shRNAs), microRNAs (miRNAs), as well as other noncoding RNAs such as piRNAs [10]. piRNAs are a large class of small non-coding RNAs, 26–31 nucleotides in length that function in the germline to silence genetic elements such as retrotransposons. The cellular origins of miRNAs and siRNA are somewhat different: miRNAs are encoded in the genome, whereas siRNAs may be endogenous or arise via viral infection or other exogenous sources [9]. miRNA genes are transcribed by RNA polymerase II into a long noncoding RNA known as the primary miRNA (pri-- miRNA) that is processed by the ribonuclease III enzyme Drosha into 70-nt pri-miRNA [9, 11]. Following its export from the nucleus by exportin-5, the pre-miRNA is further processed by cytoplasmic Dicer and TAR RNA binding protein (TRBP) into a functional miRNA duplex, which then assembles with Argonaute 2 (Ago2) protein to form a miRNA-containing ribonucleoprotein complex, after which the mature miRNA strand of 21–23 is selected (Fig. 1). Depending on the homology to mRNAs, miRNA can lead either to mRNA cleavage and degradation (complete homology) or to translation inhibition (partial homology). In mammalian cells, miRNA-RISC complexes are frequently targeted to the 3′ untranslated regions of mRNAs, whereas in plant, miRNAs exhibit complete homology with their corresponding target genes [11].

SiRNAs are very similar to miRNAs and are widely used as research tools for sequence-specific gene inactivation. Depending on the organism, siRNAs are either introduced as short dsRNAs by transfection or long dsRNAs. With respect to long dsRNAs, the process starts by breaking down the RNA into 21–23-nt shorter RNA duplexes by Dicer, a RNAse III family member. Like all RNAse III enzyme, Dicer leaves 2-nt 3′-overhangs and 5′ phosphate groups. These siRNA duplexes are then incorporated into a multiprotein complex, the RNA-induced silencing complex known as RISC. In *Drosophila*, two distinct enzymes are involved in parallel pathways for small regulatory RNA biogenesis. Dicer-2 processes long-double-stranded precursors and generates siRNAs, while Dicer-1 processes pre-miRNAs into mature miRNAs [12, 13]. Mammals possess only a single species of Dicer in association with the dsRNA-binding protein (dsRBP) such as TRBP protein, which is required for the recruitment of Ago2 to the siRNA-Dicer complex. The tree proteins (Dicer, Argonaute, and dsRBP) constitute a minimal RISC-loading complex. Subsequent to strand separation based on asymmetric unwinding (see below), the antisense strand guides the activated RISC to the complementary site in the target mRNA, resulting in mRNA cleavage (Fig. 2).

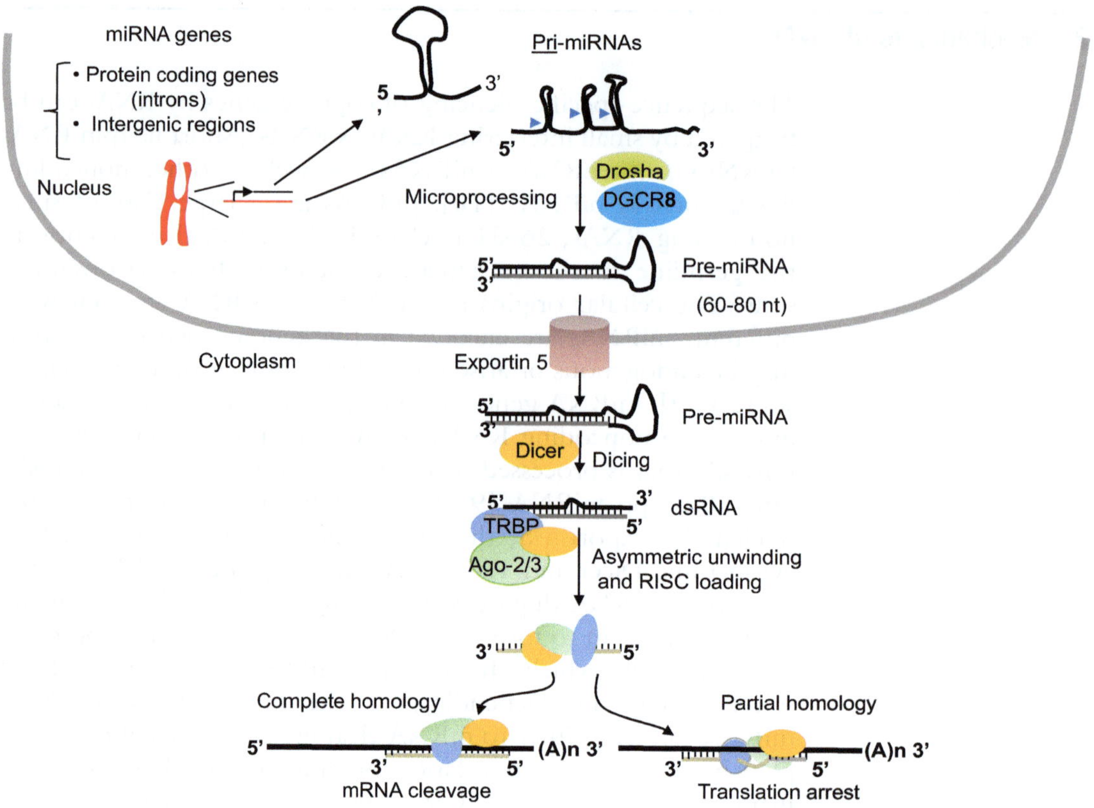

Fig. 1 Schematic representation of miRNA silencing pathway. miRNAs are encoded in the genome and are transcribed to yield primary transcripts, which are processed by Drosha with the aid of DGCR8 protein to generate short precursor miRNAs (pre-miRNAs) that are exported from the nucleus to the cytoplasm, where they are further processed by Dicer to generate short RNA duplexes containing two strands. Subsequent to processing, miRNAs are assembled into miRNA-induced silencing complex (miRISC). Only one strand of the complex is stably associated with an active miRISC complex. Unwinding of the complex seems to occur so rapidly after duplex formation due to the Ago2 presence in the complex with Dicer and TRBP. The loaded strand RNA guides the RISC complex to recognize target mRNA sequences. A perfect homology allows Ago cleavage of mRNA, whereas central mismatches promote repression of mRNA translation

The catalytic activity of RISC is mediated by Ago2 protein, which is composed of PAZ, Mid, and PIWI domains. The analysis of the crystal structure of an siRNA guide strand associated with an Ago2 PIWI domain identified a seed sequence (nucleotides 2–8) that directs target mRNA recognition by RISC [14, 15]. The 3′-end of the guide strand is recognized by the PAZ domain and is accommodated into its hydrophobic binding pocket. It is strict requirement for the siRNA to be 5′ phosphorylated to enter into RISC, and siRNA that lack a 5′ phosphate is rapidly phosphorylated by an endogenous kinase prior to loading into the RISC complex. In flies and humans, the passenger strand cleaved by AGO2 is

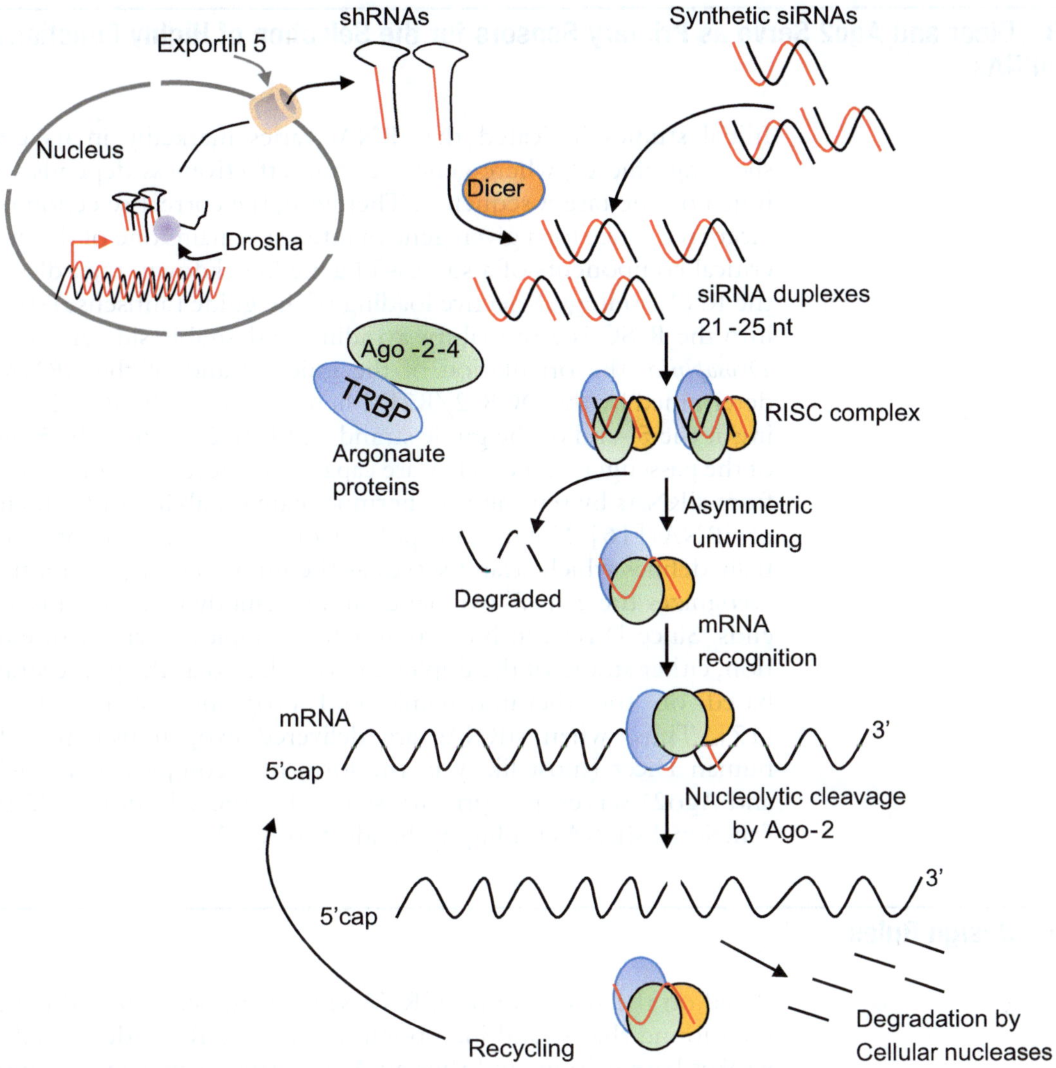

Fig. 2 Schematic representation of gene silencing by siRNAs. Synthetic siRNAs are directly loaded into a multiprotein complex termed RNA-induced silencing complex (RISC) where the sense strand with high 5′-end stability is cleaved by the nuclease Argonaute 2, resulting in strand separation. Subsequently, the RISC containing the antisense strand binds to complementary mRNA sequences. Gene silencing is a result of nucleolytic degradation of the targeted mRNA by Argonaute 2, a RNase H enzyme. Cleaved mRNA molecules are rapidly degraded by cellular nucleases. Following dissociation, the RISC is able to recycle and cleave additional mRNA molecules. Unlike chemically made siRNAs, hairpin RNAs (siRNAs) produced from plasmid vectors in cell nucleus are processed by Dicer in the cytoplasm before entering the RNAi pathway

removed from the RISC-loading complex by the endonuclease C3PO, known as translin. It should be noted that the competition of coding and noncoding RNAs for Argonaute protein binding can influence gene silencing.

3 Dicer and Ago2 Serve as Primary Sensors for the Selection of Highly Functional siRNAs

Initial studies indicated that RNAi varies markedly in its gene silencing efficacy, where gene silencing effectiveness depends very much on the target sequence. Therefore, the correct selection of a target sequence for a given gene of interest remains one of the most critical components of a successful gene knockdown, regardless of the RNAi strategy. Selective loading of the guide (antisense) strand into the RISC is essential for avoiding undesirable side effects. In *Drosophila*, the orientation of the guide strand of the siRNA is determined by the DCR-2/R2D2 heterodimer with DCR-2 binding at the 5′-end of the guide strand and R2D2 binding the 5′-end of the passenger strand. They are capable of selecting a guide strand from siRNAs by sensing the thermodynamic stability of both ends of siRNAs [16]. Notably, the polarity of Dicer processing preferentially defines which strand serves as the guide. Dicer preferentially recognizes the 2-nt 3′ overhangs and thermodynamically unstable ends. Since Dicer can bind to an siRNA duplex in either orientation, either strand of the duplex can be selected as the guide strand based on the thermodynamic end properties of the duplex [17]. Thus, when siRNAs are delivered exogenously to cells, human Dicer (most likely in the form of a complex with TRBP and Ago2) serves as a primary sensor for the selection of highly functional siRNAs leading to handoff to Ago2.

4 Design Rules

Based on the structure of miRNA sequences, Schwarz and colleagues found that strand incorporation into the RISC depended on weaker base pairing, and thus an A/U terminus may lead to more strand selection than a G/C terminus [17]. Moreover, the G/C content seems to play an important role in siRNA efficacy. Reynolds et al. performed systematic analysis on 180 siRNA targeting the mRNAs of two genes [18]. The data indicated that strand selection is biased, and the relative stability of the base pairs at the 5′-ends of the two siRNA strands is critical for strand selection. Functional siRNAs exhibited the following features: low G/C content, bias toward low internal stability at the 3′-end of sense strand, and lack of inverted repeats. The authors also suggested base preference at positions 3 and 10 of the sense strand (A and U, respectively). Ui-Tei et al. identified several features of the siRNA sequence that can discriminate highly functional siRNAs from those nonfunctional siRNAs [19]. Again, functional siRNAs contain A/U at the 5′-end of antisense strand, G/C at the 5′-end of sense strand, and more than four A/U nucleotides in the seed sequence (nucleotide

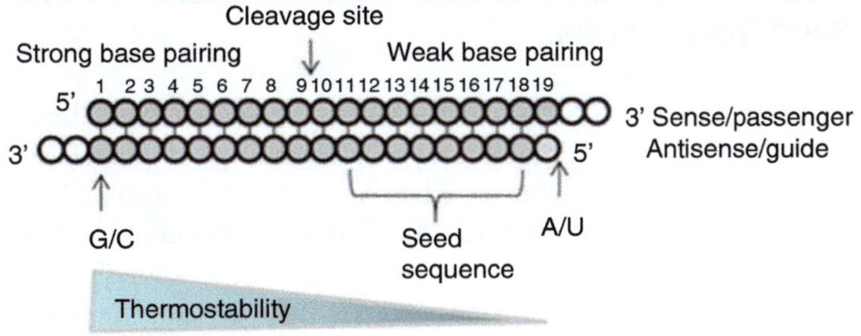

Fig. 3 Asymmetric unwinding and design rules (for details, *see* text)

2–8) and lack a long stretch in the 5′-end of sense strand. Sano et al. showed that siRNAs with unilateral 2-nt 3′ overhangs on the antisense strand (asymmetric 2-nt antisense strand overhang) are more potent than siRNAs with symmetric overhangs [20]. While there is no guarantee of effective gene silencing for a given siRNA until experimentally proven, numerous algorithms have been designed to select functional siRNA sequences. The overall guanine-cytosine content and asymmetric thermostability seem to be the most important factors for the design of effective siRNAs although the contribution of mRNA structure may play a significant role. Aside from the appropriate selection of the guide strand for incorporation into the RISC, there is the concern of off-targeting in which either the sense or antisense strand binds to non-targeted sequences triggering the inhibition of expression of these transcripts. In order to enhance the efficacy of RNAi, it is essential to better understand the mechanism of siRNA selection and strand incorporation in the RNA silencing pathway. Figure 3 summarizes the most critical features that need to be taken into consideration during siRNA design.

With respect to therapy, siRNAs should overcome the challenges of activating innate immunity and silencing off-target genes. As a seed sequence only contains seven nucleotides, it is not possible to select such a short sequence that is unique in the whole transcriptome. Thus, most, if not all, siRNAs will function as miRNAs, resulting in off-target effects. Off-target silencing cannot be ignored in developing siRNA-based therapeutics, and all potential therapeutic siRNA candidate sequences must be tested for perturbation of normal protein expression profiles. It should be noted that position-specific chemical modifications within the seed sequence reduced off-targeting ([21]; *see* Chapter 2). For each target gene, three target sequences should be designed in order to increase the likelihood that at least one sequence results in significant gene knockdown. Additionally, two or three successful knockdowns can also provide a useful control for off-target knockdowns since it is statistically unlikely that different sequences will produce the same off-target signatures.

5 Short Hairpin RNAs

Two principal approaches have been used to harness the RNAi machinery for silencing of gene expression: treatment with synthetic siRNA molecules and the expression of short hairpin (sh) RNAs that are processed intracellularly into active siRNAs (*see* Fig. 2). Notably, the silencing potency of synthetic siRNAs is transient. As the cells divide, the siRNA molecules become diluted, thereby rendering the generation of a long-term cell line with the desired target gene knockdown unfeasible. However, in nondividing cells such as mature dendritic cells, siRNA effect can be detected for several days (*see* Chapter 14). To circumvent the problem of siRNA dilution with cell division, several groups have developed DNA expression vectors for RNAi in human cells to express shRNAs from plasmid or viral vectors [22]. While synthetic siRNAs are delivered to the cytosol, shRNAs are transcribed in the nucleus and are then exported to the cytosol where they are recognized by Dicer, which processes the shRNAs into the siRNA duplexes. Dicer recognizes hairpin-structured RNAs that resemble intermediates of the microRNA pathway. Then the duplexes will be incorporated into the RISC complex for target-specific mRNA degradation. Notably, artificial miRNA expression cassettes can be used for the expression of microRNAs or shRNAs (*see* Chapters 10, 11, and 12). Although siRNAs and shRNAs ultimately use a similar cellular mechanism, the choice of which method to apply depends on several factors such as cell type and the need for transient versus stable gene silencing.

6 Enhancement of siRNA Stability via Chemical Modifications

To improve siRNA stability and pharmacokinetics, a variety of chemical modifications have been used ([23, 24]; *see* Chapters 2 and 3). Unlike antisense oligonucleotides, all siRNA chemical modifications must maintain or mimic a double-stranded, A-form RNA structure. The most promising generations of modifications which is important for siRNAs are those involving substitutions of the 2′-position of the sugar ring. These modifications of the sugar moiety are in general extremely resistant against nucleases and can be incorporated into RNA strands without altering the backbone conformation of the RNA, making them ideal candidates for use in siRNAs. Both 2’-F and 2’-O-Me modifications closely mimic the biophysical properties of 2’-OH and are highly tolerated by the RNAi machinery. 2′-O-methyl RNA is the most commonly used modification in siRNAs. This is a naturally occurring modification, and no toxicity has been reported from its use. Modification with 2′-O-methyl residues at the 5′-end of the guide strand combined with full passenger strand modification can enhance siRNA stability while retaining cleavage potency. The 2′-fluoro modification seems to be generally well tolerated in vivo, and it is usually used in combination with 2′-O-methyl modification [23]. Locked nucleic acids (LNAs) are bicyclic nucleic acids where the 2′-position of the ribose is connected to the 4′-position via a methyl bridge [25]. Although LNA modification increases duplex stability and improves nuclease resistance, it also leads to loss of siRNA potency. Therefore, this modification must be used sparingly. Unlocked nucleic acids (UNAs) can be used in combination with LNAs to improve the relative stability of siRNA termini and has also alone been proven to increase the stability and knockdown in some models [26]. Notably, the attachment of a 3′ cholesterol group to the sense strand of siRNA facilitated its uptake by human cells (*see* Chapter 3). Chemically modified self-delivered siRNAs provide an alternative RNAi strategy because they enable siRNAs to enter cells without the need of a transfection reagent.

In addition to stability, the immunostimulatory effect of siRNAs is critical and has to be taken into account in RNAi experiments, as target gene expression and function may be influenced by the activation of innate immune receptors. Indeed, immunostimulation may contribute to the therapeutic effect of siRNAs. In this respect, TLR activation was more likely responsible for the promising early clinical trial data for ocular antiangiogenic siRNAs and topically applied siRNA respiratory syncytial virus [27]. Fortunately, site-specific chemical modifications that also reduce sequence-specific off-target effects largely inhibited the activation of TLRs by siRNAs [28, 29]. It should be noted that immune activation by siRNAs may benefit patients with cancer ([31]; *see* Chapter 16).

7 Delivery Methods and Challenges

The method of choice depends on the model system and whether transient or stable expression is desired (Fig. 4). The delivery of siRNA can broadly be divided into three groups: naked siRNA delivery, lipid delivery, and conjugate delivery. These can be applied either systemically through the bloodstream or locally to the tissue of interest (i.e., tumors). To enhance siRNA potency and avoid off-target effects, nanoparticles and conjugate delivery can be targeted to the tissue of interest. With respect to shRNAs, lentiviral-mediated transduction provides a convenient method of introducing shRNAs into dividing or nondividing cells and, in general, is less toxic to the cells than adenoviral-mediated transduction [30, 31].

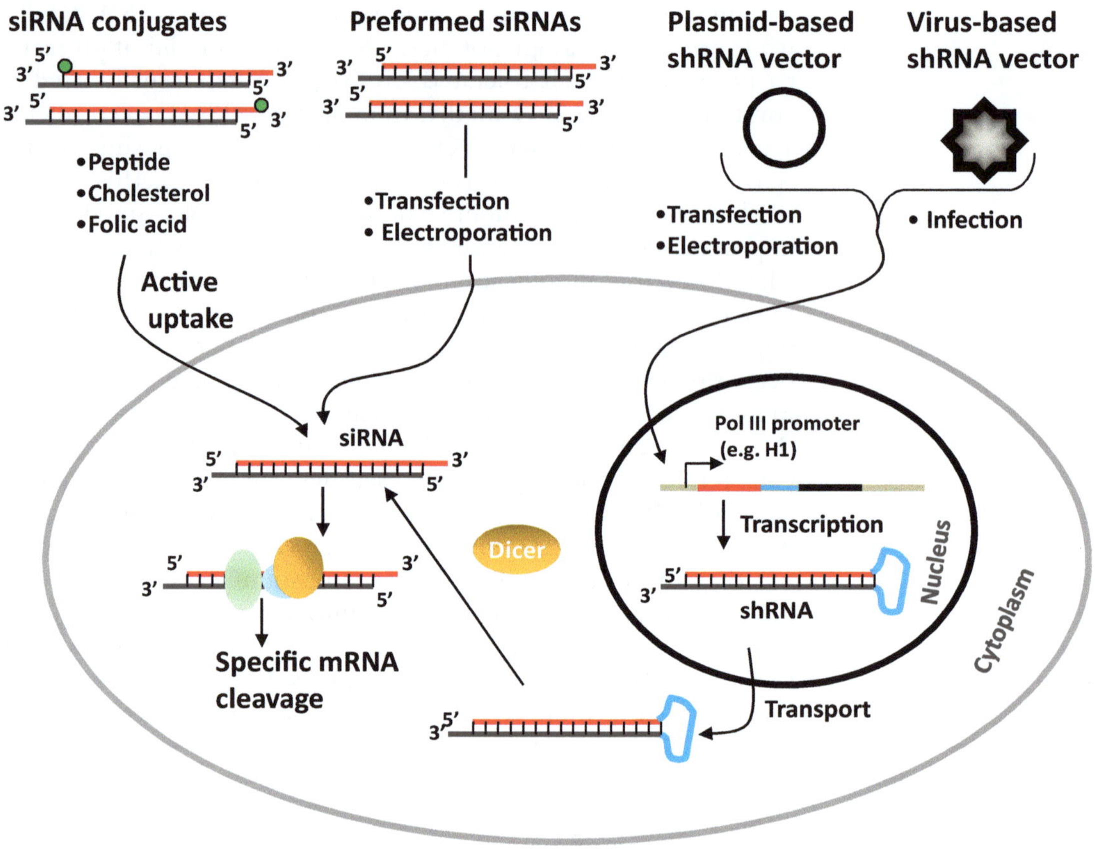

Fig. 4 siRNA delivery strategies into mammalian cells. Chemically made siRNAs can be delivered to mammalian cells through several strategies including direct conjugation to cell-surface binding ligand, encapsulation into lipids, and electroporation. Plasmid-based shRNA vectors are usually delivered via lipids or electroporation, whereas virus-based vectors are delivered via infection

7.1 Naked siRNA Delivery

Once the specific siRNAs or miRNAs are designed, they can be introduced into cells using plasmid vectors, synthetic siRNAs, PCR products, or in vitro transcribed siRNAs. Notably, the first methods of siRNA delivery in *C. elegans* were through dsRNA expressed by bacteria fed to the animal. The uptake of dsRNA into the cells involves the membrane proteins SID-1 and SID-2 and the transportation of dsRNA through the membrane by these proteins [32]. In *D. melanogaster*, no *sid* gene orthologs were found, but uptake of dsRNA probably happens through endocytosis [33]. Unfortunately, mammalian cells are deprived of this uptake and transport mechanisms. In general, naked siRNA are not always ideal to produce efficient and predictable therapeutic results [34]. Within minutes after intravenous administration, more than 99% of the injected siRNA dose is renally excreted or taken up by liver Kupffer cells from the circulation. Only a very small percentage of the administered dose remains available for the target cells or tissues. This small percentage of siRNA is additionally subject to nuclease degradation. Moreover, the efficiency of siRNA passively entering target cells is usually very low [35]. Therefore, additional delivery agents and chemical modifications are necessary to surmount siRNA instability and other barriers in the body.

7.2 Lipid Delivery of siRNAs

The plasma membrane is a significant barrier for siRNA uptake. Despite their small size, the hydrophilicity and negative charge of siRNA molecules prevent them from readily crossing biological membranes. Hence, effective siRNA delivery approaches need to overcome this limitation by facilitating cellular uptake (*see* Chapters 4, 5, and 6). A variety of approaches have been developed that promote siRNA delivery in vivo, including cationic nanoparticles, lipid, and liposome formulations. Lipid nanoparticles (LNPs) were used to enhance the cellular uptake and the pharmacological effectiveness of siRNAs in vitro and in vivo [36]. LNPs are a class of nanoparticles composed of different lipids that vary in physicochemical properties and biological activities. Several chemically optimized reagents for siRNA delivery using lipids are commercially available. For example, lipofectamine is a transfection agent that fuses with a negatively charged membrane. DOTAP is another commercially available cationic liposome using the same principle as lipofectamine. Although cationic liposomes protect siRNAs from degradation by nucleases and increase circulating half-time and cellular uptake, they are toxic to the cell and elicit hypersensitivity reactions in vivo. They form aggregates with negatively charged serum proteins, which accumulate mainly in the lungs, liver, and spleen, and often induce pulmonary toxicity associated with complement activation and inflammation [37]. Hence, several efforts are underway to make liposomes safer, such as improving their formulation by adding chemical additives to reduce cell toxicity.

A possible approach to overcome some of the problems related to toxicity was to develop sustained-release polymer formulations. Lipid polymers are compounds that protect RNA from degradation and facilitate sustained delivery to the tissues. Polymers with both hydrophobic and hydrophilic segments can self-assemble in aqueous solution to form stable polymeric particles on the nanometer scale. For siRNA delivery, these particles, with the help of cationic polymer segments, are designed to encapsulate siRNA into the particle core. A widely used strategy for making nanoparticles stable in vivo is coating with polyethylene glycol (PEG) [38, 39]. Lipid-anchored PEG is a common component in liposomes. PEG groups serve many purposes: they reduce particle size, prevent aggregation during storage, increase circulation time, and reduce uptake by unintended targets such as red blood cells and macrophages. Moreover, the hydrophilic nature of PEG provides an aqueous shield around the nanoparticle surface, thus decreasing the extent of opsonization and the subsequent recognition by the macrophages

[40]. In stable nucleic acid-lipid particles (SNALPs), the siRNA is surrounded by a lipid bilayer containing a mixture of cationic and fusogenic lipids that facilitate internalization of the SNALP and endosomal escape of its siRNA payload [41, 42]. However, it is uncertain how much of the internalized siRNAs actually get delivered to the cytosol. A phase I clinical study that involves a single injection of ApoB siRNA-containing SNALPs showed a target gene knockdown and serum cholesterol reduction and acceptable safety. Moreover, SNALP-delivered siRNA has been used to target a gene required for mitotic spindle formation (PLK1) for liver cancer treatment, resulting in a significant repression of tumor growth without secondary side effects [43]. Although PEG-based strategies have revolutionized the field of nanoparticle delivery, recent studies have shown the limitations of this strategy. Upon several administrations of PEGylated liposomes, immune response has been detected, leading to rapid blood clearance or other side effects. Also, PEGylation impairs the internalization of siRNA into target cell, as well as its escape from the endocytic pathway [40].

In general, LNPs are subject to liver accumulation, and therefore many of the successful results using this strategy target liver cells. Indeed, LNP formulations have advanced into clinical trials for delivery of siRNAs to the liver through systemic administration. Nevertheless, it is estimated that around 0.1% of injected dose reaches the cytosol of hepatocytes. Therefore, there is a need of targeting specific tissues in vivo. In this respect, Zhung et al. targeted xenograft brain cancer in murine model by PEGylated liposomes conjugated to antibodies against the transferrin receptor, transferring the particle through the blood-brain barrier, and antibodies against human insulin receptor expressed by the tumor [44]. Delivery particles with high positive surface charges can exhibit unfavorable aggregation with blood cells such as erythrocytes. However, some delivery agents make use of their interaction with particular proteins to aid uptake by target cells. For example, many liposomal delivery systems, as well as siRNA conjugated to lipophilic molecules, interact with serum lipoproteins and subsequently gain entry into hepatocytes that take up those lipoproteins [45]. Delivery agents that are not degraded, phagocytosed, or cleared by the kidney must leave the bloodstream by traversing the endothelium to reach other tissues. This occurs in tissues whose endothelia are discontinuous, such as the liver and many solid tumors. Moreover, tumors with highly permeable endothelia are characterized by poor lymphatic drainage, resulting in greater accumulation of circulating nanoparticles in malignant tissue, a process known as the enhanced permeation and retention effect. Collectively, delivery systems, such as polymer-based and lipid-based systems, significantly improve the stability of small RNAs and delivery efficiency by protecting them from degradation and enhancing the uptake by mammalian cells. However, there are several concerns of using these delivery systems, including toxicity, nonspecific uptake, and immunogenic effects.

7.3 Trafficking and Endosomal Escape of RNA Molecules

Notably, endocytosed materials are taken up into membrane-bound endocytic vesicles, which fuse with early endosomes and become increasingly acidic as they mature into late endosomes [46]. Being negatively charged and hydrophilic, siRNA molecules cannot spontaneously cross the endosomal membrane to reach the cytosol and therefore face the fate of degradation. Endosomal escape thus becomes a key step for siRNA delivery. Some delivery systems incorporate materials that are designed to respond to this low pH environment by becoming membrane disruptive in order to trigger the release of siRNA from endosomes in the cytoplasm [47]. One typical method to enhance endosomal escape is the use of cationic polymers with a *pKa* around or slightly below physiological pH. Due to the acidic nature of endosomes, these cationic polymers therefore absorb the protons and increase the osmotic pressure inside the endosomal compartment. This so-called proton sponge effect leads to the disruption of plasma membrane and siRNA cytoplasm release [48]. It has been shown that histidine residues, with *pKa* of 6.4, can enable endosomal buffering and escape. Reducible polycations containing histidine monomers are found effective to deliver siRNA targeting hepatitis B virus surface antigen and reduce the amount of viral protein shed from Alexander cells [49]. Similarly, an siRNA targeting Raf-1 was delivered to cancer cells via histidine-lysine-rich polymers [50]. The siRNA-polymer complexes markedly decreased Raf-1 mRNA and induced apoptosis in several cancer cell lines in vitro. Besides the endosomal escape, histidine has also been found to facilitate nucleic acid condensation and membrane fusion [51]. Therefore, it holds even greater potential in the field of siRNA delivery. To facilitate siRNA escape from the endosomes, Takea et al. developed PEG-detachable polyplex micelles, where the PEG palisade surrounding the particle core is conjugated via reversible disulfide bond [52]. The PEG can be detached in the reducing endosomal compartment resulting in the interaction between the exposed nanoparticle cation segments and the endosomal membrane. The PEG detachment also favors the release of nucleic acid from the polyplex core in the cytoplasm due to the removal of the surrounding PEG layer.

As indicated above, LNPs have the most potential for successful clinical use because they protect siRNA from nuclease and can be functionalized with targeting molecules for navigation to the desired tissues. In addition to endocytosis, there has been an increasing attention on the significance of exocytosis and autophagy in siRNA intracellular trafficking. These complex processes determine the retention time of cargos in intracellular compartment and directly affect the windows of opportunity for the escape and release of siRNA into the cytoplasm. Two recent studies found that certain LNPs were internalized by macropinocytosis and trafficked directly into late endosomes, bypassing early endosomes. Unexpectedly, siRNA molecules that dissociated from the LNPs were exocytosed to the extracellular milieu [53, 54]. The loss of siRNA in this manner accounted for 70% of the uptake dose 24 h after transfection. These findings indicate that escape from the endosomes is the critical, rate-limiting step in the delivery of siRNA with the used LNPs and that methods that increase retention in early endosomes might improve endosomal escape and overall delivery efficiency. Although several advances were accomplished in siRNA delivery, more research is still needed to guide the delivery system architecture and the chemical structure that is most suited for therapeutic siRNA delivery (*see* Chapter 2).

8 RNA-Based Therapies

With respect to therapy, a significant progress has been made in the fields of cancer and viral infections, and a variety of carriers including liposomes, lipid nanoparticles, aptamers, and antibodies have been used to deliver siRNAs in vitro and in vivo. These delivery carriers need to protect naked siRNA from degradation and avoid fast renal filtration as well as uptake by the mononuclear phagocyte system (macrophages and liver Kupffer cells). In the case of cancer, siRNA formulations must traffic to the site of tumors and penetrate into and distribute throughout tumor tissues. Unfortunately, several tumors are thought to be more fibrotic and more difficult to penetrate physically. After reaching the target cell/tissue, the siRNA formulations need to be internalized, escape from the endosomes into the cytoplasm, and finally release the siRNA molecules to enter the RNAi pathway. Although some progress has been made, the current clinically tested carriers showed shorter circulation properties (half-lives <2 h) [55]. Several of the more advanced therapeutic candidates in the pipeline are composed of lipid-based nanoparticles formulations or employ *N*-acetylgalactosamine conjugation to target the siRNA to the liver [19, 55]. While delivery to hepatocytes is not a major issue for the treatment of liver diseases due to the ingenious use of synthetic *N*-acetylgalactosamine (GalNAc) ligands, treatment of most diseases requires the development of clinically acceptable methods to deliver siRNAs to other tissues and organs. Once successfully inside a given target cell, additional challenges include avoiding endosomal degradation and unwanted stimulation of innate immunity [56]. To achieve accurate target gene silencing and successful therapeutic applications, it is also critical to select a target site that has very low sequence homology to human transcriptome in order to reduce off-target effects.

Alternatively, one may reduce the silencing of unintended genes by introducing few 2′ chemical modifications in the seed sequence (positions 2–8) of the siRNA guide strand (*see* Chapter 2).

9 CRISPR Interference

As indicated above, RNAi plays an essential role in many plant biological processes, including developmental timing and patterning, transposon control, DNA methylation, and chromatin modification. Although its role in human immunity is not known yet, RNAi constitutes the primary natural plant immune system against viruses [57]. The plant RNAi defense system is highly specific, because it will only target nucleic acids that are identical in sequence to the triggering dsRNAs. In addition to viral infections, RNAi protects the genome from genetic parasites such as transposons and retrotransposons. These mobile nucleic acid elements can cause diseases and alter the genome in a number of different ways.

By analogy to RNAi-mediated plant protection against pathogen attack, the clustered regularly interspaced short palindromic repeats (CRISPR) loci and CRISPR-associated (Cas) proteins provide acquired immunity against viruses and other mobile genetic elements in bacteria and archaea. These CRISPR sequences are part of the bacteria's immune system and provide endogenous adaptive immunity in approximately 40% of bacterial genomes and 70% of sequenced archaeal species [58]. Immunological memories of phages and plasmids are stored in the CRISPR array as short spacer sequences that intercalate between repeats and specify the targets of CRISPR-Cas immunity. The locus consists of CRISPR-associated (Cas) genes in operons in addition to the spacer-repeat array. These repeats were initially discovered in the 1980s in *E. coli* [59], but their function wasn't uncovered until 2007 by Barrangou and colleagues, who demonstrated that the *Streptococcus thermophilus* strains that survived the initial phage challenge were resistant to subsequent phage challenge and had incorporated sequences from the phage genome into their CRISPR arrays [60]. When a new infection occurs, the spacer-repeat array is expressed as a long precursor pre-CRISPR RNA (pre-crRNA), and the transcript is processed into individual spacer-repeat units as small guiding crRNAs. These associate with and direct RNA-guided Cas nucleases (e.g., Cas9) to cleave of foreign nucleic acids complementary to the spacer sequence. Cleavage of the viral or plasmid target DNA prevents infection (Fig. 5). Similar to the RISC complex, the specificity of the CRISPR-Cas system relies on tightly bound crRNA, which efficiently guides a nuclease to its target, a complementary DNA fragment.

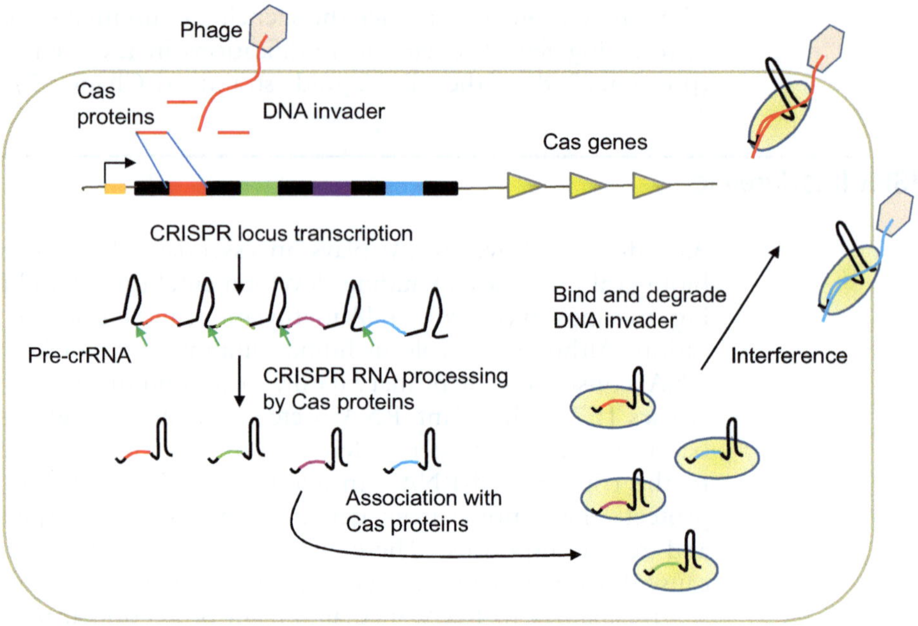

Fig. 5 Schematic representation of CRISPR-Cas systems. The CRISPR locus is transcribed and processed to generate mature CRISPR RNAs, each encoding a unique spacer sequence. After processing, each crRNA associates with Cas effector proteins that use crRNAs as guides to silence foreign genetic elements that match the crRNA sequence

9.1 Genome Editing Using CRISPR-Cas9

The CRISPR technology has been adapted for genome editing applications in eukaryotic cells and is an emerging technology because it allows for some more flexibility compared to zinc finger nucleases (ZFNs) or transcriptional activator-like effector nucleases (TALENs) (*see* Chapter 21). Moreover, the technology is incredibly precise, cheap, and easy, and it has worked on every cell or organism it is been tried on. The CRISPR-Cas system described so far falls into three major classes, each composed of different Cas gene family [61]. While the type I and III systems use various effector proteins, the type II uses primarily a single protein Cas9 to both generate crRNA and cleave corresponding target DNA and thus represent the most convenient format to use in eukaryotic cells. Indeed, in a landmark 2012 paper, Doudna, Chapentier, and Jinek showed they could use CRISPR-Cas 9 system to introduce specific mutations in any genome [62]. When loaded with both crRNA and tracrRNA (or an artificial chimera of the two), the wild-type Cas9 can site specifically introduce double-stranded DNA breaks (DSBs). The resulting DSB will either generate nonspecific mutations knocking out a gene through the error-prone NHEJ (non-homologous end joining) pathway. Alternatively, if a donor template with homology to the target locus is supplied, the DSB may be repaired by the homology-directed repair (HDR) pathway allowing for precise replacement mutations to be made [62, 63]. By directing the cells toward one of these repair mechanisms, CRISPR can be used to generate knockouts or knock-ins in desired cells. Moreover, a more refined type of CRISPR gene editing that can alter a single bit of DNA without cutting it is expected to increase CRISPR efficiency and specificity (on-target effects).

9.2 Cas-9 Specificity

For therapeutic use of genetically modified human cells, integrated sequences should be introduced specifically without the unintended disruption of other critical genome sites. Ideally, highly specific gRNAs that target only the intended genes are desired. One of the most critical steps of a CRISPR experiment is designing efficient and specific guide RNAs, which is facilitated by the availability of state-of-the-art design tools (*see* Chapter 19). Despite these sophisticated bioinformatics tools, the technology still has some sequence-specific off-target effects. The use of chemically synthesized sgRNAs without or without chemical modifications has greatly contributed to reducing the off-target effects relative to plasmids and in vitro transcribed gRNAs [64]. Notably, high-fidelity Cas9 variants including eSpCas9, SpCas9-HF1, HypaCas9, and xCas9 were developed ([65], *see* Chapter 21). These Cas9 protein variants combined with the use of chemically modified gRNAs would increase the specificity of genome editing. For therapeutic applications, unintended mutations introduced in a patient's DNA could permanently disrupt normal gene function and lead to unpredictable complications. More recently, Wienert et al. have described a method for improved identification of off-target sites [66]. The method, called DISCOVER-Seq, relies on endogenous DNA repair machinery that naturally repairs DNA breaks in the genome. By isolating the protein-DNA complexes and sequencing the DNA, the authors identified the off-target sites of an RNA guide of interest. The strategy worked well on induced pluripotent stem cells from a patient with Charcot-Marie-Tooth syndromes, as well as in mouse livers that were edited with virally delivered Cas9. Controlling Cas9 nuclease activity should also limit potential off-target cleavage. In this respect, phage inhibitors, known as anti-CRISPR, might offer a way of regulating Cas9 activity [67]. Alternatively, one might use phage display technology to select Ca9 blocking peptides or antibodies [68, 69, 70, 71].

Although the technology has important ethical implications, it has already been used to treat certain life-threatening human diseases such as beta thalassemia. In addition to DNA, certain Cas proteins such as Cas13 can edit RNAs [72, 73]. The CRISPR Cas13 system is more efficient than RNA interference and can be used in mammalian and plant cells to knock down specific mRNA transcripts. With respect to delivery, nucleofection is an easy method that usually has minimal cytotoxicity toward primary human cells (*see* Chapters 7, 26, and 27). As discussed in Chapters 26 and 27, the availability of highly purified Cas9 proteins and chemically modified guide RNA with enhanced stability and low immune-stimulatory properties from commercial vendors such as integrated DNA technologies will provide a platform for transient therapeutic delivery of CRISPR-Cas9 RNP complexes using nucleofection, a cost-effective strategy for gene disruption or editing in cell culture.

9.3 Self and Non-self Discrimination and Off-Target Effects

DNA targeting during CRISPR activation raises an important question: How the system differentiate between spacer of the CRISPR locus on the host chromosome (self) and the invading DNA (non-self). Cas9 detects genomic targets by unwinding the DNA duplex and scanning for complementarity between the genomic DNA and the spacer sequences in the gRNA. However, it will only cleave DNA if the correct protospacer-adjacent motif (PAM) is detected. While each type II system has individual PAM requirements, the *S. pyogenes* Cas 9 system requires an NGG sequence immediately downstream of the protospacer, with N being any nucleotide. The PAM sequence is crucial for the discrimination between self and non-self sequences, and only the invading DNA that is flanked by the correct PAM can be cleaved (*see* Chapter 19). It should be noted that the PAM sequence is not present in CRISPR locus, and therefore, only foreign DNA will be cleaved.

10 Conclusions

Since its discovery, RNAi has emerged as a powerful method for silencing specific genes. The technology works by cleaving messenger RNA before it is translated into a protein. As compared to other nucleic acid-based strategies, siRNA benefits from harnessing endogenous RNAi pathway to trigger gene silencing. Two main strategies have been used to harness the RNAi pathway for silencing gene expression: treatment with synthetic siRNA molecules or the expression of short hairpin RNAs that are processed intracellulary into active siRNAs or microRNAs. Chemically made siRNAs can efficiently silence gene expression without altering the host genetic material. In addition, the delivery of synthetic siRNAs can be altered based on the specific clinical needs, and the treatment can be discontinued, as warranted, without long-term effects. Although some progress has been made in delivery and chemical modifications, targeting to organs beyond the liver is still a substantial challenge in the field.

Similar to RNAi, one of the most critical steps of a CRISPR experiment is designing efficient and specific gRNAs. Efficient design tools enabled finding guide RNAs that exhibit minimal off-target effects (*see* Chapter 19). In most experiments, plasmids bearing the guide RNA sequence and nuclease encoding the Cas nuclease sequence have been used in genome engineering. In vitro transcribed RNAs have also been used widely in CRISPR experiments. However, care must be taken when using in vitro transcribed RNAs because of the activation of the interferon response. The use of chemically modified gRNAs reduced the off-target effects relative to plasmid and in vitro transcribed gRNAs. Due to its simplicity, the delivery of a ribonucleoprotein complex consisting of recombinant Cas9 protein and guide RNA is now the preferred strategy for genome engineering experiments (*see* Chapters 26 and 27). Taken together, these features make CRISPR-Cas 9 an ideal tool for gene editing.

Acknowledgments

This work was supported by grants from the Norwegian Cancer Society.

References

1. Napoli C, Lemieux C, Jorgensen R (1990) Introduction of a chimeric chalcone synthase gene into petunia results in reversible co-suppression of homologous genes in trans. Plant Cell 2:279–289
2. Guo S, Kemphues KJ (1995) par-1, a gene required for establishing polarity in C. elegans embryos, encodes a putative Ser/Thr kinase that is asymmetrically distributed. Cell 81:611–620
3. Fire A, Xu S, Montgomery MK, Kostas SA, Driver SE, Mello CC (1998) Potent and specific genetic interference by double-stranded RNA in Caenorhabditis elegans. Nature 391:806–811
4. Hamilton AJ, Baulcombe DC (1999) A species of small antisense RNA in posttranscriptional gene silencing in plants. Science 286:950–952
5. Zamore PD, Tuschl T, Sharp PA, Bartel DP (2000) RNAi: double-stranded RNA directs the ATP-dependent cleavage of mRNA at 21 to 23 nucleotide intervals. Cell 101:25–33
6. Elbashir SM, Harborth J, Lendeckel W, Yalcin A, Weber K, Tuschl T (2001) Duplexes of 21-nucleotide RNAs mediate RNA interference in cultured mammalian cells. Nature 411:494–498
7. Yan KS, Yan S, Farooq A, Han A, Zeng L, Zhou MM (2003) Structure and conserved RNA binding of the PAZ domain. Nature 426:468–474
8. Pencheva N, Tavazoie SF (2013) Control of metastatic progression by microRNA regulatory networks. Nat Cell Biol 15:546–554
9. Zamore PD, Haley B (2005) Ribogenome: the big word of small RNAs. Science 309:1519–1524
10. Sioud M (2011) Promises and challenges in developing RNAi as a research tool and therapy. Methods Mol Biol 703:173–187
11. Kim VN (2005) MicroRNA biogenesis: coordinated cropping and dicing. Nat Rev Mol Cell Biol 6:376–385
12. Liu Q, Rand TA, Kalidas S, Du F, Kim HE, Smith DP, Wang X (2003) R2d2, a bridge between the initiation and effector steps of the drosophila RNAi pathway. Science 301:1921–1925
13. Liu X, Jiang F, Kalidas S, Smith D, Liu Q (2006) Dicer-2 and R2D2 coordinately bind siRNA to promote assembly of the siRISC complexes. RNA 12:1514–1520
14. Matranga C, Tomari Y, Shin C, Bartel DP, Zamore PD (2005) Passenger-strand cleavage facilitates assembly of siRNA into Ago2-containing RNAi enzyme complexes. Cell 123:607–620
15. Wee LM, Flores-Jasso CF, Salomon WE, Zamore PD (2012) Argonaute divides its RNA guide into domains with distinct functions and RNA-binding properties. Cell 151:1055–1067
16. Tomari Y, Matranga C, Haley B, Martinez N, Zamore PD (2004) A protein sensor for siRNA asymmetry. Science 306:1377–1380
17. Schwarz DS, Hutvagner G, Du T, Xu Z, Aronin N et al (2003) Asymmetry in the assembly

of the RNAi enzyme complex. Cell 115:199–208
18. Reyolds A, Leake D, Boese Q, Scaringe S, Marchall WS et al (2004) Rational siRNA design for RNA interference. Nat Biotechnol 22:326–330
19. Ui-Tei K, Naito Y, Takahashi F, Haraguchi T, Ohki-Hamazaki H et al (2004) Guidelines for the selection of highly effective siRNA sequences for mammalian and chick RNA interference. Nucleic Acids Res 32:936–948
20. Sano M, Sierant M, Miyagishi M, Nakanishi M, Takagi Y, Sutou S (2008) Effect of asymmetric terminal structures of short RNA duplexes on the RNA interference activity and strand selection. Nucleic Acids Res 36:5812–5821
21. Jackson AL, Burchard J, Leake D, Reynolds A, Schelter J, Guo J, Johnson JM, Lim L, Karpilow J, Nichols K, Marshall W, Khvorova A, Linsley PS (2006) Position-specific chemical modification of siRNAs reduces "off-target" transcript silencing. RNA 12:1197–1205
22. Paul CP, Good PD, Winer I et al (2002) Effective expression of small interference RNA in human cells. Nat Biotechnol 20:505–508
23. Watts JK, Deleavey GF, Damha MJ (2008) Chemically modified siRNA: tools and applications. Drug Discov Today 13:842–855
24. Hoerter JA, Walter NG (2007) Chemical modification resolves the asymmetry of siRNA strand degradation in human blood serum. RNA 13:1887–1893
25. Mook OR, Baas F, de Wissel MB, Fluiter K (2007) Evaluation of locked nucleic acid-modified small interfering RNA in vitro and in vivo. Mol Cancer Ther 6:833–843
26. Laursen MB, Pakula MM, Gao S, Fluiter K, Mook OR, Baas F et al (2010) Utilisation of unlocked nucleic acid to enhance siRNA performance in vitro and in vivo. Mol BioSyst 6:862–870
27. Sioud M (2006) Single-stranded small interfering RNA are more immunostimulatory than their double-stranded counterparts: a central role for 2′-hydroxyl uridines in immune responses. Eur J Immunol 36:1222–1230
28. Rettig GR, Behlke MA (2012) Progress toward in vivo use of siRNAs-ii. Mol Ther 20:483–512
29. Furset G, Sioud M (2007) Design of bifunctional siRNAs: combining immunostimulation and gene-silencing in one single siRNA molecule. Biochem Biophy Res Commun 352:642–649
30. Barton GM, Medzhitov R (2002) Retroviral delivery of small interfering RNA into primary cells. Proc Natl Acad Sci U S A 99:14943–14945
31. Winston WM, Molodowitch C, Hunter CP (2002) Systemic RNAi in *C. elegans* requires the putative transmembrane protein SID-1. Science 295:2456–2459
32. Saleh MC, Van Rij RP, Hekele A, Gillis A, Foley E, O'Farrell PH, Andino R (2006) The endocytic pathway mediates cell entry of dsRNA to induce RNAi silencing. Nat Cell Biol 8:793–802
33. Fougerolles AD, Vornlocher HP, Maraganore J et al (2007) Interfering with disease: a progress report on siRNA-based therapeutics. Nat Rev Drug Discov 6:443–453
34. Feinberg EH, Hunter CP (2003) Transport of dsRNA into cells by the transmembrane protein SID-1. Science 301:1545–1547
35. Dowdy SF (2017) Overcoming cellular barriers for RNA therapeutics. Nat Biotechnol 35:222–227
36. Simoes S, Filipe A, Faneca H, Mano M, Penacho N, Duzgunes N, De Lima MP (2005) Cationic liposomes for gene delivery. Expert Opin Drug Deliv 2:237–254
37. Sun TM, Du JZ, Yan LF et al (2008) Self-assembled biodegradable micellar nanoparticles of amphiphilic and cationic block copolymer for siRNA delivery. Biomaterials 29:4348–4355
38. Peer D, Karp JM, Hong S et al (2007) Nanocarriers as an emerging platform for cancer therapy. Nat Nanotechnol 2:751–760
39. Gomes-Da-Silva LC, Fonseca NA, Moura V, Pedroso De Lima MC, Simoes S, Moreira JN (2012) Lipid-based nanoparticles for siRNA delivery in cancer therapy: paradigms and challenges. Acc Chem Res 45:1163–1171
40. Morrissey DV, Lockridge JA, Shaw L, Blanchard K, Jensen K, Breen W, Hartsough K, Machemer L, Radka S, Jadhav V, Vaish N, Zinnen S, Vargeese C, Bowman K, Shaffer CS, Jeffs LB, Judge A, Maclachlan I, Polisky B (2005) Potent and persistent in vivo anti-HBV activity of chemically modified siRNAs. Nat Biotechnol 23:1002–1007
41. Zimmermann TS, Lee AC, Akinc A, Bramlage B, Bumcrot D, Fedoruk MN et al (2006) RNAi-mediated gene silencing in non-human primates. Nature 441:111–114
42. Judge AD, Robbins M, Tavakoli I, Levi J, Hu L, Fronda A, Ambegia E, Mcclintock K, Maclachlan I (2009) Confirming the RNAi-mediated mechanism of action of siRNA-based cancer therapeutics in mice. J Clin Invest 119:661–673
43. Zhang Y, Zhang YF, Bryant J, Charles A, Boado RJ, Pardridge WM (2004) Intravenous RNA interference gene therapy targeting the

human epidermal growth factor receptor prolongs survival in intracranial brain cancer. Clin Cancer Res 10:3667–3677

44. Wolfrum C, Shi S, Jayaprakash KN et al (2007) Mechanisms and optimization of in vivo delivery of lipophilic siRNAs. Nat Biotechnol 25:1149–1157
45. Lundberg M, Wikstrom S, Johansson M (2003) Cell surface adherence and endocytosis of protein transduction domains. Mol Ther 8:143–150
46. Oishi M, Nagasaki Y, Itaka K et al (2005) Lactosylated poly(ethylene glycol)-siRNA conjugate through acid-labile β-thiopropionate linkage to construct pH-sensitive polyion complex micelles achieving enhanced gene silencing in hepatoma cells. J Am Chem Soc 127:1624–1625
47. Behr J-P (1997) The proton sponge: a trick to enter cells the viruses did not exploit. Chimia 51:34–36
48. Stevenson M, Ramos-Perez V, Singh S et al (2008) Delivery of siRNA mediated by histidine-containing reducible polycations. J Control Release 130:46–56
49. Leng Q, Scaria P, Lu P et al (2008) Systemic delivery of HK Raf-1 siRNA polyplexes inhibits MDA-MB-435 xenografts. Cancer Gene Ther 15:485–495
50. Kumar VV, Pichon C, Refregiers M et al (2003) Single histidine residue in head-group region is sufficient to impart remarkable gene transfection properties to cationic lipids: evidence for histidine mediated membrane fusion at acidic pH. Gene Ther 10:1206–1215
51. Takae S, Miyata K, Oba M et al (2008) PEG-detachable polyplex micelles based on disulfide-linked block catiomers as bioresponsive nonviral gene vectors. J Am Chem Soc 130:6001–6009
52. Sahay G, Querbes W, Alabi C, Eltoukhy A, Sarkar S, Zurenko C et al (2013) Efficiency of siRNA delivery by lipid nanoparticles is limited by endocytic recycling. Nat Biotechnol 31:653–658
53. Gilleron J, Querbes W, Zeigerer A, Borodovsky A, Marsico G, Schubert U et al (2013) Image-based analysis of lipid nanoparticle-mediated siRNA delivery, intracellular trafficking and endosomal escape. Nat Biotechnol 31:638–646
54. Kim HJ, Kim A, Miyata K, Kataoka K (2016) Recent progress in development of siRNA delivery vehicles for cancer therapy. Adv Drug Deliv Rev 104:61–77
55. Zimmermann TS, Karsten V, Chan A, Chiesa J, Boyce M et al (2017) Clinical Proof of concept for a novel hepatocyte-targeting GalNAc-siRNA conjugate. Mol Ther 25:71–78
56. Sioud M (2007) RNA interference and innate immunity. Adv Drug Deliv Rev 59:153–163
57. Horvath P, Barrangou R (2010) CRISPR/Cas, the immune system of bacteria and archaea. Science 327:167–170
58. Ishino Y, Shinagawa H, Makino K, Amemura M, Nakata A (1987) J Bacteriol 169:5429–5433
59. Barrangou R, Fremaux C, Deveau H et al (2007) CRISPR provides acquired resistance against viruses in prokaryotes. Science 315:1709–1712
60. Terns MP (2018) CRISPR-based technologies: impact of RNA-targeting systems. Mol Cell 72:404–412
61. Jinek M, Chylinski K, Fonfara I, Hauer M, Doudna JA, Charpentier E (2012) A programmable dual-RNA-guided DNA endonuclease in adaptive bacterial immunity. Science 337:816–821
62. Mali P, Yang L, Esvelt KM et al (2013) RNA-guided human genome engineering via Cas9. Science 339:823–826
63. Rahdar M, McMahon MA, Prakash TP, Swayze EE, Bennett CF, Cleveland DW (2015) Synthetic CRISPR RNA-Cas9-guided genome editing in human cells. Proc Natl Acad Sci U S A 112(51):E7110–E7117
64. Hu JH, Miller SM, Geurts MH, Tang W et al (2018) Evolved Cas9 variants with broad PAM compatibility and high DNA specificity. Nature 556:57–63
65. Abudayyeh OO, Gootenberg JS, Essletzbichler P et al (2017) RNA targeting with CRISPR-Cas13. Nature 550:280–284
66. Wienert B, Wyman SK, Richardson CD, Akcakaya P, Porritt MJ et al (2019) Unbiased detection of CRISPR off-targets in vivo using DISCOVER-Seq. Science 364:286–289
67. Trasanidou D, Geros AS, Mohanraju P, Nieuwenweg AC, Nobrega FL, Staals RHJ (2019) Keeping CRISPR in check: diverse mechanisms of phage-encoded anti-CRISPRs. FEMS Microbial Lett 366:fnz098. https://doi.org/10.1093/femsle/fnz098
68. Sioud M (2019) Phage display libraries: from binders to targeted drug delivery and human therapeutics. Mol Biotechnol 61:286–303
69. Dybwad A, Bogen B, Natvig JB, Førre O, Sioud M (1995) Peptide phage libraries can be an efficient tool for identifying antibody ligands for polyclonal antisera. Clin Exp Immunol 102:438–42

70. Kjeldsen-Kragh J, Rashid T, Dybwad A, Sioud M, Haugen M, Førre O, Ebringer A (1995) Decrease in anti-Proteus mirabilis but not anti-Escherichia coli antibody levels in rheumatoid arthritis patients treated with fasting and a one year vegetarian diet. Ann Rheum Dis 54:221–4

71. Sioud M, Westby P, Vasovic V, Fløisand Y, Peng Q (2018) Development of a new high-affinity human antibody with antitumor activity against solid and blood malignancie. FASEB J 32:5063–5077

72. Cox DBT, Gootenberg JS, Abudayyeh OO et al (2017) Science 358:1019–1027

73. Kleinman ME, Yamada K, Takeda A, Chandrasekaran V et al (2008) Sequence- and target-independent angiogenesis suppression by siRNA via TLR3. Nature 452:591–597

Chapter 2

Chemical Modifications in RNA Interference and CRISPR/Cas Genome Editing Reagents

Kim A. Lennox and Mark A. Behlke

Abstract

Chemically modified oligonucleotides (ONs) are routinely used in the laboratory to assess gene function, and clinical advances are rapidly progressing as continual efforts are being made to optimize ON efficacy. Over the years, RNA interference (RNAi) has become one of the main tools used to inhibit RNA expression across a wide variety of species. Efforts have been made to improve the exogenous delivery of the double-stranded RNA components to the endogenous intracellular RNAi machinery to direct efficacious degradation of a user-defined RNA target. More recently, synthetic RNA ONs are being used to mimic the bacterial-derived CRISPR/Cas system to direct specific editing of the mammalian genome. Both of these techniques rely on the use of various chemical modifications to the RNA phosphate backbone or sugar in specific positions throughout the ONs to improve the desired biological outcome. Relevant chemical modifications also include conjugated targeting ligands to assist ON delivery to specific cell types. Chemical modifications are most beneficial for therapeutically relevant ONs, as they serve to enhance target binding, increase drug longevity, facilitate cell-specific targeting, improve internalization into productive intracellular compartments, and mitigate both sequence-specific as well as immune-related off-target effects (OTEs). The knowledge gained from years of optimizing RNAi reagents and characterizing the biochemical and biophysical properties of each chemical modification will hopefully accelerate the CRISPR/Cas technology into the clinic, as well as further expand the use of RNAi to treat currently undruggable diseases. This review discusses the most commonly employed chemical modifications in RNAi reagents and CRISPR/Cas guide RNAs and provides an overview of select publications that have demonstrated success in improving ON efficacy and/or mitigating undesired OTEs.

Key words Oligonucleotides, Chemical modification, siRNA, RNAi, Guide RNA, sgRNA, crRNA, CRISPR, Cas9, Cas12a

1 Introduction

The ability to alter genomic sequences or modulate transcriptome expression levels using synthetic oligonucleotides (ONs) is instrumental in determining the function of specific DNA or RNA sequences and additionally yields much hope in the efforts to develop nucleic acid-based therapeutics to treat diseases ranging from viral infections to cancer to hereditary disorders. The idea of

Mouldy Sioud (ed.), *RNA Interference and CRISPR Technologies: Technical Advances and New Therapeutic Opportunities*, Methods in Molecular Biology, vol. 2115, https://doi.org/10.1007/978-1-0716-0290-4_2,

introducing synthetic ONs into cells to target specific RNAs for knockdown dates back to 1978, where an unmodified DNA ON complementary to Rous sarcoma virus mRNA was introduced into chicken fibroblasts to inhibit viral expression [1]. While this was an excellent proof of concept, many cell types have defense mechanisms in place to eliminate exogenously introduced nucleic acids (single-stranded (ss) or double-stranded (ds) DNAs or RNAs), which are seen as evidence of potential infectious threats. These defense mechanisms can manifest as either rapid degradation by nuclease attack or by invoking an immune response against the threat. Evading both of these defense mechanisms is key to developing safe and effective nucleic acid tools and therapeutics and can be achieved by optimal selection and placement of chemical modifications.

Antisense oligonucleotides (ASOs) are short synthetic DNA ONs complementary to a targeted RNA and employ endogenous RNase H1 which hydrolyzes the RNA component of the RNA/DNA heteroduplex. With the earliest ASO studies, such as the seminal work targeting the Rous sarcoma virus, it was realized that unmodified DNA is rapidly degraded by nucleases present in the extra- and intracellular environment and is insufficient to produce potent ASOs. Subsequently, the evolution of fine-tuning unmodified DNA ASOs with optimal chemical modification strategies to produce efficacious compounds was underway. The focus of this review is on RNA-based tools to modulate gene expression or edit the genome, specifically using RNAi (RNA interference) and CRISPR/Cas (clustered regularly interspaced short palindromic repeats/CRISPR-associated) genome editing, respectively. However, much was learned from the historic effort to produce optimally chemically modified ASOs that can be translated to these RNA-based applications.

RNAi is an endogenous posttranscriptional gene regulation pathway that typically uses 21mer asymmetric double-stranded RNAs with 2-nucleotide (nt) 3′-overhangs as sequence-specific triggers for targeted RNA degradation ([2–5]; *see* Chapter 1). These 21mer small interfering RNAs (siRNAs) can be generated endogenously by excising them from larger dsRNAs via the endonuclease DICER or chemically synthesized as single strands that are annealed together and exogenously introduced into cells. The siRNAs are loaded into the RNA-induced silencing complex (RISC), which retains the antisense (AS) (or guide) strand and discards the sense (or passenger) strand [6, 7]. This enables the AS strand to be accessible to bind to the targeted complementary RNA; if perfect (desired) or near-perfect (often undesired) complementarity is achieved, the RNA is cleaved by the endonuclease Argonaute 2 (Ago2) [8, 9]. CRISPR/Cas genome editing is a natural bacterial/archaeal pathway that employs a guide RNA and a Cas nuclease which work together to create a sequence-specific double-stranded break in the genome that is often imperfectly repaired by nonhomologous end joining (NHEJ), creating insertions or deletions (indels) and potentially disrupting gene function [10]. Alternatively, a donor DNA repair template can be exogenously introduced that can be incorporated at the cleavage site via homologous recombination (HR), resulting in precise, user-defined genome changes [11–14]. The natural guide RNA consists of two separate yet partially complementary RNA molecules: a crisprRNA (crRNA) that specifically targets the genome and a bifunctional trans-activating crRNA (tracrRNA) that both anneals to the crRNA and has structural components that recruits the Cas enzyme for DNA cleavage. Alternatively, the crRNA and tracrRNA can be fused together to form a single guide RNA (sgRNA). The RNA components of both strategies can easily be chemically synthesized. The most commonly employed Cas enzyme is *Streptococcus pyogenes* Cas9 (Class 2, type II), which requires either the two-part system of crRNA (~36–42 nts) and tracrRNA (~67–80 nts) or a fused sgRNA (~100–102 nts) and recognizes an "NGG" protospacer adjacent motif (PAM) [11, 12]. *Acidaminococcus* sp. Cas12a (Class 2, type V; also known as Cpf1) is a higher-specificity Cas enzyme that functions differently than Cas9, using a single 43 nt crRNA that recognizes a "TTTV" PAM [15]. Cas12a from *Acidaminococcus* sp. and derivatives from other species are gaining popularity, especially in applications that are sensitive to off-target editing, in regions that are AT-rich, in plant systems and other organisms that function in reduced temperatures, or when an "NGG" PAM site is unavailable around the desired editing site [16–18].

While there is universal applicability for some chemical modification strategies across a wide variety of nucleic acid-based tools, it is important to appreciate that the protein machinery involved in the some of the different applications (e.g., ASOs, RNA interference, and CRISPR/Cas9 genome editing) can greatly influence the optimal design. With this in mind, structural and length variations and how these can affect chemical modification placement will also be discussed.

2 Chemical Modifications

RNA oligonucleotides are highly sensitive to nuclease degradation; dsRNAs do confer some degree of protection, yet they are still degraded within minutes in human plasma [19]. Exogenous RNA can be immunogenic when introduced into cells, stimulating toll-like receptors (TLRs) 3, 7, and 8 localized on the cell surface, as well as protein kinase R (PKR), retinoic acid inducible gene I (RIG-I), and melanoma differentiation associated protein 5 (MDA-5) localized in the cytoplasm [20]. Unmodified siRNAs

(~14 kDa) and CRISPR/Cas9 sgRNAs (~34 kDa) are highly charged macromolecules that cannot passively diffuse across the cell membrane, requiring the need for a delivery agent [21]. These characteristics can limit their use in cell culture for knockdown or gene editing experiments and can negatively impact in vivo utility. To overcome these limitations, chemical modifications to the RNA itself (e.g., the backbone, the furanose sugar, or the nucleobase) or conjugating a targeting ligand to the RNA can improve stability, binding affinity, immunogenicity, and bioavailability. While there is an exhaustive list of chemical modifications that have been tested for RNAi applications over the years, this review will focus only on the most promising modifications from a functional and therapeutic standpoint. The majority of the discussion in this section will be the historical successes and/or failures of each chemical modification when used independently in siRNAs. Combinatorial use of modifications will be explored separately for both siRNAs and guide RNAs in Subheadings 3.1 and 3.2, respectively.

2.1 Backbone Modifications

The internucleotide phosphodiester backbone is sensitive to both exo- and endonuclease hydrolysis limiting the longevity of unprotected ONs. The first and most widely used chemical modification applied to ASOs for stabilization is the phosphorothioate (PS) linkage, where a non-bridging phosphodiester oxygen is replaced with sulfur. This modification has three distinct advantages when placed throughout an ON: (1) it desensitizes the nucleic acid to nuclease digestion which increases the half-life; (2) it binds nonspecifically to proteins, particularly serum proteins, keeping it in circulation longer and improving bioavailability; and (3) it adheres to cell surface proteins such as stabilins, promoting cellular internalization and allowing improved uptake even without the use of a delivery tool (gymnotic delivery) [22]. Disadvantageously, PS linkages additively reduce the binding affinity of an ON with each additional incorporated linkage (~−0.5 °C per modification), effectively lowering the potency for applications that benefit from having high binding affinity to a complementary nucleic acid [23, 24]. PS-modified ONs can also bind nonspecifically to immune receptors which can invoke a pro-inflammatory response [25].

PS linkages have been incorporated into siRNAs as an attempt to improve nuclease stability, yet it has been demonstrated that they are not well tolerated by Ago2. Just a single PS linkage at the scissile phosphate (between position 9 and 10 in the sense strand) prevents sense strand cleavage and can reduce RNAi activity greater than sixfold in HeLa cytoplasmic extract [7, 26]. More extensive PS modification throughout the AS strand, sense strand, or both can also reduce potency [27]. However, limited PS incorporation into siRNAs at the terminal ends of both the sense and AS strand can increase the stability from exonuclease attack while maintaining potency [28] and may improve cellular uptake [29]. Thus, strategic positioning of PS linkages is used by most in vivo and clinical applications (reviewed in [30]).

Boranophosphate internucleotide linkages can also stabilize the phosphate backbone of an siRNA while still maintaining its ability to trigger RNAi [31]. Unfortunately, boranophosphate-modified RNAs are not easily manufactured by chemical synthesis and are synthesized using enzymatic methods, which limit its widespread use.

Few studies have successfully shown the utility of phosphonoacetate (PACE) or thioPACE linkages incorporated into the phosphate backbone of ONs [32]. PACE modifications substitute the non-bridging oxygen for an acetic acid group in the internucleotide linkage that results in enhanced nuclease resistance and an additive reduction in binding affinity (~1.0–1.3 °C) per linkage incorporated [33–35]. ThioPACE linkages build onto the PACE modification by replacing the remaining oxygen in the phosphate backbone with sulfur (similar to a phosphorothioate linkage) and reduce the binding affinity slightly more (~1.3–1.5 °C) than PACE. Since both PACE and thioPACE linkages reduce ON binding affinity, they have generally been used in the context of sugar modifications that increase the binding affinity in order to off-set the destabilizing effects (e.g. 2′OMe RNAs, which are discussed in Subheading 2.2) [36]. Threlfall et al. modified the terminal ends of the sense strand of an siRNA targeting firefly luciferase with PACE or thioPACE linkages, and annealed these partially modified sense strands with an unmodified AS strand [36]. The PACE linkages reduced siRNA potency regardless of the number incorporated; even one PACE linkage on each terminal end interfered with siRNA activity. The thioPACE versions performed better, and potency was maintained with the addition of 1–4 thioPACE linkages on the terminal ends of the sense strand. While potency wasn't enhanced with these backbone modifications in this in vitro reporter assay, it would be of interest to see how thioPACE linkages compare with PS linkages in vivo, where nuclease resistance is critical for functionality.

Recently, Dowdy et al. attempted to overcome the delivery hurdle of negatively charged phosphodiester siRNAs by incorporating a neutrally charged phosphotriester backbone (short-interfering ribonucleic neutral, or siRNNs) that increases serum stability by avoiding RNase recognition [37]. This bioreversible backbone is converted by thioesterases to the native phosphodiester backbone once in the cytosol, enabling the neutralized molecules to avoid renal clearance by binding serum albumin and then subsequently transitioning into negatively charged RNA molecules that can induce a robust RNAi response.

2.2 Sugar Modifications

Modifications to the ribose sugar are a common alteration to make in therapeutic ONs. Minor changes in the furanose sugar moiety can cause considerable changes to the biophysical and biological properties of ONs, especially at the 2′hydroxyl (2′-OH) position. This position is sensitive to nucleophilic attack, and modifying it can help increase ON stability in serum and in the intracellular environment. Additionally, 2′-OH modification can influence the conformation of the furanose ring, and when an electron-withdrawing group is present (e.g., 2′-*O*-methyl (2′OMe), 2′-*O*-methoxyethyl (2′MOE), and 2′-fluoro (2′F)), the sugar takes on the C3′-endo sugar pucker conformation and increases the binding affinity [38]. This conformation can also be restricted by tethering the 2′-OH and C4′ positions on the sugar with a methylene bridge, forming the locked nucleic acid (LNA) bicyclic nucleoside which has the highest duplex stability effects [39, 40]. The advantages and disadvantages of the aforementioned modifications, as well as successful but less commonly used sugar modifications, will be discussed in this section.

The 2′-*O*-Methyl (2′OMe) nucleoside is a naturally occurring RNA analog and one of the most widely used in RNA therapeutics. The incorporation of 2′OMe groups into ONs increases binding affinity and nuclease resistance (especially against endonucleases) [41] and maintains the A-form duplex [42], making them particularly well tolerated in siRNAs. RISC loading preferences can be manipulated by modifying the two nucleotides at the 5′-end of the desired sense strand with 2′OMe residues, which blocks sense strand loading in favor of AS strand loading [43]. A common modification strategy in siRNAs to improve stability while maintaining potency is to modify every alternate sugar with 2′OMe bases [28, 44, 45]. Similarly, it has also been demonstrated that sense strands fully modified with 2′OMe RNA can still trigger potent RNAi [28, 46]. Even though extensive 2′OMe modification is well tolerated in some instances, this is not a universally applicable strategy and certain sequences or siRNA positions are less tolerant to chemical modification than others [27, 44–50]. In fact, Davis et al. discovered that the more active a chemically modified sense strand works as a steric blocking anti-miRNA oligonucleotide, the less active it functions as an siRNA [51].

2′-*O*-methoxyethyl RNA (2′MOE) is a widely and successfully used sugar modification in antisense applications. The methoxyethyl substituent in 2′MOE is bulkier than in the 2′OMe analog, and structural studies indicate that the 2′MOE side chain increases the hydration and steric hindrance in the environment near the 3′-phosphate linkage, comparatively increasing resistance to nucleases [52]. In fact, 2′MOE residues provide remarkable resistance against exonucleases present in mouse liver cell extracts and human serum, compared to 2′OMe which provide limited (if any) protection (Lennox, internal observation). Similar to other 2′-OH modifications, the methoxyethyl side chain constrains the sugar into the C3′-endo conformation, increasing the binding affinity ~2 °C per insertion [53]. The increased bulkiness of this modification makes it less tolerated by Ago2, but strategic placement of 2′MOE modifications can maintain or improve siRNA performance. Prakash et al. tested siRNAs containing three consecutive 2′MOE residues walked throughout both siRNA strands and found that these 2′MOE triplet motifs were only tolerated in the AS strand when positioned centrally (in proximity to the Ago2 cleavage site), while the sense strand was more tolerant to 2′MOE modification even at the terminal ends [50]. Structural studies indicate that a bulky 2′MOE nucleoside would be tolerated by Ago2 at the first position at the 5′-end of the AS strand, while positions 2–7 (the seed region) would be intolerant to this modification [47]. However, 2′MOE placement on the 3′-overhangs of an siRNA has been successfully employed to increase nuclease stability without compromising activity [54, 55]. Incorporating 2′MOE residues has also been shown to reduce sense strand loading when placed at positions 9 and 10 of the AS strand and can often increase potency of the siRNA compared to its unmodified counterpart [56].

One of the most tested modifications in siRNAs is the 2′-fluoro (2′F) RNA, which has a fluorine at the 2′-ribose position that replaces the natural 2′-OH group. This highly electronegative group is similar in size to a hydrogen atom and drives the conformation toward the C3′-endo sugar pucker, causing an increase in binding affinity [38, 57]. Additionally, 2′F residues also impart greater resistance to nucleases and prolonged half-life in human serum when compared to their unmodified counterparts [19, 48, 58–61]. Presumably due to its small size and minimal steric hindrance, 2′F modification is well tolerated in both the AS and sense strands of an siRNA [62, 63]. There are conflicting reports regarding the effect that 2′F modification has in an siRNA; some studies indicate that when the AS strand is fully modified with 2′F residues, there can be a reduction in potency [64], while others demonstrate that a fully 2′F-modified siRNA was nearly equipotent to the unmodified version in reducing green fluorescent protein (GFP) expression [60]. It has also been shown that 2′F pyrimidine-modified anti-luciferase siRNAs can have increased nuclease resistance [19] and bioavailability [65] without having prolonged or increased activity in cell culture or in vivo. Conversely, others report dramatic improvement in in vitro and in vivo potency with a 2′F pyrimidine siRNA against endogenous factor VII mRNA [61]. This modification seems to be optimal when used in combination with other chemical modifications (*see* Subheading 3.1).

A less commonly applied chemical modification in siRNAs is the 2′-fluoro-β-d-arabinonucleic acid (2′F-ANA), which is similar to 2′F with the exception that the 2′-fluorine orientation is equatorial (in the β-position), structurally mimicking DNA instead of RNA [66]. Like 2′F, 2′F-ANA increases nuclease resistance and binding affinity toward RNA (~1.2 °C per insertion)(reviewed in [32]). In theory, the 2′F-ANA modification should be a less than ideal candidate for use in siRNAs because of its preference for the DNA-like C2′-endo sugar pucker; however, a small number of studies have shown that 2′F-ANA is compatible for use in siRNAs, especially when positioned in the sense strand [64, 67]. For example, Dowler et al. demonstrated that a fully 2′F-ANA modified sense strand prolonged fetal calf serum stability up to about 6 h (the unmodified siRNA was rapidly degraded) and improved luciferase knockdown up to fourfold [67].

Locked nucleic acid (LNA) is a bicyclic RNA analog that covalently links the 2′- and 4′-positions on the furanose ring with a methylene bridge, locking the residue into the C3′-endo sugar pucker which increases nuclease resistance and dramatically increases the binding affinity ~5.6 °C (4–8 °C) per insertion due to improved base stacking [39, 68]. As previously highlighted, one important step in siRNA functionality is the removal of the sense strand from the siRNA duplex, so incorporating multiple high binding affinity modifications such as LNAs may reduce the unwinding capability of the duplex and decrease siRNA efficacy. Minimal use of LNA modifications at the siRNA termini or at a select internal positions can still give a robust RNAi response [48] and can substantially enhance serum half-life when compared to an unmodified siRNA [69, 70]. One study systematically substituted LNA insertions throughout the antisense strand, pinpointing three sensitive positions that reduced RNAi activity [70]. It was also shown that LNA residues placed at the 5′-end of the sense strand increased the binding affinity, which introduces strand loading bias and thus reduces off-target effects [71]. Similar reports using minimally LNA-modified siRNAs formulated with chitosan demonstrated effective knockdown of GFP in bronchoepithelium when dosed intravenously in mice [72]. The methylated analog to LNA, the 2′,4′-constrained ethyl nucleic acid (cEt), shows LNA-like increases in duplex binding affinity and even further improves nuclease resistance [73, 74]. While this bridged nucleic acid (BNA) is less widely used, substitution of cEt modifications has been successful in ASOs with better toxicology profiles (less hepatotoxicity) than their LNA counterparts [75–78]. As of now, there are no reports using cEt modifications in siRNAs; however, they have been successfully substituted in crRNAs which is discussed in Subheading 3.2.

A less studied conformationally constrained sugar modification that has been used in siRNAs is tricyclo-DNA (tcDNA). Based on a tricyclic structure, contiguous stretches of this chemical modification can increase serum stability as well as binding affinity. Its structural similarity to RNA enables compatibility with the RNAi machinery [79]; however, not every position in the siRNA is amenable to tcDNA substitution. In one study, it was observed that select sense strand pyrimidine modification generally increased potency in an siRNA targeting EGFP, while AS strand activity was less tolerant to tcDNA modification [79]. Combining the most active tcDNA-modified AS and sense strands together further increased potency than when either modified strand was combined with their unmodified RNA complement.

The even less conformationally constrained unlocked nucleic acid (UNA), a non-nucleotide residue whereby the covalent C2′–C3′ bond of the ribose ring has been cleaved, increases the flexibility of the ON while reducing duplex T_m by 5–10 °C per insertion [80–82]. The incorporation of UNA residues into the 3′-overhangs of an siRNA can both reduce toxicity in cell culture and increase biostability in serum or when delivered IV in mouse [83, 84]. While UNA modification has been demonstrated to be tolerated by RISC, it has been employed mostly as a tool to reduce off-target effects by positioning UNAs in the seed region (particularly position 7 from the 5′-end of the AS strand) and/or at the 5′-end of the sense strand which favors AS strand loading into RISC [85, 86].

Recently, glycol nucleic acid (GNA) has been tested in the context of siRNAs by Alnylam Pharmaceuticals. GNA is a destabilizing flexible three-carbon, acyclic nucleic acid analog where the (S)-isomer will specifically hybridize with A/T-rich RNA sequences [87, 88]. GNA modifications positioned at the 3′-end of a phosphodiester DNA ON were resistant to 3′-exonuclease activity by snake venom phosphodiesterase (SVP). GNA substitutions in siRNAs are tolerated by RISC in a position-specific manner, with the most amenable positions being the seed region of the AS strand as well as positions 12–16 from the 5′-end of the AS strand. Modifications at these locations are able to maintain siRNA potency in cell culture; however, the duration and activity of GNA-modified siRNAs did not hold up for in vivo studies [89].

2.3 Termini Modifications

RNAi-mediated gene silencing requires a 5′-phosphate group on the AS strand for Ago2 recognition [90]. Unmodified siRNAs containing a 5′-OH will be intracellularly phosphorylated (e.g., by Clp1 kinase) and incorporated into RISC. Alternatively, a 5-′-phosphate group can easily be chemically added during siRNA synthesis. Both of these methods of 5′-phosphorylation of an siRNA AS strand are effective when delivering an siRNA in cell culture. However, the 5′-phosphate group is rapidly removed by phosphatases when delivered systemically in vivo, lowering siRNA efficacy [91]. This is especially problematic when the 5′-end of the AS strand is modified (e.g., with GalNAc, described below) which may reduce intracellular 5′-phosphorylation efficiency [86]. To prevent the dephosphorylation and subsequent (and potentially inefficient) rephosphorylation events from occurring in vivo, a nonhydrolyzable 5′-phosphate analog, such as 5′-(*E*)-vinyl-phosphonate (5′-*E*-VP, where the bridging 5′-oxygen is replaced with a double bond carbon group), can be introduced to improve siRNA efficacy [92–94]. This modification strategy is particularly valuable when using naked delivery of a chemically modified siRNA in vivo. The 5′-C-malonyl group is another nonhydrolyzable phosphate mimic that is recognized by Ago2 and can stabilize the 5′-end of the AS strand from exonuclease and phosphatase attack [95].

2.4 Conjugation of Targeting Ligands and Fatty Acids

In vivo efficacy of ONs is dependent on multiple factors, starting with introduction into the body, migration to the desired targeted tissue, cellular internalization, endosomal escape, and localizing to the appropriate intracellular compartment. While ASOs generally contain chemical modifications that facilitate increased bioavailability, cellular uptake, and intracellular trafficking (e.g., a fully phosphorothioate backbone), we have discussed that extensive modification of siRNAs isn't always amenable to producing a potent RNAi reagent. Simply injecting an siRNA in vivo with no delivery tool (gymnotically) is ineffective—they don't bind well to proteins, limiting their circulation time as well as protein-assisted cellular internalization. Over the years, the use of targeting ligands conjugated to ONs has been extensively studied to try and overcome the delivery challenges in model organisms from rodents all the way up to humans (reviewed in [96]). However, targeting ligands have varied levels of cell-specific selectivity. Ideally, the ultimate delivery achievement would include identification of a cell type-specific receptor that binds the ligand (and attached cargo) with high affinity combined with rapid internalization. Unfortunately, it is challenging to identify cell surface receptors that meet all of these criteria. Here, we discuss a handful of the more successful ligand conjugations for in vivo receptor-mediated delivery of siRNAs.

Hydrophobic lipid-based conjugates have an attractive appeal for ON delivery due to their association with serum lipoproteins which increases bioavailability, as well as their ability to facilitate receptor-mediated membrane permeability (*see* Chapter 3). Cholesterol conjugation was one of the first ligand-assisted delivery successes for delivering siRNAs in vivo to the liver [97, 98]. The seminal in vivo study using this approach employed an siRNA targeting apolipoprotein B (*apoB*) that was stabilized at the 3-′-termini with 2′OMe (AS strand) and PS (AS and sense strands) and conjugated to cholesterol (3′-end of the sense strand). This chemically modified siRNA was injected IV into mice and specifically reduced *apoB* mRNA levels and lowered total cholesterol levels, while the unconjugated control showed no activity [98]. Cholesterol can also be used to deliver ON cargo to extrahepatic tissue, including small intestine, kidney, adipose, lung, and brain [97–101]. Recently, it was demonstrated that a cholesterol-conjugated siRNA targeting myostatin (Mstn) could be delivered IV by tail vein injection to whole-body muscle in mouse [102]. In this study, a single 50 mg/kg dose reduced Mstn mRNA levels by 85–95% in multiple muscle tissues and improved muscle functionality.

Similar successes have also been reported with siRNA conjugation to α-tocopherol (vitamin E), a fat-soluble molecule that also binds serum lipoproteins and is taken up into hepatocytes [103]. In this example, the α-tocopherol was conjugated on the 5′-end of the AS strand of a 27/29mer chemically modified Dicer-substrate siRNA (see Subheading 3.3), with the intent that the α-tocopherol would be intracellularly removed by Dicer during the conversion of the 27/29mer into the active 21mer siRNA. This target-specific α-tocopherol-conjugated siRNA was able to markedly reduce *apoB* mRNA and reduce serum triglyceride and cholesterol levels when dosed at 2 mg/kg in mouse, compared with the unconjugated siRNA that showed no effect [103].

Recently, Ionis Pharmaceuticals screened a series of fatty acids conjugated to ASOs to see if they could identify a derivative that would facilitate ASO transport from circulation to muscle cells via albumin binding and increase ASO functionality in skeletal and cardiac muscles [104]. Prakash et al. found that palmitate was the most effective fatty acid for delivery to muscle in mouse, increasing knockdown of the nuclear long noncoding RNA MALAT1 with an ASO more than fivefold in the heart and quadricep [104]. In a follow-up study, Østergaard et al. compared cholesterol-, α-tocopherol-, and palmitate-conjugated ASOs targeting dystrophia myotonica protein kinase (DMPK) in mouse, rat, and monkey and importantly found that the effectiveness of the hydrophobic moiety correlated with the level of relevant plasma lipoproteins in each species [105]. They had to eliminate testing the cholesterol-modified siRNA in monkey due to toxicity from this ligand in mouse at higher doses. With the other two promising fatty acids, palmitate and α-tocopherol, they observed only a modest potency improvement in monkey. While these studies used ASOs as their platform, information from these in vivo delivery results may be translatable to other ON therapeutic technologies. In fact, another recent study by Biscans et al. looked at the tissue distribution of siRNAs conjugated to 15 different lipids (including cholesterol) and found that the saturated fatty acid docosanoic acid (DCA, with or without phosphocholine), when conjugated with fully chemically modified siRNAs, efficiently accumulated in extrahepatic tissues such as muscle, heart, fat, and lung in mice [106]. By optimizing the lipid chemical composition, it may be possible to bias the delivery of a conjugated siRNA to the targeted organ or alter the distribution to certain cell types; however, the majority of the injected dose will still be delivered to clearance organs such as the liver, kidney, and spleen.

Several key characteristics are important when trying to select successful ligand/receptor candidates for receptor-mediated uptake of ONs in vivo. The ligand should be amenable to conjugation onto an ON and not interfere with ON functionality and should deliver the ON to productive compartments in the cell. The receptor should be specific to the desired targeted cell type; it should be abundantly expressed and rapidly recycled following cargo internalization. The most successful targeting ligand to-date for productive uptake of an siRNA is the triantennary N-acetylgalactosamine (GalNAc) conjugate, which specifically and tightly binds to the asialoglycoprotein receptor (ASGPR) on the surface of hepatocytes for efficient delivery of ONs to the liver [107]. GalNAc is a particularly attractive ligand since its receptor is abundant on hepatocytes (~0.5–1.0 million copies per cell), and it is rapidly recycled approximately every 15 min [108, 109]. Initial studies using chemically stabilized siRNA-GalNAc conjugates demonstrated that GalNAc could efficiently deliver the ON cargo to hepatocytes both in vitro and in vivo and elicit a robust RNAi response against the targeted transthyretin (*TTR*) mRNA [107]. Subsequent studies looked at optimizing siRNA chemical composition to increase in vivo stability without negatively impacting on-target potency, which are discussed in Subheading 3.1 [110, 111]. Another successful ligand, eGPL1, has been used to deliver ASOs to pancreatic β-cells via the GLP1R receptor [112]. However, since ASOs work more effectively in the nucleus and siRNAs work more effectively in the cytoplasm [113], it is currently unknown whether this ligand/receptor strategy will also promote productive uptake of an siRNA to the cytoplasm of pancreatic β-cells.

3 Design Strategies for ON Optimization

The most commonly used chemical modifications in ON therapeutics have been outlined above and are instrumental in protecting ONs from nuclease degradation, modulating ON binding affinity, improving pharmacokinetic properties, and reducing off-target effects. Here, we discuss the challenges and successes in combinatorial use of chemical modifications for both siRNAs and CRISPR/Cas guide RNAs that promote all of these desired characteristics and generate ONs more applicable for clinical use. For example, siRNAs intended for in vivo applications may require extensive modification of both the AS and sense strand. This presents challenges, as the chemical modification strategy must be compatible with multiple protein interactions, and the thermodynamic properties of the duplex must be balanced for desired strand loading bias and proper unwinding of the sense strand to occur for siRNA functionality [6, 47, 51]. With this in mind, suboptimal modification (e.g., too extensive, too uniform, or of sensitive positions) can have a negative impact on siRNA efficacy. Additionally, the main focus of this review so far has been on standard canonical 21-mer duplexed siRNAs. Other strategies modifying the structure or length from the canonical design have also been successful in producing active, efficacious RNAi knockdown reagents and will also be discussed in this section.

CRISPR/Cas genome editing is a relatively new nucleic acid-based technique that relies on high performing nucleic acids to produce highly efficient and specific alterations to the genome. Coming in late to the game has distinct advantages—everything learned from the failures and successes of optimizing chemical modification strategies for ASOs and siRNAs, which is a wealth of information, can be applied to this new application. However, translatability of chemical modification successes isn't perfect, as different protein interactions are involved in each application. While the biophysical properties of commonly used chemical modifications are established (as previously discussed), optimizing the placement of the various chemical modifications in the context of the Cas proteins to maximize functionality of the guide RNAs is still necessary. This section includes discussion of the recent literature looking specifically at optimizing the chemical modification strategies for guide RNAs.

3.1 Chemical Modification Strategies for siRNAs

The most effective siRNA design strategies to date have been to simultaneously employ several different chemical modifications, particularly a combination of 2′OMe, 2′-F, and phosphorothioate linkages [62, 109, 114]. By strategically using a combination of modifications, siRNAs that are entirely chemically modified can still be potent triggers of RNAi. This was exemplified early on with an siRNA targeting hepatitis B virus in mice using a combination of 2′-F pyrimidines, 2′OMe purines (AS strand), DNA purines (sense strand), 3′-terminal PS linkage (AS strand), and 5′ and 3′ inverted abasic end caps (sense strand). Serum half-life of this duplex was 2–3 days compared to just minutes with the unmodified control siRNA, which translated into increased in vivo efficacy [115]. Today, the most commonly employed modification strategy for preparing siRNAs for effective in vivo use includes blocking the duplex termini from exonuclease attack and alternating 2′OMe residues with or without 2′-F pyrimidines. Each strand has a different chemical modification tolerability profile based on its function. The AS strand functions as an RNA, so chemical modifications that best mimic the natural 2′-OH conformation, such as 2′-F (similar size and charge), are generally well tolerated. The sense strand is less sensitive to architectural constraints and thus is often heavily modified with 2′OMe residues (reviewed in [109]). It is necessary for the siRNA duplex to be unwound for proper RISC function, so careful thermodynamic consideration needs to be taken into account to prevent siRNAs from having a binding affinity too high for unwinding to occur. Also, having structural malformation can also impede RNAi which can occur with duplex overwinding (full 2′-F) or underwinding (full 2′OMe); however, alternating these chemical modifications can balance out the duplex geometry and create an active, stable siRNA [109, 116]. In fact, all current siRNAs in clinical trials are extensively modified with a combination of 2′-F and 2′OMe residues [30, 96, 117]. There have been other successes with completely modified siRNAs, such as combining 2′F-ANA with 2′-F and/or LNA to create RNAi reagents more potent than their unmodified counterparts in cell culture [64].

Over the years, Alnylam has prolifically published on optimizing the chemical composition of their therapeutic-bound GalNAc-siRNAs [92, 107, 110, 111, 118]. Most recent studies by Alnylam demonstrate the importance of PS modifications on the 5′-end of both strands to confer stability, increase liver exposure, improve potency, and prolong the duration of gene silencing in both rodents and nonhuman primates [111]. A subsequent study looked at the design of a fully 2′OMe/2′-F GalNAc-conjugated duplex with the aim to maximize 2′OMe and minimize 2′-F content; this was achieved by identifying critical positions throughout the duplex where 2′OMe residues were not well tolerated and instead substituting the smaller 2′-F residues, while leaving the remaining positions 2′OMe modified. By this methodology, an optimized chemical composition was developed that reduced the 2′-F content by >50% (by increasing 2′OMe content), enhancing the efficacy of a GalNAc-siRNA targeting antithrombin (AT) in both mice and nonhuman primates [110]. After many years of development, Alnylam successfully managed to secure the first FDA-approved RNAi drug, Onpattro (patisiran), in treatment of peripheral nerve disease caused by hereditary transthyretin-mediated amyloidosis (hATTR) in adults [119]. Onpattro is modestly modified with 2′OMe pyrimidines in the sense strand and two 2’OMe uridines in the AS strand, with both strands having 3′-overhangs of two 2-′-deoxythymidine residues, while all other bases are unmodified RNA. This liver-bound siRNA is formulated as a lipid nanoparticle (LNP) instead of delivered via GalNAc conjugation, obviating the need for extensive chemical modification as the LNP will help protect the cargo from nucleases.

The major challenge to developing widespread adoption of therapeutic siRNAs is the ability to effectively deliver the siRNA to target cells in vivo. Extracellular vesicles (ELVs) are membrane-enclosed vesicular structures that transport nucleic acids, such as mRNA and microRNA, to recipient cells at various distances throughout the body [120, 121]. Exploiting ELVs as a nucleic acid drug delivery tool has recently become of great interest, due to their inherent ability to deliver endogenous RNAs as well as their in vivo biocompatibility [122]. However, one of the most challenging issues with widespread use of ELVs as delivery vesicles is optimizing reproducible and efficient loading of nucleic acid cargo, as ELVs are not well adapted to encapsulate exogenous hydrophilic macromolecules [123]. Early methods used to load hydrophilic synthetic siRNAs into ELVs include electroporating or transfecting; however, it was discovered that these methods can disrupt vesicle integrity or aggregate siRNAs, causing loading variability and poor performance [122, 124, 125]. To try and overcome these inadequacies in loading siRNAs *into* ELVs, hydrophobic cholesterol moieties were conjugated to chemically stabilized siRNA cargo and tested for reproducible loading *onto* ELVs (i.e., using cholesterol to anchor the siRNA to the surface of the lipid bilayer). Stremersch et al. had limited success with this strategy; while efficient ELV loading of chemically stabilized (siSTABLE from Dharmacon) siRNAs was achieved, there was no functional knockdown of the target mRNAs [122]. In a different study, Didiot et al. demonstrated that a cholesterol-conjugated chemically stabilized siRNA (2′-F, 2′OMe, 3′-PS) targeting the huntingtin (HTT) gene was efficiently loaded onto ELVs, internalized in vitro into primary cortical neurons or in vivo into mouse brain, and was functionally active in silencing HTT mRNA and protein [126]. As an extension to this study, Biscans et al. included other hydrophobic lipid moieties conjugated to a chemically modified siRNA targeting HTT (15-mer duplex with a 5′-*E*-VP and a 3′ single-stranded PS-modified tail on the AS strand, the lipid on the 3′-end of the sense strand, and alternating 2′OMe and 2′-F residues on both strands) to compare loading density onto ELVs and subsequent productive delivery of the siRNAs [127]. Fifteen different lipids (fatty acids, sterols, and vitamins) with a broad range of hydrophobicity were conjugated to the siRNA, and a direct correlation was observed between increased lipid-conjugate hydrophobicity, increased siRNA loading efficiency onto ELVs, and increased productive gene knockdown in primary mouse neurons, with α-tocopherol outperforming all other lipids tested [127]. Optimizing loading efficiency of siRNAs onto ELVs and thus increasing productive knockdown capabilities make ELV-siRNA delivery an attractive and viable option for nucleic acid therapeutics.

3.2 Chemical Modification Strategies for CRISPR/Cas Genome Editing

CRISPR/Cas genome editing is a relatively new nucleic acid-based technique that relies on high performing nucleic acids to produce highly efficient and specific alterations to the genome. Coming in late to the game has distinct advantages—everything learned from the failures and successes of optimizing chemical modification strategies for ASOs and siRNAs, which is a wealth of information, can be applied to this new application. However, translatability of chemical modification successes isn't perfect, as different protein interactions are involved in each application. While the biophysical properties of commonly used chemical modifications are established (as previously discussed), optimizing the placement of the various chemical modifications in the context of the Cas proteins to maximize functionality of the guide RNAs is still necessary.

The CRISPR/Cas9 system is a ribonucleoprotein (RNP) complex containing the Cas9 protein commonly from *Streptococcus pyogenes* and the guide RNA. The native nucleic acid components of the Cas9 guide RNA consist of two RNAs: a crRNA (42 nt) that directs the RNP complex via sequence homology to the targeted region of the genome and a universal tracrRNA (80 nt) that mediates the interaction between the crRNA and Cas9 protein, by binding to both and bringing them into proximity for genome editing to occur. Early attempts to recreate the CRISPR/Cas9 RNP in mammalian systems employed a fused version of the crRNA and tracrRNA, a ~102 nt single-guide RNA (sgRNA) that could be chemically synthesized or generated enzymatically by in vitro transcription (IVT). Not unexpectedly, similar problems that were encountered when introducing unmodified siRNAs into mammalian systems were also applicable to these guide RNAs. Unmodified RNAs are rapidly degraded in serum and the intracellular environment and can activate an undesired innate immune response. It is difficult to control the chemical modification patterns when generating sgRNAs by IVT, and universal or extensive modification of guide RNAs can interfere with Cas9 functionality [128]. Therefore, early reports attempting to increase genome editing efficiency through RNA stabilization used chemically synthesized sgRNAs to enable control of introducing specific chemical modifications. To improve stability and reduce immunostimulatory responses, Hendel et al. introduced either 2′OMe, 2′OMe 3′PS, or 2′OMe 3′thioPACE at the three terminal nucleotides on the flanking ends of 100-mer sgRNAs [129]. When looking at genome editing efficiency via NHEJ indel formation or HDR insertion (donor template consisting of ~0.8 kb homology arms with a GFP expression cassette), the chemically modified versions all outperformed the unmodified sgRNAs targeting three different genes (IL2RG, HBB, and CCR5). The 2′OMe end-modified sgRNAs only showed modest improvement, while the 2′OMe with the additional backbone modifications (PS or thioPACE) showed a substantial improvement over the unmodified sgRNA in editing efficiency (e.g., IL2RG was edited with modified sgRNA at 68–76% frequency compared to unmodified at 2.4% via NHEJ). Similar improvements in genome editing were seen when comparing the backbone-modified 2′OMe sgRNAs with the unmodified sgRNAs when co-delivering Cas9 either as an mRNA or as a recombinant protein (RNP) into K562, human primary T cells, or CD34+ hematopoietic stem and progenitor cells (HSPCs) [129]. In a follow-up study, Ryan et al. examined sgRNAs end modified with a single 2′OMe PACE for stability against exonucleases and then systematically substituted another destabilizing single 2′OMe PACE modification at each position of the 20 nt DNA recognition domain at the 5′-end of sgRNAs [35]. Destabilization of the DNA recognition domain was predicted to reduce off-target editing, and it was found that modifying position 5 or 11 from the 5′-end of the sgRNA with 2′OMe PACE did in fact reduce the off-target indels (in K562 cells, induced pluripotent stem cells (iPSCs) and HSPCs) and HDR template misincorporation (in K562 cells) without altering on-target activity when delivered as RNPs [35].Yin et al. examined the crystal structure of the Cas9-sgRNA complex to identify intolerant positions within the sgRNA that make contact with the Cas9 protein in order to maximize the chemical modifications at tolerant positions [128]. With this rational design strategy, they developed enhanced sgRNA (e-sgRNA) that contains 70/101 modified positions (2′OMe or 2′-F) with several PS linkages. The finalized e-sgRNA design was validated and compared with unmodified or 2′OMe PS end-modified (Hendel et al. design strategy) sgRNAs in vitro targeting GFP or human HBB and was found to outperform both versions when delivered with Cas9 mRNA (unmodified, 2′OMe PS end-modified, and e-sgRNA had ~0%, ~15%, and ~50% indels, respectively) [128, 129]. However, this difference in potency between 2′OMe PS end-modified sgRNA and e-sgRNA was leveled out when using high doses in combination with recombinant Cas9 protein (RNP), as the Cas9 protein can protect the RNA from degradation. When these three sgRNA variants were encapsulated in lipid nanoparticles (LNPs) and compared in vivo by IV injection in mice targeting PCSK9, Fah, or ROSA26, the genome editing trends correlated with in vitro data with e-sgRNA >2′OMe PS end-modified sgRNA > unmodified sgRNA (e.g., PCSK9 showed 65%, 25%, and 5% indel formation, respectively) [128]. A subsequent study evaluated the benefit of using a chemically modified sgRNA with 2′OMe PS ends and additional 2′OMe modifications in the sgRNA loop structures, co-delivered with Cas9 mRNA encapsulated in an LNP targeting murine transthyretin (Ttr) in vivo [130]. This more heavily modified sgRNA had higher genome editing efficiency than the sgRNA with only 2′OMe PS ends and the unmodified version. In fact, when IV administered in mice using a 3 mg/kg dose, ~70% genome editing was observed in the liver (compared to ~50% for the 2′OMe PS ends) for up to a year with a corresponding >97% knockdown of serum TTR levels [130].

A different approach to artificially recreating the CRISPR/Cas9 complex in mammalian systems is to introduce the guide RNA as a separate crRNA and tracrRNA two-part system, mimicking the natural bacterial components. There are advantages to this approach, such as simplifying chemical synthesis by using shorter oligonucleotides and being able to bulk produce the universal tracrRNA. Rahdar et al. first reported on the use of a synthetic crRNA that was modified with a combination of PS linkages, 2′OMe, 2′-F, and cEt substitutions [131]. The higher binding affinity chemical modifications enabled a truncated functional crRNA that could be shortened from the native 43 nt Cas9 crRNA down to 29 nt length (17 nt DNA recognition domain, 12 nt tracrRNA recognition domain). Used in combination with plasmids expressing Cas9 and the tracrRNA, they remarkably identified a chemical modification strategy (full-PS backbone with seven 2′-F in the 5′-end of the DNA recognition domain and five cEt in the tracrRNA recognition domain to improve intramolecular folding) where the 29 nt crRNA slightly outperformed an unmodified 102 nt sgRNA that was continually overexpressed from a plasmid. The shortening of the DNA recognition domain also reduced the off-target hit rate of the modified crRNA compared with the unmodified sgRNA [131]. Reasoning that optimal extensive chemical modification of the crRNA and tracrRNA may further enhance in vivo efficacy and provide better CRISPR/Cas9 reagents with increased metabolic stability for clinical applications, Mir et al. took lessons learned from Alnylam's heavily chemically modified siRNAs (*see* Subheading 3.1) and applied this rationale to CRISPR guide RNAs [132]. Similar to the e-sgRNA strategy, they examined the crystal structure of the Cas9 protein and crRNA:tracrRNA complex, identified the positions that do not come in contact with Cas9, and modified these positions with the bulkier 2′OMe residues. Other positions were rationally modified and remodified with 2′OMe, 2′-F, and PS linkages or left as unmodified 2′-OH to identify the optimal pattern that still maintained efficacy. The most potent combination of modified crRNA and tracrRNA (exhibiting higher genome editing that the unmodified counterpart) had >80% substitutions at the 2′-OH position on the ribose with the crRNA containing at least one modification (backbone or sugar) at each position. These highly modified RNAs maintained genome editing efficiency in HEK293T cells and human embryonic stem cells and were able to direct HDR of a GFP cassette with comparable activity. Of note, they identified a functional crRNA:tracrRNA combination where every position was chemically modified. A ~twofold reduction in genome editing was measured for this fully modified crRNA:tracrRNA when compared to the unmodified version; however, the value as an improved stabilized complex may be better appreciated when used in in vivo applications where enhanced nuclease resistance is required for functionality [132].

Like Cas9, the Cas12a (Cpf1) protein from *Acidaminococcus* sp. *BV3L6* functions to create double-stranded breaks in the genome. However, there are a few key differences between the requirements of these enzymes to fulfill their genome editing capabilities. Cas12a utilizes a single 43 nt crRNA (no universal tracrRNA) and instead recognizes a "TTTV" PAM adjacent to the targeted sequence in the genome [15]. The 5′-end of the crRNA forms a pseudoknot structure stabilized by intramolecular base pairing and is critical for recruiting Cas12a, while the 23 nt DNA specificity domain on the 3′-end of the crRNA contains a seed region required for the recognition of the targeted genomic sequence [133]. Similar to Cas9 guide RNAs, the Cas12a crRNA was predicted to benefit from incorporating chemical modifications to promote stabilization and reduce immune stimulation. McMahon et al. systematically studied length requirements and the positioning of a variety of chemical modifications throughout the Cas12a crRNA with guidance from the Cas12a-crRNA crystal structure, using the same experimental approach as they had previously done with Cas9 gRNA optimization [131, 134]. Truncation studies determined that gene editing activity could be mostly maintained with a 37 nt crRNA, generated by removing 1 nt from the 5′-end and 5 nts from the 3′-end. Due to chemical modification sensitivity of key nucleotide positions, the crRNA was extended with a 3 nt linker on the 5′-end (now a 40 nt crRNA) to effectively enable the crRNA to be end blocked with modifications without interfering with Cas12a activity. It was found that several positions in the crRNA pseudoknot structure as well as the seed region are intolerant to chemical modification. Rationally positioning PS linkages, 2′OMe, 2′-F, cEt, or DNA nucleotides at other positions throughout the crRNA (particularly at the terminal ends) were tolerated and mediated improved nuclease resistance with high levels of genome editing in HEK293T cells with a Cas12a expression plasmid [134].

3.3 Structure and Length Variations of siRNAs

Longer, asymmetric dsRNAs consisting of a 25 nt sense strand and a 27 nt AS strand with a 2 nt 3′-overhang can function as substrates for Dicer (Dicer-substrate RNAs, DsiRNAs), engaging the RNAi pathway prior to the Dicer processing step [135]. The asymmetric structure combined with two DNA bases at the 3′-end of the passenger strand introduces polarity during processing and influences strand selection, favoring AS strand loading into RISC and enforcing predictable dicing patterns that ensure the most active 21-mer is being produced [135]. As with the canonical 21-mer siRNAs, DsiRNAs can be chemically modified to enhance stability and reduce immune activation (discussed in Subheading 4.2), and universal optimized modification patterns for in vivo use are developed (internal observation) with the recognition that the Dicer cleavage domain needs to remain unmodified for proper DsiRNA processing [45, 103, 136, 137]. Dicerna Pharmaceuticals recently initiated a DsiRNA phase I clinical trial (first participant dosed in May 2018) using a GalNAc-conjugated universal tetraloop structure appended to a chemically modified DsiRNA, targeting lactate dehydrogenase A (LDHA) in treatment of primary hyperoxaluria type 1 (PH1) and type 2 (PH2). Data from this clinical trial was just released, demonstrating that a single dose was well tolerated and produced the expected phenotype by reducing urinary oxalate for participants with PH1 by an average of 71%, bringing participant urinary oxalate levels into the normal range (ClinicalTrials.gov: NCT03392896). Dicerna Pharmaceuticals also initiated another phase I clinical trial (first participant dosed in January 2019) using a similar compound to target hepatitis B virus (HBV) mRNA to treat HBV infection in adults (ClinicalTrials.gov: NCT03772249).

The small segmented interfering RNA (sisiRNA) is another RNAi construct that employs an intact antisense strand annealed to two shorter 10–12 nt LNA-modified sense strands, forming a tri-molecule construct [138]. This sense strand nicking strategy precludes undesired sense strand loading into RISC to mitigate off-target event that may occur from sense strand activity. The sense strand LNA modifications are required to increase binding affinity of the smaller sense strands to prevent premature strand separation, as well as increase sisiRNA serum stability. In fact, with this design, the AS strand was able to be more heavily modified than the canonical intact siRNA and still maintain potency, which may be beneficial if used in in vivo applications [138]. Similar to this concept are asymmetric interfering RNAs (aiRNAs), a strategy where the sense strand is shorter (down to 15 nt) than the traditional 21 nt sense strand, reducing sense strand RISC loading and lowering off-target effects [139–142]. However, aiRNA instability is problematic, and this strategy may be improved by the use of optimal chemical modifications. Petrova et al. took a different approach to bias AS strand loading by designing 2′OMe-modified fork-siRNAs that have thermodynamic asymmetry between the 5′- and 3′-end, promoting AS strand loading into RISC [143]. The reported fork-siRNA showed improved nuclease resistance and increased duration of silencing of the targeted ABCB1 gene.

Circular siRNAs are another unique structure recently developed by ligating the sense strand ends of a 21-mer canonical siRNA and subsequently annealing the AS strand to the circularized sense strand [144]. Circularizing the sense strand eliminates the free ends of this strand that are required for RISC loading and exonuclease degradation, forming a more stabilized siRNA with a reduced sense strand-related off-target effects. This siRNA construct was validated to have removed sense strand activity in a luciferase reporter assay and increased longevity when exposed to RNase A [144]. While the pilot study used unmodified RNA for validation, it is presumed that modifications could be incorporated to further protect the AS strand and reduce immune stimulation (discussed in Subheading 4.2). Circularization has also been employed to create an siRNA dumbbell structure which is more potent with prolonged activity when compared to the cognate open-ended siRNA [145, 146]. Similar to the circular siRNAs, this performance improvement is most likely due to their enhanced stability from eliminating exonuclease degradation and can be further enhanced by substituting DNA bases into the dumbbell loops.

Chemical modification strategies for synthetically manufactured short small hairpin RNAs (sshRNAs), RNAi triggers that have a 19 bp stem with a loop up to 10 nt between the AS and sense strand, have also been evaluated [147–149]. SshRNAs are too small for Dicer processing, and potent compounds can be designed with either the loop on the 3′-end of the AS strand (L sshRNAs) or the 5′-end of the AS strand (R sshRNAs) by loading into RISC intact. The addition of 2′OMe chemical modification benefits sshRNAs similarly to siRNAs (improved stability and reduced immune stimulation) and can also similarly impede function (over-modification of sensitive positions can reduce potency). Defining the biochemical properties of sshRNAs was accelerated through the use of 2′OMe chemical modification by substituting key positions to define if/how the hairpin was being processed by Dicer and Ago2 [148]. Of note, modification tolerance in the loop is different between the two structural variants (L versus R sshRNAs), due to requirements of needing a free 5′-phosphate available on the AS strand for RISC loading (whether synthetically appended or intracellularly phosphorylated; see Subheading 2.3). The loop of L sshRNAs can be chemically modified as the AS strand inherently has a free 5′-end available in this construct. However, the loop of R sshRNAs needs to remain an unmodified substrate for endogenous RNase cleavage to occur and open up the loop, providing the required free 5′-end on the AS strand for proper RISC loading [148].

Heavily chemically modified single-stranded siRNAs (ss-siRNAs) can also be used to successfully silence genes via an RNAi mechanism of action [150–154]. Similar to ASOs, chemical modification strategies can enable ss-siRNAs to be delivered in vivo to peripheral tissues gymnotically (without a delivery tool). Lima et al. performed a chemical modification survey to optimize the combination of sugar and backbone modifications that improved ss-siRNA potency by increasing stability and pharmacokinetics in vivo [150]. The most important modification for ss-siRNA activity is the inclusion of the metabolically stabilized 5′-*E*-VP, as a 5′-phosphate group is required for RNAi activation and intracellular phosphorylation can be impeded if the 5′-end is chemically modified. Along with the 5′-*E*-VP, the most potent ss-siRNA compound identified in this study also included 14 PS linkages positioned to be compatible with RISC, alternating 2′-F and 2′OMe residues, and two terminal 2′MOE analogs on the 3′-end; for in vivo use, the addition of a lipophilic C16 modification improved serum protein binding and cellular uptake [150]. This strategy was also successfully employed for allele-specific knockdown of the expanded CAG-repeat allele of huntingtin (HTT) in the mouse brain via intraventricular infusion, demonstrating the potential for ss-siRNAs to specifically target genes in the central nervous system [151, 153]. This allele-specific inhibition of a CAG repeat proof-of-principle study was bolstered by an in vitro *study* showing that ss-siRNAs could be used to specifically inhibit the CAG repeat alleles for Ataxin-3 and Atrophin-1 [154]. Pendergraff et al. evaluated different commercially available modifications to substitute at the 3′-end and found that 2′OMe and sometimes LNA residues can further enhance ss-siRNA potency in vitro [151].

4 Off-Target Effects (OTEs)

Physiological changes that occur in addition to the phenotype caused by degradation of the target RNA (RNAi) or desired genome editing event (CRISPR/Cas genome editing) are considered off-target effects (OTEs). No drug is entirely devoid of potential side effects, especially nucleic acid-derived drugs that rely on specific hybridization to a target nucleic acid for the mechanism of action. Binding affinity of ONs toward other nontargeted yet identical or near-identical sequences in the transcriptome (siRNAs) or genome (guide RNAs) make them particularly prone to OTEs. Additionally, their size, negative charge, and chemical modification strategy (e.g., PS backbone) can cause a variety of OTEs by binding nonspecifically to cell surface or intracellular proteins that inherently interact with polyanionic macromolecules or naturally function to bind nucleic acids. This can be particularly problematic with proteins tasked to elicit an innate immune response against foreign invaders. Understanding the underlying cause of the OTE and taking careful measure to avoid triggering it are critical for both interpretation of experimental results and ensuring that therapeutically relevant nucleic acid drugs are specific, safe, and tolerable.

4.1 Sequence-Specific OTEs

Widespread sequence-specific off-target changes in gene expression have been reported with the use of siRNAs [155]. These OTEs can result when an siRNA has near complementarity to a closely related sequence resulting in Ago2 slicing of a nontargeted gene. Efforts to reduce off-target cleavage using allele-specific ss-siRNAs were previously discussed in Subheading 3.3. Alternatively, extensive BLAST searching for specific sequences and avoiding problematic off-target sites can reduce OTEs of this nature. More difficult to address are OTEs resulting from complementarity of the 5′-end of the AS strand, which is analogous to the microRNA (miRNA) "seed region" that includes nucleotides 2–8 (herein called "AS seed region-equivalent"). If complementarity of this region occurs in the 3′UTR of an off-target gene, unintended translational inhibition and/or mRNA destabilization can occur [156–160]. Various strategies using chemical modifications in the AS seed region-equivalent domain of an siRNA have been shown to help mitigate these miRNA-like OTEs [43, 161–164]. Jackson et al. identified that 2′OMe residues placed at position 2 of the siRNA AS strand could alleviate miRNA-related OTEs an average of 66% compared to the unmodified counterpart, but for some sequences, the on-target potency was also reduced [43]. Ui-Tei et al. also demonstrated that full substitution of the siRNA AS seed region-equivalent with DNA nucleotides reduced miRNA-related OTEs without substantial decreases in siRNA activity, presumably by reducing the thermostability of seed-target interactions in the 3′UTR of off-target genes [164]. Similarly, Bramsen et al. studied ten different modification types in the AS seed region-equivalent and found that their best candidate was a destabilizing UNA at position 7, which consistently attenuated miRNA-like OTEs without compromising siRNA activity [163]. In a recent study, Lee et al. optimized the position in the AS seed region-equivalent to place an abasic dSpacer and found that substitution at position 6 conserved on target activity (80–100%) with no significant miRNA-related OTEs [162]. When compared to siRNAs containing a 2′OMe (position 2) or UNA (position 7), the abasic dSpacer (position 6) outperformed both previously validated chemical modification strategies with on target potency and OTE mitigation. Additionally, an siRNA targeting PCSK9 containing a position 6 dSpacer showed good in vivo activity with a desired reduction in plasma cholesterol levels and no deleterious off target phenotypes [162].

4.2 Immune-Related OTEs

Innate immunity is a rapid defensive response to potential biological threats that may be detrimental to the host. The intracellular presence of unmodified dsRNA is often an indicator of a viral infection and can cause the cell to respond by invoking an innate immune response, especially when the dsRNAs are longer than 30 base pairs (bps) [165]. It has also been recognized that shorter dsRNAs as well as ssRNA (such as CRISPR/Cas gRNAs) can activate mammalian innate immune responses by stimulating TLRs 3, 7, and 8, RIG-I, PKR, and MDA5 by both sequence-dependent and sequence-independent mechanisms [166–169]. It is important to understand the nature of these immune responses in order to evade or manipulate them when intracellularly employing exogenous synthetic nucleic acids. This understanding is critical, especially when using siRNAs in the clinic as some of the early siRNA successes may have been the result of unrecognized immune activation instead of RNAi-mediated target degradation [170]. Chemical modifications, such as 2′OMe and pseudouridine, are commonly found in mammalian tRNAs and rRNAs and can be used to distinguish "self" from "non-self" RNAs, avoiding undesired autoimmune responses. Foreign RNAs, such as viral RNAs, are recognized by the innate immune system by the absence of these naturally occurring chemical modifications [171]. Modifying siRNAs with as few as two 2′OMe modifications at any position in the siRNA (with the exception of two cytidine residues) can eliminate the immunogenicity that can occur with unmodified siRNAs in sensitive cells [171–175]. Even though chemically modifying one strand of the siRNA has been shown to abrogate immune activation, it is recommended that both strands be chemically modified in tolerant positions as a precaution to eliminate any subtle immune responses that could occur in vivo [176]. Other 2′-ribose modifications, such as 2′-F and LNA, can also help reduce immune activation; however, the effects are not as pronounced as 2′OMe RNA. Additionally, certain siRNA sequences are more immunostimulatory than others and mitigating the immune response in these instances may rely more on the position and extent of the chemical modification [166, 175]. It is general practice to incorporate at minimum 2′OMe modifications into therapeutically relevant siRNAs, but most are heavily modified at strategic positions with a combination of 2′OMe, 2′-F, PS linkages, and a stabilized 5-′-phosphate for clinical utility (reviewed in [109]). Less work has been published characterizing the immune effects that CRISPR/Cas gRNAs can invoke in mammalian cells; however, it has been empirically shown that unmodified sgRNAs can induce type I interferon production in vitro in peripheral blood mononuclear cells (PBMCs) and that substitution of 2′OMe residues can eliminate this response [177].

5 Concluding Remarks

Developing siRNAs as therapeutics has been a roller coaster of near successes followed by failure over the years, but the FDA's recent approval of patisiran has made the pipe dream a clinical reality. Now there are numerous RNAi drugs in various stages in clinical trials for a wide variety of diseases such as cancer, HIV/AIDS, age-related macular degeneration, hypercholesterolemia, and sickle cell disease (reviewed in [178]). None of these successes would have been possible without the diligence of so many scientists working on optimizing chemical modification strategies, figuring out how to improve the toxicology profile and reduce OTEs to ensure that RNAi drugs are safe and tolerable and searching out tricks and tools to deliver them effectively into the targeted disease-associated cells in the body. The recent success of RNAi therapeutics and all the knowledge gained from years of optimization will hopefully be translatable to some extent to CRISPR/Cas genome editing and help expedite safe reagents to the clinic to help treat undruggable diseases. This review touches on the most commonly used strategies that have for the most part been successful minimally in vitro; however, there are multiple other chemical modifications and structural variations in the literature (or that failed at the bench) that weren't included here but also contributed to the overall repertoire of knowledge necessary to hone in on the therapeutically viable constructs. One of the biggest hurdles still in ON therapeutics is targeting specific organs or cell populations, yet recent advancements seen with conjugated targeting ligands such as GalNAc are providing hope for a promising future of having an arsenal of cell type-specific delivery tools. It is also the hope of many scientists in the field to enlist the expertise of nucleic acid chemists to expand the chemical modification tool kit by developing new modifications with unique properties that could be further improve the field of ON therapeutics.

References

1. Stephenson ML, Zamecnik PC (1978) Inhibition of Rous sarcoma viral RNA translation by a specific oligodeoxyribonucleotide. Proc Natl Acad Sci U S A 75:285–288
2. Fire A, Xu S, Montgomery MK, Kostas SA, Driver SE, Mello CC (1998) Potent and specific genetic interference by double-stranded RNA in caenorhabditis elegans. Nature 391:806–811
3. Elbashir SM, Harborth J, Lendeckel W, Yalcin A, Weber K, Tuschl T (2001) Duplexes of 21-nucleotide RNAs mediate RNA interference in cultured mammalian cells. Nature 411:494–498
4. Zamore PD, Tuschl T, Sharp PA, Bartel DP (2000) RNAi: double-stranded RNA directs the ATP-dependent cleavage of mRNA at 21 to 23 nucleotide intervals. Cell 101:25–33
5. Meister G, Tuschl T (2004) Mechanisms of gene silencing by double-stranded RNA. Nature 431:343–349
6. Matranga C, Tomari Y, Shin C, Bartel DP, Zamore PD (2005) Passenger-strand cleavage facilitates assembly of siRNA into Ago2-containing RNAi enzyme complexes. Cell 123:607–620
7. Leuschner PJ, Ameres SL, Kueng S, Martinez J (2006) Cleavage of the siRNA passenger strand during RISC assembly in human cells. EMBO Rep 7:314–320
8. Liu J, Carmell MA, Rivas FV, Marsden CG, Thomson JM, Song JJ, Hammond SM, Joshua-Tor L, Hannon GJ (2004) Argonaute2 is the catalytic engine of mammalian RNAi. Science 305:1437–1441
9. Meister G, Landthaler M, Patkaniowska A, Dorsett Y, Teng G, Tuschl T (2004) Human Argonaute2 mediates RNA cleavage targeted by miRNAs and siRNAs. Mol Cell 15:185–197
10. Kim H, Kim JS (2014) A guide to genome engineering with programmable nucleases. Nat Rev Genet 15:321–334
11. Jinek M, Chylinski K, Fonfara I, Hauer M, Doudna JA, Charpentier E (2012) A programmable dual-RNA-guided DNA endonuclease in adaptive bacterial immunity. Science 337:816–821
12. Cong L, Ran FA, Cox D, Lin S, Barretto R, Habib N, Hsu PD, Wu X, Jiang W, Marraffini LA et al (2013) Multiplex genome engineering using CRISPR/Cas systems. Science 339:819–823
13. Cho SW, Kim S, Kim JM, Kim JS (2013) Targeted genome engineering in human cells with the Cas9 RNA-guided endonuclease. Nat Biotechnol 31:230–232
14. Mali P, Yang L, Esvelt KM, Aach J, Guell M, DiCarlo JE, Norville JE, Church GM (2013) RNA-guided human genome engineering via Cas9. Science 339:823–826
15. Zetsche B, Gootenberg JS, Abudayyeh OO, Slaymaker IM, Makarova KS, Essletzbichler P, Volz SE, Joung J, van der Oost J, Regev A et al (2015) Cpf1 is a single RNA-guided endonuclease of a class 2 CRISPR-Cas system. Cell 163:759–771
16. Kim H, Kim ST, Ryu J, Kang BC, Kim JS, Kim SG (2017) CRISPR/Cpf1-mediated DNA-free plant genome editing. Nat Commun 8:14406
17. Tang X, Lowder LG, Zhang T, Malzahn AA, Zheng X, Voytas DF, Zhong Z, Chen Y, Ren Q, Li Q et al (2017) A CRISPR-Cpf1 system for efficient genome editing and transcriptional repression in plants. Nat Plants 3:17103
18. Moreno-Mateos MA, Fernandez JP, Rouet R, Vejnar CE, Lane MA, Mis E, Khokha MK, Doudna JA, Giraldez AJ (2017) CRISPR-Cpf1 mediates efficient homology-directed repair and temperature-controlled genome editing. Nat Commun 8:2024
19. Layzer JM, McCaffrey AP, Tanner AK, Huang Z, Kay MA, Sullenger BA (2004) In vivo activity of nuclease-resistant siRNAs. RNA 10:766–771

20. Gantier MP, Williams BR (2007) The response of mammalian cells to double-stranded RNA. Cytokine Growth Factor Rev 18:363–371
21. Dowdy SF (2017) Overcoming cellular barriers for RNA therapeutics. Nat Biotechnol 35:222–229
22. Miller CM, Tanowitz M, Donner AJ, Prakash TP, Swayze EE, Harris EN, Seth PP (2018) Receptor-mediated uptake of phosphorothioate antisense oligonucleotides in different cell types of the liver. Nucleic Acid Ther 28:119–127
23. Freier SM, Altmann KH (1997) The ups and downs of nucleic acid duplex stability: structure-stability studies on chemically-modified DNA:RNA duplexes. Nucleic Acids Res 25:4429–4443
24. Lennox KA, Behlke MA (2010) A direct comparison of anti-microRNA oligonucleotide potency. Pharm Res 27:1788–1799
25. Yanai H, Chiba S, Ban T, Nakaima Y, Onoe T, Honda K, Ohdan H, Taniguchi T (2011) Suppression of immune responses by nonimmunogenic oligodeoxynucleotides with high affinity for high-mobility group box proteins (HMGBs). Proc Natl Acad Sci U S A 108:11542–11547
26. Schwarz DS, Tomari Y, Zamore PD (2004) The RNA-induced silencing complex is a Mg2+–dependent endonuclease. Curr Biol 14:787–791
27. Chiu YL, Rana TM (2003) siRNA function in RNAi: a chemical modification analysis. RNA 9:1034–1048
28. Choung S, Kim YJ, Kim S, Park HO, Choi YC (2006) Chemical modification of siRNAs to improve serum stability without loss of efficacy. Biochem Biophys Res Commun 342:919–927
29. Detzer A, Sczakiel G (2009) Phosphorothioate-stimulated uptake of siRNA by mammalian cells: a novel route for delivery. Curr Top Med Chem 9:1109–1116
30. Rettig GR, Behlke MA (2012) Progress toward in vivo use of siRNAs-II. Mol Ther 20:483–512
31. Hall AH, Wan J, Shaughnessy EE, Ramsay Shaw B, Alexander KA (2004) RNA interference using boranophosphate siRNAs: structure-activity relationships. Nucleic Acids Res 32:5991–6000
32. Deleavey GF, Damha MJ (2012) Designing chemically modified oligonucleotides for targeted gene silencing. Chem Biol 19:937–954
33. Dellinger DJ, Sheehan DM, Christensen NK, Lindberg JG, Caruthers MH (2003) Solid-phase chemical synthesis of phosphonoacetate and thiophosphonoacetate oligodeoxynucleotides. J Am Chem Soc 125:940–950
34. Sheehan D, Lunstad B, Yamada CM, Stell BG, Caruthers MH, Dellinger DJ (2003) Biochemical properties of phosphonoacetate and thiophosphonoacetate oligodeoxyribonucleotides. Nucleic Acids Res 31:4109–4118
35. Ryan DE, Taussig D, Steinfeld I, Phadnis SM, Lunstad BD, Singh M, Vuong X, Okochi KD, McCaffrey R, Olesiak M et al (2018) Improving CRISPR-Cas specificity with chemical modifications in single-guide RNAs. Nucleic Acids Res 46:792–803
36. Threlfall RN, Torres AG, Krivenko A, Gait MJ, Caruthers MH (2012) Synthesis and biological activity of phosphonoacetate- and thiophosphonoacetate-modified 2'-O-methyl oligoribonucleotides. Org Biomol Chem 10:746–754
37. Meade BR, Gogoi K, Hamil AS, Palm-Apergi-C, van den Berg A, Hagopian JC, Springer AD, Eguchi A, Kacsinta AD, Dowdy CF et al (2014) Efficient delivery of RNAi prodrugs containing reversible charge-neutralizing phosphotriester backbone modifications. Nat Biotechnol 32:1256–1261
38. Wan WB, Seth PP (2016) The medicinal chemistry of therapeutic oligonucleotides. J Med Chem 59:9645–9667
39. Koshkin AA, Nielsen P, Meldgaard M, Rajwanshi VK, Singh SK, Wengel J (1998) LNA (locked nucleic acid): an RNA mimic forming exceedingly stable LNA:LNA duplexes. J Am Chem Soc 120:2
40. Obika S, Nanbu D, Hari Y, Andoh J, Morio K, Doi T, Imanishi T (1998) Stability and structural features of the duplexes containing nucleoside analogues with a fixed N-type conformation, 2′-O,4′-C-methyleneribonucleosides. Tetrahedron Lett 39:4
41. Lennox KA, Owczarzy R, Thomas DM, Walder JA, Behlke MA (2013) Improved performance of Anti-miRNA oligonucleotides using a novel non-nucleotide modifier. Mol Ther Nucleic Acids 2:e117
42. Nishizaki T, Iwai S, Ohtsuka E, Nakamura H (1997) Solution structure of an RNA.2'-O-methylated RNA hybrid duplex containing an RNA.DNA hybrid segment at the center. Biochemistry 36:2577–2585
43. Jackson AL, Burchard J, Leake D, Reynolds A, Schelter J, Guo J, Johnson JM, Lim L, Karpilow J, Nichols K et al (2006) Position-specific chemical modification of siRNAs reduces “off-target” transcript silencing. RNA 12:1197–1205

44. Czauderna F, Fechtner M, Dames S, Aygun H, Klippel A, Pronk GJ, Giese K, Kaufmann J (2003) Structural variations and stabilising modifications of synthetic siRNAs in mammalian cells. Nucleic Acids Res 31:2705–2716
45. Collingwood MA, Rose SD, Huang L, Hillier C, Amarzguioui M, Wiiger MT, Soifer HS, Rossi JJ, Behlke MA (2008) Chemical modification patterns compatible with high potency dicer-substrate small interfering RNAs. Oligonucleotides 18:187–200
46. Kraynack BA, Baker BF (2006) Small interfering RNAs containing full 2'-O-methylribonucleotide-modified sense strands display Argonaute2/eIF2C2-dependent activity. RNA 12:163–176
47. Egli M, Manoharan M (2019) Re-engineering RNA molecules into therapeutic agents. Acc Chem Res 52:1036–1047
48. Braasch DA, Jensen S, Liu Y, Kaur K, Arar K, White MA, Corey DR (2003) RNA interference in mammalian cells by chemically-modified RNA. Biochemistry 42:7967–7975
49. Elbashir SM, Martinez J, Patkaniowska A, Lendeckel W, Tuschl T (2001) Functional anatomy of siRNAs for mediating efficient RNAi in drosophila melanogaster embryo lysate. EMBO J 20:6877–6888
50. Prakash TP, Allerson CR, Dande P, Vickers TA, Sioufi N, Jarres R, Baker BF, Swayze EE, Griffey RH, Bhat B (2005) Positional effect of chemical modifications on short interference RNA activity in mammalian cells. J Med Chem 48:4247–4253
51. Davis S, Lollo B, Freier S, Esau C (2006) Improved targeting of miRNA with antisense oligonucleotides. Nucleic Acids Res 34:2294–2304
52. Teplova M, Minasov G, Tereshko V, Inamati GB, Cook PD, Manoharan M, Egli M (1999) Crystal structure and improved antisense properties of 2'-O-(2-methoxyethyl)-RNA. Nat Struct Biol 6:535–539
53. Manoharan M (1999) 2′-carbohydrate modifications in antisense oligonucleotide therapy: importance of conformation, configuration and conjugation. Biochim Biophys Acta 1489:117–130
54. Hemmings-Mieszczak M, Dorn G, Natt FJ, Hall J, Wishart WL (2003) Independent combinatorial effect of antisense oligonucleotides and RNAi-mediated specific inhibition of the recombinant rat P2X3 receptor. Nucleic Acids Res 31:2117–2126
55. Dorn G, Patel S, Wotherspoon G, Hemmings-Mieszczak M, Barclay J, Natt FJ, Martin P, Bevan S, Fox A, Ganju P et al (2004) siRNA relieves chronic neuropathic pain. Nucleic Acids Res 32:e49
56. Song X, Wang X, Ma Y, Liang Z, Yang Z, Cao H (2017) Site-specific modification using the 2′-methoxyethyl group improves the specificity and activity of siRNAs. Mol Ther Nucleic Acids 9:242–250
57. Purser S, Moore PR, Swallow S, Gouverneur V (2008) Fluorine in medicinal chemistry. Chem Soc Rev 37:320–330
58. Muhonen P, Tennila T, Azhayeva E, Parthasarathy RN, Janckila AJ, Vaananen HK, Azhayev A, Laitala-Leinonen T (2007) RNA interference tolerates 2′-fluoro modifications at the Argonaute2 cleavage site. Chem Biodivers 4:858–873
59. Harborth J, Elbashir SM, Vandenburgh K, Manninga H, Scaringe SA, Weber K, Tuschl T (2003) Sequence, chemical, and structural variation of small interfering RNAs and short hairpin RNAs and the effect on mammalian gene silencing. Antisense Nucleic Acid Drug Dev 13:83–105
60. Blidner RA, Hammer RP, Lopez MJ, Robinson SO, Monroe WT (2007) Fully 2′-deoxy-2′-fluoro substituted nucleic acids induce RNA interference in mammalian cell culture. Chem Biol Drug Des 70:113–122
61. Manoharan M, Akinc A, Pandey RK, Qin J, Hadwiger P, John M, Mills K, Charisse K, Maier MA, Nechev L et al (2011) Unique gene-silencing and structural properties of 2-′-fluoro-modified siRNAs. Angew Chem Int Ed Engl 50:2284–2288
62. Watts JK, Deleavey GF, Damha MJ (2008) Chemically modified siRNA: tools and applications. Drug Discov Today 13:842–855
63. Deleavey GF, Watts JK, Damha MJ (2009) Chemical modification of siRNA. Curr Protoc Nucleic Acid Chem 16:13
64. Deleavey GF, Watts JK, Alain T, Robert F, Kalota A, Aishwarya V, Pelletier J, Gewirtz AM, Sonenberg N, Damha MJ (2010) Synergistic effects between analogs of DNA and RNA improve the potency of siRNA-mediated gene silencing. Nucleic Acids Res 38:4547–4557
65. Viel T, Boisgard R, Kuhnast B, Jego B, Siquier-Pernet K, Hinnen F, Dolle F, Tavitian B (2008) Molecular imaging study on in vivo distribution and pharmacokinetics of modified small interfering RNAs (siRNAs). Oligonucleotides 18:201–212
66. Trempe JF, Wilds CJ, Denisov AY, Pon RT, Damha MJ, Gehring K (2001) NMR solution structure of an oligonucleotide hairpin with a

2'F-ANA/RNA stem: implications for RNase H specificity toward DNA/RNA hybrid duplexes. J Am Chem Soc 123:4896–4903

67. Dowler T, Bergeron D, Tedeschi AL, Paquet L, Ferrari N, Damha MJ (2006) Improvements in siRNA properties mediated by 2′-deoxy-2′-fluoro-beta-D-arabinonucleic acid (FANA). Nucleic Acids Res 34:1669–1675
68. Eichert A, Behling K, Betzel C, Erdmann VA, Furste JP, Forster C (2010) The crystal structure of an 'All Locked' nucleic acid duplex. Nucleic Acids Res 38:6729–6736
69. Mook OR, Baas F, de Wissel MB, Fluiter K (2007) Evaluation of locked nucleic acid-modified small interfering RNA in vitro and in vivo. Mol Cancer Ther 6:833–843
70. Elmen J, Thonberg H, Ljungberg K, Frieden M, Westergaard M, Xu Y, Wahren B, Liang Z, Orum H, Koch T et al (2005) Locked nucleic acid (LNA) mediated improvements in siRNA stability and functionality. Nucleic Acids Res 33:439–447
71. Khvorova A, Reynolds A, Jayasena SD (2003) Functional siRNAs and miRNAs exhibit strand bias. Cell 115:209–216
72. Glud SZ, Bramsen JB, Dagnaes-Hansen F, Wengel J, Howard KA, Nyengaard JR, Kjems J (2009) Naked siLNA-mediated gene silencing of lung bronchoepithelium EGFP expression after intravenous administration. Oligonucleotides 19:163–168
73. Pallan PS, Allerson CR, Berdeja A, Seth PP, Swayze EE, Prakash TP, Egli M (2012) Structure and nuclease resistance of 2′,4-′-constrained 2'-O-methoxyethyl (cMOE) and 2'-O-ethyl (cEt) modified DNAs. Chem Commun (Camb) 48:8195–8197
74. Seth PP, Vasquez G, Allerson CA, Berdeja A, Gaus H, Kinberger GA, Prakash TP, Migawa MT, Bhat B, Swayze EE (2010) Synthesis and biophysical evaluation of 2′,4′-constrained 2'O-methoxyethyl and 2′,4′-constrained 2'O-ethyl nucleic acid analogues. J Org Chem 75:1569–1581
75. Seth PP, Siwkowski A, Allerson CR, Vasquez G, Lee S, Prakash TP, Wancewicz EV, Witchell D, Swayze EE (2009) Short antisense oligonucleotides with novel 2′-4′ conformationaly restricted nucleoside analogues show improved potency without increased toxicity in animals. J Med Chem 52:10–13
76. Hong D, Kurzrock R, Kim Y, Woessner R, Younes A, Nemunaitis J, Fowler N, Zhou T, Schmidt J, Jo M et al (2015) AZD9150, a next-generation antisense oligonucleotide inhibitor of STAT3 with early evidence of clinical activity in lymphoma and lung cancer. Sci Transl Med 7:314ra185
77. Burel SA, Han SR, Lee HS, Norris DA, Lee BS, Machemer T, Park SY, Zhou T, He G, Kim Y et al (2013) Preclinical evaluation of the toxicological effects of a novel constrained ethyl modified antisense compound targeting signal transducer and activator of transcription 3 in mice and cynomolgus monkeys. Nucleic Acid Ther 23:213–227
78. Pandey SK, Wheeler TM, Justice SL, Kim A, Younis HS, Gattis D, Jauvin D, Puymirat J, Swayze EE, Freier SM et al (2015) Identification and characterization of modified antisense oligonucleotides targeting DMPK in mice and nonhuman primates for the treatment of myotonic dystrophy type 1. J Pharmacol Exp Ther 355:310–321
79. Ittig D, Luisier S, Weiler J, Schumperli D, Leumann CJ (2010) Improving gene silencing of siRNAs via tricyclo-DNA modification. Artif DNA PNA XNA 1:9–16
80. Mangos MM, Min KL, Viazovkina E, Galarneau A, Elzagheid MI, Parniak MA, Damha MJ (2003) Efficient RNase H-directed cleavage of RNA promoted by antisense DNA or 2'F-ANA constructs containing acyclic nucleotide inserts. J Am Chem Soc 125:654–661
81. Campbell MA, Wengel J (2011) Locked vs. unlocked nucleic acids (LNA vs. UNA): contrasting structures work towards common therapeutic goals. Chem Soc Rev 40:5680–5689
82. Langkjaer N, Pasternak A, Wengel J (2009) UNA (unlocked nucleic acid): a flexible RNA mimic that allows engineering of nucleic acid duplex stability. Bioorg Med Chem 17:5420–5425
83. Bramsen JB, Laursen MB, Nielsen AF, Hansen TB, Bus C, Langkjaer N, Babu BR, Hojland T, Abramov M, Van Aerschot A et al (2009) A large-scale chemical modification screen identifies design rules to generate siRNAs with high activity, high stability and low toxicity. Nucleic Acids Res 37:2867–2881
84. Laursen MB, Pakula MM, Gao S, Fluiter K, Mook OR, Baas F, Langklaer N, Wengel SL, Wengel J, Kjems J et al (2010) Utilization of unlocked nucleic acid (UNA) to enhance siRNA performance in vitro and in vivo. Mol BioSyst 6:862–870
85. Vaish N, Chen F, Seth S, Fosnaugh K, Liu Y, Adami R, Brown T, Chen Y, Harvie P, Johns R et al (2011) Improved specificity of gene silencing by siRNAs containing unlocked

nucleobase analogs. Nucleic Acids Res 39:1823–1832
86. Kenski DM, Cooper AJ, Li JJ, Willingham AT, Haringsma HJ, Young TA, Kuklin NA, Jones JJ, Cancilla MT, McMasters DR et al (2010) Analysis of acyclic nucleoside modifications in siRNAs finds sensitivity at position 1 that is restored by 5′-terminal phosphorylation both in vitro and in vivo. Nucleic Acids Res 38:660–671
87. Zhang L, Peritz A, Meggers E (2005) A simple glycol nucleic acid. J Am Chem Soc 127:4174–4175
88. Schlegel MK, Peritz AE, Kittigowittana K, Zhang L, Meggers E (2007) Duplex formation of the simplified nucleic acid GNA. Chembiochem 8:927–932
89. Schlegel MK, Foster DJ, Kel'in AV, Zlatev I, Bisbe A, Jayaraman M, Lackey JG, Rajeev KG, Charisse K, Harp J et al (2017) Chirality dependent potency enhancement and structural impact of glycol nucleic acid modification on siRNA. J Am Chem Soc 139:8537–8546
90. Martinez J, Patkaniowska A, Urlaub H, Luhrmann R, Tuschl T (2002) Single-stranded antisense siRNAs guide target RNA cleavage in RNAi. Cell 110:563–574
91. Weitzer S, Martinez J (2007) The human RNA kinase hClp1 is active on 3′ transfer RNA exons and short interfering RNAs. Nature 447:222–226
92. Parmar R, Willoughby JL, Liu J, Foster DJ, Brigham B, Theile CS, Charisse K, Akinc A, Guidry E, Pei Y et al (2016) 5′-(E)-vinylphosphonate: a stable phosphate mimic can improve the RNAi activity of siRNA-GalNAc conjugates. Chembiochem 17:985–989
93. Prakash TP, Lima WF, Murray HM, Li W, Kinberger GA, Chappell AE, Gaus H, Seth PP, Bhat B, Crooke ST et al (2015) Identification of metabolically stable 5′-phosphate analogs that support single-stranded siRNA activity. Nucleic Acids Res 43:2993–3011
94. Prakash TP, Lima WF, Murray HM, Elbashir S, Cantley W, Foster D, Jayaraman M, Chappell AE, Manoharan M, Swayze EE et al (2013) Lipid nanoparticles improve activity of single-stranded siRNA and gapmer antisense oligonucleotides in animals. ACS Chem Biol 8:1402–1406
95. Zlatev I, Foster DJ, Liu J, Charisse K, Brigham B, Parmar RG, Jadhav V, Maier MA, Rajeev KG, Egli M et al (2016) 5'-C-Malonyl RNA: small interfering RNAs modified with 5′-monophosphate bioisostere demonstrate gene silencing activity. ACS Chem Biol 11:953–960
96. Juliano RL (2016) The delivery of therapeutic oligonucleotides. Nucleic Acids Res 44:6518–6548
97. Lorenz C, Hadwiger P, John M, Vornlocher HP, Unverzagt C (2004) Steroid and lipid conjugates of siRNAs to enhance cellular uptake and gene silencing in liver cells. Bioorg Med Chem Lett 14:4975–4977
98. Soutschek J, Akinc A, Bramlage B, Charisse K, Constien R, Donoghue M, Elbashir S, Geick A, Hadwiger P, Harborth J et al (2004) Therapeutic silencing of an endogenous gene by systemic administration of modified siRNAs. Nature 432:173–178
99. Wolfrum C, Shi S, Jayaprakash KN, Jayaraman M, Wang G, Pandey RK, Rajeev KG, Nakayama T, Charrise K, Ndungo EM et al (2007) Mechanisms and optimization of in vivo delivery of lipophilic siRNAs. Nat Biotechnol 25:1149–1157
100. Krutzfeldt J, Rajewsky N, Braich R, Rajeev KG, Tuschl T, Manoharan M, Stoffel M (2005) Silencing of microRNAs in vivo with 'antagomirs'. Nature 438:685–689
101. DiFiglia M, Sena-Esteves M, Chase K, Sapp E, Pfister E, Sass M, Yoder J, Reeves P, Pandey RK, Rajeev KG et al (2007) Therapeutic silencing of mutant huntingtin with siRNA attenuates striatal and cortical neuropathology and behavioral deficits. Proc Natl Acad Sci U S A 104:17204–17209
102. Khan T, Weber H, DiMuzio J, Matter A, Dogdas B, Shah T, Thankappan A, Disa J, Jadhav V, Lubbers L et al (2016) Silencing myostatin using cholesterol-conjugated siRNAs induces muscle growth. Mol Therapy Nucleic Acids 5:e342
103. Nishina K, Unno T, Uno Y, Kubodera T, Kanouchi T, Mizusawa H, Yokota T (2008) Efficient in vivo delivery of siRNA to the liver by conjugation of alpha-tocopherol. Mol Ther 16:734–740
104. Prakash TP, Mullick AE, Lee RG, Yu J, Yeh ST, Low A, Chappell AE, Ostergaard ME, Murray S, Gaus HJ et al (2019) Fatty acid conjugation enhances potency of antisense oligonucleotides in muscle. Nucleic Acids Res 47:6029–6044
105. Ostergaard ME, Jackson M, Low A, A EC, R GL, Peralta RQ, Yu J, Kinberger GA, Dan A, Carty R et al (2019) Conjugation of hydrophobic moieties enhances potency of antisense oligonucleotides in the muscle of rodents and non-human primates. Nucleic Acids Res 47:6045–6058

106. Biscans A, Coles A, Haraszti R, Echeverria D, Hassler M, Osborn M, Khvorova A (2018) Diverse lipid conjugates for functional extra-hepatic siRNA delivery in vivo. Nucleic Acids Res 47:1082–1096
107. Nair JK, Willoughby JL, Chan A, Charisse K, Alam MR, Wang Q, Hoekstra M, Kandasamy P, Kel'in AV, Milstein S et al (2014) Multivalent N-acetylgalactosamine-conjugated siRNA localizes in hepatocytes and elicits robust RNAi-mediated gene silencing. J Am Chem Soc 136:16958–16961
108. Spiess M (1990) The asialoglycoprotein receptor: a model for endocytic transport receptors. Biochemistry 29:10009–10018
109. Khvorova A, Watts JK (2017) The chemical evolution of oligonucleotide therapies of clinical utility. Nat Biotechnol 35:238–248
110. Foster DJ, Brown CR, Shaikh S, Trapp C, Schlegel MK, Qian K, Sehgal A, Rajeev KG, Jadhav V, Manoharan M et al (2018) Advanced siRNA designs further improve in vivo performance of GalNAc-siRNA conjugates. Mol Ther 26:708–717
111. Nair JK, Attarwala H, Sehgal A, Wang Q, Aluri K, Zhang X, Gao M, Liu J, Indrakanti R, Schofield S et al (2017) Impact of enhanced metabolic stability on pharmacokinetics and pharmacodynamics of GalNAc-siRNA conjugates. Nucleic Acids Res 45:10969–10977
112. Ammala C, Drury WJ 3rd, Knerr L, Ahlstedt I, Stillemark-Billton P, Wennberg-Huldt C, Andersson EM, Valeur E, Jansson-Lofmark R, Janzen D et al (2018) Targeted delivery of antisense oligonucleotides to pancreatic beta-cells. Sci Adv 4:eaat3386
113. Lennox KA, Behlke MA (2016) Cellular localization of long non-coding RNAs affects silencing by RNAi more than by antisense oligonucleotides. Nucleic Acids Res 44:863–877
114. Dar SA, Thakur A, Qureshi A, Kumar M (2016) siRNAmod: a database of experimentally validated chemically modified siRNAs. Sci Rep 6:20031
115. Morrissey DV, Blanchard K, Shaw L, Jensen K, Lockridge JA, Dickinson B, McSwiggen JA, Vargeese C, Bowman K, Shaffer CS et al (2005) Activity of stabilized short interfering RNA in a mouse model of hepatitis B virus replication. Hepatology 41:1349–1356
116. Allerson CR, Sioufi N, Jarres R, Prakash TP, Naik N, Berdeja A, Wanders L, Griffey RH, Swayze EE, Bhat B (2005) Fully 2′-modified oligonucleotide duplexes with improved in vitro potency and stability compared to unmodified small interfering RNA. J Med Chem 48:901–904
117. Fitzgerald K, Kallend D, Simon A (2017) A highly durable RNAi therapeutic inhibitor of PCSK9. N Engl J Med 376:e38
118. Rajeev KG, Nair JK, Jayaraman M, Charisse K, Taneja N, O'Shea J, Willoughby JL, Yucius K, Nguyen T, Shulga-Morskaya S et al (2015) Hepatocyte-specific delivery of siRNAs conjugated to novel non-nucleosidic trivalent N-acetylgalactosamine elicits robust gene silencing in vivo. Chembiochem 16:903–908
119. Adams D, Gonzalez-Duarte A, O'Riordan WD, Yang CC, Ueda M, Kristen AV, Tournev I, Schmidt HH, Coelho T, Berk JL et al (2018) Patisiran, an RNAi therapeutic, for hereditary transthyretin amyloidosis. N Engl J Med 379:11–21
120. Valadi H, Ekstrom K, Bossios A, Sjostrand M, Lee JJ, Lotvall JO (2007) Exosome-mediated transfer of mRNAs and microRNAs is a novel mechanism of genetic exchange between cells. Nat Cell Biol 9:654–659
121. Pitt JM, Kroemer G, Zitvogel L (2016) Extracellular vesicles: masters of intercellular communication and potential clinical interventions. J Clin Invest 126:1139–1143
122. Stremersch S, Vandenbroucke RE, Van Wonterghem E, Hendrix A, De Smedt SC, Raemdonck K (2016) Comparing exosome-like vesicles with liposomes for the functional cellular delivery of small RNAs. J Control Release 232:51–61
123. van der Meel R, Fens MH, Vader P, van Solinge WW, Eniola-Adefeso O, Schiffelers RM (2014) Extracellular vesicles as drug delivery systems: lessons from the liposome field. J Control Release 195:72–85
124. Wahlgren J, De LKT, Brisslert M, Vaziri Sani F, Telemo E, Sunnerhagen P, Valadi H (2012) Plasma exosomes can deliver exogenous short interfering RNA to monocytes and lymphocytes. Nucleic Acids Res 40:e130
125. Kooijmans SAA, Stremersch S, Braeckmans K, de Smedt SC, Hendrix A, Wood MJA, Schiffelers RM, Raemdonck K, Vader P (2013) Electroporation-induced siRNA precipitation obscures the efficiency of siRNA loading into extracellular vesicles. J Control Release 172:229–238
126. Didiot MC, Hall LM, Coles AH, Haraszti RA, Godinho BM, Chase K, Sapp E, Ly S, Alterman JF, Hassler MR et al (2016) Exosome-mediated delivery of

hydrophobically modified siRNA for huntingtin mRNA silencing. Mol Ther 24:1836–1847

127. Biscans A, Haraszti RA, Echeverria D, Miller R, Didiot MC, Nikan M, Roux L, Aronin N, Khvorova A (2018) Hydrophobicity of lipid-conjugated siRNAs predicts productive loading to small extracellular vesicles. Mol Ther 26:1520–1528
128. Yin H, Song CQ, Suresh S, Wu Q, Walsh S, Rhym LH, Mintzer E, Bolukbasi MF, Zhu LJ, Kauffman K et al (2017) Structure-guided chemical modification of guide RNA enables potent non-viral in vivo genome editing. Nat Biotechnol 35:1179–1187
129. Hendel A, Bak RO, Clark JT, Kennedy AB, Ryan DE, Roy S, Steinfeld I, Lunstad BD, Kaiser RJ, Wilkens AB et al (2015) Chemically modified guide RNAs enhance CRISPR-Cas genome editing in human primary cells. Nat Biotechnol 33:985–989
130. Finn JD, Smith AR, Patel MC, Shaw L, Youniss MR, van Heteren J, Dirstine T, Ciullo C, Lescarbeau R, Seitzer J et al (2018) A single administration of CRISPR/Cas9 lipid nanoparticles achieves robust and persistent in vivo genome editing. Cell Rep 22:2227–2235
131. Rahdar M, McMahon MA, Prakash TP, Swayze EE, Bennett CF, Cleveland DW (2015) Synthetic CRISPR RNA-Cas9-guided genome editing in human cells. Proc Natl Acad Sci U S A 112:E7110–E7117
132. Mir A, Alterman JF, Hassler MR, Debacker AJ, Hudgens E, Echeverria D, Brodsky MH, Khvorova A, Watts JK, Sontheimer EJ (2018) Heavily and fully modified RNAs guide efficient SpyCas9-mediated genome editing. Nat Commun 9:2641
133. Swarts DC, van der Oost J, Jinek M (2017) Structural basis for guide RNA processing and seed-dependent DNA targeting by CRISPR-Cas12a. Mol Cell 66(221–233):e224
134. McMahon MA, Prakash TP, Cleveland DW, Bennett CF, Rahdar M (2018) Chemically modified Cpf1-CRISPR RNAs mediate efficient genome editing in mammalian cells. Mol Ther 26:1228–1240
135. Rose SD, Kim DH, Amarzguioui M, Heidel JD, Collingwood MA, Davis ME, Rossi JJ, Behlke MA (2005) Functional polarity is introduced by Dicer processing of short substrate RNAs. Nucleic Acids Res 33:4140–4156
136. Kubo T, Zhelev Z, Ohba H, Bakalova R (2007) Modified 27-nt dsRNAs with dramatically enhanced stability in serum and long-term RNAi activity. Oligonucleotides 17:445–464
137. Kubo T, Zhelev Z, Ohba H, Bakalova R (2008) Chemically modified symmetric and asymmetric duplex RNAs: an enhanced stability to nuclease degradation and gene silencing effect. Biochem Biophys Res Commun 365:54–61
138. Bramsen JB, Laursen MB, Damgaard CK, Lena SW, Babu BR, Wengel J, Kjems J (2007) Improved silencing properties using small internally segmented interfering RNAs. Nucleic Acids Res 35:5886–5897
139. Yuan Z, Wu X, Liu C, Xu G, Wu Z (2012) Asymmetric siRNA: new strategy to improve specificity and reduce off-target gene expression. Hum Gene Ther 23:521–532
140. Sun X, Rogoff HA, Li CJ (2008) Asymmetric RNA duplexes mediate RNA interference in mammalian cells. Nat Biotechnol 26:1379–1382
141. Chang CI, Yoo JW, Hong SW, Lee SE, Kang HS, Sun X, Rogoff HA, Ban C, Kim S, Li CJ et al (2009) Asymmetric shorter-duplex siRNA structures trigger efficient gene silencing with reduced nonspecific effects. Mol Ther 17:725–732
142. Hong SW, Park JH, Yun S, Lee CH, Shin C, Lee DK (2014) Effect of the guide strand 3′-end structure on the gene-silencing potency of asymmetric siRNA. Biochem J 461:427–434
143. Petrova Kruglova NS, Meschaninova MI, Venyaminova AG, Zenkova MA, Vlassov VV, Chernolovskaya EL (2010) 2'-O-methyl-modified anti-MDR1 fork-siRNA duplexes exhibiting high nuclease resistance and prolonged silencing activity. Oligonucleotides 20:297–308
144. Zhang L, Liang D, Chen C, Wang Y, Amu G, Yang J, Yu L, Dmochowski IJ, Tang X (2018) Circular siRNAs for reducing off-target effects and enhancing long-term gene silencing in cells and mice. Mol Ther Nucleic Acids 10:237–244
145. Abe N, Abe H, Ito Y (2007) Dumbbell-shaped nanocircular RNAs for RNA interference. J Am Chem Soc 129:15108–15109
146. Abe N, Abe H, Nagai C, Harada M, Hatakeyama H, Harashima H, Ohshiro T, Nishihara M, Furukawa K, Maeda M et al (2011) Synthesis, structure, and biological activity of dumbbell-shaped nanocircular RNAs for RNA interference. Bioconjug Chem 22:2082–2092
147. Ge Q, Dallas A, Ilves H, Shorenstein J, Behlke MA, Johnston BH (2010) Effects of chemical

modification on the potency, serum stability, and immunostimulatory properties of short shRNAs. RNA 16:118–130

148. Dallas A, Ilves H, Ge Q, Kumar P, Shorenstein J, Kazakov SA, Cuellar TL, McManus MT, Behlke MA, Johnston BH (2012) Right- and left-loop short shRNAs have distinct and unusual mechanisms of gene silencing. Nucleic Acids Res 40:9255–9271
149. Ge Q, Ilves H, Dallas A, Kumar P, Shorenstein J, Kazakov SA, Johnston BH (2010) Minimal-length short hairpin RNAs: the relationship of structure and RNAi activity. RNA 16:106–117
150. Lima WF, Prakash TP, Murray HM, Kinberger GA, Li W, Chappell AE, Li CS, Murray SF, Gaus H, Seth PP et al (2012) Single-stranded siRNAs activate RNAi in animals. Cell 150:883–894
151. Yu D, Pendergraff H, Liu J, Kordasiewicz HB, Cleveland DW, Swayze EE, Lima WF, Crooke ST, Prakash TP, Corey DR (2012) Single-stranded RNAs use RNAi to potently and allele-selectively inhibit mutant huntingtin expression. Cell 150:895–908
152. Pendergraff HM, Debacker AJ, Watts JK (2016) Single-stranded silencing RNAs: hit rate and chemical modification. Nucleic Acid Ther 26:216–222
153. Hu J, Liu J, Yu D, Aiba Y, Lee S, Pendergraff H, Boubaker J, Artates JW, Lagier-Tourenne C, Lima WF et al (2014) Exploring the effect of sequence length and composition on allele-selective inhibition of human huntingtin expression by single-stranded silencing RNAs. Nucleic Acid Ther 24:199–209
154. Hu J, Liu J, Narayanannair KJ, Lackey JG, Kuchimanchi S, Rajeev KG, Manoharan M, Swayze EE, Lima WF, Prakash TP et al (2014) Allele-selective inhibition of mutant atrophin-1 expression by duplex and single-stranded RNAs. Biochemistry 53:4510–4518
155. Jackson AL, Bartz SR, Schelter J, Kobayashi SV, Burchard J, Mao M, Li B, Cavet G, Linsley PS (2003) Expression profiling reveals off-target gene regulation by RNAi. Nat Biotechnol 21:635–637
156. Jackson AL, Burchard J, Schelter J, Chau BN, Cleary M, Lim L, Linsley PS (2006) Widespread siRNA "off-target" transcript silencing mediated by seed region sequence complementarity. RNA 12:1179–1187
157. Birmingham A, Anderson EM, Reynolds A, Ilsley-Tyree D, Leake D, Fedorov Y, Baskerville S, Maksimova E, Robinson K, Karpilow J et al (2006) 3' UTR seed matches, but not overall identity, are associated with RNAi off-targets. Nat Methods 3:199–204
158. Vickers TA, Lima WF, Wu H, Nichols JG, Linsley PS, Crooke ST (2009) Off-target and a portion of target-specific siRNA mediated mRNA degradation is Ago2 'Slicer' independent and can be mediated by Ago1. Nucleic Acids Res 37:6927–6941
159. Jackson AL, Linsley PS (2010) Recognizing and avoiding siRNA off-target effects for target identification and therapeutic application. Nat Rev Drug Discov 9:57–67
160. Wilczynska A, Bushell M (2015) The complexity of miRNA-mediated repression. Cell Death Differ 22:22–33
161. Janas MM, Schlegel MK, Harbison CE, Yilmaz VO, Jiang Y, Parmar R, Zlatev I, Castoreno A, Xu H, Shulga-Morskaya S et al (2018) Selection of GalNAc-conjugated siRNAs with limited off-target-driven rat hepatotoxicity. Nat Commun 9:723
162. Lee HS, Seok H, Lee DH, Ham J, Lee W, Youm EM, Yoo JS, Lee YS, Jang ES, Chi SW (2015) Abasic pivot substitution harnesses target specificity of RNA interference. Nat Commun 6:10154
163. Bramsen JB, Pakula MM, Hansen TB, Bus C, Langkjaer N, Odadzic D, Smicius R, Wengel SL, Chattopadhyaya J, Engels JW et al (2010) A screen of chemical modifications identifies position-specific modification by UNA to most potently reduce siRNA off-target effects. Nucleic Acids Res 38:5761–5773
164. Ui-Tei K, Naito Y, Zenno S, Nishi K, Yamato K, Takahashi F, Juni A, Saigo K (2008) Functional dissection of siRNA sequence by systematic DNA substitution: modified siRNA with a DNA seed arm is a powerful tool for mammalian gene silencing with significantly reduced off-target effect. Nucleic Acids Res 36:2136–2151
165. Minks MA, West DK, Benvin S, Baglioni C (1979) Structural requirements of double-stranded RNA for the activation of 2′,5-′-oligo(A) polymerase and protein kinase of interferon-treated HeLa cells. J Biol Chem 254:10180–10183
166. Hornung V, Guenthner-Biller M, Bourquin C, Ablasser A, Schlee M, Uematsu S, Noronha A, Manoharan M, Akira S, de Fougerolles A et al (2005) Sequence-specific potent induction of IFN-alpha by short interfering RNA in plasmacytoid dendritic cells through TLR7. Nat Med 11:263–270

167. Hornung V, Ellegast J, Kim S, Brzozka K, Jung A, Kato H, Poeck H, Akira S, Conzelmann KK, Schlee M et al (2006) 5-′-Triphosphate RNA is the ligand for RIG-I. Science 314:994–997
168. Marques JT, Williams BR (2005) Activation of the mammalian immune system by siRNAs. Nat Biotechnol 23:1399–1405
169. Robbins M, Judge A, MacLachlan I (2009) siRNA and innate immunity. Oligonucleotides 19:89–102
170. Robbins M, Judge A, Ambegia E, Choi C, Yaworski E, Palmer L, McClintock K, MacLachlan I (2008) Misinterpreting the therapeutic effects of small interfering RNA caused by immune stimulation. Hum Gene Ther 19:991–999
171. Kariko K, Buckstein M, Ni H, Weissman D (2005) Suppression of RNA recognition by Toll-like receptors: the impact of nucleoside modification and the evolutionary origin of RNA. Immunity 23:165–175
172. Robbins M, Judge A, Liang L, McClintock K, Yaworski E, MacLachlan I (2007) 2'-O-methyl-modified RNAs act as TLR7 antagonists. Mol Ther 15:1663–1669
173. Judge AD, Bola G, Lee AC, MacLachlan I (2006) Design of noninflammatory synthetic siRNA mediating potent gene silencing in vivo. Mol Ther 13:494–505
174. Whitehead KA, Dahlman JE, Langer RS, Anderson DG (2011) Silencing or stimulation? siRNA delivery and the immune system. Annu Rev Chem Biomol Eng 2:77–96
175. Cekaite L, Furset G, Hovig E, Sioud M (2007) Gene expression analysis in blood cells in response to unmodified and 2-′-modified siRNAs reveals TLR-dependent and independent effects. J Mol Biol 365:90–108
176. Judge AD, Robbins M, Tavakoli I, Levi J, Hu L, Fronda A, Ambegia E, McClintock K, MacLachlan I (2009) Confirming the RNAi-mediated mechanism of action of siRNA-based cancer therapeutics in mice. J Clin Invest 119:661–673
177. Schubert MS, Cedrone E, Neun B, Behlke MA, Dobrovolskaia MA (2018) Chemical modification of CRISPR gRNAs eliminate type I interferon responses in human peripheral blood mononuclear cells. J Cytokine Biol 3:121
178. Setten RL, Rossi JJ, Han SP (2019) The current state and future directions of RNAi-based therapeutics. Nat Rev Drug Discov 18:421–446

Chapter 3

Preparation, Determination of Activity, and Biodistribution of Cholesterol-Containing Nuclease-Resistant siRNAs In Vivo

Ivan V. Chernikov, Mariya I. Meschaninova, and Elena L. Chernolovskaya

Abstract

RNA interference (RNAi) is a powerful tool for suppressing gene expression associated with various diseases that are not amenable to treatment with low molecular weight drugs. Despite significant progress in this area, the potential for therapeutic use of RNAi in humans is limited due to the lack of efficient delivery systems. Bioconjugation is one of the most promising methods for delivering siRNA to cells and tissues, since conjugation of siRNA with molecules capable of penetrating cells through natural transport mechanisms can provide specificity of delivery without toxic effects and unwanted immunostimulation. Here we describe the design, preparation, and in vivo evaluation of cholesterol-containing siRNA conjugates able to accumulate in the tumor, penetrate into cells without a carrier, and suppress the expression of the target genes.

Key words Lipophilic siRNA, Nuclease-resistant siRNA, In vivo imaging, Mice models

1 Introduction

The introduction of molecules of natural origin into siRNAs is one of the promising methods for nonviral delivery. Unlike other approaches (delivery using cationic lipids and polymers, high pressure injections), the advantages of using bioconjugates include the specificity of their penetration into target cells and the absence of toxic effects [1]. A feature of this approach is the need to select the appropriate modification to solve a specific problem due to the specificity of ligand-receptor interactions. The ability of cholesterol [2], α-tocopherol [3], aptamers [4], antibodies [5], and cell-penetrating peptides [6] to alter the bioavailability and biodistribution of siRNA was demonstrated in several studies. The use of cholesterol conjugates of siRNA in vivo is not strictly specific because LDL receptors are expressed at a high level by various

Mouldy Sioud (ed.), *RNA Interference and CRISPR Technologies: Technical Advances and New Therapeutic Opportunities*, Methods in Molecular Biology, vol. 2115, https://doi.org/10.1007/978-1-0716-0290-4_3,

types of cells but more advantageous if siRNA delivery to a specific cell type only is not within the scope of the task.

Previously, we proposed an experimental algorithm for the site-specific modification of siRNAs, based on mapping of their nuclease-sensitive sites in the presence of serum followed by incorporation of 2′-O-methyl analogs of ribonucleotides at the identified positions of cleavage [7]. Based on RNA resistant to nucleases, we designed conjugates of anti-*MDR1* siRNA with cholesterol attached through an optimized linker that are capable of penetrating efficiently into cells in a carrier-free mode, to silence the expression of *P*-glycoprotein and to restore the sensitivity of drug-resistant cancer cells to vinblastine [8]. A convenient combination of H-phosphonate and phosphoramidite methods was developed for the synthesis of 5′-lipophilic conjugates of siRNAs [8]. It was found that lipophilic siRNAs are able to effectively penetrate into cancer cells when used in a micromolar concentration range [8]. The efficiency of the uptake is dependent upon the type of lipophilic moiety, the length of the linker between the moiety and the siRNA, and cell type. Among all the conjugates tested, the cholesterol-conjugated siRNAs with linkers containing from six to ten carbon atoms demonstrate the optimal uptake and gene silencing properties [8]. We have demonstrated that the attachment of cholesterol to siRNA provided its efficient accumulation in the liver and reduced its retention in the kidneys, both in healthy and tumor-bearing mice after IV and IP administration. Cholesterol-containing selectively 2′-O-methyl-modified siRNA after IV, IP, or PT injection was able to overcome all barriers from the injection site to the cytoplasm of tumor cells, accumulated there in sufficient amounts, and suppressed the expression of the target gene [2]. Here we provide detailed protocols for the design, preparation, and in vivo evaluation of cholesterol-containing siRNA conjugates able to accumulate in the tumor, penetrate into cells without a carrier, and suppress the expression of the target genes.

2 Materials

Prepare all solutions using ultrapure water obtained by Simplicity® Water Purification System (Millipore) or equivalent, to attain a sensitivity of 18 MΩ cm at 25 °C and molecular biology grade reagents.

Store all oligonucleotides and their conjugates at −20 °C.

2.1 Synthesis of H-Phosphonate of Cholesterol-3-(carboxyamino-hexan-6-ol) (See Note 1)

1. Dichloromethane (CH_2Cl_2), ≥99.8%, ~20 ppm amylene as stabilizer, water (H_2O) <0.02%.
2. Pyridine (Py), water (H_2O) <0.01%.
3. Cholesteryl chloroformate, 95%.
4. Triethylamine (TEA), ≥99.5%, water (H_2O) ≤0.1%.
5. 6-Aminohexan-1-ol, 97%.
6. Imidazole anhydrous, ≥99%, water (H_2O) ≤0.2%.
7. Phosphorus trichloride (PCl_3), 99%.
8. Ethyl acetate (EtOAc), ≥99.7%, water (H_2O) ≤0.03%.
9. *n*-Hexane, water (H_2O) <0.01%.
10. Methanol (CH_3OH), ≥99.9%.
11. Sodium bicarbonate ($NaHCO_3$), ≥95.0%.
12. Sodium sulfate (Na_2SO_4), ≥99.0%, anhydrous, granular.
13. Kieselgel F_{254} thin-layer chromatography plate.
14. Silica gel, pore size 60 Å, 230–400 mesh.
15. Rotary evaporator system, Rotavapor R-200.
16. Standard organic laboratory glassware.
17. UV lamp (254 nm).
18. Spectrophotometer AV-300.

2.2 Synthesis of Components of siRNA and Their Conjugates

1. Acetonitrile (AcN), water (H_2O) <0.02%.
2. Tetrahydrofuran (THF), ~300 ppm butylhydroxytoluene as inhibitor, water (H_2O) <0.02%.
3. Dichloromethane (CH_2Cl_2), ≥99.8%, ~20 ppm amylene as stabilizer, water (H_2O) <0.02%.
4. Pyridine (Py), water (H_2O) <0.01%.
5. 5′,2′,*N*-protected phosphoramidites (Glen Research): 5′-O-dimethoxytrityl-2′-O-*tert*-butyldimethylsilyl(TBDMS)-N^6-benzoyl-adenosine-3′-*N*,*N*-diisopropyl-(2-cyanoethyl)phosphoramidite (A(bz)-RNA-CE); 5′-O-dimethoxytrityl-2′-O-TBDMS-N^4-acetyl-cytidine-3′-*N*,*N*-diisopropyl-(2-cyanoethyl)phosphoramidite (C(ac)-RNA-CE); 5′-O-dimethoxytrityl-2′-O-TBDMS-N^2-acetyl-guanosine-3′-*N*,*N*-diisopropyl-(2-cyanoethyl)phosphoramidite (G(ac)-RNA-CE); 5′-O-dimethoxytrityl-2′-O-TBDMS-uridine-3′-*N*,*N*-diisopropyl-(2-cyanoethyl)phosphoramidite (U-RNA-CE); 5′-O-dimethoxytrityl-2′-O-methyl-N^4-acetyl-cytidine-3′-*N*,*N*-diisopropyl-(2-cyanoethyl)phosphoramidite (2′-OMe-C(ac)-RNA-CE); 5′-O-dimethoxytrityl-2′-O-methyl-uridine-3′-*N*,*N*-diisopropyl-(2-cyanoethyl)phosphoramidite (2′-OMe-U-RNA-CE). Phosphoramidite solutions were prepared in AcN at concentrations of 0.1 M for 2′-O-TBDMS-phosphoramidites and 0.05 M for 2′-O-methyl-

phosphoramidites and were generally stable for up to 2 weeks when stored under inert atmosphere (*see* **Notes 2–4**).

6. Solid support (Glen Research): 5′-O-dimethoxytrityl-2′-O-acetyl-N^6-acetyl-adenosine-3′-succinoyl-long chain alkylamino-CPG, 5′-dimethoxytrityl-2′-O-acetyl-N^2-acetyl-guanosine-3′-succinoyl-long chain alkylamino-CPG or *N*-(6-(O-dimethoxytrityl)-hexyl)-(2-carboxamide)-phthalimidyl-lcaa-CPG 500 Å pore size and ~24–35 μmol loading.
7. Activator for coupling: 0.25 M solution of ethylthiotetrazole (ETT) in AcN (*see* **Note 3**).
8. Capping reagent A: THF/acetic anhydride/2,6-lutidine, 8/1/1, v/v/v (Glen Research).
9. Capping reagent B: THF/*N*-methylimidazole, 84/16, v/v (Glen Research) (*see* **Note 3**).
10. Oxidizing reagent A: A 0.02 M solution of iodine in THF/H_2O/Py, 45/4.5/0.5, v/v/v.
11. Detritylating reagent: 3% dichloroacetic acid (DCA) (w/v) in CH_2Cl_2.
12. DNA/RNA synthesizer ASM-800 (*see* **Notes 5 and 6**).
13. Columns and filters for synthesizer.
14. Pivaloyl chloride, 99%.
15. Oxidizing reagent B: A 0.2 M solution of iodine in Py/H_2O, 98/2, v/v (*see* **Note 7**).
16. Tris–HCl buffer (0.5 M, pH 8.3).
17. *N*-Hydroxysuccinimide esters of Cyanine5.5 dye.
18. Dimethyl sulfoxide (DMSO), ≥99.5%.
19. Solution for precipitated: 2% (w/v) $NaClO_4$ (≥98.0%) solution in acetone.
20. Acetone, water (H_2O) <0.2%.

2.3 Cleavage and Deprotection

1. Cleavage reagent (AMA solution): 40% methylamine in water/concentrated ammonium hydroxide solution, 1/1, v/v, freshly prepared (*see* **Notes 8** and **9**).
2. Washing reagent: AcN/H_2O/ethanol, 1/1/1, v/v/v.
3. TBDMS deprotecting cocktail: *N*-methylpyrrolidinone/triethylamine/triethylamine tris(hydrogen fluoride), 3/1.5/2, v/v/v, freshly prepared (*see* **Note 10**).
4. Ethoxytrimethylsilane, 98%.
5. Diethyl ether.
6. Acetone, water (H_2O) <0.2%.
7. Speedvac Concentrator Plus.

8. Thermomixer Comfort.
9. Centrifuge 5424 R.

2.4 Analysis and Purification of Oligoribonucleotides and Their Conjugates

1. Ethanol, 94–96%.
2. Acetone, water (H_2O) <0.2%.
3. *N,N,N,N'*-Tetramethylethylenediamine (TEMED). Store at 4 °C.
4. 10× TBE buffer: 0.5 M Tris–HCl, pH8.3, 0.5 M boric acid, 10 mM EDTA.
5. Ammonium persulfate (APS): 10% (w/v) solution in water, freshly prepared (*see* **Note 11**).
6. Denaturing polyacrylamide gel (stock) for oligonucleotide isolation and analysis: 12% solution of acrylamide/*N,N'*-methylenebisacrylamide (30/0.5), 7 M urea with 1× TBE buffer (*see* **Note 12**).
7. Loading buffer: 7 M urea, 0.025% bromphenol blue, 0.025% xylene cyanol FF.
8. Stains-All (Sigma-Aldrich): 0.05% (w/v) solution in formamide/water, 1/1, v/v.
9. Elution solution A: 0.3 M NaOAc (≥99.0%) solution in water, pH 5.5. Store at 4 °C.
10. Elution solution B: 0.3 M $NaClO_4$ (≥98.0%) solution in water.
11. Solution for precipitated: 2% (w/v) $NaClO_4$ solution in acetone.
12. 50% (v/v) solution of AcN in water.
13. Vertical electrophoresis apparatus.
14. Kieselgel F_{254} thin-layer chromatography plate for visualizing oligonucleotides and their conjugates on gels by UV shadowing covered with polyethylene or polypropylene film.
15. UV lamp (254 nm).
16. Sep Pak C18 Cartridges.
17. Amicon 3 K®.
18. Thermomixer Comfort.
19. Centrifuge 5424 R.
20. Speedvac Concentrator Plus.
21. Spectrophotometer NanoDrop 1000.
22. Mass spectrometer ESI LC/MS/MSD XCT.

2.5 siRNA Biodistribution In Vivo

1. Annealing buffer: 30 mM HEPES-KOH, pH 7.5, 100 mM potassium acetate, 2 mM magnesium acetate. Store at +4°C or at −20°C for long-term storage.

2. Sterile 0.9% saline solution.
3. PBS.
4. 10× TBE buffer: 0.5 M Tris–HCl, pH 8.3, 0.5 M boric acid, 10 mM EDTA.
5. Loading buffer N: 15% Ficoll-400, 0.05% bromphenol blue, 0.05% xylene cyanol.
6. Stains-All solution: 0.1% stains in 50% deionized formamide/ water.
7. Native polyacrylamide gel: 15% acrylamide, 0.5% *N*,*N*-′-methylene-bis-acrylamide, TBE×1.
8. Mice: 8- to 10-week-old SCID (SHO-Prkdc^{scid}Hr^{hr}).
9. Cells: KB-8-5 (or other cell line expressing target gene).
10. Anesthesia Systems series 21100 with induction chamber and Multiple Delivery System, for four animals or equivalent Anesthesia System for Isoflurane.
11. In-Vivo MS FX PRO Imaging System (Carestream, USA).
12. Amplificator iCycler iQ5 (Bio-Rad, USA).
13. Tissue-Tek O.C.T. (Sacura Finetek, USA).

2.6 siRNA Activity In Vivo

1. RIPA lysis buffer: 150 mM NaCl, 1% Triton X-100, 0.5% sodium deoxycholate, 0.1% SDS, 50 mM Tris–HCl, (pH 7.5).
2. 5× Laemmli sample buffer 60 mM Tris–HCl pH 6.8, 2% SDS, 10% glycerol, 5% β-mercaptoethanol, 0.01% bromophenol blue.
3. PageRuler Prestained Protein Ladder.
4. TGB Buffer: 25 mM Tris–HCl pH 8.3, 0.25 M glycine, 0.1% SDS.
5. Mini-PROTEAN® TGX™ Precast Gels.
6. Mini-PROTEAN Tetra Vertical Electrophoresis Cell.
7. A Mini Trans-Blot® cell.
8. PVDF membrane.
9. Monoclonal anti-P-glycoprotein and anti-β-actin antibodies or appropriate antibodies for your target.
10. Secondary rabbit anti-mouse antibodies in conjugated with horseradish peroxidase.
11. Chemiluminescent Reagent Kit.
12. Versadoc 4000 MP.
13. GelPro 4.0. software (Media Cybernetics, USA).

3 Methods

3.1 Preparation of H-Phosphonate of Cholesterol-3-(Carboxyamino-hexan-6-ol) Triethylammonium Salt (3) (Scheme 1)

H-phosphonates are stable, non-hygroscopic, resistant to oxidation by air, and can be easily synthesized in preparative amounts.

3.1.1 Silica Gel Column Chromatography

Pack the column using the appropriate eluent EtOAc/hexane (10/90) or CH_3OH/CH_2Cl_2 (10/90). Dissolve the crude product into a minimum volume of eluent and place the resulting oily mixture on top of the silica bed. Identify fractions containing pure product by TLC. Combine fractions containing pure product and remove all solvents using a rotary evaporator.

3.1.2 Cholesterol-3-(Carboxyamino-hexan-6-ol) (2)

1. Dissolve cholesteryl chloroformate (**1**, 0.5 g, 1.1 mmol) and dry TEA (0.15 mL, 1.1 mmol) in dry dichloromethane (4 mL) and add to 6-amino-1-hexanol (0.129 g, 1.1 mmol). Stir the mixture at room temperature for 2 h, and then evaporate under reduced pressure.
2. Add dichloromethane (50 mL) to residue, and wash the resulting solution with saturated aqueous $NaHCO_3$ (2 × 50 mL) and water (50 mL), dry over Na_2SO_4, filter, and then evaporate under reduced pressure.
3. Dissolve the residue in EtOAc/hexane (10/90) and purify via silica gel column chromatography (*see* Subheading 3.1.1) (elution with EtOAc/hexane, 10/90 to 50/50) to obtain **2** (92%, R_f (TLC, EtOAc/hexane 50/50): 0.2).
4. ^{1}H NMR (300 MHz, $CDCl_3$, δ, ppm): 0.69 (s, 3H, H-18/19, cholesterol); 0.87 (d, 3H, H-26/27, cholesterol); 0.89 (d, 3H, H-26/27, cholesterol); 0.93 (d, 3H, H-21, cholesterol); 1.025 (s, 3H, H-18/19, cholesterol); 3.19 (dd, 2H, -C$\underline{H}_2$-NH-, linker); 3.67 (t, 1H, -C$\underline{H}_2$-OH, linker); 4.52 (m, 1H, H-3, cholesterol); 4.62 (t, 1H, -NH-, linker); 5.4 (d, 1H, H-6, cholesterol).

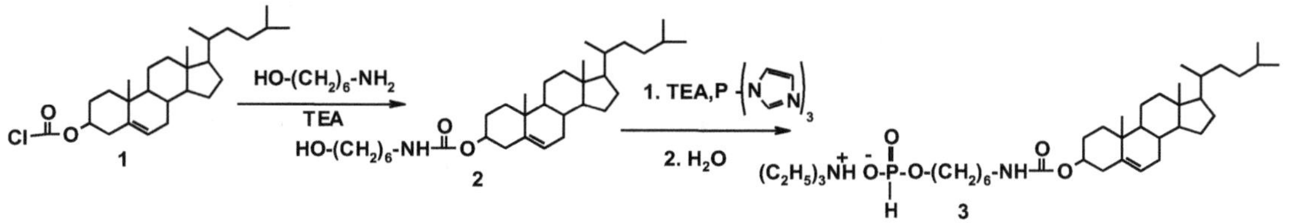

Scheme 1 Synthesis of H-phosphonate of cholesterol-3-(carboxyaminohexan-6-ol)

3.1.3 H-Phosphonate of Cholesterol-3-(Carboxyamino-hexan-6-ol), Triethylammonium Salt (3)

1. Dissolve imidazole (1 g, 14.5 mmol) in dry AcN (20 mL) and then cool to 0 °C. Add TEA (2 mL, 14.0 mmol) and PCl_3 (0.4 mL, 4.5 mmol) to this solution and subsequently stir for 25 min at 0 °C.
2. Dissolve cholesterol-3-(carboxyaminohexan-6-ol) (**2**, 0.53 g, 1 mmol) in dry dichloromethane (10 mL) and add dropwise into the mixture for 40 min under argon. Stir the reaction mixture for an additional 1 h at room temperature followed by the addition of water (3 mL) and subsequent evaporation.
3. Dissolve the residue in dichloromethane (30 mL), wash with water (2 × 30 mL), dry over Na_2SO_4, filter, and then evaporate under reduced pressure. Dissolve the crude product in CH_3OH/CH_2Cl_2 (10/90) and then subject to silica gel chromatography (*see* Subheading 3.1.1) (elution with CH_3OH/CH_2Cl_2, 10/90 to 70/30, with 0.1% TEA), in order to afford the corresponding H-phosphonate **3** (triethylammonium salt, 83%, R_f (TLC, CH_3OH/CH_2Cl_2, 50/50): 0.24).
4. ^{31}P NMR (300 MHz, $CDCl_3$, δ, ppm): 5.0 ($J_{P\text{-}H}$ 628.5 Hz).
5. ESI-MS: calculated for $C_{40}H_{76}PO_5N$ $[M + H]^+$: 695.6; found: 695.0.

3.2 Synthesis of Sense and Antisense Strands of siRNA and Their Conjugates

(*see* **Note 4** for sequence selection and modification patterns)

Oligonucleotides were synthesized on an automatic ASM-800 synthesizer (or any other solid-phase NA synthesizer) at 0.4 μmol scale using solid-phase phosphoramidite synthesis protocols optimized for the instrument (*see* **Notes 5** and **6**) or order oligonucleotides commercially.

3.2.1 Synthesis of Sense and Antisense Strands of siRNA

1. The 16 mg of required solid support was packed in the appropriate column and placed on the automated DNA/RNA synthesizer.
2. Start the synthesis cycle (Scheme 2) with detritylating reagent to remove 5′-O-dimethoxytrityl (DMT) groups from the first support-bound nucleoside for 120 s. After that, wash the column with AcN followed by addition of a mixture (1/1, v/v) of phosphoramidite monomer solution and 0.25 M ETT, with a 10 min coupling step for 2′-O-TBDMS-protected phosphoramidites and 6 min coupling step for 2′-O-methylated phosphoramidites. Use a mixture of capping reagents A and B (1/1, v/v) for 30 s for capping of unreacted 5′-hydroxyl groups. At the last step, oxidize the phosphite-triester to phosphorotriester with oxidizing reagent A for 65 s.
3. After the entire sequence synthesis is completed synthesized, remove the 5′-O-DMT group of the last monomer.
4. Wash the column with AcN and dry thoroughly.

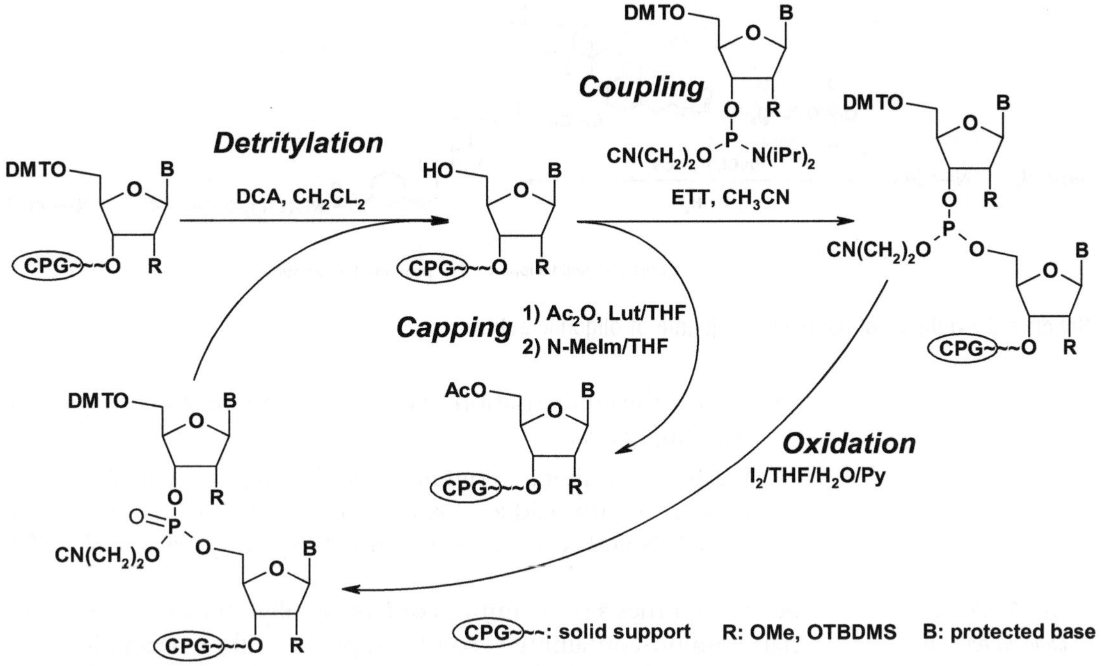

Scheme 2 Synthetic cycle of automated synthesis of oligonucleotides

3.2.2 Synthesis of Lipophilic Conjugates of Oligonucleotides

For the synthesis of 5′-lipophilic RNA conjugates, the solid-phase phosphoramidite method is usually used, introducing a lipophilic residue in the form of a phosphoramidite in the last stage of the synthesis. Here we present another variant of preparation of lipophilic siRNA derivatives, based on a combination of methods of phosphoramidite and H-phosphonate chemistry. The essence of this approach is using a polymer-bound protected oligonucleotide, synthesized by the phosphoramidite method, and a lipophilic residue in the form of an H-phosphonate (Scheme 3). It should be noted that this approach is universal and suitable for the synthesis of conjugates of various functional nucleic acids (such as antisense oligonucleotides, aptamers, and others) with ligands of different chemical nature bearing hydroxyl groups suited for transformation to H-phosphonates.

1. Dissolve H-phosphonate of cholesterol-3-(carboxyamino-hexan-6-ol) (2.4 mg, 4.5 μmol) in dichloromethane (150 μL).
2. Transfer the 5′-O-detritylated, support-bound protected oligonucleotide (15 mg, 0.45 μmol) from the synthesis column to a 0.6 mL tube and subsequently blend with a solution of the corresponding H-phosphonate (50 μL), followed by treatment with pivaloyl chloride (0.9 μL, 7.5 μmol in 150 μL Py/AcN, 2/1, v/v) for 5 min.
3. Wash the support-bound conjugate with AcN in order to remove an uncoupled lipophilic compound.

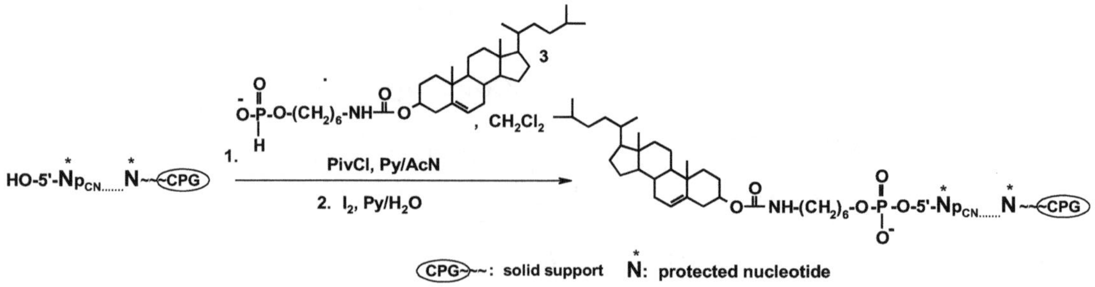

Scheme 3 Synthesis of lipophilic conjugate of oligonucleotide

4. Repeat the condensation twice in order to increase the yield of the conjugates.
5. Wash the conjugate-containing solid support with AcN, and then treat with oxidizing reagent B (0.2 mL) for 20 min, wash with AcN and acetone, and then subsequently dry (*see* **Note 7**).

3.2.3 Synthesis of Labeled Antisense Strand of siRNA

For the synthesis of 3′-amino-containing oligonucleotide, use special amino-containing solid support, for example, *N*-(6-(O-dimethoxytrityl)-hexyl)-(2-carboxamide)-phthalimidyl-lcaa-CPG.

1. Add 10 volume of 2% $NaClO_4$ in acetone to the solution of fully deblocked (*see* Subheading 3.3.2) 3′-amino-containing oligonucleotide, mix well, and centrifuge for 3 min at 14,000 × *g*.
2. After centrifugation, remove the supernatant; wash the pellet with acetone, and dry in a SpeedVac (*see* **Note 13**).
3. Redissolve the pellet of crude reaction mixture of 3′-amino-containing oligonucleotide (up to 20 OD_{260}) in 20 μL of 0.5 M Tris–HCl (pH 8.3).
4. Add Cy5.5 solution (1 mg in 80 μL DMSO) (*see* **Note 14**) to the oligonucleotide-containing solution and mix well.
5. Carry out the reaction at room temperature with shaking.
6. Upon completion of the reaction (usually 2 h, *see* **Note 15**), precipitate with 2% $NaClO_4$ in acetone.
7. Analyze and purify the labeled antisense strand of siRNA as described in Subheading 3.3.4.

3.3 Deprotection, Purification, and Analysis of Strands of siRNAs and Their Conjugates

To prevent the nuclease degradation of oligoribonucleotides and their analogs and conjugates, you may follow several rules: water used in experiments should be RNase-free; all containers (bottles, tubes) and pipette tips should be autoclaved or sterilized; wear clean gloves and use caution when handling RNA samples.

3.3.1 Cleaving from the Support and Removing Acetyl, Benzoyl, and Cyanoethyl Protecting Groups

1. Transfer solid support from the column to a 0.6 mL tube.
2. Cleave the oligonucleotides and their conjugates from the support and remove acetyl, benzoyl, and cyanoethyl protecting groups by freshly prepared AMA solution (400 μL) at 65 °C for 15 min and shaking (*see* **Note 9**).
3. Cool the tube to 0 °C before opening (*see* **Note 16**).
4. Transfer solution to a 1.5 mL tube, wash the solid support with AcN/H_2O/ethanol (200 μL) twice, combine the solutions, and dry completely in a Speedvac Concentrator Plus to obtain the dry pellet (*see* **Note 17**).

3.3.2 Removing TBDMS Protecting Groups

1. Add freshly prepared TBDMS deprotecting cocktail (300 μL) to the pellet and incubate the reaction mixture at 65 °C for 90 min (*see* **Note 10**).
2. Add ethoxytrimethylsilane (200 μL) to quench the reaction, mix well, and then add the diethyl ether and centrifuge for 3 min at 14,000 × *g* (*see* **Note 18**).
3. After centrifugation, remove the supernatant and rinse the pellet with acetone.
4. Redissolve the pellet (color from white to light yellow) in 100 μL water (*see* **Note 19**).

3.3.3 Analysis of Crude Oligonucleotides and Their Conjugates

1. Measure the amount of crude oligonucleotides and their conjugates by spectrophotometer (*see* **Note 20**).
2. Analyze crude oligonucleotides and their conjugates by analytical electrophoresis (about ~0.1 OD_{260} of the reaction mixture with the loading buffer per 0.5 cm width well) as detailed below; visualize the bands by Stains-All.

*3.3.4 The Purification of Oligonucleotides and Their Conjugates by Electrophoresis (*See **Note 21**)

1. Prepare 12% denaturing polyacrylamide gel for the purification of oligonucleotides and their conjugates (*see* **Note 22**) by mixing 12% denaturing polyacrylamide gel solution, TEMED, and freshly prepared APS solution (*see* **Note 11**); swirl the mixture and quickly pour into glass plates cassette and insert comb before polymerization occurs.
2. Allow the stacking gel to polymerize for at least 10–15 min for obtaining gel slab with the thickness of 0.8 mm; after that, pre-run the gel at 5 W in TBE × 1 buffer for 30 min.
3. Apply 10–15 OD_{260} of the reaction mixture in the loading buffer per the 5 cm width well.
4. Perform electrophoresis at constant 5–10 W and maintain the gel temperature during the run <40 °C to avoid bowing of the bands.

5. Detect the products by a UV shadowing method on UV fluorescent screen upon irradiation at 254 nm in the form of a violet fluorescence bands.
6. Excise the product band and put it to the tube followed by addition of elution solution A or B and then precipitate with ethanol or desalt with Amicon 3 K® or Sep Pak C18 cartridge as detailed below (*see* **Note 22**).

3.3.5 Elution with 0.3 M NaOAc and Precipitation with Ethanol

1. Add 600–1000 μL of 0.3 M NaOAc (pH 5.5) to the gel slice with oligonucleotide and shake well for 16 h at 25 °C.
2. Transfer the supernatant to the new tube, add 4 volume of ethanol, and incubate for 4–16 h at −20 °C.
3. Collect the pellet by centrifugation, remove the supernatant; wash the pellet twice with chilled 70% ethanol (until no acetic acid smell), and dry in a SpeedVac.
4. Redissolve the pellet in water (100 μL) and measure the amount by spectrophotometer (*see* **Note 23**).
5. Analyze the purity of oligonucleotides and their conjugates by analytical electrophoresis (about ~0.05 OD_{260} of sample per 0.5 cm width well) as detailed above.

3.3.6 Elution with 0.3 M $NaClO_4$ and Desalting with Amicon (Alternative Method)

1. Add 600–1000 μL of 0.3 M $NaClO_4$ to the gel slice with oligonucleotide and shake well for 16 h at 25 °C.
2. Aspirate the eluate and desalt it with Amicon® according to the supplier's recommended protocol.
3. Collect the ultrafiltrate and add 10 volumes of 2% $NaClO_4$ in acetone, mix well, and centrifuge for 3 min at 14,000 × *g*.
4. After centrifugation, remove the supernatant; wash the pellet with acetone, and dry in a SpeedVac.
5. Redissolve the pellet in water (100 μL) and measure the amount by spectrophotometer.
6. Check the purity of oligonucleotides and their conjugates by analytical electrophoresis (about ~0.05 OD_{260} of sample per 0.5 cm width well) as detailed above.

3.3.7 Elution with 0.3 M $NaClO_4$ and Desalting with Sep Pak C18

1. 600–1000 μL of 0.3 M $NaClO_4$ was added to the gel slice with oligonucleotide and shaken well for 16 h at 25 °C.
2. Wash Sep Pak cartridge (*see* **Note 24**) consistently with 5 mL of 50% acetonitrile in water, 5 mL water, and 5 mL 0.3 M $NaClO_4$.
3. Load the supernatant to Sep Pak cartridge.
4. Wash the Sep Pak cartridge with 5 mL 0.3 M $NaClO_4$.
5. Add 1.0 mL of 50% acetonitrile to the cartridge and collect the eluent in a 1.5 mL tube.

6. Evaporate the eluent by a SpeedVac to a volume of 100–150 μL, add 10 volumes of 2% $NaClO_4$ in acetone, mix well, and centrifuge for 3 min at 14,000 × *g*.
7. After centrifugation, remove the supernatant; wash the pellet with acetone, and dry in a SpeedVac.
8. Redissolve the pellet in water (100 μL) and measure the amount by spectrophotometer.
9. Analyze the purity of oligonucleotides and their conjugates by analytical electrophoresis (about ~0.05 OD_{260} of sample per 0.5 cm width well) as detailed above.

3.4 siRNA Biodistribution In Vivo

3.4.1 siRNA Duplex Preparation

1. Prepare a mixture of the antisense and the sense strands of siRNA (100 μM each) in annealing buffer.
2. Heat the mixture at 85°C for 3 min, and then cool down slowly to room temperature (e.g., leave in a switched off thermostat for 1 h).
3. Analyze siRNA duplexes by electrophoresis in a 15% native PAAG, followed by staining of the gel with Stains-All solution.
4. Store siRNA frozen in aliquots at −20°C, and thaw when needed in ice.

3.4.2 Mice Model, Imaging, Sample Collection (See **Notes 25** and **26** for Mice Models and Ethical Issues)

1. Cultivate KB-8-5 cells (or other cell line tumorigenic in mice) in appropriate cell culture media (e.g., DMEM supplemented with 10% FBS, 100 U/mL penicillin, 100 μg/mL streptomycin, and 0.25 μg/mL amphotericin and 300 nM vinblastine) at 37 °C in a humidified atmosphere containing 5% CO_2/95% air (*see* **Note 27**).
2. Remove and discard the spent cell culture media from the culture vessel.
3. Wash cells with PBS twice, add the pre-warmed dissociation reagent such as 0.025% trypsin or to the side of the flask; use enough reagent to cover the cell layer.
4. Incubate the culture flask at 37 °C for approximately 2 min. Add the pre-warmed complete growth medium and disperse the medium by pipetting.
5. Transfer the cells to a 15 mL conical tube and centrifuge at 200 × *g* for 5–10 min.
6. Resuspend the cell pellet in a minimal volume of 0.9% saline solution and determine the total number of cells and percent viability using a hemocytometer or automated cell counter.
7. Prepare the suspension of 10^6 KB-8-5 cells in 100 μL 0.9% saline solution and then inject the solution subcutaneously into the right flank of mice (*see* **Note 28**).

8. In 1–3 weeks, when tumors will reach a palpable volume of at least 50 mm^3, randomly assign mice to experimental or control groups.
9. Inject intravenously 2–10 μg/g Cy7- or Cy5.5-labeled siRNA for bioimaging studies or non-labeled siRNA for steam-loop assay in 200 μL of Opti-MEM medium.
10. At various time points (15 min, 1, 2, 4, 6, and 24 h), anesthetize mice in induction chamber and put four mice at a time on the heated tray in In-Vivo MS FX PRO Imaging System, attach individual masks connected to Anesthesia Systems series 21100 via Multiple Delivery System (*see* **Note 29**).
11. Scan mice's fluorescence (10 s exposure) and X-ray (15 s exposure) by In-Vivo MS FX PRO Imaging System (*see* **Note 30**).
12. Kill mice by dislocation of the cervical vertebrae or carbon dioxide chamber 24 h after siRNA injection, remove their organs, put them on plastic wrap, and visualize their fluorescence by In-Vivo MS FX PRO Imaging System.
13. Calculate organ fluorescence intensity, measured in relative fluorescent units (RFU), and the percentage of organ fluorescence intensity relative to the total fluorescent intensity of all organs.
14. Immediately prepare samples for steam-loop PCR (detailed in Subheading 3.4.4) to determine siRNA concentration in organs.
15. Kill mice 4–7 days after siRNA injection and immediately prepare samples for Western blot (detailed in Subheading 3.4.5) to measure biological activity of siRNA.

3.4.3 siRNA Concentration in Blood Serum Assay

1. Take 50 μL of blood from retro-orbital sine of mice injected with siRNA in heparinized tube at various time points (15 min, 1, 2, 4, and 24 h) to measure siRNA concentration in blood plasma (*see* **Note 31**).
2. Centrifuge blood sample for 5 min at 500 × *g*.
3. Transfer 20 μL of supernatant in the new tube and add 180 μL of 0.25% Triton X-100.
4. Prepare samples for calibration to get quantitative results. Add 10 μL of 10, 1, and 0.1 pmole siRNA into three new tubes with 20 μL of supernatant obtained from control mice (intact) and add 180 μL of 0.25% Triton X-100 to each sample.
5. Heat all samples for 10 min at 95°C with shaking each for 5 min.
6. Cool lysates on ice for 10 min.
7. Centrifuge for 20 min at 12,000 × *g* at 4 °C.

8. Heat the supernatant for 10 min at 95°C and take 2 μL of supernatant for cDNA synthesis reaction. Set the RT reaction in 40 μL volume using appropriate master mix for reverse transcription (RT) (*see* **Note 32** *for primer design*).
9. Incubate RT reaction for 1 h at 42 °C (*see* **Note 33**).
10. Add 5 μL of RT reaction to each 20 μL of master mix for real-time PCR and amplify samples.
11. Build the calibration curves using specially prepared samples (*see* ***step 5*** *of this Subheading 3.4.3*) for quantification.

3.4.4 siRNA Concentration in Organ Assay (See **Note 34**)

1. Cut several 50 mg pieces of each organs and put each one in the tube with 500 μL of 0.25% Triton X-100.
2. Homogenize tissue thoroughly using dispergator for small volume, for example, T 10 basic ultra-turrax (IKA).
3. Prepare calibration samples for each type of tissue to get quantitative results. Add 10 μL of 10, 1, and 0.1 pmole siRNA into Eppendorf tubes containing 200 μL of 0.25% Triton X-100 with 20 mg of corresponding tissue obtained from control mice (intact) and homogenized as described above (*see* **Note 35**).
4. Proceed to **steps 5–11** of Subheading 3.4.3.

3.4.5 Microscopy

1. Immediately put 50–70 mg pieces of organs into cryomold with O.C.T. compound and freeze in liquid nitrogen. Keep the containers with samples at −80 °C.
2. Cut 7 μm cryosections at −20 °C with a cryotome, place on poly-L-lysine coated slides, and let it melt quickly by pressing finger to the bottom side of the slide.
3. Carefully rinse slides with PBS.
4. Fix section in 2% paraformaldehyde for 10 min pouring on the slide the amount of solution needed to cover the section at room temperature.
5. Aspirate fixation solution and quench excess of aldehyde with 0.1 M glycine in PBS for 5 min at room temperature.
6. Dilute 5 μL methanolic stock solution of the phalloidin-TRITC into 200 μL PBS with 1% BSA for each slide and place the staining solution on the slide for 10 min at room temperature.
7. Mount coverslips on slides using mounting medium with DAPI and antifade agent of your choice (e.g., Vectashield Antifade Mounting Medium (Vector Laboratories)) and seal with nail polish (*see* **Note 36**).
8. Visualize with confocal fluorescent microscope.

3.5 siRNA Activity In Vivo Assay

3.5.1 Samples Preparation

1. Homogenize tissue samples in RIPA buffer supplemented with freshly added 1 mM phenylmethylsulfonyl fluoride or cocktail of protease inhibitors (*see* **Note 37**).
2. Wait for 10 min, and then centrifuge for 10 min at 10,000 × *g* at 4 °C, collect the supernatant, and store it at −80 °C or use immediately for the analysis by Western blot.

3.5.2 Western Blot

1. Mix 10 μL sample in RIPA buffer with 20 μL Laemmli loading buffer and heat for 15 min at 95 °C.
2. Assemble SDS-PAGE precast gel into Vertical Electrophoresis Cell filed with TG buffer, load samples, and pre-stained protein molecular weight standards, and then separate them at 10 V/cm for 1 h.
3. Transfer proteins from polyacrylamide gel to polyvinylidene fluoride (PVDF) membrane using Mini Trans-Blot® cell.
4. Block membrane for 1 h in 2% non-fat dried milk with 0.1% Tween 20 in PBS.
5. Incubate membrane overnight with monoclonal anti-P-glycoprotein and anti-β-actin antibodies at 1:1000 and 1:7500 dilution, respectively, at 4 °C (*see* **Note 38**).
6. Wash the membrane with 0.1% Tween 20 in PBS three times for 10 min.
7. Incubate membrane for 1 h with secondary rabbit anti-mouse antibodies conjugated with horseradish peroxidase.
8. Use a chemiluminescent reagent kit and Versadoc 4000 MP for bands detection.
9. Analyze data using GelPro 4.0. software (Media Cybernetics).

4 Notes

1. Always wear personal protection such as lab coat, safety glasses, and gloves when performing experiments. Work with all chemicals in a ventilated chemical hood. All reactions should be handled under the inert atmosphere.
2. You may use phosphoramidites with the other base-protecting groups (e.g., phenoxyacetyl, isobutyryl) from different manufacturers (Glen Research, ChemGen, Sigma-Aldrich, TriLink, etc.).
3. To improve synthesis quality and yield by removing trace amounts of moisture, add molecular sieves (3 Å, 8, to 12 mesh) to phosphoramidite solutions, activator solution, capping reagent B, and acetonitrile.

4. Use the experimentally validated nuclease-resistant siRNA sequences to design the lipophilic siRNA conjugates. When selecting the sequence from the very beginning, find mRNA target sequence number at GenBank at http://www.ncbi.nlm.nih.gov/nucleotide/, use siRNA sequences design tools, for example, BLOCK-iT™ RNAi Designer (Invitrogen, https://rnaidesigner.thermofisher.com/rnaiexpress/) or siDESIGN Center (Darmacon, https://dharmacon.horizondiscovery.com/design-center/), or any other appropriate tool to select several sequences per target gene. After that, test their biological activity and action specificity on a suitable cell line to select the most active sequence. Use chemical modification to increase the nuclease resistance of siRNAs using one of the published patterns. For example, replace the pyrimidine nucleotides at the 5′-end of 5′CpA3′, 5′UpA3′, and 5′UpG3′ motives with their 2′-O-methyl analogs and check the stability of the obtained siRNA in the presence of serum. siRNAs with high silencing activity and stability can be used to design lipophilic conjugates. In each experiment, appropriate specificity controls, including negative sequence control and scramble sequence control, should be used in accordance with [9].
5. Use helium gas for operating ASM-800 DNA/RNA synthesizer, and for the other synthesizers, use gas recommended by the manufacturer.
6. You may synthesize oligonucleotides using the other automatic DNA/RNA synthesizer (e.g., from Applied Biosystems or Mermade) with a protocol optimized for this instrument.
7. Do not use *N*-dimethylformamidine-protected guanosine phosphoramidite for the synthesis. The reaction of the *N*-dimethylformamidine (dmf) derivatives of guanosine with iodine and aqueous ammonia gives the corresponding *N*-cyano nucleosides. This reaction also occurs in oligonucleotides if the iodine (concentrations of oxidizing reagent more than 0.02 M) is retained on the solid support/in the column before cleavage with aqueous ammonia [10]. When using *N*-dimethylformamidine-protected guanosine phosphoramidite for the synthesis, wash the solid support well from traces of an oxidizing agent B.
8. Store the concentrated (28–30%) ammonium hydroxide solution at −20 °C. Ammonium hydroxide is the most conventional deblocking reagent for complete cleavage from the support and removing cyanoethyl- and base-protecting groups (incubation for 16 h at 55 °C or 3 days at room temperature). Frequent gentle shake or vortexing during the cleavage/deprotection will improve the reaction yield.

9. If you choose the AMA solution as a deblocking reagent, use only N^4-acetyl protected cytidine phosphoramidite, but not N^4-benzoyl protected one. Otherwise, the deprotection by AMA will lead to the transamidation side reaction of benzoyl-C, giving the mutant N^4-Me-C [11].
10. Use the TBDMS deprotecting cocktail only freshly prepared and hot (65 °C).
11. The APS solution can be stored at 5 °C no more than 1 week.
12. The stock gel can be stored no more than 3 weeks.
13. Be careful upon handling the tube with dry pellet of oligonucleotides and their conjugates. Centrifuge the tube with the pellet before opening and open slowly; otherwise, the pellet can fly out of the tube upon opening.
14. You may use *N*-hydroxysuccinimide esters of any dye suitable for this reaction.
15. After 2 h, you may take 0.3 μL aliquot from the solution and add to 10 μL of H_2O. Next, precipitate with the 10× volume of 2% $NaClO_4$ in acetone, mix well, and centrifuge for 3 min at 14,000 × *g*. Analyze the reaction mixture by analytical electrophoresis. If the reaction is not complete, leave for another 2 h at room temperature or overnight at 5 °C.
16. If you open the tube without cooling the hot ammonia/methylamine, the vapors can burst from the tube which results in a loss of oligonucleotide material.
17. Additional washes of solid support will increase the yield of crude reaction mixture.
18. Adding the diethyl ether and incubating the tube with crude reaction mixture 30 min at −20 °C before centrifuging help to form a good precipitate.
19. Some sequences are less soluble in water after 2′-deprotection. The addition of acetonitrile improves the solubility.
20. For sufficient accuracy of the measurement, do not dilute the stock solution of oligonucleotides more than 20 times.
21. PAGE purification is suitable for oligonucleotides of any length but is strongly recommended particularly for long oligonucleotides and RNA oligonucleotides. PAGE purification typically gives oligonucleotides of very high purity but with a somewhat lower yield.
22. Do not use the Sep Pak C18 cartridge for desalting of lipophilic conjugates (especially cholesterol-containing ones), because of the loss of the conjugate due to the irreversible adsorption of cholesterol on the C18 phase.

23. The solution may contain small pieces of the gel that whirl up when collecting the eluate. To remove them, the solution can be passed through a membrane (0.45 μm pore size) filter.
24. Soak the new Sep Pak cartridge with 50% acetonitrile in water at least for 2 h before use.
25. All animal procedures should be carried out in strict accordance with the recommendations for proper use and care of laboratory animals (e.g., ECC Directive 86/609/EEC or according to the specific country requirements). The protocol must be approved by the Local Bioethics Commission. The number of animals in each experimental group should be sufficient to obtain reliable results.
26. Certified laboratory animals from certified vivarium should be used.
27. Cells should be in good condition, no more than 6 weeks after defrosting. Avoid using cells that were overgrown or contain less than 90–95% of living cells.
28. To implant tumor cells in animals, always use freshly prepared cell suspensions, do not allow cells to settle to the bottom of the tube, and use resuspension from time to time. When cells are implanted into a large number of animals, control the time of manipulation with cell: depending on the cell type, use the cell suspension for 30–40 min after its preparation, discard the rest, and prepare a new batch if required.
29. Alternatively, use Avertin (150 μg/kg) IP injection to anesthetize mice, wait for 5 min, and then put them on tray.
30. Use fluorophores with near-infrared emission spectrum, since light with a longer wavelength passes through the body more easily. Other fluorophores and their combinations with similar corresponding properties can be used for these purposes. Cy7 could be successfully used due to low background; however, Cy5.5 can also be used both for in vivo imaging and for FACS analysis. Fluorophores with shorter wavelength such as Cy3, or Alexa 488 could be used for microscopy. Select exposure time depending on the spectral characteristics of the fluorophore, the dose administered, and the efficiency of its accumulation.
31. We recommend to take blood from the same mouse no more than once every 2 h and no more than three times in 24 h, and if you need to use shorter time intervals between experimental points, it is better to have several mice and take blood from them one by one. Always have additional mice in the group, as blood sampling is an invasive procedure, and some animals may die during the experiment.

32. To design siRNA-specific stem-loop RT-qPCR primers, use the instructions available online at http://genomics.dote.hu:8080/mirnadesigntool.
33. At this stage, samples can be stored at −20 °C overnight or up to week.
34. The critical point for determining the concentration of siRNA in an organ is the effective conduct of the reverse transcription reaction, since total RNA and tissue components present in the sample inhibit the reaction. Therefore, it is important to use a sufficient dilution of the sample with the triton solution and the determination of the linear range of determination in the preliminary experiment.
35. Alternatively, relative determination of the level of RNA in a tissue can be used as compared to sno-miRNA whose expression in cells is fairly stable. Nevertheless, we recommend using calibration curves, which not only provide quantitative data but also monitor the adequacy of the procedure.
36. For some tissues, DAPI staining in mountain medium gives unsatisfactory results, and in this case, it is possible to stain the nucleus by treatment of the section with 0.5 μg/mL DAPI solution for 10 min at room temperature before incubation with phalloidin-TRIZ solution. Use mountain medium without DAPI in this case.
37. Homogenize 100 μg or less tissue in 300 μL RIPA buffer to ensure high quality of the probe.
38. Use antibodies suitable for Western blot application. Use the dilutions indicated in the documentation for the antibodies or select them experimentally.

Acknowledgments

The authors thank Dr. Alya G. Venyaminova for her valuable help and comments regarding the synthetic part. This work was supported by Russian Scientific Foundation grant 19-14-00251.

References

1. Benizri S, Gissot A, Martin A, Vialet B, Grinstaff MW et al (2019) Bioconjugated oligonucleotides: recent developments and therapeutic applications. Bioconjug Chem 30:366–383
2. Chernikov IV, Gladkikh DV, Meschaninova MI, Ven'yaminova AG, Zenkova MA et al (2017) Cholesterol-containing nuclease-resistant siRNA accumulates in tumors in a carrier-free mode and silences MDR1 gene. Mol Ther Nucleic Acids 6:209–220
3. Biscans A, Coles A, Haraszti R, Echeverria D, Hassler M et al (2018) Diverse lipid conjugates for functional extra-hepatic siRNA delivery in vivo. Nucleic Acids Res 47:1082–1096
4. Esposito CL, Nuzzo S, Catuogno S, Romano S, de Nigris F et al (2018) STAT3

gene silencing by aptamer-siRNA chimera as selective therapeutic for glioblastoma. Mol Ther Nucleic Acids 10:398–411

5. Cuellar TL, Barnes D, Nelson C, Tanguay J, Yu SF et al (2014) Systematic evaluation of antibody-mediated siRNA delivery using an industrial platform of THIOMAB-siRNA conjugates. Nucleic Acids Res 43:1189–1203
6. Meade BR, Gogoi K, Hamil AS, Palm-Apergi-C, van den Berg A et al (2014) Efficient delivery of RNAi prodrugs containing reversible charge-neutralizing phosphotriester backbone modifications. Nat Biotechnol 32:1256–1261
7. Volkov AA, Kruglova NS, Meschaninova MI, Venyaminova AG, Zenkova MA et al (2009) Selective protection of nuclease-sensitive sites in siRNA prolongs silencing effect. Oligonucleotides 19:191–202
8. Petrova NS, Chernikov IV, Meschaninova MI, Dovydenko IS, Venyaminova AG et al (2012) Carrier-free cellular uptake and the gene-silencing activity of the lipophilic siRNAs is strongly affected by the length of the linker between siRNA and lipophilic group. Nucleic Acids Res 40:2330–2344
9. Gagnon KT, Corey DR (2019) Guidelines for experiments using antisense oligonucleotides and double-stranded RNAs. Nucleic Acid Ther 29:116–122
10. Mullah B, Andrus A, Zhao H, Jones RA (1995) Oxidative conversion of N-dimethylformamidine nucleosides to N-cyano nucleosides. Tetrahedron Lett 36:4373–4376
11. Reddy MP, Hanna NB, Farooqui F (1997) Ultrafast cleavage and deprotection of oligonucleotides synthesis and use of C-Ac derivatives. Nucleosides Nucleotides 16:1589–1598

Chapter 4

Multifunctional Nanodelivery Platform for Maximizing Nucleic Acids Combination Therapy

Seung Koo Lee, Benedict Law, and Ching-Hsuan Tung

Abstract

The silencing of an oncogene with a small interfering RNA (siRNA) is a promising way for cancer therapy. Its efficacy can be further enhanced by integrating with other therapeutics; however, transporting siRNA and other active ingredients to the same location at the same time is challenging. Here, we report a novel multifunctional nanodelivery platform by sequentially layering several functional ingredients, such as siRNAs, microRNAs, peptides, and targeting ligands, onto a core through charge–charge interaction. The prepared nanovectors effectively and programmably delivered multiple active components to maximize therapeutic combination with minimal off-targeting effects.

Key words Small interfering RNA, KLA peptide, Hyaluronic acid, microRNA, Nucleic acids, Layer-by-layer, Gold nanoparticles, Poly-L-lysine, Electrostatic interactions, Combination therapy

1 Introduction

RNA interference (RNAi), which includes the usage of small interfering RNAs (siRNAs) and microRNAs (miRNAs), is an important natural gene regulatory process (*see* Chapter 1). siRNAs specifically recognize their complementary sequences on mRNAs and result in a single gene silencing in cells [1–3]. miRNAs, which are small evolutionarily conserved and noncoding RNAs, concurrently target multiple effectors of pathways involved in cell differentiation, proliferation, and survival [4, 5]. Until now, only few successful in vivo cases were demonstrated, despite promising cellular results. This rarity in success is mainly caused by their poor delivery [6].

To overcome this drawback, we have previously developed a layer-by-layer (LbL) method for the delivery of siRNA [7]. The LbL fabrication is a straightforward and gentle assembly approach that is based on the electrostatic interactions between negatively and positively charged polyelectrolytes [8–11]. Multiple layers of siRNA were sequentially coated onto a gold nanoparticle (AuNP)

Mouldy Sioud (ed.), *RNA Interference and CRISPR Technologies: Technical Advances and New Therapeutic Opportunities*, Methods in Molecular Biology, vol. 2115, https://doi.org/10.1007/978-1-0716-0290-4_4,

core, which was chosen for its uniform size, feasibility of surface modification, and electronic feature [12–14]. The assembled nanoparticle (NP) effectively carried siRNA into cells and gradually released the embedded siRNA. The LbL assembly protected siRNA from degradation, enhanced the delivery, and prolonged the gene silencing effect [7].

siRNA and miRNA are chemically the same; they bear the same length (19–23 nucleotides) and charge; hence the developed LbL construct could be applied to deliver miRNA as well [15]. miRNA-708 was previously identified as a "metastasis suppressor" due to its significant downregulation in metastatic lesions when compared to the matched primary tumors in human and mouse triple-negative breast cancer (TNBC) [16]. We thus tested a miRNA-708 NP, which possessed two layers of miRNA-708 mimetic, in a metastatic TNBC animal model. The study showed that miRNA-708 NP impaired breast cancer metastasis by targeting the cancer stem cells population [15].

In addition to monotargeting, gene silencing potency can be enhanced with the combinations of two different siRNAs or miRNAs [17, 18]. Therefore, the modular LbL assembly was used as a multifunctional nanodelivery platform (MNP) to hold siRNAs and miRNAs. In one of our early developments, two siRNAs against two different regions of the same luciferase gene were layered separately onto one MNP. In vitro cellular validation showed that the two siRNA combination NP is more effective in gene silencing than the single siRNA NP [19].

In addition to the single siRNA or single miRNA strategy, miRNA with siRNA combination strategies was reported to block cancer growth [20, 21]. To validate the therapeutic potential of the miRNA/siRNA combination in our system, a MNP carrying a siRNA targeting neurotrophin receptor ($p75^{NTR}$) and a miRNA-141 mimetic was prepared using the same protocol. $p75^{NTR}$ was known to be responsible for the promotion of tumor growth [22, 23], and miRNA-141 was known to inhibit tumor growth, invasion, and metastasis in various tumors [24, 25]. When miRNA-141 mimetic and $p75^{NTR}$ siRNA (Table 1) were co-delivered by MNP into MDA-MB231 cells, a superb cancer cell killing effect, >80%, was attainable (Fig. 1). In contrast, the therapeutic effect of MNPs with only miRNA or siRNA was less than a half of miRNA/siRNA combination.

With respect to siRNA assemble, a positively charged poly-L-lysine (PLL) was placed in between to glue the two negatively charged siRNA layers. The positively charged spacing layer is essential for MNP assemble, and PLL is one of the options. It can be replaced by other positively charged polymers. When positively charged bioactive peptides are used, the choice of drug combinations would be more versatile beyond nucleic acids.

Table 1
The sequence of siRNAs, miRNA mimetic, and KLA peptide

Name	Sense strand sequence (5′ → 3′)	Antisense strand sequence (5′ → 3′)
siRNA-p75^{NTR} [23]	AUGCCUCCUUGGCACCUCCdTdT	GGAGGUGCCAAGGAGGCAUdTdT
siRNA-scr [23]	GCUGACCCUGAAGUUCAUCdTdT	GAUGAACUUCAGGGUCAGCdTdT
miRNA-141 mimetic [24]	UAACACUGUCUGGUAAAGAUGG dTdT	CCAUCUUUACCAGACAGUGUUA dTdT
KLA [26]	KLAKLAKKLAKLAKKLAKLAKKLAKLAK	

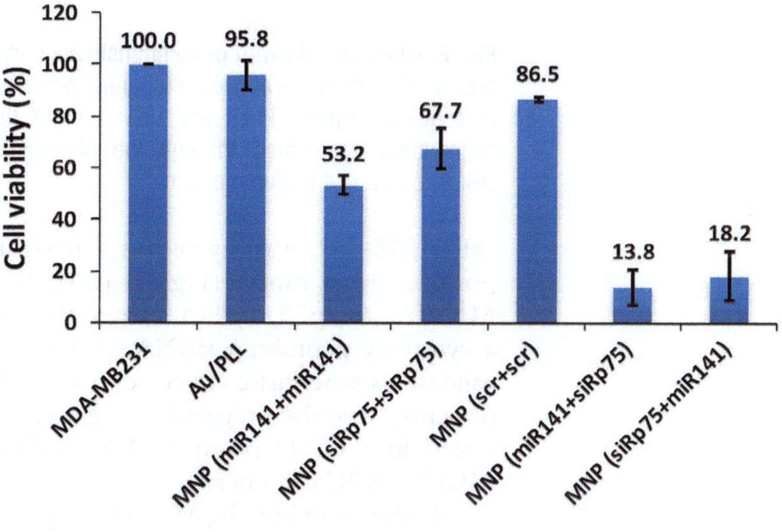

Fig. 1 Synergistic cytotoxic effect of the nanodelivery system induced by the miRNA-141/siRNA-p75^{NTR} combination therapy. Cell viability is evaluated 2 days after treatment with various nanocomplexs (each miRNA-141 mimetic, siRNA-p75^{NTR}, and siRNA-scr: 0.12 μM) in MDA-MB231. The cell viability of the untreated MDA-MB231 cells is set as 100%. *miR141* miRNA-141 mimetic, *siRp75* siRNA-p75^{NTR}, *scr* scramble

An amphipathic antimicrobial peptide, named KLA peptide, is a positively charged peptide (Table 1). The KLA peptide was reported to induce apoptosis in mammalian cells by disrupting the mitochondrial membrane [27, 28]. However, KLA could not permeate the plasma membrane of cells; thus, its mitochondrial lysing activity was only observed when a proper delivery vector was used [29–31]. The developed MNP is an ideal delivery system to transport these unconventional pair of drugs, the KLA peptide and an siRNA against a pro-survival gene, p75^{NTR}. The MNP was fabricated by sequentially layering KLA, siRNA, and PLL to the AuNP core. Lastly, a ligand for CD44, hyaluronic acid

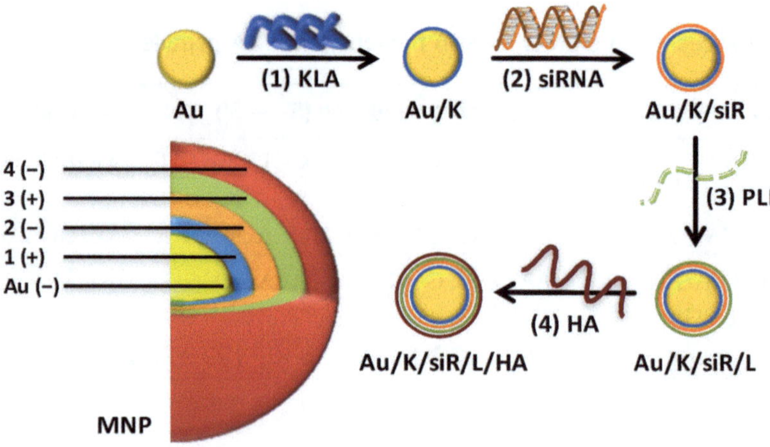

Fig. 2 Schematic diagram of multifunctional nanodelivery platform (MNP) preparation. Two layers of active therapeutic ingredients (KLA and siRNA), one layer of spacing polymer (PLL), and one layer of the targeting ligand (HA) were sequentially assembled through the electrostatic interactions. *K* KLA, *siR* siRNA, *L* PLL, *HA* hyaluronic acid

(HA) [32–34], was layered as a navigator to target the CD44-positive breast cancer (Fig. 2) [26]. Notably, the fully equipped MNP, Au/K/p75/L/HA, carried three major active components, a cytotoxic peptide, a siRNA, and a CD44 targeting ligand. To validate its synergistic effect, several MNPs, with fewer active components, were also prepared as negative controls. These MNPs were tested in a CD44-positive MDA-MB231 and a CD44-negative MCF7 TNBC cell lines.

As shown in Fig. 3a, MNPs incorporated with either a KLA or a $p75^{NTR}$ siRNA showed only moderate cell toxicity (63% or 70% viable, respectively). However, the KLA and $p75^{NTR}$ siRNA-combined MNP, Au/K/p75/L, demonstrated a significant killing effect (22% viable) in MDA-MB231 cells. This data confirms the synergistic therapeutic effect of KLA and $p75^{NTR}$ siRNA and the advantage of combining these two non-typical anticancer agents into a single platform. Importantly, the cytotoxic effect of the MNP with scramble (scr) siRNA (Table 1) which bears the same nucleotide composition but with a random sequence of $p75^{NTR}$ (Au/K/scr/L) has dramatically bounced to 85% viable. This data supports the critical inhibition effect of the siRNA targeting $p75^{NTR}$ gene. The HA-coated MNP, Au/K/p75/L/HA, further lowered the amount of viable cells to 14% in CD44+ MDA-MB231 cells. Conversely, the combined effect was not achieved (83% viable) in MCF7 (CD44−) cells with Au/K/p75/L/HA, due to a lack of HA and CD44 binding (Fig. 3b) [26].

Thereafter, MNP containing one layer of KLA and two layers of $p75^{NTR}$ siRNA was systemically injected into triple-negative

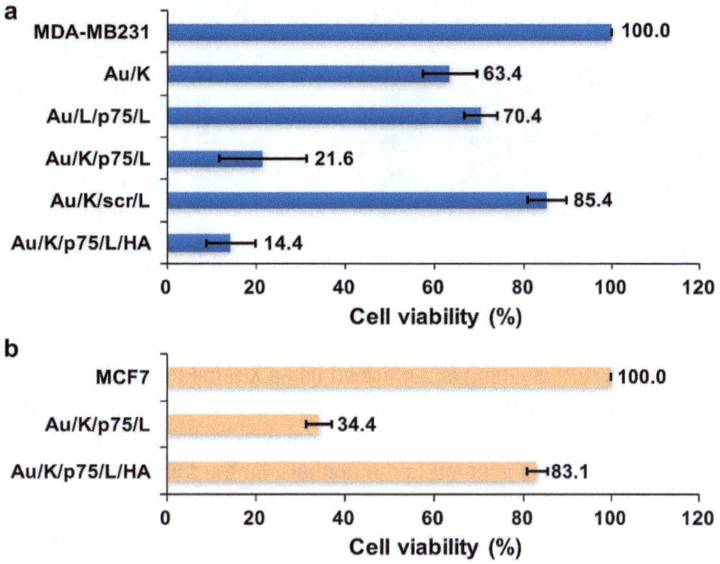

Fig. 3 Specific and synergistic therapeutic effect of MNP in CD44+ TNBC cells. Cell viability is evaluated 2 days after treatment with various MNPs (KLA, 1.6 μM; siRNA-p75^{NTR} or siRNA-scr, 0.12 μM) in (**a**) CD44+ MDA-MB231 and (**b**) CD44- MCF7 cells. The viability of the untreated MDA-MB231 and MCF7 cells is set as 100%. *p75* siRNA-p75^{NTR}, *K* KLA, *L* PLL, *scr* scramble

MDA-MB468 tumor-bearing mice to further investigate its in vivo therapeutic potential. The biodistribution of MNP was noninvasively examined by near-infrared fluorescence (NIRF) imaging. One day postinjection, MNP circulated throughout the body including the tumor. After 3 days, the signal from MNP mainly localized at the tumor site, and the majority of background noise has disappeared (Fig. 4a). A significant antitumor effect was examined in the MNP-treated group, while the PBS-treated tumors continued to grow (Fig. 4b).

Because all siRNAs have the same size and charge, it is easy to replace one siRNA with another to silence different genes. The modulation capability of the LbL technology further allows for flexibility in the number and types of multi-components integrated into one MNP. The applications of MNP are more than a nucleic acid transporter. For example, by incorporating fluorescently labeled polymers, we have designed MNPs as a cell fate tracker [35] and a long-lasting cell imaging probe [36].

In this chapter, we describe the preparation and functional analysis of KLA and siRNA-p75^{NTR} containing MNP (Au/K/p75/L/HA) in vitro and in vivo. For more details, also *see* ref. 37.

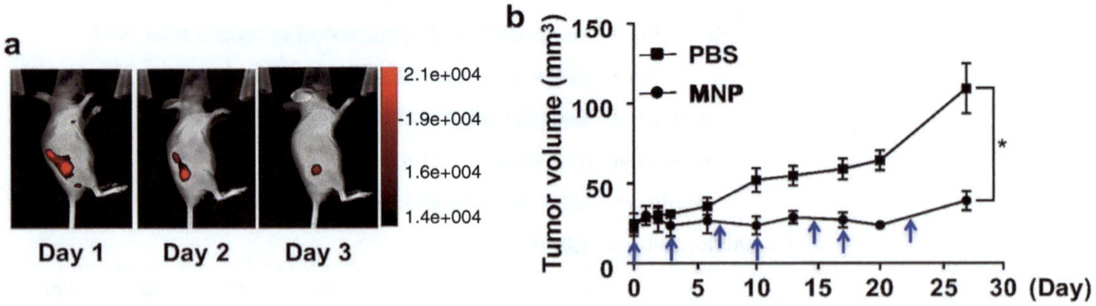

Fig. 4 Synergistic antitumor effect induced by KLA/siRNA-p75^{NTR} combined MNP in a TNBC xenograft model. (**a**) Noninvasive NIRF imaging is performed using Xtreme during 3 days after systemic injection of MNP to the MDA-MB468 tumor-bearing mice. The last layer of PLL is conjugated with cy5.5 dye for the imaging purpose. (**b**) Tumors are measured at the indicated time with digital calipers. The arrows indicate the time when MNP (*see* **Note 14**) or PBS is injected to the animals via the tail veins ($n = 3$ per group). $*p < 0.05$

2 Materials

Prepare all solutions using sterile filtered pure water and store all reagents at room temperature unless otherwise specified.

2.1 Synthesis of KLA

1. Rink Amide MBHA low loading resin (100–200 mesh) (Millipore Sigma, Burlington, MA).
2. Prepacked Fmoc-protected amino acid and activator (HBTU) including Fmoc-L-Lys(Boc)-OH/HBTU, Fmoc-L-Leu-OH/HBTU, and Fmoc-L-Ala/HBTU in vials for PS3 peptide synthesizer (Gyros Protein Technologies, Tucson, AZ) (*see* **Note 1**).
3. Dimethylformamide (DMF).
4. 20% piperidine in DMF.
5. 0.4 M *N*-methylmorpholine in DMF.
6. Thioanisole.
7. Ethanedithiol.
8. Anisole.
9. Trifluoracetic acid.
10. Methyl tert-butyl ether.
11. Acetonitrile HPLC Grade.

2.2 Fabrication of Multifunctional Nanodelivery Platform (MNP)

1. Bare AuNPs (size, 40 nm) (BB International, Cardiff, UK) (*see* **Note 2**).
2. Poly-L-Lysine (PLL, 15,000–30,000 g/mol of molecular weight) (*see* **Note 3**).
3. siRNA (Sigma-Aldrich) against p75^{NTR} and control nonsense scramble (scr) siRNA (*see* **Note 4**).

4. Purified KLA peptide.
5. Sodium hyaluronate HA: 100–150 KDa of molecular weight (*see* **Note 5**).
6. Diethyl pyrocarbonate (DEPC)-treated water.

2.3 Cell Culture

1. Cell maintenance: MDA-MB231 (triple-negative), MDA-MB468 (triple-negative), and MCF7 (ER+) human breast cancer cell lines were purchased from ATCC (Manassas, VA). All cell lines are cultured in DMEM medium supplemented with 2 mM L-glutamine, 100 U/mL penicillin, 100 mg/mL streptomycin, and 10% heat-inactivated fetal bovine serum in a humidified atmosphere of 5% CO_2 at 37 °C.
2. Therapeutic effect of MNP in vitro: A cell proliferation assay is performed using a CellTiter solution (Promega, Madison, WI). Cell viability is assessed after treatment with various MNPs on CD44+ MDA-MB231 and CD44- MCF7 cells.

2.4 In Vivo Antitumor Effect of MNP

1. BALB/c Nu/Nu female nude mice (5–6 weeks) (*see* **Note 6**).
2. TNBC MDA-MB468 cells.
3. Digital calipers.
4. Small animal optical imaging system, Xtreme (Bruker, Billerica, MA).

3 Methods

3.1 Preparation and Characterization of KLA

1. Load Rink Amide MBHA low loading resin (0.05 mmol) into the reaction vessel provided by the manufacturer (*see* **Note 7**).
2. Load the prepacked Fmoc-L-Lys(Boc)-OH/HBTU, Fmoc-L-Leu-OH/HBTU, and Fmoc-L-Ala/HBTU amino acids (0.4 mmol, 8 equivalent) to the peptide synthesizer (*see* **Note 8**).
3. Place the reagents, including 20% (w/w) piperidine, in DMF for Fmoc deprotection, 0.4 M *N*-methylmorpholine in DMF as a base for amino acid coupling, and DMF for washing, to the allocated bottles provided.
4. Start peptide synthesis according to the manufacturer's instruction.
5. Cleave the peptide from the resin using a cocktail of thioanisole, ethanedithiol, anisole, and TFA (5 mL) in a ratio of 5:3:2:90 in volume (*see* **Note 9**).
6. Precipitate the peptide with an equal volume (5 mL) of methyl tert-butyl ether to remove the cleaving cocktail.

7. Wash the peptide three times with ether, and then dry it under vacuum overnight (*see* **Note 10**).
8. Purify the peptide with HPLC (equipped with UV detector at 220 nm), using preparative column (Higgins, 250 × 20 mm, PROTO 300 C18, 10 μM).
9. Freeze-dry the collected elute.
10. Characterize the final product for the molecular weight with Matrix-Assisted Laser Desorption/Ionization-Time of Flight (MALDI-TOP) mass spectrometry.
11. Confirm the purity of the peptide (>98% purity) with HPLC using analytical column (Higgins, 250 × 4.6 mm, PROTO 200 C18, 5 μM).

3.2 Preparation and Characterization of MNP

1. Add AuNPs solution (3.15×10^9 particles in 0.7 mL) dropwisely onto KLA solution (0.5 mL of 62.5 nmol stock solution) in DEPC-treated water (*see* **Note 11**).
2. Incubate for 30 min in the dark with gentle shaking.
3. Centrifuge the incubated solutions for 30 min at 16,100 × *g* using a micro-centrifuge (Eppendorf, Hauppauge, NY).
4. Remove the supernatant.
5. Resuspend the gel-like deep red pellet with DEPC-treated water and centrifuge for 30 min at 16,100 × *g* using a micro-centrifuge (Eppendorf).
6. Remove the supernatant.
7. Wash one more time with centrifugation (repeat **steps 5–6**).
8. Resuspend KLA-coated AuNPs in 0.5 mL of DEPC-treated water.
9. Add KLA-coated AuNPs dropwisely onto siRNA solution (4.0 nmol, 0.5 mL DEPC-treated water).
10. Incubate the reaction solution in the dark for 30 min with gentle shaking.
11. Wash the particles three times with DEPC-treated water (repeat **steps 3–8**).
12. Repeat the deposition procedures sequentially with PLL (0.5 mL of 5 mg/mL stock solution) and HA (0.5 mL of 8 mg/mL stock solution).
13. Acquire MNP with a total of four layers of polyelectrolytes (KLA, siRNA, PLL, and HA, respectively) (Fig. 2).
14. Measure the sizes and zeta potentials of bare and multilayered AuNPs in DEPC-treated water by Zetasizer Nano-ZS according to the manufacturer's instruction (*see* **Note 12**).

3.3 Therapeutic Effect of MNP In Vitro

1. Seed MDA-MB231 cells in a 96-well black clear bottom culture plate at a density of 0.5×10^4 (or 1.0×10^4 of MCF7) cells per well.
2. Incubate cells for 1 day at 37 °C incubator.
3. Replace culture medium with various MNPs (miRNA-141 mimetic, siRNA-p75^{NTR}, and siRNA-scr, 0.12 μM; KLA, 1.6 μM) containing medium.
4. Incubate cells for 12 h at 37 °C incubator.
5. Wash cells twice with phosphate-buffered saline (PBS) and culture them in complete medium for 2 days at 37°C incubator.
6. On 2 days post incubation, replace cells on each well with fresh medium, add 20 μL of CellTiter solution, and incubate for an additional 3 h.
7. Measure the absorbance of the solution at 490 nm using a plate reader.
8. Cell viability was calculated after setting the absorbance of untreated MDA-MB231 and MCF7 cells as 100% (Figs. 1 and 3).

3.4 In Vivo Antitumor Effect of MNP

1. Collect MDA-MB468 cells by trypsinization and count cells.
2. Implant the cells (1.0×10^7 cells in 200 μL of PBS) via subcutaneous into both flanks of mice (*see* **Note 13**).
3. On 4 weeks postimplantation of MDA-MB468 cells, inject MNP intravenously, two times per week, for 27 days (*see* **Note 14**).
4. Perform real-time noninvasive NIRF imaging by Xtreme (Bruker) during 3 days after systemic injection of MNP to MDA-MB468 tumor-bearing mice (Fig. 4a) (*see* **Note 15**).
5. Continue measuring tumor size daily with digital calipers until the end of experiment (Fig. 4b) (*see* **Note 16**).

4 Notes

1. The Fmoc-protected amino acids, in the presence of activators (HBTU), are prepacked in vials that are specific for the PS3 peptide synthesizer. If using other peptide synthesizers, the protected amino acids can be purchased from other vendors such as Millipore Sigma or Peptide International (Louisville, KY).
2. 20 mL of AuNPs from BB International contains 9.0×10^{10} particles.
3. PLL is dissolved with DEPC-treated water and stored with the concentration of 5 mg/mL at −20 °C.

4. All siRNAs and miRNA mimetic are ordered as the yield of 50 OD (250 nmol) with HPLC purification and stored with the concentration of 100 μM at −80 °C.
5. HA is dissolved with DEPC-treated water and stored with the concentration of 8 mg/mL at 4 °C.
6. All animal studies are performed in compliance with the approved animal protocols and guidelines of the Institutional Animal Care and Use Committee of Weill Cornell Medicine.
7. Use a low loading resin for synthesizing peptides with longer than 15 amino acid residues. This can increase the purity of the final products.
8. The use of excessive amino acids increases the purity of peptides.
9. This is a universal cocktail that can be applied to cleave peptide of any sequence and should be freshly prepared. Other cocktails such as a mixture of triisopropylsilane, water, and TFA in a ratio of 2:5:93 in volume also work as effective.
10. The use of vacuum for drying the peptide helps to remove the trace amount of the leftovered cleaving cocktail.
11. The number of AuNPs from BB International is 4.5×10^9 particles/mL; thus, the number of prepared AuNPs in 0.7 mL is 3.15×10^9.
12. The nanoparticle solution, if not in use, should be best stored at 4 °C.
13. For the optimization of cell growth, 5–6-week-old female BALB/c Nu/Nu mice are used, and *n* number of mice is three.
14. Injected MNP is Au/PLL/siRNA-p75^{NTR}/KLA/siRNA-p75^{NTR}/PLL-cy5.5 (total siRNA, 1.34 mg/kg; KLA, 2.84 mg/kg).
15. Real-time NIRF noninvasive imaging is performed at cy5.5 channel (Ex: 650 nm/Em: 700 nm) using Xtreme (Bruker).
16. The tumor volumes are calculated according to the formula as (length × width × height × 0.5).

Acknowledgments

This study was supported in part by NIH CA135312 and DOD W81XWH-11-1-0442.

References

1. Elbashir SM, Harborth J, Lendeckel W, Yalcin A, Weber K, Tuschl T (2001) Duplexes of 21-nucleotide RNAs mediate RNA interference in cultured mammalian cells. Nature 411:494–498
2. Bumcrot D, Manoharan M, Koteliansky V, Sah DW (2006) RNAi therapeutics: a potential new class of pharmaceutical drugs. Nat Chem Biol 2:711–719
3. Burnett JC, Rossi JJ (2012) RNA-based therapeutics: current progress and future prospects. Chem Biol 19:60–71
4. Ambros V (2004) The functions of animal microRNAs. Nature 431:350–355
5. Garzon R, Marcucci G, Croce CM (2010) Targeting microRNAs in cancer: rationale, strategies and challenges. Nat Rev Drug Discov 9:775–789
6. Conde J, Artzi N (2015) Are RNAi and miRNA therapeutics truly dead? Trends Biotechnol 33:141–144
7. Lee SK, Han MS, Asokan S, Tung CH (2011) Effective gene silencing by multilayered siRNA-coated gold nanoparticles. Small 7:364–370
8. Peyratout CS, Dahne L (2004) Tailor-made polyelectrolyte microcapsules: from multilayers to smart containers. Angew Chem Int Ed Engl 43:3762–3783
9. Jewell CM, Lynn DM (2008) Multilayered polyelectrolyte assemblies as platforms for the delivery of DNA and other nucleic acid-based therapeutics. Adv Drug Deliv Rev 60:979–999
10. Reum N, Fink-Straube C, Klein T, Hartmann RW, Lehr CM, Schneider M (2010) Multilayer coating of gold nanoparticles with drug-polymer coadsorbates. Langmuir 26:16901–16908
11. Chanana M, Gliozzi A, Diaspro A, Chodnevskaja I, Huewel S, Moskalenko V et al (2005) Interaction of polyelectrolytes and their composites with living cells. Nano Lett 5:2605–2612
12. Daniel MC, Astruc D (2004) Gold nanoparticles: assembly, supramolecular chemistry, quantum-size-related properties, and applications toward biology, catalysis, and nanotechnology. Chem Rev 104:293–346
13. Jans H, Jans K, Lagae L, Borghs G, Maes G, Huo Q (2010) Poly(acrylic acid)-stabilized colloidal gold nanoparticles: synthesis and properties. Nanotechnology 21:455702
14. Liz-Marzan LM (2006) Tailoring surface plasmons through the morphology and assembly of metal nanoparticles. Langmuir 22:32–41
15. Ramchandani D, Lee SK, Yomtoubian S, Han MS, Tung CH, Mittal V (2019) Nanoparticle delivery of miR-708 mimetic impairs breast cancer metastasis. Mol Cancer Ther 18:579–591
16. Ryu S, McDonnell K, Choi H, Gao D, Hahn M, Joshi N et al (2013) Suppression of miRNA-708 by polycomb group promotes metastases by calcium-induced cell migration. Cancer Cell 23:63–76
17. Wang S, Shi Z, Liu W, Jules J, Feng X (2006) Development and validation of vectors containing multiple siRNA expression cassettes for maximizing the efficiency of gene silencing. BMC Biotechnol 6:50
18. Peter ME (2010) Targeting of mRNAs by multiple miRNAs: the next step. Oncogene 29:2161–2164
19. Lee SK, Tung CH (2013) A fabricated siRNA nanoparticle for ultra-long gene silencing. Adv Funct Mater 23:3488–3493
20. Nishimura M, Jung EJ, Shah MY, Lu C, Spizzo R, Shimizu M et al (2013) Therapeutic synergy between microRNA and siRNA in ovarian cancer treatment. Cancer Discov 3:1302–1315
21. Xue W, Dahlman JE, Tammela T, Khan OF, Sood S, Dave A et al (2014) Small RNA combination therapy for lung cancer. Proc Natl Acad Sci U S A 111:E3553–E3561
22. Reis-Filho JS, Steele D, Di Palma S, Jones RL, Savage K, James M et al (2006) Distribution and significance of nerve growth factor receptor (NGFR/p75NTR) in normal, benign and malignant breast tissue. Mod Pathol 19:307–319
23. Vanhecke E, Adriaenssens E, Verbeke S, Meignan S, Germain E, Berteaux N et al (2011) Brain-derived neurotrophic factor and neurotrophin-4/5 are expressed in breast cancer and can be targeted to inhibit tumor cell survival. Clin Cancer Res 17:1741–1752
24. Zuo QF, Zhang R, Li BS, Zhao YL, Zhuang Y, Yu T et al (2015) MicroRNA-141 inhibits tumor growth and metastasis in gastric cancer by directly targeting transcriptional co-activator with PDZ-binding motif, TAZ. Cell Death Dis 6:e1623
25. Finlay-Schultz J, Cittelly DM, Hendricks P, Patel P, Kabos P, Jacobsen BM et al (2015) Progesterone downregulation of miR-141 contributes to expansion of stem-like breast cancer cells through maintenance of progesterone receptor and Stat5a. Oncogene 34:3676–3687

26. Lee SK, Law B, Tung CH (2017) Versatile nanodelivery platform to maximize siRNA combination therapy. Macromol Biosci 17:1600294
27. Javadpour MM, Juban MM, Lo WC, Bishop SM, Alberty JB, Cowell SM et al (1996) De novo antimicrobial peptides with low mammalian cell toxicity. J Med Chem 39:3107–3113
28. Ellerby HM, Arap W, Ellerby LM, Kain R, Andrusiak R, Rio GD et al (1999) Anti-cancer activity of targeted pro-apoptotic peptides. Nat Med 5:1032–1038
29. Agemy L, Friedmann-Morvinski D, Kotamraju VR, Roth L, Sugahara KN, Girard OM et al (2011) Targeted nanoparticle enhanced proapoptotic peptide as potential therapy for glioblastoma. Proc Natl Acad Sci U S A 108:17450–17455
30. Cieslewicz M, Tang J, Yu JL, Cao H, Zavaljevski M, Motoyama K et al (2013) Targeted delivery of proapoptotic peptides to tumor-associated macrophages improves survival. Proc Natl Acad Sci U S A 110:15919–15924
31. Chen WH, Xu XD, Luo GF, Jia HZ, Lei Q, Cheng SX et al (2013) Dual-targeting pro-apoptotic peptide for programmed cancer cell death via specific mitochondria damage. Sci Rep 3:3468
32. Aruffo A, Stamenkovic I, Melnick M, Underhill CB, Seed B (1990) CD44 is the principal cell surface receptor for hyaluronate. Cell 61:1303–1313
33. Oh EJ, Park K, Kim KS, Kim J, Yang JA, Kong JH et al (2010) Target specific and long-acting delivery of protein, peptide, and nucleotide therapeutics using hyaluronic acid derivatives. J Control Release 141:2–12
34. Dreaden EC, Morton SW, Shopsowitz KE, Choi JH, Deng ZJ, Cho NJ et al (2014) Bimodal tumor-targeting from microenvironment responsive hyaluronan layer-by-layer (LbL) nanoparticles. ACS Nano 8:8374–8382
35. Lee SK, Mortensen LJ, Lin CP, Tung CH (2014) An authentic imaging probe to track cell fate from beginning to end. Nat Commun 5:5216
36. Lee SK, Han MS, Tung CH (2012) Layered nanoprobe for long-lasting fluorescent cell label. Small 8:3315–3320
37. Lee SK, Tung CH (2016) siRNA nanoparticles for ultra-long gene silencing in vivo. Methods Mol Biol 1372:113–120

Chapter 5

PAMAM Dendrimers as a Delivery System for Small Interfering RNA

Shiva Kheiriabad, Maryam Ghaffari, Jafar Ezzati Nazhad Dolatabadi, and Michael R. Hamblin

Abstract

Polyamidoamine dendrimers (PAMAM) form positively charged nanoparticles that function as nonviral delivery vectors for gene therapy. They protect nucleic acids from enzymatic degradation and facilitate endocytosis and endosomal escape. In this chapter, we describe the preparation and in vitro evaluation of small interfering RNA (siRNA)-PAMAM dendrimers. The physicochemical properties of the designed formulations were evaluated by size and zeta potential assessment and atomic force microscopy (AFM). The binding and release of the siRNA molecules from the PAMAM dendrimers were also assessed. Visualization and quantitative analysis of the siRNA-PAMAM dendrimers in live cells were analyzed by fluorescence microscopy and flow cytometry, respectively. Improving siRNA delivery to human cells through PAMAM dendrimers should accelerate the clinical applications of RNA interference.

Key words PAMAM dendrimers, siRNA, Gene delivery, Nonviral vector, RNA interference, RNA-induced silencing complexes, Plasmid-based short hairpin RNA

1 Introduction

RNA interference (RNAi) is a highly specific method to silence or knock down matching mRNA sequences using formation of a double-stranded structure (*see* Chapter 1). RNAi can be used as a therapeutic method by two different methods including small interfering RNA (siRNA) and plasmid-based short hairpin RNA (shRNA) [1, 2]. Double-stranded shRNA can penetrate the nucleus of targeted cells, leading to more efficient and continuous effects and avoid off-target gene silencing. Like other plasmids, shRNA has difficulty in crossing the cellular membrane and penetrating into the nucleus and has therefore attracted less attention as a pharmaceutical agent in gene therapy [1, 3]. Small (20–25 base pairs) interfering RNAs (siRNAs) are used to silence posttranscriptional gene expression and target mRNA with sequence-

Mouldy Sioud (ed.), *RNA Interference and CRISPR Technologies: Technical Advances and New Therapeutic Opportunities*, Methods in Molecular Biology, vol. 2115, https://doi.org/10.1007/978-1-0716-0290-4_5,

specific degradation [4–7]. Their potential as therapeutic agents has been investigated in various diseases such as cancer and infections like malaria. The silencing mechanism of siRNA inside the cells involves self-assembly with RNA-induced silencing complexes (RISC), which function as an endonuclease. In this process, double-stranded siRNA is cleaved to single strands (sense and antisense) and are then able to guide the RISC to bind to complementary RNA strands, which are cleaved into discarded segments [2, 8–11]. MicroRNAs (miRNAs) are another way to silence gene expression in the posttranscriptional phase. miRNAs are small endogenous noncoding single-stranded RNAs, containing 20–24 nucleotides. The mechanism of action of miRNA involves binding to the 3′ untranslated regions (UTRs) of mRNAs, resulting in the degradation and suppression of translation. However, some obstacles regarding the delivery of RNA-targeting agents such as degradation, by several enzymatic activities, and also their pronounced hydrophilic nature, negative charge, and high molecular weights lead to difficulties in penetration through the cell membrane, and these are considered to be significant drawbacks for transition to clinical application. In order to achieve successful therapeutic use of siRNA, it is necessary to design transfection carriers with optimized cellular uptake and low cytotoxicity and to avoid any immunological responses [3, 12]. Recent approaches to overcome these problems have involved developing new nanodelivery systems. Generally, there are two types of vectors to deliver therapeutic genetic material such as DNA or siRNA to cells, which can be divided into viral and nonviral vectors. Although viral vectors can be highly efficient, they have disadvantages of triggering immune responses, cytotoxicity, mutagenicity, and even carcinogenesis. These side effects are a matter of concern for widespread clinical uses. Nonviral vectors can avoid these problems due to their biocompatibility, stability, and surface functionality. One of the nonviral vectors is PAMAM dendrimers, which are nanoparticles produced by a synthetic polymerization process, which possess a promising capability for gene delivery. Several unique features of PAMAM dendrimers make it a desirable alternative for gene delivery, consisting of the controlled and defined size and molecular weight, non-immunogenicity, branched three-dimensional architecture, positively charged surface, easy surface modification, a low polydispersity index, and a high ratio of surface area to volume [4, 5]. Nucleic acid binding to the cationic surface of dendrimers occurs via ionic interactions and results in the formation of condensed particles below several hundred nanometers in size. The PAMAM nanocomplex protects the nucleic acids against degradation and helps delivery into cells. Furthermore, PAMAM dendrimers (in common with other cationic dendrimers) have a large number of tertiary amine groups with pKa values around 6.0. Deprotonation of these amine groups in lysosomes and endosomes that have lower pH values leads to the "proton-sponge effect" which allows the endosomal escape of the nucleic acids. The formation of stable complexes with siRNA, physical protection, and efficient transfection across the cell membrane are the most important advantages PAMAM dendrimers as gene delivery nanodevices. The formation of suitable complexes between siRNA and PAMAM dendrimers is based on surface charge density, the size of the cationic polymer, and the structural morphology of this type of nanocarrier [4, 5]. Efficient transfection of siRNA requires the proper weight ratio of PAMAM dendrimers to balance the positive and negative charges. Endocytosis into cells is facilitated by a high weight ratio (excessive positive charges), while increased cytotoxicity and the tendency to absorb various negatively charged proteins in vivo are significant shortcomings of using a high weight ratio. Two important factors governing the transfection of these nanocomplexes into cells are the size and zeta potential of the nanoparticles. In common with other cationic polymeric carriers, we need to understand more about the physicochemical properties of polymeric vectors on the formation of complexes with nucleic acids and subsequent adsorption by the cellular membrane [10]. Modification of the PAMAM dendrimer surface with various ligands can be performed to enhance the specificity and gene delivery efficiency and reduce the cytotoxicity of these nanocarriers. Alkylation, PEGylation, peptide conjugation, and acetylation are different ways to modify PAMAM dendrimers [13, 14]. Figure 1 describes the mechanism of action of siRNA-mediated gene silencing and how it can be potentiated by the use of PAMAM dendrimers as delivery vehicles.

In this chapter, we will first describe the characterization of the PAMAM dendrimer complexes with siRNA, including the charge, size, and structure via dynamic light scattering (DLS), gel electrophoresis (GE), and atomic force microscopy (AFM). These measurements confirm the PAMAM/siRNA complex formation and determine the stability of the nanocomplex and also the adsorption to a negatively charged surface. Finally, we describe the delivery of siRNA-PAMAM dendrimers to human cells in vitro.

2 Methods

2.1 Agarose Gel Electrophoresis (GE)

2.1.1 Equipment

1. 250 mL flask.
2. Two 1 L orange cap bottles.
3. Spatula.
4. Volumetric cylinders.
5. Gel combs (*see* **Note 1**).
6. Gel casting tray and tape (*see* **Note 2**).
7. Power supply and cables.
8. Electrophoresis tank.

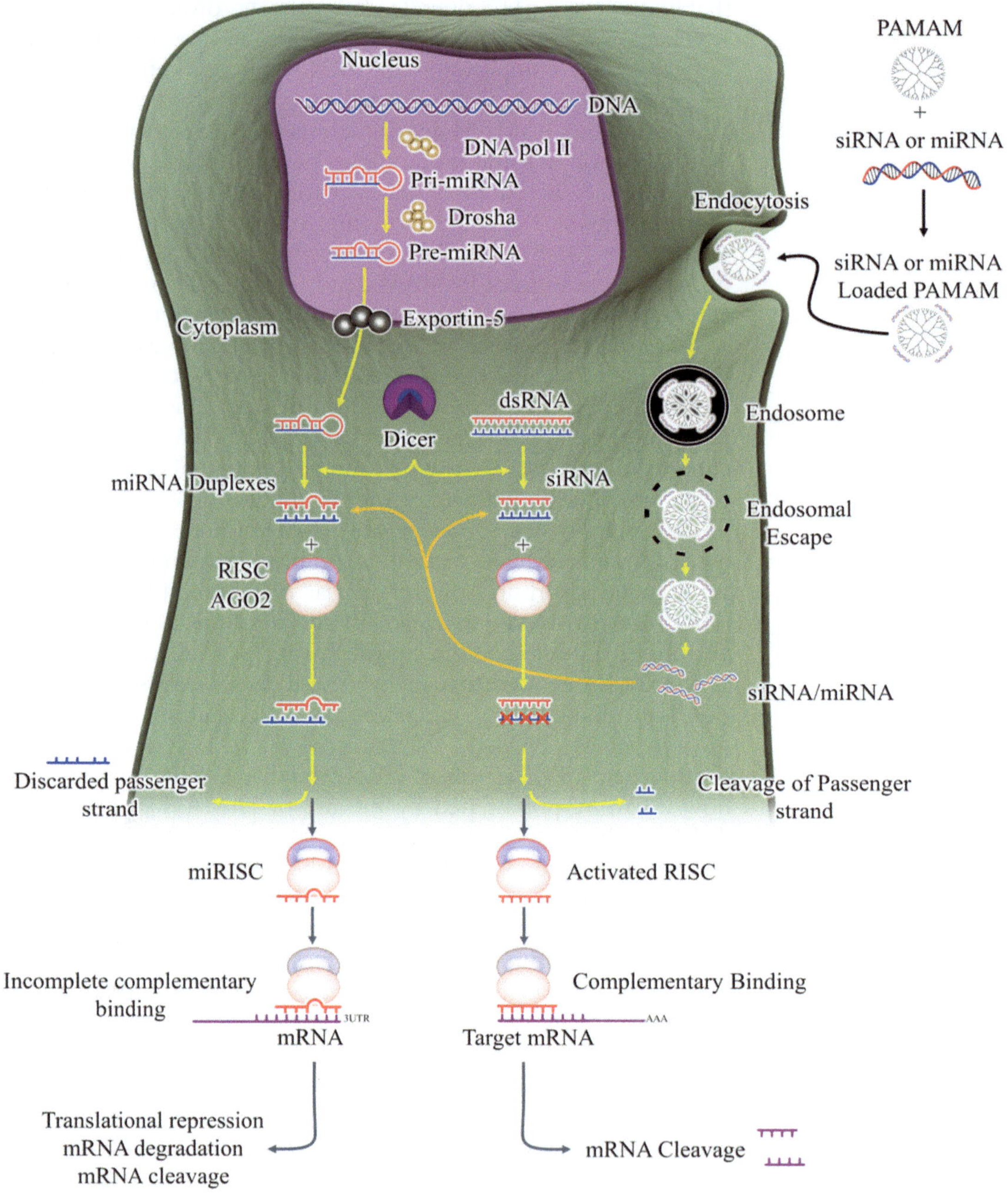

Fig. 1 Schematic illustration of the use of PAMAM dendrimers to potentiate the use of siRNA or miRNA for gene silencing

2.1.2 Buffers (See ***Note 3****)*

1. 50× TAE: 57.1 mL acetic acid, glacial, 100 mL 0.5 M EDTA, 242 g Tris base.
2. 6× loading dye: 0.25% xylene cyanol FF—XC (or same as BB), 15% Ficoll, 120 mM EDTA (240 μL of 0.5 M EDTA in 1 mL total 6× loading dye), 0.25% bromophenol blue—BB (tiny amount on spatula tip).

2.1.3 Reagents

1. Agarose and powdered agar.
2. YTA or SYBR safe gel stains.

2.2 Cell Culture and Transfection

1. MCF-7 breast cancer cell line: Culture cells in RPMI-1640 supplemented with 10% fetal bovine serum and 100 units/mL penicillin–streptomycin.
2. Phosphate-buffered saline (PBS): 140 mmol/l of NaCl, 10 mmol/l of Na phosphate buffer, pH 7.4.
3. Trypsin/EDTA.
4. Fetal bovine serum (FBS).
5. Dimethyl sulfoxide (DMSO).
6. Purified water (*see* **Note 4**).
7. RPMI-1640 medium.
8. Mission® siRNA duplex: The siRNA targets the Bcl-2 gene (sense/antisense, M_w = 13,273 g/mol) and was obtained from Sigma-Aldrich. Nonspecific scrambled siRNA (si-control) was also used.
9. Poly(amidoamine) (PAMAM) dendrimer solution: Different generation of commercially available PAMAM dendrimers (i.e., PAMAM G5 and PAMAM G4, etc.) in methanol with terminal amino groups and ethylenediamine core were used, molecular weight (PAMAM generation 4 (M_w, 14,214 Da), PAMAM generation 5 (M_w, 28,824), etc.).
10. Fluorescein isothiocyanate (FITC).
11. MTT: 3-[4, 5-dimethylthiazol-2-yl]-2, 5-diphenyl tetrazolium bromide.
12. Dialysis membrane: cutoff 12 kDa.

2.3 Equipment

1. Dynamic light scattering (DLS) and zeta plus analyzer (Zetasizer, Malvern Nano ZS, UK).
2. UV transilluminator.
3. Atomic force microscopy, AFM.
4. UV-Vis spectrometer.
5. Enzyme Immunoassay Instrument.
6. BD FACSCalibur Cell Sorting System.

2.4 Reagents and Equipment for RNA Isolation and RT-PCR

1. SuperScript™ II Reverse Transcriptase kit.
2. 5× First-Strand Buffer: 250 mM Tris–HCl, pH 8.3 at room temperature; 375 mM KCl; 15 mM $MgCl_2$ (*see* **Note 26**).
3. 0.1 M DTT (*see* **Note 26**).
4. 10× PCR buffer: 200 mM Tris–HCl (pH 8.4), 500 mM KCl.
5. Random hexamers.
6. Trizol reagent.
7. Taq DNA polymerase.
8. 1.5 mL Eppendorf tubes.
9. Isopropyl alcohol.
10. Ethanol.
11. RNase-free water.
12. Nuclease-free microcentrifuge tubes.
13. RNaseOUT.
14. 50 mM $MgCl_2$.
15. 10 mM dNTP mix.
16. 10 μM Bcl-2 forward primer (5-′-TGCACCTGACGCCCTTCAC-3′) (*see* **Note 27**).
17. 10 μM Bcl-2 reverse primer (5-′-AGACAGCCAGGAGAAATCAAACAG-3′).
18. DNA-free™ kit.
19. Silicone oil.
20. Thermocycler.

3 Methods

3.1 Preparation of Dendrimer Complexed with siRNA (See Note 5)

1. Dilute the freshly synthesized PAMAM dendrimer in distilled water at pH 7.4.
2. Dilute the lyophilized siRNA stock in RNAse-free water at a desired final concentration prior to the association with the PAMAM dendrimer.
3. Calculate the charge ratio of N/P which is accounted for molar ratio of primary amines on PAMAM dendrimer to phosphate group of siRNA.
4. Prepare the complex at different range of N/P including 1:1, 2:1, 4:1, 6:1, and 8:1, respectively. For this purpose, add siRNA solution dropwise to dendrimer solution in the above-mentioned ratios.
5. Then, vortex the mixture quickly for 30 s at room temperature.

6. To prove the stability and formation of PAMAM/siRNA complex, perform agarose gel electrophoresis and atomic force microscopy (AFM) analysis and indicate below.

3.2 Characterization of Synthesized Nanocomplex (PAMAM/siRNA)

3.2.1 Size and Zeta Potential

1. Prepare a fresh solution of nanocomplex at N/P according to Subheading 3.1 and dilute the samples properly in distilled water.
2. Determine the hydrodynamic diameter of the particles, polydispersity index, and the zeta potential (particle surface charge) of the freshly prepared samples using dynamic light scattering (DLS) and zeta plus analyzer (Zetasizer, Malvern nanozs, UK) (*see* **Note 6**).
3. For the measurements, use the refractive index (1.33) and the viscosity (0.89) of ultrapure water at 25 °C .
4. Record the size and zeta potential values using a 90-degree scattering angle and calculate the average values from the data for each sample done in triplicate at a total concentration of 0.1 μg/μL in a volume of 2 mL for each sample.

3.2.2 Analysis of the Dendrimer/siRNA Nanocomplexes by Agarose Gel Electrophoresis

Agarose gel electrophoresis is carried out to evaluate both siRNA binding ability of the PAMAM dendrimers and validate the formation of nanoparticle/siRNA complexes (*see* **Note 7**). Follow the following steps for gel preparation and electrophoresis.

1. Make 1 L of TAE buffer in concentration of 1× (20 mL into 1 L).
2. Pour 100 mL of prepared TAE buffer into a clean 250 Erlenmeyer flask and use another 900 mL as a running buffer (*see* **Note 8**).
3. Agarose gel is prepared with (2.0%, w/v) weighing 2 g of agarose gel powder and adding it into 100 mL of TAE buffer (40 mmol/L Tris–acetate, 1 mmol/L EDTA). Heat the solution in a microwave until agarose gel powder dissolves. Undissolved agarose appears as floating "lenses" (*see* **Note 9**).
4. Remove the solution from the microwave and let it cool to about 60 °C (*see* **Note 10**).
5. After, add a safe stain like YTA or SYBR solution to a final concentration of 0.05 μg/mL. For example, add 5 μL of 10 mg/mL stock for a 100 mL gel. Alternatively, the gel can be stained after electrophoresis (*see* **Note 11**).
6. Pour the liquid agarose and take the sides into the notches and then put the comb into the casting tray.
7. Wait until the gel polymerizes (*see* **Note 12**).
8. Remove the comb carefully to expose the sample wells.
9. Place the prepared gel into the gel chamber.

10. Add safe stain to the ~900 mL 1× TAE buffer at a concentration of 0.05 μg/mL. Pour the buffer into the chamber until it covers the gel.
11. Add 20 μL from each sample, containing 100 base pairs RNA ladder, free siRNA (control), and PAMAM-siRNA complexes in different N/P rate to 1 μL of RNA loading buffer (0.4 M formaldehyde, 0.02% bromophenol blue, 62.5% formamide, 0.02% xylene cyanol FF, and MOPS buffer) (*see* **Note 13**).
12. Prior to sample loading, add 1 μL of loading dye (containing EDTA, glycerol, and bromophenol blue) per 5 μL sample and then load the samples into the agarose gel.
13. Connect the tank to the power supply and set the voltage at ~70 V (*see* **Note 14**).
14. Run for about 30 min at 70 V.
15. After electrophoresis, visualize the migration of nucleic acid bands using UV transilluminator (*see* **Notes 15** and **16**).

3.2.3 Atomic force Microscopy (AFM) Analysis of the Complexes (See Note 17)

1. With respect to AFM evaluation, form the PAMAM /siRNA complexes at defined N/P ratios.
2. Place 1–5 aliquots of 100-fold diluted siRNA into 10 mL solution to coat the surface of freshly cleaved mica sheet (1×1 cm^2) for 10 min.
3. Dry out at room temperature and then wash twice with ddH_2O (*see* **Note 18**).
4. Expose to mild air stream for 10 min and then do the analysis by means of AFM, using a DME DualScope/Rasterscope™ SPM (DME Co., Denmark).

3.3 In Vitro siRNA Release Assay

1. Dissolve the nanodelivery system (PAMAM/siRNA) in 4 mL PBS buffer.
2. Carry out dialysis (cutoff 12 kDa, Sigma-Aldrich) against PBS (50 mL) pH 5.5 and 7.4 for 120 h on a shaker incubator (37 °C, 100 rpm).
3. Take out 2 mL of PBS from the container and then replace with 2 mL fresh PBS at various time points (1, 3, 6, 12, 24, 48, 72, 96, and 120 h).
4. Measure the extent of siRNA release using UV-Vis spectrometry at 260 nm wavelength.
5. Use the following equation to measure the released siRNA based on percentage.

$$\text{Released siRNA\%} = [(\text{OD of siRNA in PBS/initial total content of siRNA})] \times 100.$$

3.4 Cell Culture

1. Acquire the MCF-7 breast cancer cells from the National Cell Bank.
2. Culture cells in RPMI-1640 medium supplemented with 10% fetal bovine serum (FBS) and 1% antibiotics (streptomycin 100 g/mL and penicillin 100 U/mL).
3. Maintain the cells in a humidified cell incubator at 37 °C with 5% CO_2/95% air.
4. Monitor the cell culture flasks daily. When the cells reach confluence of 80–90%, passage the cells at ratio of 1:4 twice a week (*see* **Note 19**).

3.5 Cytotoxicity Evaluation of Blank Nanocarriers

MTT assay is used to determine the nontoxic concentrations of the PAMAM dendrimer reagent (*see* **Note 20**).

1. Seed the cells at a density of 4×10^3 cells/well in 96-well plates and incubate overnight.
2. After 24 h, treat the cells with various concentrations of PAMAM in serum-free media and then incubate the treated cells for 6 h at 37 °C.
3. After 6 h, replace the serum-free cell culture media with 200 mL fresh complete medium.
4. Add 20 μL MTT solution (5 mg/mL in PBS) and incubate the cells for another 4 h.
5. Remove the MTT solution and replace with 200 μL of DMSO.
6. Agitate the plate for 5 min and read the absorbance via an Enzyme Immunoassay Instrument (ELx800, BioTek Instruments, Winooski, VT) at 490 nm.
7. Analyze the cell viability.

3.6 Cellular Uptake Studies

Label PAMAM dendrimer with FITC according to the published procedure [15] (*see* **Note 21**) and then investigate the uptake of the complexes as described below.

3.6.1 Flow Cytometry Analysis (Quantitative Analysis)

1. Seed the cells at density of 1×10^5 cells/well on 24-well plate with complete RPMI-1640 medium.
2. After 24 h, replace the growth medium with 200 μL serum-free culture medium, containing desirable concentration of FITC-PAMAM, and then incubate for 6 h.
3. Remove the supernatant and wash cells with PBS (pH = 7.4) three times and then detach the cells with trypsin treatment.
4. Centrifuge cell suspensions at $1000 \times g$ for 5 min, resuspend in PBS (pH = 7.4), and then analyze the uptake by flow cytometry.

3.6.2 Fluorescence-Activated Cell Sorter (FACS) Analysis: Qualitative Analysis

1. Treat the cells using the same method described above for fluorescence-activated cell sorter (FACS) analysis.
2. After incubation, wash the cells with PBS buffer for three times and detach the cells with trypsin/EDTA treatment.
3. Analyze the fluorescence signal intensities of the samples on BD FACSCalibur Cell Sorting System (*see* **Note 22**).

3.7 In Vitro Transfection of siRNA/ PAMAM Complexes

1. Culture cells in RPMI-1640 complete medium (*see* **Note 23**).
2. Prior to transfection, prepare siRNA/PAMAM complexes according to the followin
3. Dilute the desired amount of siRNA and PAMAM in 200 μL of OPTIMEM transfection medium.
4. Add the PAMAM dendrimers to the siRNA solution, homogenize by vortexing for 10 s, and then incubate for 30 min at room temperature.
5. Add 1.6 mL serum-free medium to the prepared complexes and make final volume to 2 mL. The final concentration should be around 50 nM siRNA with PAMAM in N/P ratio of 10 (*see* **Note 24**).
6. For the in vitro transfection, seed cells at density of 2×10^4 cells per well in 24-well plates using complete medium.
7. Incubate for 24 h to reach confluence of almost 70–80%.
8. After incubation, remove the complete medium and wash cells with PBS once. Then add 100 nM siRNA, complexed with PAMAM at weight ratio of 1:100 (siRNA to polymer) in OPTIMEM medium (*see* **Note 25**).
9. Incubate the cells for 6 h in OPTIMEM media.
10. Replace the OPTIMEM medium with complete RPMI medium.
11. Maintain the cells under normal growth condition until analysis of gene expression.

3.8 RNA Isolation and RT-PCR Assay (See Note 28)

Seed MCF-7 cells at density of 1×10^6 cells per well in six-well plates, and transfect with siRNA/PAMAM complex and non-silencing control siRNA as described in Subheading 3.7. After 48 h of transfection time, wash the cells with cold PBS, and prepare the total cellular mRNA as described below (*see* **Note 29**).

3.8.1 RNA Extraction

1. Lyse the cells in1 mL Trizol reagent, and, through a pipette, pass the lysate several times.
2. Transfer the lysed cells into a 1.5 mL Eppendorf tube.
3. Allow the homogenized samples to stand at 15–30 °C for 5 min (*see* **Note 30**).

4. Add 0.2 mL of chloroform per mL of Trizol.
5. Shake the tubes vigorously by hand for 15 s and then allow to stand for 2–3 min at 15–30 °C.
6. Centrifuge the resulting mixture at 12,000 × *g* for 15 min at 2–8 °C (*see* **Note 31**).
7. Transfer the colorless upper aqueous phase to a fresh and clean 1.5 mL Eppendorf tube (*see* **Note 32**).
8. Precipitate the RNA from the aqueous phase by adding 0.5 mL isopropyl alcohol to the aqueous phase, mix gently, and incubate the samples at 15–30 °C for 10 min.
9. Centrifuge at 12,000 × *g* for 10 min at 2–8 °C (*see* **Note 33**).
10. Remove the supernatant. Wash the RNA pellet once by adding at least 1 mL of 75% ethanol (*see* **Note 34**). Mix the sample by vortexing gently and centrifuge at 7500 × *g* at 2–8 °C for 5 min.
11. Pour off the ethanol and briefly air-dry the RNA pellet for 5–10 min (*see* **Note 35**).
12. Dissolve RNA pellet in RNase-free water and incubate for 10 min at 55–60 °C (*see* **Note 36**).

3.8.2 Reverse Transcription (See ***Note 37****)*

1. To prepare the RNA/primer mixture, transfer 5 μg prepared total RNA (x μL) into a nuclease-free microcentrifuge tube and add 1 μL of random hexamers (50 ng/μL) and 1 μL of 10 mM dNTP mix.
2. Add nuclease-free water to bring the volume up to 12 μL.
3. Incubate the mixture at 65 °C for 5 min and then chill on ice for 2 min.
4. Microcentrifuge briefly to collect the solution at the bottom of the tube (*see* **Note 38**).
5. To prepare the reaction mixture, add the following components to a separate tube (*see* **Note 39**):

5× First-Strand Buffer	4 μL
0.1 M DTT	2 μL
RNaseOUT (optional) (*see* **Note 40**)	1 μL

6. Add the reaction mixture to the RNA/primer mixture (**step 4**), mix briefly, and then incubate at 25 °C for 2 min.
7. Add 1 μL (200 units) of SuperScript II Reverse Transcriptase (*see* **Note 41**) to the tube, mix, and incubate at 25 °C for 10 min.
8. Incubate the reaction mixture at 42 °C for 50 min.

9. Inactivate the enzyme by heating at 70 °C for 15 min, and then place on ice for 5 min.
10. Add 1 μL (2 units) RNase H (*see* **Notes 41** and **42**) and incubate at 37 °C for 20 min.
11. Microcentrifuge briefly, and then freeze the reaction at −20 °C until PCR amplification.

3.8.3 PCR Amplification

1. Prepare the following components in a PCR tube (*see* **Note 43**):

cDNA (from RT reaction)	2 μL
10× PCR buffer	5 μL
Taq DNA polymerase (5 U/μL)	0.4 μL
50 mM $MgCl_2$	1.5 μL
10 mM dNTP mix	1 μL
10 μM Bcl-2 forward primer	1 μL
10 μM Bcl-2 reverse primer	1 μL
Autoclaved, distilled water	up to 50 μL

2. Mix gently (*see* **Note 44**).
3. Incubate in thermocycler using the following program:
   (a) Initial denaturation: 94 °C for 4 min.
   (b) 15 to 40 cycles of PCR: Denature at 94 °C for 30 s, anneal at 55 °C for 20–30 s, and extend at 72 °C for 45 s (*see* **Note 45**).
   (c) Final extension: 72 °C for 5 min (*see* **Note 46**).

4 Notes

1. Select the proper type of combs.
2. Tape both sides of the casting tray to hold the liquid gel.
3. Make a stock of 50× TAE running buffer.
4. It was used for all the experiments.
5. siRNA stock solution, which is required for each transfection, was calculated with the following equation:

$$V_{\text{siRNA}} = \left(C_{\text{siRNA,final}} \times V_{\text{final}}\right) / \left(C_{\text{siRNA,stock}}\right),$$

where $C_{\text{siRNA, final}}$ and V_{final} are remarked as final desired concentration of siRNA and volume, respectively. The concentration of siRNA stock solution is named C_{siRNA}.

The stock solution of polymer, which is used for preparing samples, was calculated using the following equation:

$$V_{\text{polymer}} = \left(C_{\text{siRNA,final}} \times M_{\text{w,siRNA}}/M_{\text{w}}, P \times \Phi_{\text{N:P}} \times \Phi V_{\text{final}}\right)/\left(C_{\text{polymer,stock}}/M_{\text{w}}, F\right)$$

where $M_{\text{w siRNA}}$ is the molecular weight of the siRNA, P is the molecular weight per phosphate group of siRNAs, $\Phi_{\text{N:P}}$ is the charge ratio (N/P), and F is the molecular weight of the polymer's functional group [3].

6. Here, Zetasizer is used to depict the transfection efficiency appropriately and study the precise physical and chemical characteristics of the synthesized nanocomplexes.
7. SiRNA stability was analyzed by incubating the PAMAM/siRNA complex and blank siRNA in 50% FBS up to 48 h at 37 °C. To detect degradation, aliquots taken at various time points were analyzed by gel electrophoresis.
8. Instead of flask, any other container that fits well into the microwave can be used.
9. Please use insulated mitts to transport the flask.
10. Gel solution can be cooled under running water. It will solidify if you cool it too much.
11. YTA safe stain is a new nucleic acid stain, which is used instead of conventional ethidium bromide (EB). EB is carcinogenic; hence, it must be handled with gloves.
12. The gel should look uniform and opaque.
13. Ladder can be defined as one of the samples that include RNA fragments with known lengths in the range of your samples.
14. In the first minutes, monitor the gel carefully to make sure that the dyes are migrating in the correct direction.
15. As the gel is running by shining UV light on it, the progress of the RNA running can be detected directly (if you have already added the safe stain).
16. Wear gloves and carry the gel in the casting tray. To protect yourself from UV, use a plastic shield, face mask, etc. At the end of the process, before opening the plastic cover, turn off the UV.
17. Atomic force microscope (Digital Instrument, Santa Barbara, CA) is considered to image PAMAM dendrimer/siRNA complex (N/P) and also analyze morphology and size of established complex (N/P) under identical condition. Utilize AFM in tapping mode [16].
18. Prior to AFM imaging, rinse samples mildly with 10 mM ddH_2O or HEPES buffer to ensure removal of unadsorbed or excess polymers from the surface of mica sheets.

19. Detach cells from flasks by washing twice with PBS and incubate with trypsin-EDTA for certain minutes. After, centrifuge the separated cells to obtain cellular pellet. Suspend the pellet in complete medium and culture the appropriate cell number at 37 °C, 5% CO_2/95% air.
20. MTT assay is carried out as a classical method to measure the cytotoxicity of nanocarriers.
21. Conjugate PAMAM dendrimers to FITC dye by reacting 2000 μg of dendrimer with 100 μg of FITC in pH 7.4 at room temperature for 12 h. After, purify the conjugates.
22. Calculate the uptake percentage by subtracting the number of florescent cells from the number of total cells.
23. The cells are allowed to grow for 2–3 days to 65–70% confluence before transfection.
24. The volume of the transfection sample was estimated around 20% of total volume in each well.
25. Transfection efficiency can be measured by flow cytometry.
26. Just prior to use, thaw 0.1 M DTT and 5× First-Strand Buffer at 25 °C and refreeze immediately.
27. To make a primer stock at 100 μM concentration, dissolve the oligonucleotide in 10 mM Tris, pH 7.5. Dilute from this stock 1:10 (in water) to make a solution at 10 μM.
28. In this assay, a two-step reverse transcription PCR (RT-PCR), mRNA is used as a template to assess the gene expression level of Bcl-2. The reverse transcription reaction is performed using SuperScript™ II Reverse Transcriptase kit (Invitrogen/Life Technologies).
29. Use the untreated cells as a control. Also, use the β-actin housekeeping gene as an internal normalization standard (positive control) and measure the mRNA level of this gene in each sample. The forward and reverse primers targeting the β-actin sequence were 5′-CCATCATGAAGTGTGACGTGG-3′ and 5′-GTCCGCCTAGAAGCATTTGCG-3′, respectively.
30. This step is to ensure the complete dissociation of nucleoprotein complexes.
31. After centrifugation, there are three phases in the mixture, which include a colorless upper aqueous phase (containing RNA), a white interphase (containing the DNA), and a red organic phase (containing the protein).
32. Transfer up to 700 μL of the colorless phase to the clear tube. Be careful to avoid *DNA* contamination from *interphase* during the transferring colorless aqueous phase.
33. Before centrifugation, the RNA precipitate forms a gel-like pellet on the bottom and side of the tube.

34. Use RNase-free water to prepare the 75% ethanol. Store samples in ethanol for at least 1 week at 4 °C and up to 1 year at −20 °C.
35. It is important not to let the RNA pellet dry out too much as this will greatly decrease its solubility.
36. After this step, treat the sample with DNaseI to remove DNA contamination. The Ambions DNA-free™ kit (#1906) is used. This kit contains a DNaseI enzyme (2 U/μL), a buffer for the enzyme activity (100 mM Tris–Cl pH 7.5, 1 mM $CaCl_2$, 25 mM $MgCl_2$), and an inactivation reagent (required at the end of the reaction for inactivation of the enzyme).
37. To avoid RNase contamination, wear gloves at all times.
38. During the preparation of the reaction mixture, leave the RNA mixture on ice.
39. Do not add these components to your RNA solution and use a new tube.
40. If using <50 ng starting RNA, RNaseOUT is required.
41. Keep the samples on ice.
42. The RNase H enzyme is used to remove RNA complementary to the cDNA.
43. It is better to use only 10% of the first-strand reaction for PCR. Higher volumes may result in decreased amounts of PCR product and may not increase the amplification.
44. After this step, layer with 1–2 drops (~50 μL) of silicone oil. It should be mentioned that in thermal cyclers with a heated lid, silicone oil is unnecessary.
45. The recommended extension and annealing conditions for the Taq polymerase should be used. For this purpose, a reasonable starting point for annealing temperature is about 55 °C which used with the cycling times, but the optimal temperature and cycling times for the template combination and primer can be determined experimentally.
46. After PCR, perform electrophoresis using 2% agarose gel containing ethidium bromide at 100 V for 15 min and visualize the PCR products of each sample by a UV illuminator. Normalize the relative amount of mRNA expression to β-actin (loading control) [16].

References

1. Ayatollahi S, Salmasi Z, Hashemi M, Askarian S, Oskuee RK, Abnous K, Ramezani M (2017) Aptamer-targeted delivery of Bcl-xL shRNA using alkyl modified PAMAM dendrimers into lung cancer cells. Int J Biochem Cell Biol 92:210–217
2. Gu J, Fang X, Hao J, Sha X (2015) Reversal of P-glycoprotein-mediated multidrug resistance by CD44 antibody-targeted nanocomplexes for short hairpin RNA-encoding plasmid DNA delivery. Biomaterials 45:99–114
3. Chang PK, Prestidge CA, Bremmell KE (2017) Interfacial analysis of siRNA complexes with poly-ethylenimine (PEI) or PAMAM dendrimers in gene delivery. Colloids Surf B: Biointerfaces 158:370–378
4. Abedi-Gaballu F, Dehghan G, Ghaffari M, Yekta R, Abbaspour-Ravasjani S, Baradaran B, Dolatabadi JEN, Hamblin MR (2018) PAMAM dendrimers as efficient drug and gene delivery nanosystems for cancer therapy. Appl Mater Today 12:177–190
5. Ghaffari M, Dehghan G, Abedi-Gaballu F, Kashanian S, Baradaran B, Ezzati Nazhad Dolatabadi J, Losic D (2018) Surface functionalized dendrimers as controlled-release delivery nanosystems for tumor targeting. Eur J Pharm Sci 122:11–330
6. Ezzati Nazhad Dolatabadi J, Omidi Y (2016) Solid lipid-based nanocarriers as efficient targeted drug and gene delivery systems. TrAC Trends Anal Chem 77:100–108
7. Ezzati Nazhad Dolatabadi J, Omidi Y, Losic D (2011) Carbon nanotubes as an advanced drug and gene delivery nanosystem. Curr Nanosci 7:297–314
8. Pandi P et al (2018) Dendrimer as a new potential carrier for topical delivery of siRNA: comparative study of dendriplex vs. lipoplex for delivery of TNF-α siRNA. Int J Pharm 550 (1–2):240–250
9. Perez A, Romero E, Morilla M (2009) Ethylendiamine core PAMAM dendrimers/siRNA complexes as in vitro silencing agents. Int J Pharm 380(1–2):189–200
10. Wang Y, Fang J, Cheng D, Wang Y, Shuai X (2014) A pH-sensitive micelle for codelivery of siRNA and doxorubicin to hepatoma cells. Polymer 55:3217–3226
11. Zhou J, Neff CP, Liu X, Zhang J, Li H, Smith DD, Swiderski P, Aboellail T, Huang Y, Du Q, Liang Z, Peng L, Akkina R, Rossi JJ (2011) Systemic administration of combinatorial dsiRNAs via nanoparticles efficiently suppresses HIV-1 infection in humanized mice. Mol Ther 19:2228–2238
12. Zhang Y, Wang Z, Gemeinhart RA (2013) Progress in microRNA delivery. J Control Release 172:962–974
13. Shan Y, Luo T, Peng C, Sheng R, Cao A, Cao X, Shen M, Guo R, Tomás H, Shi X (2012) Gene delivery using dendrimer-entrapped gold nanoparticles as nonviral vectors. Biomaterials 33:3025–3035
14. Tariq I, Pinnapireddy SR, Duse L, Ali MY, Ali S, Amin MU, Goergen N, Jedelská J, Schäfer J, Bakowsky U (2018) Lipodendriplexes: A promising nanocarrier for enhanced gene delivery with minimal cytotoxicity. Eur J Pharm Biopharm 135:72–82
15. Ma D, Lin QM, Zhang LM, Liang YY, Xue W (2014) A star-shaped porphyrin-arginine functionalized poly(L-lysine) copolymer for photo-enhanced drug and gene co-delivery. Biomaterials 35:4357–4367
16. Zarebkohan A, Najafi F, Moghimi HR, Hemmati M, Deevband MR, Kazemi B (2015) Synthesis and characterization of a PAMAM dendrimer nanocarrier functionalized by SRL peptide for targeted gene delivery to the brain. Eur J Pharm Sci 12:19–30

Chapter 6

Delivery of Functional Small RNAs via Extracellular Vesicles In Vitro and In Vivo

Duo Zhang, Heedoo Lee, and Yang Jin

Abstract

Extracellular vesicles (EVs) are naturally generated nanovesicles which potentially mediate the intercellular communication and interorgan crosstalk. EVs have recently gained significant interest as a promising material for delivery of therapeutics. Small RNAs, including small interfering RNA (siRNA) and microRNA (miRNA), provide a great therapeutic strategy for treating human diseases. However, it remains a challenge to deliver unconjugated small RNAs to the target tissue or cells. The delivery of small RNAs in an EV-encapsulating manner has a number of advantages, such as enhancing the concentration of small RNAs, improving the uptake of small RNAs by the recipient cells, and potentially achieving a cell-specific delivery. In this chapter, a protocol is provided for EV preparation and loading with small RNAs. Additionally, a detailed experimental protocol for tracking and validating small RNA delivery into the lungs is described. Overall, the described protocols are valuable for delivering functional small RNAs both in vitro and in vivo.

Key words microRNA, siRNA, Microvesicles, Exosomes, Small RNA-based therapeutics

1 Introduction

Extracellular vesicles (EVs) are a group of heterogeneous, membrane-surrounded structures released by all cells [1]. Although a consensus nomenclature has not been developed, currently, the term EVs refers to exosomes (EXOs), microvesicles (MVs), and apoptotic bodies (ABs), classified based on their sizes, components, and mechanisms of formation [1, 2]. These evolutionally conserved vesicles carry out important functions in intercellular communication and interorgan crosstalk, due to their capacity to transfer proteins, lipids, and nucleic acids [1, 2]. EVs also exert essential physiological and pathological effects on both recipient and parent cells. Emerging evidence in recent years suggests that EVs potentially serve as novel targets for the development of therapeutic and diagnostic reagents.

Mouldy Sioud (ed.), *RNA Interference and CRISPR Technologies: Technical Advances and New Therapeutic Opportunities*, Methods in Molecular Biology, vol. 2115, https://doi.org/10.1007/978-1-0716-0290-4_6,

Small RNAs, particularly small interfering RNA (siRNA) and microRNA (miRNA), have great potentials to serve as a therapeutics of human diseases [3]. SiRNA is a chemically synthesized, short double-stranded RNA (dsRNA) molecule with 20–25 base pairs in length [4]. In contrast, miRNAs are endogenously encoded small noncoding RNAs with about 22 nucleotides in length, which serve as crucial regulators in numerous biological processes such as proliferation and differentiation, development, and cell death ([5], *see* Chap. 1. For years, a number of pharmaceutical companies have invested in the RNAi field and aim to bring siRNA drug candidates into clinical applications (*see* Chaps. 1 and 2). For example, the Food and Drug Administration (FDA) has approved an RNAi-based therapy for the rare hereditary disease transthyretin-mediated amyloidosis in adult patients [6]. However, despite the scientific investigation and pharmaceutical investment, the delivery of small RNA remains a challenge. The major issues are stability and nonspecific uptake. Although chemical modification improves the stability (*see* Chaps. 2 and 3), unconjugated free small RNAs still has poor permeability across cell membrane due to their charge and molecular weight, which leads to low delivery efficiency [3]. Delivery systems, such as polymer-based and lipid-based systems, significantly improve the stability of small RNAs and delivery efficiency by protecting them from degradation and enhancing the uptake by cells ([7, 8]; *see* Chaps. 4 and 5). However, there are several concerns of using these delivery systems, including toxicity, nonspecific uptake, and immunogenic effects [9, 10].

Compared with man-made delivery systems, EVs provide many advantages as mentioned above, such as protecting EV-encapsulated miRNAs from enzymatic digestion and degradation, as well as providing improved efficiency of cell uptake [11]. In addition, EVs are naturally generated by most mammalian cells. MVs and EXOs are both released by healthy cells, and thus they are less toxic and less immunogenic compared to exogenous delivery vehicles [11, 12]. It has been shown that EVs are able to cross the blood-brain barrier (BBB) and transmit biological signals from the circulation to the central nervous system [13]. The original components of EVs comprised of RNAs (such as tRNA, miRNA, lncRNA), lipids, and proteins [1]. These components that are transported by EVs potentially serve as novel therapeutic agents [14]. Furthermore, engineered EVs were designed to target specific tissue/cell in vivo. For example, Alvarez Erviti and colleagues displayed neuron-specific RVG peptide on the surface of EXOs [15]. RVG-displayed EVs could deliver small RNA specifically into neurons, microglia, and oligodendrocytes in the mouse brain [15]. Taken together, these features make EV an ideal carrier for small nucleotide delivery.

Here, we present protocols for the EV isolation and preparation for small RNA formulations. In this protocol, we use MVs and EXOs as small RNA carrier and exclude ABs due to their size and heterogeneity. Two methods for encapsulating small RNA in EVs are described. We also describe the labeling of EVs using PKH26 dye. For in vivo delivery, two routes of EV administration are described, including intravenous injection (IV) and intratracheal inhalation (IH). Finally, we provide experimental protocols for visualizing the EV biodistribution and validating the silencing potency of siRNAs.

2 Materials

2.1 EV Preparation

Figure 1 illustrates the steps used to prepare EVs. Sterile solutions and supplies, as well as aseptic technique, must always be applied in the following protocol. Results could be affected by microbial contamination both in vivo and in vitro.

2.1.1 Equipment and Reagents

1. Ultracentrifuge (Beckman).
2. Benchtop centrifuge.
3. Phosphate-buffered saline (PBS, pH 7.4; no calcium, no magnesium; store at room temperature).
4. Polyethylene glycol average Mn 6000.
5. Centrifuge/ultracentrifuge tubes (Beckman).
6. 20G × 1″ safelet catheter (EXEL): sterile, nonpyrogenic, and nontoxic.
7. 2 and 15 mL collection tubes.
8. Forane (isoflurane): store at room temperature.
9. Thread (Thermo Fisher).
10. 1 mL syringe.

2.2 Loading of Small RNAs into EVs

Sterile solutions and supplies, as well as aseptic technique, must always be applied in the following protocol. Results could be affected by microbial contamination both in vivo and in vitro.

2.2.1 Equipment and Reagents

1. Neon® Transfection System 100 μL Kit (Thermo Fisher Scientific).
2. Ultracentrifuge.
3. Water bath.
4. Benchtop centrifuge.
5. Phosphate-buffered saline (PBS) (Gibco): pH 7.4; no calcium, no magnesium; store at room temperature.

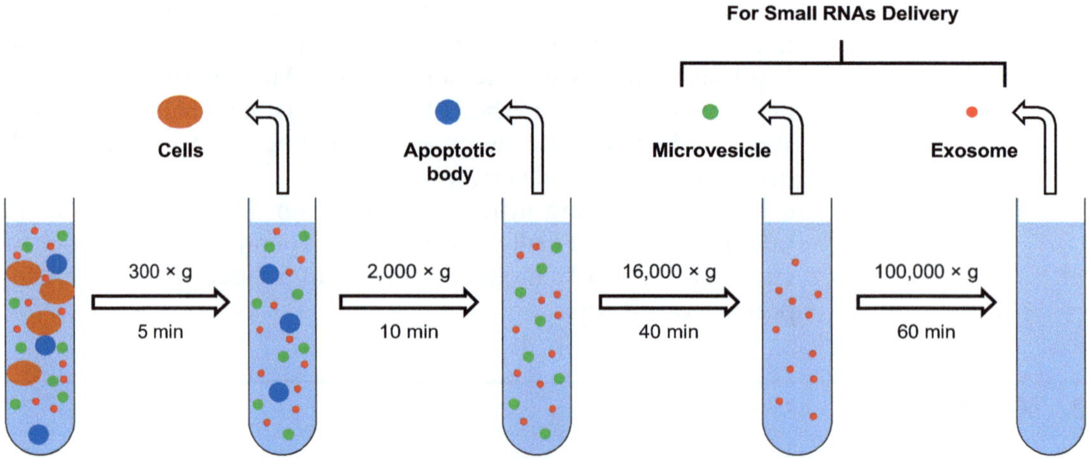

Fig. 1 Schematic diagram: preparation of extracellular vesicles used for small RNAs delivery

6. Calcium chloride ($CaCl_2$) (Sigma-Aldrich): a 10× stock solution (1 M $CaCl_2$) can be prepared in ddH_2O; store at room temperature.
7. Small interfering RNAs (siRNAs) and microRNA mimics/inhibitors: dissolve the small RNAs (20 μM) and prepare aliquots in RNase-free water (20 μM) to avoid freeze-thaw cycles. Store at −20 °C for up to 6 months. For extended storage, freeze at −80 °C.
8. 1.5 mL Eppendorf microtubes.
9. Centrifuge/ultracentrifuge tubes.

2.3 EV-Mediated Delivery of Small RNA Molecules

Sterile solutions and supplies, as well as aseptic technique, must always be applied in the following protocol. Results could be affected by microbial contamination both in vivo and in vitro.

2.3.1 Equipment and Reagents

1. Tailveiner Restrainer for Mice (Braintree Scientific).
2. Benchtop Eppendorf centrifuge.
3. Centrifuge cryostat.
4. Fluorescence microscope.
5. Phosphate-buffered saline (PBS): pH 7.4; no calcium, no magnesium; store at room temperature.
6. PKH26 Red Fluorescent Cell Linker (Sigma-Aldrich): store at room temperature.
7. Optimal cutting temperature compound (VWR): store at room temperature.
8. 1× trypsin/EDTA solution: store at −20 °C. Freshly dilute to 0.2× trypsin/EDTA with PBS before use.

9. Paraformaldehyde solution 4% in PBS: store at 4 °C. Aliquot and freeze at −20 °C for long-term storage.
10. Forane (isoflurane): store at room temperature.
11. 1.5 mL microtubes.
12. Exosome Spin Columns (Thermo Fisher Scientific).
13. Microscope slides.
14. Gauze pads.
15. 1 mL syringe.

3 Methods

3.1 Isolation of EVs from Mouse BALF (See Note 1)

1. Isoflurane euthanasia: expose the mouse to the isoflurane (5% or greater) until 1 min after breathing stops (*see* **Note 2**).
2. Use scissors to expose the thoracic cage and neck. Dissect tissue from the neck to expose the trachea.
3. Proceed to open the diaphragm by cutting the rib cage to expose both the heart and lungs. Take caution not to pierce the heart or lungs.
4. Use forceps to slide a 1-inch-long piece of thread underneath the trachea.
5. Carefully insert the needle catheter into the trachea and tie thread into a single knot around tubing in the trachea.
6. Slowly inject 1 mL cold PBS using 1 mL syringe (it takes 15–20 s): watch the lungs inflating and do not overinflate.
7. Slowly collect BAL fluid (BALF) from the lungs using 1 mL syringe (it takes 15–20 s) and transfer the BALF to a pre-chilled tube on ice.
8. Repeat **steps 6–7**.
9. Centrifuge the BALF at 300 × *g* for 5 min.
10. Carefully collect BALF supernatant.
11. Centrifuge the BALF supernatant at 2000 × *g* for 10 min.
12. Carefully collect AB pellets and BALF supernatant.
13. Centrifuge the BALF supernatant at 16,000 × *g* for 40 min.
14. Carefully collect MV pellets and BALF supernatant.
15. Centrifuge the BALF supernatant at 100,000 × *g* for 60 min.
16. Carefully collect Exo pellets.
17. Wash the EV pellets with cold PBS using centrifugation.
18. Resuspend the washed EV pellets with cold PBS.
19. Store at −80 °C.

3.1.1 Isolation of EVs from Serum

1. Isoflurane euthanasia: expose the mouse to the isoflurane (5% or greater) until 1 min after breathing stops.
2. Collect the blood from the mouse heart.
3. Incubate the collected blood at RT for 30 min.
4. Centrifuge the sample for 20 min at 3000 rpm (1500 × *g*).
5. Carefully transfer the supernatant (serum) to new tubes.
6. Centrifuge the supernatant at 16,000 × *g* for 40 min.
7. Carefully collect MV pellets and the remaining supernatant.
8. Centrifuge the remaining supernatant at 100,000 × *g* for 60 min.
9. Carefully collect Exo pellets.
10. Wash the EV pellets with cold PBS using centrifugation.
11. Resuspend the washed EV pellets with cold PBS.
12. Store at −80 °C.

3.1.2 Alternative Method of EV Isolation

1. Add 5% PEG to the collected BALF supernatant or serum.
2. Incubate the mixture at 4 °C overnight.
3. Centrifuge the mixture at 3000 × *g* for 40 min.
4. Resuspend the EV pellets with cold PBS.
5. Store at −80 °C.

3.2 Loading of Small RNAs into EVs

Figure 2 illustrates the introduction of siRNAs into EVs using electroporation or calcium chloride transfection method.

3.2.1 Small RNA Loading via Electroporation

1. After pelleting EVs, resuspend the EVs using pre-cold E2 electrolytic buffer provided in Neon™ Transfection System 100 μL Kit (*see* **Note 3**).
2. To achieve a 110% reaction volume, add 5.5 μL 20 μM small RNA (total, 110 pmol) with 110 μg resuspend EVs (measured by EV protein) to 1.5 mL tube. Add E2 electrolytic buffer to make the final reaction volume to 110 μL.
3. Gently mix the tube by inverting 8–10 times and briefly centrifuge the tube.
4. Aspirate the EV/small RNA mixture into the 100 μL Neon™ Tip using Neon™ pipette. Avoid air bubbles during pipetting.
5. Perform the electroporation at 0.5 kV using 10 μs pulse five times using the Neon™ Transfection System (*see* **Note 4**).
6. After electroporation, transfer 100 μL of the small RNA loaded EVs into a new tube and proceed to wash steps below.

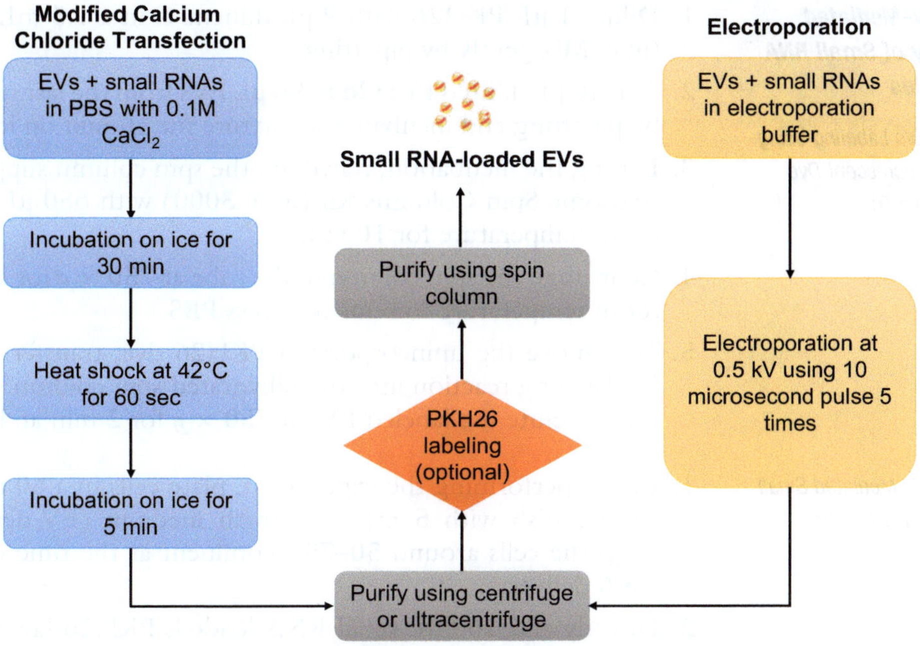

Fig. 2 Flowchart illustrating two methods to introduce small RNA molecules into extracellular vesicles via modified calcium chloride transfection or electroporation

3.2.2 Small RNA Loading via Calcium-Mediated Transfection

1. After pelleting EVs, resuspend the EVs using pre-cold PBS (*see* **Note 5**).
2. Add 5 μL 20 μM small RNA (total, 100 pmol) with 100 μg resuspend EVs (measured by EV protein) to 1.5 mL tube.
3. Add pre-cold PBS to make the final reaction volume to 450 μL.
4. Add 50 μL 1 M $CaCl_2$ to the final concentration of 0.1 M (*see* **Note 6**).
5. Gently mix the tube by inverting 8–10 times and briefly centrifuge the tube.
6. Place the tube on ice for 30 min.
7. Place the bottom 2/3 of the tube into a 42 °C water bath for 60 s (*see* **Note 7**).
8. Put the tubes back on the ice for 5 min.
9. After transfection, proceed to wash steps below.

3.2.3 Wash and Purify Small RNA-Loaded EVs

1. To remove the unloaded small RNA molecules, add 10 mL pre-cold PBS to the electroporated mixture and transfer to Beckman™ centrifuge tubes.
2. Pellet the EVs: for MVs, centrifugation is at 16,000 × *g* for 40 min at 4 °C; for EXOs, ultracentrifugation is at 100,000 × *g* for 120 min at 4 °C.
3. Carefully remove the supernatant and resuspend the EVs in 90 μL PBS (*see* **Note 8**).

3.3 EV-Mediated Delivery of Small RNA Molecules

3.3.1 EVs Labeling Using PKH26 Fluorescent Dye (See **Note 9**)

1. Dilute 1 μL PKH26 with 9 μL diluent C in a 1.5 mL microtube. Mix gently by pipetting.
2. Add 10 μL diluted PKH26 to 90 μL EVs from the last step. Mix by pipetting and incubate the mixture for 10 min on ice.
3. During the incubation, rehydrate the spin column supplied by Exosome Spin Columns Kit (MW 3000) with 650 μL PBS at room temperature for 10 min.
4. Centrifuge the spin column in the tube at 750 × *g* for 2 min at room temperature to remove excess PBS.
5. To remove the unincorporated PKH26 dye, transfer 100 μL EV labeling reaction into the rehydrated spin column. Centrifuge to elute the labeled EVs at 750 × *g* for 2 min at 4 °C.

3.3.2 EV-Mediated Small RNA Delivery In Vitro

1. Before performing the experiment, plate cells in a 60 mm cell culture dish with 5 mL of growth medium (EV-depleted). Keep the cells around 50–70% confluent at the time of small RNA delivery.
2. Directly add 100 μL small RNA-loaded, PKH26-labeled EVs prepared in Subheading 2 into the recipient cells (*see* **Note 10**). Incubate the cells at 37 °C in a CO_2 incubator for several hours until you can observe clear red fluorescence under a microscope. The medium may be changed after 2–6 h.
3. To check the internalization of EVs, after incubation, treat the cells with 500 μL 0.2× trypsin/EDTA buffer for 1 min and then wash the cells extensively with 5 mL citric acid buffer to remove non-internalized EVs. The internalization of PKH26-labeled EVs can be viewed under a fluorescence microscope (*see* **Note 11**).
4. After 24–48 h after incubation, collect the cells and proceed to RNA or protein purification. Small RNA delivery efficiency can be evaluated using real-time qRT-PCR or Western blot analysis (*see* **Note 12**).

3.3.3 EV-Mediated Small RNA Delivery In Vivo

A schematic representation of the assessment of EV-siRNA molecules to regulate gene expression in vivo is shown in Fig. 3.

1. Intravenous administration: place the mouse in the restraining device and soak the tail in warm water for 5 min to cause vasodilation of the tail vein. Locate one of the two lateral tail veins and slowly press the plunger to inject 5 μL/g (body weight of mouse) small RNA-loaded, PKH26-labeled EVs prepared in Subheading 2 into the vein. Then, remove the needle and slightly pressure the puncture site with a dry piece of gauze until the bleeding has stopped. Return the mouse to the cage.

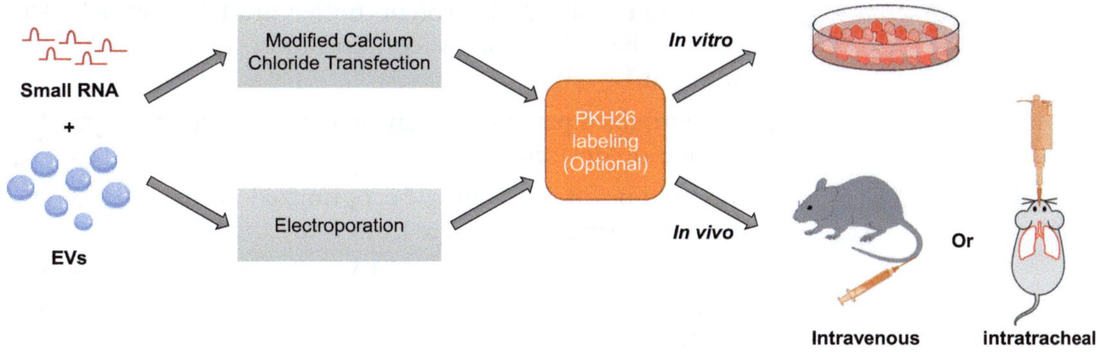

Fig. 3 Schematic diagram: experimental steps to deliver small RNA loaded EVs in vivo and in vitro

2. Intratracheal administration: after the mouse is deeply anesthetized by inhalation anesthesia with 2% isoflurane (confirmed by the absence of reflexes on the footpad), the mouse is held by its teeth, and 2.5 μL/g (body weight of mouse) of the small RNA-loaded EVs prepared in Subheading 2 is gradually released into the mouth and throat with the help of a micropipette. Adjust the rate of release so as to allow the mouse to inhale the solution without forming bubbles. Hold the mouse in the hanging position for another couple of minutes till its breathing gradually returns to normal. Return the mouse to the cage.
3. To check the biodistribution of EVs, 2 h after administration, fix the tissues from the treated mice using 4% PFA. Freeze the fixed tissue in OCT in a suitable tissue mold. The tissue is typically cut into thin sections (5–15 μm thick) in the cryostat at −20 °C and mounted onto microscope slides for the observation under a fluorescent microscope (*see* **Note 13**).
4. Mice are euthanized via isoflurane inhalation as described in Subheading 1 24–48 h after treatment. To determine the efficiency of small RNA delivery, collect the tissues/organs from the treated mouse and proceed to RNA or protein purification. Efficiency can be evaluated using real-time qRT-PCR or Western blot analysis (*see* **Note 14**).

4 Notes

1. EVs here include MVs and EXOs. In this protocol, both MVs and EXOs can be used as carriers for small RNA delivery.
2. In our experiments, no difference was observed among mice species, such as C57BL/6 and BALB/c.

3. The volume of E2 electrolytic buffer must be optimized to achieve the EV protein concentration range from 5 to 10 mg/mL.
4. The optimized parameters may be varied among different electroporation devices.
5. The volume of PBS must be optimized to achieve the EV protein concentration range from 5 to 10 mg/mL.
6. Depending on the origin of EVs, the mixture may appear cloudy after adding $CaCl_2$. It does not affect the small RNA transfection.
7. Small RNA loading efficiency can be improved by longer incubation at 42 °C, which also causes EV degradation. This step might be optimized for the best performance of EV-mediated small RNA delivery.
8. Depending on the experiment, the volume of PBS for resuspension can be optimized.
9. This is an optional step. Proceed to the EV delivery if the tracking of EVs is not necessary.
10. Depending on the origin of EVs, the dose of EVs can be optimized.
11. For long-term storage, the cells can be fixed using 4% formalin. Do not use menthol or ethanol for the fixation, which can wash PKH26 dye away.
12. For the control group, small RNA control loaded EV is recommended. Other groups, such as original EVs or PBS, could also be used as a negative control.
13. Do not use the paraffin-embedded tissue section if you want to perform immunofluorescence. Procedure for deparaffinization can eliminate PKH26 fluorescence signal.
14. For the control group, small RNA control loaded EV is recommended. Other groups, such as original EVs or PBS, could also be used as a negative control.

Acknowledgments

This work was supported by National Institutes of Health (NIH) grants R33 AI121644, R01 GM111313, R01 GM127596, Wing Tat Lee award (all to Y.J.), by NIH grant K99HL141685 (to D.Z.).

References

1. van Niel G, D'Angelo G, Raposo G (2018) Shedding light on the cell biology of extracellular vesicles. Nat Rev Mol Cell Biol 19 (4):213–228
2. Raposo G, Stoorvogel W (2013) Extracellular vesicles: exosomes, microvesicles, and friends. J Cell Biol 200(4):373–383
3. Lam JK, Chow MY, Zhang Y, Leung SW (2015) siRNA versus miRNA as therapeutics for gene silencing. Mol Ther Nucleic Acids 4: e252
4. Dana H, Chalbatani GM, Mahmoodzadeh H, Karimloo R, Rezaiean O, Moradzadeh A, Mehmandoost N, Moazzen F, Mazraeh A, Marmari V, Ebrahimi M, Rashno MM, Abadi SJ, Gharagouzlo E (2017) Molecular mechanisms and biological functions of siRNA. Int J Biomed Sci 13(2):48–57
5. Zhang D, Li Y, Liu S, Wang YC, Guo F, Zhai Q, Jiang J, Ying H (2017) microRNA and thyroid hormone signaling in cardiac and skeletal muscle. Cell Biosci 7:14
6. Mullard A (2018) FDA approves landmark RNAi drug. Nat Rev Drug Discov 17(9):613
7. Dilnawaz F (2017) Polymeric biomaterial and lipid based nanoparticles for oral drug delivery. Curr Med Chem 24(22):2423–2438
8. Yingchoncharoen P, Kalinowski DS, Richardson DR (2016) Lipid-based drug delivery systems in cancer therapy: what is available and what is yet to come. Pharmacol Rev 68 (3):701–787
9. Lv H, Zhang S, Wang B, Cui S, Yan J (2006) Toxicity of cationic lipids and cationic polymers in gene delivery. J Control Release 114 (1):100–109
10. Wu SY, McMillan NA (2009) Lipidic systems for in vivo siRNA delivery. AAPS J 11 (4):639–652
11. Lee H, Zhang D, Rai A, Jin Y (2017) The obstacles to current extracellular vesicle-mediated drug delivery research. J Pharm Pharm 4(2):156–158
12. Zhang D, Lee H, Wang X, Rai A, Groot M, Jin Y (2018) Exosome-mediated small RNA delivery: a novel therapeutic approach for inflammatory lung responses. Mol Ther 26 (9):2119–2130
13. Matsumoto J, Stewart T, Banks WA, Zhang J (2017) The transport mechanism of extracellular vesicles at the blood-brain barrier. Curr Pharm Des 23(40):6206–6214
14. Wang J, Sun X, Zhao J, Yang Y, Cai X, Xu J, Cao P (2017) Exosomes: a novel strategy for treatment and prevention of diseases. Front Pharmacol 8:300
15. Alvarez-Erviti L, Seow Y, Yin H, Betts C, Lakhal S, Wood MJ (2011) Delivery of siRNA to the mouse brain by systemic injection of targeted exosomes. Nat Biotechnol 29 (4):341–345

References

Chapter 7

Optimized siRNA Delivery into Primary Immune Cells Using Electroporation

Mouldy Sioud

Abstract

Effective RNA delivery strategies for primary human monocytes and dendritic cells (DCs) are useful tools for both basic research and cancer immunotherapy applications. Compared to viral delivery, electroporation is a relatively safe and simple technique that has been established for most immune cells. This chapter describes the feasibility of introducing small interfering RNAs into human primary monocytes and DCs using either nucleofection or standard electroporation techniques. DC cancer vaccines that integrate siRNA targeting relevant DC-intrinsic immunosuppressive signals induced robust and durable anti-tumor immune responses.

Key words RNA interference, siRNAs, Dendritic cells, Electroporation, Nucleofection, Immunosuppressive factors, Cancer vaccines

1 Introduction

RNA interference is currently the most used reverse genetics technique to study gene function in mammalian cells [1, 2]. A major advantage of RNAi is that the silencing machinery is present in practically every mammalian somatic cell. Hence, no prior genetic manipulation of the target cells is required, and a simple siRNA transfection can result in a loss-of-function phenotype in vitro [3, 4]. However, nonviral methods are often inefficient for most primary cells, particularly non-adherent blood cells such as T cells, B cells and monocytes. Notably, primary cells are a valuable tool for researchers and are biologically more relevant than their transformed or immortalized cell lines.

Based on their potent antigen presentation and T cell activation, dendritic cells (DCs) derived from blood monocytes were tested as cancer vaccines (*see* Chapter 15). Ex vivo-generated DC-based preparations have been tested in cancer patients for more than a decade. While the DC-based vaccines exhibited a

Mouldy Sioud (ed.), *RNA Interference and CRISPR Technologies: Technical Advances and New Therapeutic Opportunities*, Methods in Molecular Biology, vol. 2115, https://doi.org/10.1007/978-1-0716-0290-4_7,

good safety profile, objective clinical responses have been seen only in very few patients [5]. One main reason for the failure of DC cancer vaccines is believed to be the presence of immunosuppressive factors that hamper DC activation [6–8]. These factors are expressed by DCs themselves and/or tumor cells. Hence, the development of agents that stimulate DC function and/or suppress the expression of immunosuppressive factors, such as interleukin (IL)-10, PD-L1, and indoleamine 2,3-dioxygenase (IDO), will enhance the vaccine potency. IL-10 has been associated with the development of tolerogenic DCs, and IDO is an enzyme involved in inhibiting the activation of alloreactive T cells, leading to tumor evasion of the host immune system [6, 7]. IDO-positive DCs converted naïve T cells into regulatory T cells, thus inducing tolerance instead of T cell activation. Additionally, IDO-positive DCs produce various tryptophan metabolites such as *N*-formylkynurenine with a general immunosuppressive effect on T cell activation [7]. Thus, blocking IDO expression with siRNAs might enhance the efficacy of cancer vaccines.

Electroporation is an efficient technique to deliver siRNAs into human cells that are refractory to lipid-based transfection strategies, and it was used to test the recognition of siRNA by cytosolic RNA-sensing receptors expressed by monocytes and DCs [8]. Also, the technique has been validated by different groups for loading DCs with antigen-encoding mRNA leading to higher transfection efficiency in comparison to simple co-incubation or lipofection [9–14]. *Nucleofection* is also an electroporation-based transfection method which enables transfer of nucleic acids such as DNA and RNA into cells by applying a specific voltage and reagents. Here, optimized protocols for siRNA delivery into human monocytes and DCs via electroporation and nucleofection are provided.

2 Materials

2.1 Isolation of Blood Monocytes

1. Buffy coat from healthy adult donors.
2. RMPI 1640 medium.
3. RMPI 1640 supplemented with 10% fetal calve serum (FCS), penicillin (100 μg/mL), and streptomycin (100 μg/mL). Complete medium.
4. Lymphoprep ™.
5. Pasteur pipettes.
6. Z1 Coulter Particle Counter.
7. Cell scraper.
8. Granulocyte macrophage colony-stimulating factor (GM-CSF).

Table 1
siRNA sequences

siRNA	Target site
IDO siRNA-1	5Ç-GCCUCCUAUUUUGGUUUAU-3′
IDO siRNA-2	5Ç-GCAGCGUCUUUCAGUGCUU-3′
IDO siRNA-3	5Ç-CCCUUCAAGUGUUUCACCA-3Ç
IDO siRNA-4	5Ç-CCGUGAGUUUGUCCUUUCA-3Ç
Scrambled IDO siRNA-4	5-GAUUUCGUACUCUGUGCUC-3′
Control siRNA β-Galactosidase	5-UUGAUGUGUUUAGUCGCUA-3′
IL10 siRNA	5-AGGAUCAGCUGGACAACUU-3

9. Interleukin-4.
10. Tumor necrosis factor-α (TNF-α).

2.2 Electroporation of siRNA into Monocytes and Dendritic Cells

1. Electroporation apparatus BTX ECM 830.
2. 4 mm gap cuvettes.
3. Chemically synthesized siRNAs (*see* Table 1).
4. 6-well cell culture plates.
5. RMPI 1640 complete medium.
6. Human monocytes and dendritic cells generated from buffy coats.

2.3 Nucleofection of siRNA into DCs

1. Nucleofactor kit (Amaxa, Biosciences).
2. Transfection solutions supplemented with Amaxa transfection kit.
3. Transfection cuvettes supplemented with Amaxa transfection kit.
4. RMPI 1640 complete medium.
5. Human monocytes and dendritic cells generated from buffy coats.
6. 6-well cell culture plates.

2.4 Preparation of Protein Extracts

1. Phosphate-buffered saline (PBS): dissolve 8.g NaCl, 1.16 g Na_2HPO_4, 0.2 g KH_2PO_4, and 0.2 g KCl into 1 L distilled water. Adjust the pH to 7.0.
2. Lysis buffer: 10% NP40 in PBS and 10 μg/mL each of aprotinin, pepstatin, and leupeptin.

3. Sample buffer (2×): 125 mM Tris–HCl, pH 6.8, 4% SDS, 20% glycerol, 0.1% bromophenol blue, 5% β 2-mercaptoethanol.
4. Eppendorf centrifuge.
5. γ-globulin (10 mg/mL).
6. Protein assay kit.
7. Disposable cuvettes.

2.5 Polyacrylamide Gel Electrophoresis of Proteins

1. Vertical electrophoresis plate assembly with comb.
2. Electrophoresis power supply EPS 600.
3. 30% acrylamide mix, 29% (w/v) acrylamide, 1% bisacrylamide, prepared in distilled water.
4. 10% SDS, 10% ammonium sulfate (APS), and TEMED.
5. Stacking gel buffer: 1 M Tris–HCl, pH 6.8.
6. Resolving gel buffer: 1 M Tris–HCl, pH 8.8.
7. Electrophoresis buffer 10×: dissolve 144 g glycine, 30 g Tris–Base, and 10 g SDS into 1 L distilled water.
8. Colored protein markers.

2.6 Western Blots

1. BioRad Trans-Blot.
2. Transfer buffer 10×: dissolve144 g glycine and 30 g Tris–Base into 1 L distilled water.
3. Transfer buffer 1×: 100 mL 10× transfer buffer, 200 mL methanol, and 700 mL distillated water.
4. Nitrocellulose transfer membrane.
5. Chromatography Whatman paper.
6. Phosphate-buffered saline with 0.1% tween (PBS-T).
7. Blocking buffer: 5% milk in PBS-T.
8. Antibody dilution buffer: 1% milk in PBS-T.
9. Anti-indoleamine 2,3-dioxygenase (IDO) antibody.
10. Anti-mouse immunoglobulin-HRP conjugated.
11. Antihuman β actin polyclonal antibodies.
12. Enhanced chemiluminescent (ECL) reagent.
13. High-performance chemiluminescence film.
14. Stripping buffer: 1×: Tris 7.57 g/L, SDS 20 g/L, pH 6.8. Autoclave.
15. β-Mercaptoethanol.

3 Methods

3.1 Isolation of Mononuclear Cells

1. Transfer the buffy coat into a 75 cm^3 cell culture flask.
2. Add 50 mL RPMI medium and mix gently.
3. Carefully layer 20 mL of the diluted blood over 10 mL Lymphoprep™ in a 50 mL falcon tube.
4. Centrifuge at 800 × *g* for 30 min in swing-out rotor.
5. Subsequent to centrifugation, mononuclear cells form a distinct band at the sample/medium interface.
6. Remove the cells from the interface using a Pasteur pipette.
7. Dilute the harvested cells with medium and pellet the cells by centrifugation for 5 min at 800 × *g*.
8. Wash the cells twice with serum-free RPMI 1640 medium or PBS.
9. Subsequent to washing, resuspend cells in 50–100 mL complete RPMI 1640 medium.
10. Dispense 30–50 × 10^6 cells per 25 mL complete medium into 75 cm^3 cell culture flasks and then incubate at 37 °C for 1–2 h.
11. Carefully aspirate the medium and wash adherent cells with 10 mL warmed RPMI 1640 complete medium.
12. After washing, add 10–15 mL complete medium and carefully harvest cells by scraping (*see* **Note 1**).
13. Transfer cells to 50 mL falcon tubes, and then centrifuge at 800 × *g* for 5 min.
14. Remove the supernatant, resuspend cells in complete RPMI 1640 medium, and then determine cell number.
15. Disperse the appropriate cell number into 25 cm3 or 75 cm3 tissue culture flasks.

3.2 Generation of Immature and Mature Dendritic Cells

Immature monocyte-derived dendritic cells (iDCs) are generated by culturing monocytes in the presence of GM-CSF (25 ng/mL) and IL-4 (50 ng/mL) for 4–5 days. To generate mature monocyte-derived DCs (mDCs), add TNF-α (50 ng/mL) to the culture and incubate for 2 more days.

3.3 Electroporation of siRNAs into Human Monocytes and Dendritic Cells

Prior to electroporation, wash the cells twice with RMPI medium and then resuspend at 5–10 × 10^6 cells/0.5 mL RPMI medium (serum-free).

1. Dispense RMPI complete medium into 6-well tissue culture plate (2.5 mL/well) and incubate at 37 °C.
2. Prepare Eppendorf tubes with the desired siRNA concentrations.

3. Wash the electroporation cuvettes with RPMI medium.
4. Mix cells (500 μL) with desired siRNA concentrations (0.1–3 μg/transfection), transfer the mixtures into the electroporation cuvette, and then incubate at 4 °C for 2–5 min.
5. During the incubation time, set up the electroporation parameters:
   - Volt:500.
   - Pulse length:2 ms.
   - Pulse:01.
   - Interval: 01.0 s.
   - Polarity: unipolar.
6. After, place the metal walls in the cuvette against the metal walls in the electroporator and then pulse (*see* **Note 2**).
7. Transfer the electroporated cells immediately to 2. 5 mL pre-warmed RMPI complete medium and incubate cells at 37 °C in humidified 5% CO_2 for at least 1 h.
8. To induce IDO gene expression, add INF-γ (500 ng/mL).
9. Incubate at 37 °C in humidified 5% CO_2 for 15–20 h prior to analyzing siRNA effects.

3.4 siRNA Delivery to Monocytes Using the Nucleofection Method

1. Before transfection, dispense RMPI complete medium into 6-well tissue culture plate (2. 5 mL/well) and incubate at 37 °C.
2. Count cells, calculate the number of cell needed, and spin down. We usually use 5–10^6 cells per transfection.
3. In the Eppendorf tubes, dispense the siRNA concentrations needed. Usually we use 0.1 μg to 3 μg per transfection.
4. Select program Y-001 on the nucleofector apparatus.
5. For reach transfection, resuspend the cell pellet in 100 μL transfection solution provided in the kit (*see* **Note 3**).
6. Transfer 100 μL cell solution to the Eppendorf tube containing siRNA, mix gently, and then transfer the mixture into the transfection cuvette.
7. Insert the cuvette in the nucleofector instrument, the metal walls facing you, and press the parameter X.
8. When the transfection is completed, it is essential to add 0.5 mL complete medium to the cells as soon as possible.
9. Transfer the cells from the cuvette (≈0.6 mL) to the 6-well plate using the white plastic pipette provided with the electroporation cuvettes.
10. The machine has to be reset between each transfection by pressing the parameter X.

3.5 Preparation of Protein Extracts

1. Subsequent to overnight transfection, harvest the cells by centrifugation for 5 min at 12,000 × *g*.
2. Wash the cells twice with PBS.
3. Add to the cell pellet lysing buffer (≈30 μL per 10^6 cells), mix well, and then place the cell on ice for 30 min. Vortex three times during the incubation.
4. Subsequently, centrifuge the cells at 12,000 × *g* for 10 min.
5. Save the supernatant which contains the cytoplasmic proteins, and then measure the protein concentration using, for example, the Bio-Rad protein assay kit.

3.6 Hand Casting Polyacrylamide Gels

Several systems exist for performing polyacrylamide gel electrophoresis (PAGE), and consideration should be given to which system best suits a given sample prior to running the samples.

1. Assemble plates according to the manufacturer's instructions. Prepare 10% sodium dodecyl sulfate-polyacrylamide gel (SDS-PAGE) using standard protocols (*see* **Note 4**).
2. Pour the resolving gel, leaving space for the stacking gel (around 2 cm).
3. Gently overlay the acrylamide with water-saturated isobutanol.
4. Allow the gel to polymerize for 30–60 min. A very sharp liquid-gel interface will be visible when the gel has polymerized.
5. Pour off the overlay and rinse the top of the gel several times with water.
6. Blot any remaining water with a paper towel.
7. Prepare a 5% stacking gel.
8. Pour the stacking gel directly onto the resolving gel.
9. Insert the comb immediately.
10. Allow the gel to polymerize for at least 15–30 min.
11. Prepare protein samples while the gel is polymerizing.
12. Place the gel into the electrophoresis chamber.
13. Add 1x electrophoresis buffer to cover the wells.
14. Carefully remove the comb, and rinse each well with the electrophoresis buffer.

3.7 Sample Preparation and Electrophoresis

1. To prepare the protein samples, mix protein extract (up to 60 μg) 10 μL and 10 μL of 2× sample buffer.
2. Heat samples for 5 min at 95 °C in a dry bath heat block.
3. Centrifuge the samples at 300 × *g* for 20 s, and then gently load into the gel.
4. Electrophorese at 80 V for 5 min, and then at 100 V until the blue dye has migrated to the bottom of the gel.

3.8 Transfer onto Nitrocellulose Membranes

Protein transfer efficiency in blotting applications is dependent upon multiple factors, including the gel percentage, gel thickness, protein size, transfer conditions, and the type the membrane. Therefore, transfer conditions must be adjusted to address these varying factors. The protocol given below is adjusted for the siRNA experiments done with human monocytes and DC cells. All washes are done at room temperature using TBS-T buffer, unless otherwise indicated.

1. Equilibrate the nitrocellulose or polyvinylidene fluoride (PVDF) membrane for 2–5 min in 10 mL 1× transfer buffer. While equilibrating, assemble filter pad and filter paper in the cathode side of the transfer module and soak in 1× transfer buffer.
2. Place the gel on large plate down, break open, and remove the small plate, being careful to leave the gel on large plate. Cut off and remove wells (stacking gel).
3. Place wet filter paper on the gel, remove bubbles, and flip over.
4. Assemble the transfer cassette in the following order, from the bottom (cathode unit) side-1 filter pad, Wattman paper, polyacrylamide gel, nitrocellulose membrane, Wattman paper, and then filter pad (anode unit).
5. Transfer the assembled cassette to the inner electrophoresis chamber.
6. Place an ice block and a magnetic stir bar in the outer chamber, and then fill the chamber with 1× transfer buffer to just cover the top of the assembled cassette.
7. Transfer at 4 °C for 1 h at 100 V, 400 mA.
8. When the transfer is completed, disassemble the cassette, and remove the membrane from the gel. The colored molecular weight marker should be visible on the membrane (*see* **Note 5**).
9. Block the membrane in a 5% milk solution prepared in 1× TBST for 1–2 h at room temperature.
10. Wash the membrane once with 1× TBST, and then incubate the membrane with anti-IDO antibody (1/1000 in 1× PBST) overnight at 4°Cwith gentle mixing (*see* **Note 6**).
11. After incubation, wash the membrane for 30 min with TBST in 5–10 min intervals.
12. Remove the primary antibody and wash the membrane three times with PBS-T.
13. Incubate the membrane with secondary antibody (goat anti-rabbit IgG-HRP conjugated) for 1 h at room temperature with gentle rocking.
14. After incubation, wash the membrane for 30 min in 5–10 min intervals.

15. Develop the membrane with, for example, Super Signal West Dura Extended Duration Substrate for 1–5 min.
16. Drain the membrane with a paper, and then place it in a plastic wrap. Make sure that there are no air bubbles between the membrane and the plastic wrap.
17. Detect the signal with BioRad *ChemDoc Imaging System which* provides fast, reliable, and sensitive imaging and documentation of chemiluminescence Western blots. Alternatively, high-performance chemiluminescence films can be used to detect the emitted light.

3.9 Membrane Stripping and Actin Detection

When a satisfactory exposure has been obtained for the desired protein target (e.g., IDO), strip the membrane in order to analyze actin protein levels (an internal control).

1. Soak the nitrocellulose membrane in 45 mL stripping buffer supplemented with 350 μL β-mercaptoethanol. Use a 50 mL falcon tube.
2. Incubate at 55 °C in a water bath for 30 min with gentle rocking.
3. After incubation, wash the membrane with TBST for 10 min, and then block in 5% milk for 1 h at room temperature with gentle rocking.
4. Incubate the membrane with anti- β actin antibody (C4, sc47778, Santa Cruz), and then process the membrane as indicated above (Subheading 3.8, **steps 12–17**).

Under our experimental conditions, the transfection efficiency was up to 95%, consistent with the high percentage of IDO knockdown in transfected cells of 95–99% (Figs. 1 and 2). Also, transfected cells showed that 75–85% of total cells are viable cells as assessed by trypan blue exclusion test. Using Gene-Pulser II (pulsed conditions, 300 V, 150 μF, 100 Ω), and 5 μg siRNA (375 pmol), Prechtel and colleagues achieved only 60% reduction of lamin A/C expression in DC [13]. On the basis of these results, the electroporation of siRNA into DCs using BTX instrument should be adapted in the clinic. Any gene involved in immunosuppression such as IDO, IL10, TGF-β, and SOCS-1 can be silenced by siRNAs (*see* Chapter 15). Figure 3a, b shows that siRNA maintained its silencing activity after freezing and thawing of DCs, further supporting the inclusion of siRNAs in today DC cancer vaccines.

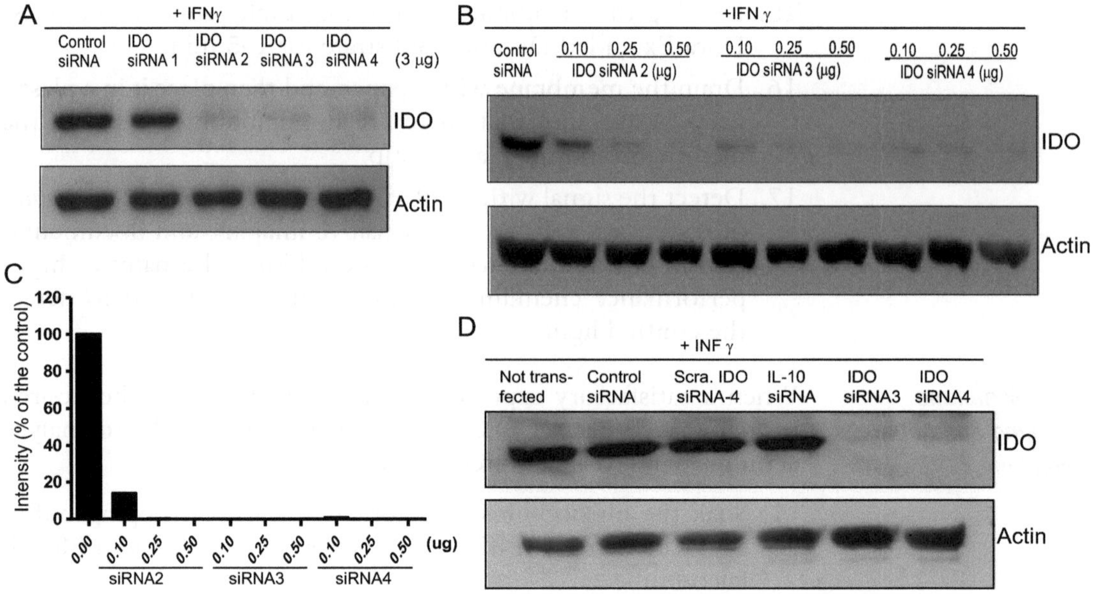

Fig. 1 Silencing potency of IDO siRNAs in human monocytes. (**a**) Downregulation of IDO gene expression by siRNAs. Freshly isolated monocytes were transfected with anti-IDO siRNAs (3 μg/225 pmol each) or control siRNA targeting β-galactosidase (3 μg/ 225 pmol) using nucleofection technology (*see* Subheading 3.4). After, IDO expression was induced with IFN-γ (500 U/mL) for 18 h, and IDO protein levels were determined by Western blots. (**b**) Concentration-dependent downregulation of IDO gene expression by siRNAs. Freshly isolated monocytes were transfected with various concentrations of IDO siRNAs and then processed as in A. (**c**) Quantification of the IDO signals shown in B. The control intensity is defined as 100%. (**d**) Anti-IDO siRNAs are specific. Monocytes were transfected with the indicated siRNA duplexes (1 μg/75 pmol) and processed as in A. Sca = a scrambled IDO siRNA 4. Cells are not transfected, but IFN-γ-stimulated cells were included in these experiments as an additional control

4 Notes

1. Differentiation factors can be added directly to adherent monocytes. Hence, there is no need for harvesting the cells by scraping.
2. If the electroporation was successful, you should see some air bubbles immediately after pulsing.
3. The cells should not stay in the transfection solution more than 15 min, so when having many transfections, it is better to process the samples in groups.
4. Commercially, precasted gels are available from several commercial sources.
5. Subsequent to transfer, it is also possible to color the membrane with Ponceau stain before blocking with milk in order to check the transfer efficiency. Wash the Ponceau stain off with PBS-T prior to blocking.

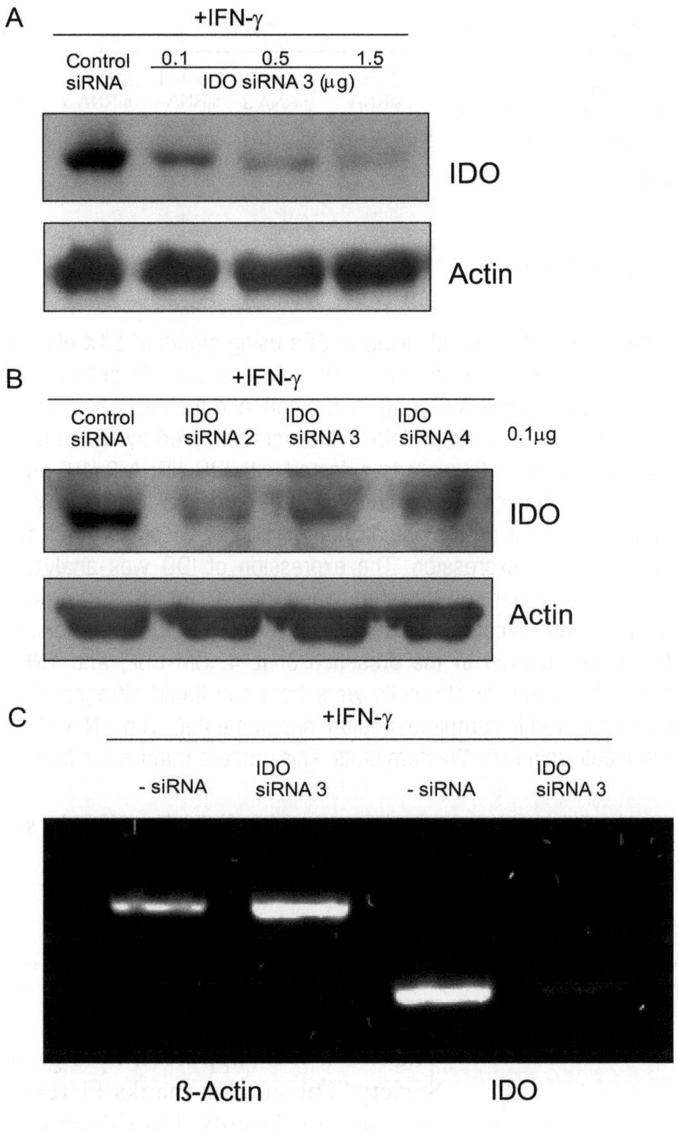

Fig. 2 Silencing potency of IDO siRNA in human dendritic cell. (**a**) Downregulation of IDO gene expression by siRNAs in immature (i) DCs. To generate iDCs, monocytes were treated with GM-CSF (25 ng/mL) and IL-4 (50 ng/mL) for 5 days (*see* Subheading 3.2). Then cells were transfected with various concentrations of IDO siRNA3 (0.1–1.5 μg/ 7.5–112.5 pmol) or control siRNA (1.5 μg/112.5 pmol) using nucleofection technology (*see* Subheading 3.4). After, IDO gene expression was induced by IFN-γ (500 U/mL) for 18 h, and IDO protein levels were determined by Western blots. (**b**) Downregulation of IDO gene expression by siRNAs in mature (m) DCs. To generate mDCs, iDCs were stimulated with TNF-α (50 ng/mL) for 2 days, and then they were washed, transfected with various IDO siRNAs (0.1 μg/7.5 pmol each), and processed as in A. (**c**) RT-PCR analysis of IDO mRNA levels. iDCs were transfected with siRNA-3 (0.4 μg/30 pmol) or control siRNA (0.4 μg/30 pmol), and then IDO expression was induced with IFN-γ for 18 h. After, total RNA was prepared, and IDO and β-actin mRNA levels were analyzed by RT-PCR

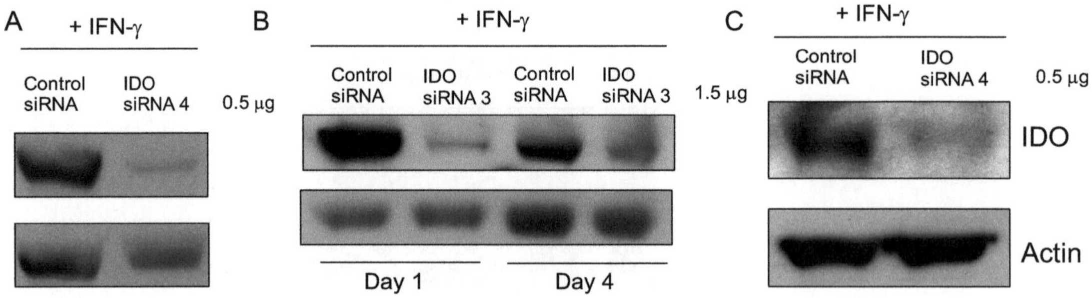

Fig. 3 Efficient siRNA gene silencing in DCs using standard BTX electroporation technique (*see* Subheading 3.3). (**a**) Inhibition of IDO expression in iDCs. Around 5×10^6 cells were transfected with IDO siRNA4 (0.5 μg/37.5 pmol) or control siRNA (0.5 μg/37.5 pmol), and then IDO expression was induced with IFN-γ (500 U/mL) for 18 h. After, IDO and actin protein levels were analyzed by Western blots. (**b**) IDO siRNA remains functional during DC maturation. iDCs were transfected with IDO siRNA-3 (1.5 μg/112.5 pmol) or control siRNA (1.5 μg/112.5 pmol), and then cells were cultured in the presence of IL-4, GM-CSF, and TNF-α (50 ng/mL) to induce DC maturation, mimicking the preparation DC cancer vaccines. After, the cells were incubated with IFN-γ for 18 h to induce IDO expression. The expression of IDO was analyzed by Western blots at day 1 or day 4 subsequent to interferon IFN-γ stimulation (day 0). (**c**) IDO siRNA remains functional after DC freezing and thawing. iDCs were transfected with IDO siRNA4 (0.5 μg/37.5 pmol) or control siRNA (0.5 μg/37.5 pmol), and then they were cultured in the presence of IL-4, GM-CSF, and TNF-α (50 ng/mL) for 2 days to induce maturation. Subsequently, the cells were frozen in liquid nitrogen. On day 2 after freezing, the cells were thawed and cultured in complete medium supplemented with IFN-γ (500 U/mL) for 18 h, and then IDO protein levels were determined by Western blots. This process mimics the freezing and thawing of DC cancer vaccines

6. In general, diluted antibody solution can be used up to four times or more. Therefore, store the samples at −20 °C until use.

Acknowledgments

This work was supported by a grant from the Norwegian Cancer Society. The author thanks Flatekval G Furset for performing the experiments described in this chapter.

References

1. Fire A, Xu S, Montgomery MK, Kostas SA, Driver SE, Mello CC (1998) Potent and specific genetic interference by double-stranded RNA in Caenorhabditis elegans. Nature 391:806–811
2. Elbashir SM, Harborth J, Lendeckel W, Yalcin A, Weber K, Tuschl T (2001) Duplexes of 21-nucleotides RNAs mediate RNA interference in cultured mammalian cells. Nature 411:494–498
3. Sioud M (2004) Therapeutic siRNAs. Trends Pharmacol Sci 25:22–28
4. Sioud M (2011) Promises and challenges in developing RNAi as a research tool and therapy. Methods Mol Biol 703:173–187
5. Berzofsky JA, Terabe M, Trepel JB, Pastan I, Stroncek DF, Morris JC, Wood LV (2018) Cancer vaccine strategies: translation from mice to human clinical trials. Cancer Immunol Immunother 67:1863–1869
6. Grütz G (2005) New insights into the molecular mechanism of interleukin-10-mediated immunosuppression. J Leukoc Biol 77:3–15

7. Munn DH, Mellor AL (2004) IDO expression by dendritic cells: tolerance and tryptophan catabolism. Nat Rev Immunol 4:762–774
8. Sioud M (2005) Induction of inflammatory cytokines and interferon responses by double-stranded and single-stranded siRNAs is sequence-dependent and requires endosomal localization. J Mol Biol 348:1079–1090
9. Van Tendeloo VF, Ponsaerts P, Lardon F, Nijs G, Lenjou M, Van Broeckhoven C, Van Bockstaele DR, Berneman ZN (2001) Highly efficient gene delivery by mRNA electroporation in human hematopoietic cells: superiority to lipofection and passive pulsing of mRNA and to electroporation of plasmid cDNA for tumor antigen loading of dendritic cells. Blood 98:49–56
10. Flatekval GF, Sioud M (2009) Modulation of dendritic cell maturation and function with mono- and bifunctional small interfering RNAs targeting indoleamine 2,3-dioxygenase. Immunology. https://doi.org/10.1111/j.1365-2567.2009.03093.x
11. Michiels A, Tuyaerts S, Bonehill A, Corthals J, Breckpot K, Heirman C, Van Meirvenne S, Dullaers M, Allard S, Brasseur F, van der Bruggen P, Thielemans K (2005) Electroporation of immature and mature dendritic cells: implications for dendritic cell-based vaccines. Gene Ther 12:772–782
12. Prechtel AT, Turza NM, Theodoridis AA, Kummer M, Steinkasserer A (2006) Small interfering RNA (siRNA) delivery into monocyte-derived dendritic cells by electroporation. J Immunol Methods 311:139–152
13. Furset G, Sioud M (2007) Design of bifunctional siRNAs: combining immunostimulation and gene-silencing in one single siRNA molecule. Biochem Biophys Res Commun 352:642–649
14. Sioud M, Kjeldsen-Kragh J, Suleyman S, Vinje O, Natvig JB, Førre O (1992) Limited heterogeneity of T cell receptor variable region gene usage in juvenile rheumatoid arthritis synovial T cells. Eur J Immunol 22(9):2413–2418

Chapter 8

Synthesis and Evaluation of Caged siRNAs with Single cRGD Modification for Photoregulating RNA Interference

Lijia Yu, Duanwei Liang, Nannan Jing, Changmai Chen, and Xinjing Tang

Abstract

We designed and synthesized caged siRNAs with photolabile linker and single cRGD peptide modifications for the photoregulation of gene expression. Photolabile linker and cRGD were inserted at 5′ terminus of siRNAs to obtain cRGD-modified caged siRNAs. All these caged siRNAs could be activated through light activation to release the native siRNAs and further achieve the photoregulation of gene silencing of two exogenous reporter genes (firefly luciferase and green fluorescent protein, GFP) and one endogenous gene (the mitosis motor protein, Eg5). The intracellular distribution and cellular uptake pathways of these caged siRNAs were also investigated. Tumor-bearing mice were further used to demonstrate the photoregulation of gene silencing with cRGD-modified caged siRNAs in vivo. Overall, the data support the use of this new generation of caged siRNAs in cancer therapy.

Key words Caged siRNA, Photoactivation, RNAi, Photochemical biology, cRGD modification

1 Introduction

Small interfering RNAs (siRNAs) have attracted interest for the treatment of various diseases due to their high sequence specificity and gene silencing potency (*see* Chapters 1 and 2). However, unmodified RNAs are negatively charged and often could not be effectively delivered to target cells and/or tissues [1–4]. Chemical modification of siRNAs was one of the important methods to improve stability, intracellular delivery, and pharmacokinetic properties [5], *see* Chapters 2 and 3. The covalent conjugation of cell-specific targeting moieties to siRNAs improved their delivery efficiency to target cells (*see* Chapter 2). However, this conjugation often causes negative effects, such as the decrease of siRNA gene silencing efficiency [6]. Introducing stimulus-responsive groups to siRNAs is an important method to solve this shortcoming [7]. In this respect, photolabile moieties have been widely applied in RNA modifications for gene silencing studies due to its easy modification

Mouldy Sioud (ed.), *RNA Interference and CRISPR Technologies: Technical Advances and New Therapeutic Opportunities*, Methods in Molecular Biology, vol. 2115, https://doi.org/10.1007/978-1-0716-0290-4_8,

and precise spatiotemporal manipulation [8]. A photolabile linker could be inserted between the targeting ligands and cargo oligonucleotides. These siRNA conjugates were temporarily inactive. However, upon light irradiation, both targeting ligand and photolabile linker were removed, leading to the restoration of siRNA activities and gene silencing. To block Dicer/siRNA binding or processing, previous caging strategies were mainly focused on modifications targeting the ribose or phosphate backbone of siRNAs. However, heavily caged siRNAs might result in the incomplete restoration of siRNA activity [9]. Modification of siRNAs with large bulky steric cyclo-dodecyl DMNPE caging group at four terminus of siRNAs achieved good photoregulation activity of gene silencing [10]. We developed a new strategy by caging the 5′ terminus of antisense strand of siRNAs with a targeting ligand, which virtually block the gene silencing activity of siRNA due to the large hindrance of siRNA interaction with the RISC proteins. Upon light activation, targeting ligands together with its receptor protein were removed, and the siRNA activity is restored [11–14].

In the present chapter, we described the synthesis and purification of caged siRNAs with ligand modification at 5′ terminus of antisense strand of siRNA. Caged siRNAs with single cyclic cyclo (Arg-Gly-Asp-D-Phe-Lys) (cRGD) peptide were designed by sequentially inserting photolabile group, alkyl linker, and cRGD at the terminus of antisense strand RNA. Photoregulation of gene silencing in cells or in mice was evaluated using these cRGD-modified caged siRNAs. Intracellular distribution and cellular uptake pathways of cRGD-modified caged siRNA were also investigated. Overall, the data confirm the universal use of caged siRNAs to block gene expression.

2 Materials

2.1 Synthesis of Phosphoramidites of Photolabile Group 1 (PL) and Two Alkyl Linkers (Linker 1 and Linker 2)

1. 2’-Nitroacetophenone.
2. 1,4-Dioxane and bromine (Br_2).
3. Sodium borohydride ($NaBH_4$).
4. Methanol (MeOH).
5. Sodium hydroxide (NaOH).
6. Sodium sulfate (Na_2SO_4).
7. Potassium carbonate (K_2CO_3).
8. Hydrochloric acid (HCl, 37%).
9. Nitrogen or argon gas (N_2 or Ar_2, respectively).
10. 3-Hydroxy-2,2-dimethyl-propanoic acid.
11. 1-Ethyl-3-(3-dimethylaminopropyl)carbodiimide hydrochloride (EDC·HCl).

12. Methylene chloride (CH_2Cl_2).
13. *N*-Hydroxysuccinimide (NHS).
14. Ethyl acetate.
15. Petroleum ether.
16. Tetrazole.
17. Calcium hydride (CaH_2).
18. *N*,*N*,*N*′,*N*′-tetraisopropyl-[1-(2-nitrophenyl)ethyl] phosphorodiamidite.
19. Trimethylamine (TEA).
20. $CDCl_3$.
21. DMSO-*d6*.
22. Aluminum oxide (Al_2O_3).
23. 1,6-Hexanediol.
24. 4,4′-Dimethoxytriphenylmethyl chloride (DMTr-Cl).
25. 100 mL and 150 mL three-neck round-bottom flask.
26. 25 mL, 50 mL, and 100 mL round-bottom flask.
27. Magnetic stir bar and plate.
28. Merck silica gel.
29. Purification columns.
30. 10 mL constant pressure funnel.
31. Thin-layer chromatography (TLC) plates (Silica gel 60F, Merck).
32. 10 mL or 5 mL syringe.
33. 500 mL separator funnel.
34. Büchner funnel.
35. Filter paper.
36. Allihn condenser.
37. Precision pH test paper.
38. Vacuum pump.
39. Rubber stopper or Parafilm.

2.2 Synthesis, Purification, and Characterization of Caged RNA Oligoribonucleotides

1. Photolinker phosphoramidite 1 (PL).
2. Alkyl liner phosphoramidite 1 (**linker 1**).
3. Alkyl linker phosphoramidite 2 (**linker 2**).
4. Commercially available nucleoside 3′-phosphoramidtes adenosine.
5. *N*-benzoyl-5’-*O*-[bis(4-methoxyphenyl)phenylmethyl]-2’-*O*-[(1,1dimethylethyl) dimethylsilyl], 3′-[2-cyanoethyl *N*,*N*-bis (1-methylethyl)phosphoramidite] cytidine.

6. *N*-benzoyl-5'-*O*-[bis(4-methoxyphenyl)phenylmethyl]-2'-*O*-[(1,1-dimethylethyl) dimethylsilyl], 3′-[2-cyanoethyl *N,N*-bis (1-methylethyl)phosphoramidite].guanosine.
7. *N*-2-methyl-1-oxopropyl-5'-*O*-[bis(4-methoxyphenyl)phenylmethyl]2'-*O*-[(1,1-dimethylethyl)dimethylsilyl]3′-[2-cyanoethyl *N,N*-bis(1-methylethyl)phosphoramidite] uridine.
8. 5'-*O*-bis(4-methoxyphenyl)phenylmethyl]-2'-*O*-[(1,1-dimethylethyl)dimethylsilyl]3′-[2- cyanoethyl *N,N*-bis (1-methylethyl)phosphoramidite] thymidine.
9. 5'-*O*-[bis(4-methoxyphenyl)phenylmethyl] 3′-[2-cyanoethyl *N,N*-bis(1-methylethyl) phosphoramidite].
10. Acetonitrile (CH_3CN, anhydrous 10 ppm).
11. Nitrogen.
12. 3′-dT CPG.
13. Acetonitrile (CH_3CN, HPLC grade).
14. Cyclo(Arg-Gly-Asp-D-Phe-Lys) (cRGD) peptide.
15. Dimethyl formamide (DMF).
16. *N,N*-Diisopropylethylamine (DIPEA).
17. Ammonium hydroxide.
18. Dimethyl sulfoxide (DMSO).
19. Triethylamine trihydrofluoride (TEA·3HF).
20. Sodium acetate (3 M aqueous solution).
21. n-Butyl alcohol.
22. Diethyl pyrocarbonate (DEPC).
23. Pure water.
24. Triethylamine (TEA).
25. TEAB, triethylammonium bicarbonate (2M aqueous solution, prepared by bubbling CO_2 to 2M TEA aqueous solution and adjusting pH to 8.0).
26. 1× PBS buffer.
27. 40% acrylamide/bisacrylamide (19:1).
28. 1× Tris-borate-EDTA(TBE) running buffer .
29. Ammonium persulfate (APS).
30. *N,N,N′,N′*-Tetramethylethylenediamine (TEMED).
31. 6× RNA loading buffer.
32. SYBR gold.
33. Diethyl pyrocarbonate (DEPC) water.
34. Automated oligonucleotide synthesizer (Applied Biosystems 394).

35. 200 μL and 1.5 mL microcentrifuge tubes.
36. Parafilm.
37. Oscillator.
38. SpeedVac evaporator.
39. Vacuum pump and water aspirator.
40. Vortex mixer.
41. Shaker.
42. Refrigerated microcentrifuge.
43. HPLC system (Waters, Alliance e2695).
44. Semi-preparative column (Waters XBridgeTM OST C18).
45. UV/vis spectrophotometer (NanoDrop ND-1000).
46. Conical vials.
47. Mass spectrometer.
48. Gel plates.
49. 0.75 mm spacers.
50. 0.75-mm-thick combs.
51. Syringe filter (PES, 13 mm, 0.22 μm).
52. Gel electrophoresis equipment.
53. LED lamp (Lamplic, UVDRV-8-A-0150-30).
54. Chemiluminescence gel imaging system (BIO-RAD, Chemi-Doc XRS System).

2.3 Photoregulation of Gene Silencing of Exogenous and Endogenous Gene Expression in Cell and In Vivo

1. Firefly plasmid, pGL-3.
2. Renilla plasmid, pRL-TK.
3. Nuclease-free water.
4. Human glioblastoma cells, U87MG.
5. Human glioblastoma cells with stable expression of enhanced green fluorescence protein, U87MG-GFP.
6. Dulbecco's modified Eagle's medium (DMEM).
7. Defined fetal bovine serum (FBS).
8. Penicillin-streptomycin (100 μg/mL).
9. 1× phosphate buffer saline (PBS).
10. Trypsin-EDTA.
11. Lipofectamine2000.
12. Lipofectamine RNAiMAX.
13. OptiMEM.
14. 1× passive lysis buffer.
15. Dual-luciferase reporter assay system.

16. TRIZOL reagent.
17. Chloroform.
18. Isopropanol.
19. Ethanol.
20. NanoDrop2000.
21. DEPC-water.
22. HiScript QRT SuperMix for qPCR kit.
23. GoTaq$^{@}$ qPCR Master Mix kit.
24. Eg5 primers.
25. GAPDH primers.
26. Propidium iodide staining solution.
27. Formaldehyde.
28. Hoechst 33342.
29. Rhodamine-cyclopeptide.
30. Cell culture flasks.
31. Cell counting chamber.
32. Cell culture plates.
33. 1.5 mL microcentrifuge tubes.
34. Fluoroskan Ascent FL.
35. LED lamp.
36. Well luminescence test plate.
37. Mesh screen.

2.4 Intracellular Distribution and Cellular Uptake Pathways

1. Human glioblastoma cells, U87MG.
2. Dulbecco's modified Eagle's medium (DMEM).
3. Fetal bovine serum (FBS).
4. Penicillin-streptomycin (100 μg/mL each).
5. 1× phosphate buffer saline (PBS).
6. Trypsin-EDTA.
7. Lipofectamine2000.
8. OptiMEM.
9. Firefly plasmid, pGL-3.
10. Renilla plasmid, pRL-TK.
11. Hoechst33342.
12. Lysotracker green.
13. Chlorpromazine.
14. Amiloride.
15. Methyl-β-cyclodextrin.

16. Genistein.
17. Dimethyl sulfoxide.
18. 35 mm laser confocal culture dish.
19. Tissue culture incubator.
20. 1.5 mL microcentrifuge tubes
21. LED lamp.
22. Vortex mixer.
23. Laser scanning confocal microscopy.
24. 12-well tissue culture plates.
25. Eppendorf centrifuge.
26. 400-mesh screen.
27. Cell counting chamber.
28. Flow cytometer (BD FACSAriaII).

2.5 Photoregulation of GFP Gene Expression In Vivo

1. Human glioblastoma cells with stable expression of enhanced green fluorescence protein, U87MG-GFP.
2. Dulbecco's modified Eagle's medium (DMEM).
3. Fetal bovine serum (FBS).
4. Penicillin-streptomycin (100 μg/mL each).
5. 1× phosphate buffer saline (PBS).
6. BALB/c nude mice.
7. Transfection regent, Entranster-in vivo.
8. Cell counting chamber.
9. 75 cm^2 cell culture flasks.
10. 1.5 mL microcentrifuge tubes.
11. LED lamp.
12. Vortex mixer.

3 Methods

3.1 Synthesis of Phosphoramidites of Photolabile Group 1 (PL) and Two Alkyl Linkers (Linker 1 and Linker 2)

The phosphoramidite of photolabile **linker 1** (PL) and two alkyl linker (**linker 1** and **linker 2**) were synthesized as shown in Fig. 1. These reactions were monitored by thin-layer chromatography (TLC). The products were purified and determined by NMR and/or ESI-MS as described below.

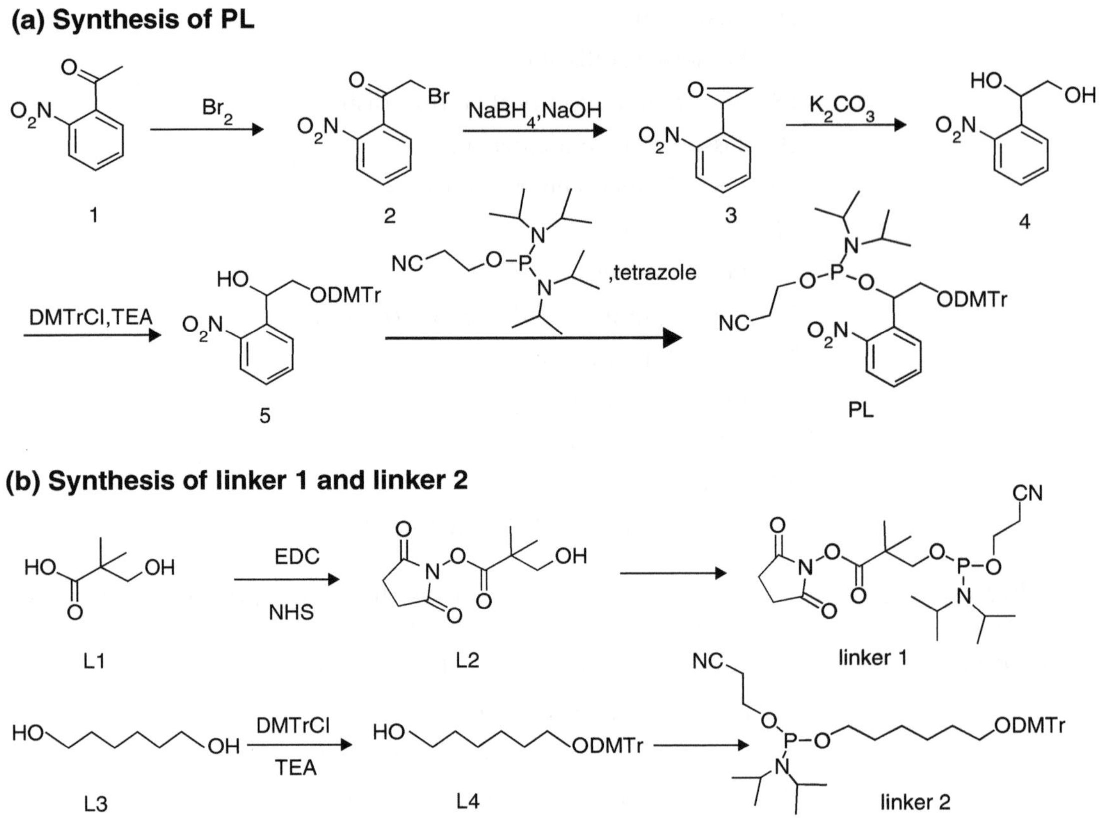

Fig. 1 Synthetic route of (**a**) photolinker phosphoramidite 1(PL) and (**b**) **linker 1** and **linker 2**. (Figure adapted from Yu. et al. [11])

3.1.1 Synthesis of PL

Synthesis of Compound 2 (*See* Note 1)

1. Add 2′-nitroacetophenone (20.0 g, 12.1 mmol) to 1,4-dioxane (50 mL) in a 150 mL round-bottom flask containing a magnetic stir bar.
2. Add 6.2 mL Br_2 (19.3 g, 12.1 mmol) to the reaction mixture. Stir the reaction until the color changes from purple to light yellow.
3. Evaporate 1,4-dioxane in a rotary evaporator. Purify the residue using silica gel column chromatography to obtain compound **2** using 20:1 (v/v) petroleum ether/ethyl acetate as the eluent.

Approximately 25 g (85.2% yield) of the compound **2** can be obtained. The compound was determined by ^{1}H-NMR and ^{13}C-NMR. ^{1}H-NMR (400 MHz, $CDCl_3$) δ8.20 (d, J = 8.3 Hz, 1H), 7.84–7.72 (m, 1H), 7.72–7.64 (m, 1H), 7.50 (d, J = 8.6 Hz, 1H), 4.29 (s,2H).^{13}C NMR (400 MHz, $CDCl_3$) δ 147.84, 140.96, 133.63, 128.12, 127.59, 124.31, 65.57, 24.15.

Synthesis of Compound **3**

1. Dissolve compound **2** (19.5 g,79.0 mmol) in 50 mL 1,4-dioxane, and place the solution in an ice bath. Stir for 10 min at room temperature.
2. Dissolve $NaBH_4$ (5.9 g, 158 mmol) in 12 mL MeOH. Transfer it to a 10 mL constant pressure funnel, and add it to the reaction mixture slowly. Then stir in ice bath for 30 min.
3. Remove the ice bath and incubate the solution at room temperature.
4. When TLC results show the disappearance of compound **2**, add 10% aqueous NaOH solution (20 mL) and continue stirring for another 0.5 h in ice bath.
5. Transfer the mixture to the separator funnel (500 mL). Add 50 mL ethyl acetate. Shake thoroughly and then allow the mixture to stand for phase separation. Extract aqueous layer with 50 mL ethyl acetate three times. Collect the organic layer together.
6. Dry the organic solution using anhydrous Na_2SO_4 overnight. Filter off Na_2SO_4 using a Büchner funnel, filter paper, and vacuum pump. After, condense the solution in a rotary evaporator.
7. Purify the residue by silica gel column chromatography to obtain compound **3** using 20:1 (v/v) petroleum ether/ethyl acetate as the eluent.

Approximately 11.8 g (88.9% yields) of the compound **3** was obtained. The compound was determined byl H-NMR. ^{1}H-NMR (400 MHz, $CDCl_3$) δ 8.14 (dd, J = 8.2, 1.0 Hz, 1H), 7.73–7.58 (m, 2H), 7.52–7.42 (m, 1H), 4.47 (dd, J = 4.3, 2.7 Hz, 1H), 3.29 (dd, J = 5.5, 4.5 Hz, 1H), 2.67 (dd, J = 5.5, 2.6 Hz, 1H).

Synthesis of Compound **4**

1. Dissolve compound **3** (2.0, 12.1 mmol) in 10 mL 1,4-dioxane using a 100 mL round-bottom flask containing magnetic stir bar.
2. Add 10% aqueous K_2CO_3 (2.5 g, 18.1 mmol) to reaction mixture. Reflux at 80 °C for 24 h with an Allihn condenser.
3. Cool the reaction at room temperature. Adjust pH of the mixture to around 5 using 10% aqueous HCl. Use precision pH test paper to check the pH.
4. Add 25 mL ethyl acetate and transfer the mixture to a 500 mL separatory funnel. Re-extract aqueous layer with 50 mL ethyl acetate three times. Combine the organic layers.
5. Dry the organic solution using anhydrous Na_2SO_4 overnight. Filter off Na_2SO_4 using a Büchner funnel, filter paper, and vacuum pump. Condense the solution in a rotary evaporator.

6. Purify the residue as above step by silica gel column chromatography to obtain compound **4** using 5:1 (v/v) petroleum ether/ethyl acetate as the eluent.

Approximately 2.1 g (79.0% yields) of compound **4** was obtained. The compound was determined by ^{1}H-NMR. ^{1}H-NMR (400 MHz, $CDCl_3$) δ 7.98 (d, J = 8.2 Hz, 1H), 7.88 (d, J = 7.9 Hz, 1H), 7.70 (t, J = 7.6 Hz, 1H), 7.50 (t, J = 7.7 Hz, 1H), 5.48 (d, J = 6.3 Hz, 1H), 3.97–3.81 (m, 1H), 3.58–3.45 (m, 1H).

Synthesis of Compound **5**

1. Dissolve compound **4** (0.5 g, 2.7 mmol) in 10 mL dry CH_2Cl_2 using a 100 mL round-bottom flask containing a magnetic stir bar.
2. Add 0.76 mL trimethylamine (0.5 g, 5.4 mmol) to the mixture.
3. Dissolve 4,4′-dimethoxytriphenylmethyl chloride (0.8 g, 2.36 mmol) in 5 mL dry CH_2Cl_2. Transfer mixture to 10 mL constant pressure funnel. Drip it slowly to the reaction mixture.
4. Seal the system with a group-glass stopper and stir for 1 h at room temperature.
5. Remove the solvent using a rotary evaporator, and purify the residue by silica gel column chromatography to obtain compound **5** using10:1 (v/v) petroleum ether/ethyl acetate as the eluent.

Approximately 0.8 g (27.4% yields) of compound **5** was obtained. The compound was determined by ^{1}H-NMR. ^{1}H-NMR (400 MHz, $CDCl_3$) δ 7.93 (dd, J = 8.2, 1.3 Hz, 1H), 7.82 (dd, J = 7.9, 1.4 Hz, 1H), 7.62 (td, J = 7.6, 1.3 Hz, 1H), 7.49–7.38 (m, 3H), 7.32 (tt, J = 7.9, 2.5 Hz, 6H), 7.29–7.20 (m, 1H), 6.91–6.80 (m, 4H), 5.57 (dd, J = 7.4, 3.4 Hz, 1H), 3.81 (s, 6H), 3.66 (dd, J = 9.4, 3.5 Hz, 1H), 3.28–3.16 (m,2H).

Synthesis of Compound **6**

1. Add to compound **5** (0.6 g, 1.2 mmol) and tetrazole (0.3 g, 4.7 mmol) in a 25 mL round-bottom flask. Remove oxygen in the flask by vacuum and fill with nitrogen three times. Add anhydrous CH_2Cl_2 (2 mL), which is freshly distilled over Ca_2H_2 to the flask under nitrogen.
2. Add *N,N,N′,N′*-tetraisopropyl-[1-(2-nitrophenyl)ethyl] phosphorodiamidite (1.1 g, 3.7 mmol in 2 mL CH_2Cl_2) to reaction mixture under nitrogen condition using a 5 mL syringe. Stir at room temperature for 2 h.
3. Condense the reaction solution using rotary evaporator. Purify the residue using dry Merck silica gel column chromatography to obtain product using 6:1:0.1(v/v/v) petroleum ether/ethyl acetate/triethylamine as eluent.

Approximately 0.76 g (72.3% yields) of compound **6** was obtained. The compound was determined by^1 H-NMR and ^{31}P-NMR. ^{1}H NMR (400 MHz, $CDCl_3$) δ 7.88 (dt, J = 8.2, 1.5 Hz, 1H), 7.84–7.74 (m, 1H), 7.62 (dt, J = 12.1, 7.5 Hz, 1H), 7.46–7.36 (m, 3H), 7.30 (dd, J = 8.9, 3.0 Hz, 6H), 7.19 (dt, J = 6.9, 3.7 Hz, 1H), 6.87–6.72 (m, 4H), 5.50–5.32 (m, 1H), 3.79 (d, J = 1.8 Hz, 8H), 3.59 (ttd, J = 13.4, 6.7, 3.0 Hz, 2H), 3.45 (dp, J = 10.5, 6.7 Hz, 1H), 3.04 (dt, J = 10.6, 5.7 Hz, 2H), 2.52 (td, J = 6.4, 3.2 Hz, 1H), 1.16 (ddd, J = 19.6, 9.3, 6.8 Hz, 9H), 0.84 (d, J = 6.8 Hz, 3H).^{31}P NMR (400 MHz, $CDCl_3$) δ 150.35, 149.83.

*3.1.2 Synthesis of **Linker 1***

Synthesis of Compound **L2**

1. Mix 3-hydroxy-2,2-dimethyl-propanoic acid (**L1**, 3.0 g, 25.4 mmol) and 1-ethyl-3-(3-dimethylaminopropyl)carbodiimide hydrochloride (EDC·HCl) (5.4 g, 30.2 mmol) in dry methylene chloride (25 mL) using a 100 mL round-bottom flask containing a magnetic stir bar. Add *N*-hydroxysuccinimide (NHS) (3.3 g, 28.6 mmol) to the reaction, and stir the mixture on magnetic plate at room temperature overnight.
2. Evaporate methylene chloride in a rotary evaporator.
3. Purify the residue by silica gel column chromatography to obtain compound **L2** using 1:1 ethyl acetate/petroleum ether as the eluent.

Approximately 4.3 g (81%, yield) of compound **L2** was obtained. The compound was identified by ^{1}H-NMR. ^{1}H-NMR (400 mHz, $CDCl_3$) δ 3.63 (s, 2H), 3.12 (s, 1H), 2.78 (s, 4H), 1.28 (ss, 6H).

Synthesis of **Linker 1**

1. Add compound **L2** (1.5 g, 5.0 mmol), tetrazole (0.44 g, 6.28 mmol), and a magnetic stir bar to a 25 mL round-bottom flask. Remove oxygen in the flask by vacuum and fill with nitrogen three times. Add anhydrous CH_2Cl_2 (2 mL), which is freshly distilled over Ca_2H_2 to the flask under nitrogen.
2. Add *N*,*N*,*N'*,*N'*-tetraisopropyl-[1-(2-nitrophenyl)ethyl] phosphorodiamidite (1.5 g, 5.0 mmol in 2 mL CH_2Cl_2) to reaction mixture under nitrogen condition with 5 mL syringe. Stir at room temperature for 2 h.
3. Remove half of solvent using rotary evaporator. Purify the residue by dry Merck silica gel column chromatography to obtain compound using 10:1:0.1(v/v/v) petroleum ether/ethyl acetate/triethylamine as eluent. Characterize the compound by ESI-MS and ^{31}P- NMR.

Approximately 0.76 g (27%, yields) of the **linker 1** was obtained. ESI-MS *m/z* Calcd. 415.19, Found 416.22, $[M+Na]^+$ 438.16. ^{31}P-NMR (400 MHz, $CDCl_3$). δ, 145.9.

3.1.3 Synthesis of ***Linker 2***

Synthesis of Compound **L4**

1. Add 1,6-hexanediol (**L3**, 0.5 g, 4.2 mmol), triethylamine (0.43 g, 4.2 mmol), and a magnetic stir bar to a 100 mL round-bottom flask with 10 mL CH_2Cl_2 (*see* **Note 2**).
2. Dissolve 4,4′-dimethoxytriphenylmethyl chloride (0.64 g, 1.9 mmol) in 10 mL CH_2Cl_2. Transfer solution to 10 mL constant pressure funnel. Drip the solution slowly to the above flask.
3. Seal the system with a ground-glass stopper and stir for 4 h at room temperature.
4. Remove the solvent using a rotary evaporator, and purify the residue to obtain the compound **L4** by silica gel column chromatography using 5:1 (v/v) petroleum ether/ethyl acetate as the eluent.

Approximately 0.67 g (38%) of compound **L4** was obtained. The compound was identified by ^{1}H-NMR. ^{1}H NMR (400 MHz, DMSO-d_6) δ, 7.40 (d, J = 7.8 Hz, 2H), 7.34–7.23 (m, 6H), 7.20 (t, J = 7.1 Hz, 1H), 6.88 (d, J = 7.8 Hz, 4H), 3.72 (s, 6H), 3.39 (q, J = 5.7 Hz, 2H), 2.98 (t, J = 6.2 Hz, 2H), 1.99 (s, 1H), 1.54 (dd, J = 13.0, 6.4 Hz, 2H), 1.47–1.36 (m, 2H), 1.37–1.20 (m, 4H).

Synthesis of **Linker 2** Phosphorodiamidite (*See* **Note 3**)

1. Add compound **L4** (0.45 g, 1.1 mmol), tetrazole (0.18 g, 2.6 mmol), and a magnetic stir bar to a 25 mL round-bottom flask. Remove oxygen in the flask by vacuum and fill with nitrogen three times. Add anhydrous CH_2Cl_2 (2 mL), which is freshly distilled over Ca_2H_2 to the flask under nitrogen.
2. Add *N,N,N′,N′*-tetraisopropyl-[1-(2-nitrophenyl)ethyl] phosphorodiamidite (0.64, 2.1 mmol in 2 mL CH_2Cl_2) to reaction mixture under nitrogen condition with 5 mL syringe. Stir at room temperature for 2 h.
3. Remove solvent using rotary evaporator. Purify the residue by dry Merck silica gel column chromatography to obtain compound using 10:1:0.1(v/v/v) petroleum ether/ethyl acetate/triethylamine as eluent. Characterize the compound by ESI-MS, ^{1}H-NMR, and ^{31}P- NMR.

Approximately 0.48 g (72.7%) of **linker 2** was obtained. The compound was characterized by ESI-MS, ^{1}H-NMR and ^{31}P-NMR. ESI-MS m/z Calcd. for **linker 2**, 620.34; Found $[M+H]^+$ 621.34, $[M+Na]^+$ 643.26, [DMTr] 303.17. ^{1}H-NMR (400 MHz, DMSO-d_6) δ, 7.38 (d, J = 7.7 Hz, 2H), 7.33–7.17 (m, 7H), 6.87 (d, J = 8.3 Hz, 4H), 3.70 (d, J = 17.4 Hz, 8H), 3.54 (dt, J = 14.3, 7.0 Hz, 4H), 3.35 (s, 1H), 2.95 (t, J = 6.1 Hz, 2H), 2.73 (t, J = 5.7 Hz, 2H), 2.51 (s, 1H), 1.59–1.45 (m, 4H), 1.39 (s, 2H), 1.29 (d, J = 10.5 Hz, 4H), 1.12 (dd, J = 11.2, 6.8 Hz, 12H). ^{31}P NMR (400 MHz, DMSO-d_6) δ 146.39.

3.2 Synthesis and Purification of cRGD-Modified RNA Oligonucleotides

3.2.1 RNA Synthesis Using an ABI394 DNA/RNA Synthesizer

1. Dissolve commercially available nucleoside phosphoramidites in the appropriate amount of dry acetonitrile under nitrogen, and use a 10 mL syringe to add the required amount to obtain 0.1 M phosphoramidites.
2. Calculate the required amount of CPG support according to the synthesis scale. Load an empty synthesis column with dT CPG (1 μmol scale, 33 mg).
3. Connect the oligonucleotide synthetic column, reagents, and phosphoramidites to DNA/RNA synthesizer.
4. Create synthetic methods of the target sequence. Use 10-min coupling time of RNA monomer.
5. Choose the option "DMTr on" mode on standard phosphoramidite chemistry.
6. Click the start button. Stay near the DNA synthesizer for synthesis of the first several nucleotides by observing the appearance of red color when "DMTr" was deprotected, which indicates the successful running of DNA synthesizer.
7. Check the synthesis yield on the monitor. If one step of coupling yield is less than 85%, this synthesis column would be given up.

3.2.2 Synthesis of cRGD-Modified Non-caged RNA Oligonucleotides

1. Dissolve **linker 1** phosphoramidite in the appropriate amount of dry acetonitrile to obtain 0.1 M solution.
2. Use the corresponding DNA solid support (T CPG). Synthesize **linker 1**-modified RNA strands on a 1 μmol scale using an ABI394 DNA/RNA synthesizer on DMT-ON mode based on standard phosphoramidite chemistry. Use 30-min coupling time for the **linker 1** phosphoramidite.
3. Air-dry the support thoroughly in centrifuge tube. Remove the synthesis column from the synthesizer. Open the synthesis column and transfer the CPG support to a 1.5 mL centrifuge tube. 1 mg cRGD, 1 mL dry DMF, and 10 μL DIPEA were added to the centrifuge tube. Seal tube and shake it for 24 h at room temperature.
4. Remove the supernatant solution from the support. Rinse the support with 1 mL of anhydrous ethanol three times, and rinse the support with 1 mL of ether three times. Thoroughly air-dry the support in the tube.

3.2.3 Synthesis of cRGD-Modified Caged RNA Oligonucleotides

1. Dissolve photolinker (**PL**) phosphoramidite, **linker 1** phosphoramidite, and **linker 2** phosphoramidite in the appropriate amount of dry acetonitrile to obtain 0.1 M solutions.
2. Use corresponding synthesis column solid support to synthesize the caged RNAs according to modified sequences on a 1 μmol scale on DMT-ON mode using ABI394 DNA/RNA

synthesizer. Extend the coupling time to 30 min for photolinker phosphoramidite (**PL**), **linker 1** phosphoramidite, and **linker 2** phosphoramidite.

3. Remove the synthesis column from the synthesizer. Open the synthesis column and pour the support into a 1.5 mL centrifuge tube. Air-dry the support thoroughly in centrifuge tube. Add 1 mg cRGD, 1 mL dry DMF, and 10 μL DIPEA to centrifuge tube. Seal the tube and shake it for 24 h at room temperature.
4. Remove the supernatant solution from the support. Rinse the support with 1 mL of anhydrous ethanol three times and then 1 mL of ether three times. Thoroughly air-dry the support in the tube.

3.2.4 Deprotection of RNA Oligoribonucleotides

1. After the completion of native and modified RNA synthesis. Transfer the solid from the synthesis column to 1.5 mL centrifuge tube.
2. Add AMA solution (1:1 mixture (v/v) of aqueous ammonium hydroxide (28%) and aqueous methylamine (40%)) and seal tubes tightly with Parafilm.
3. Shake it for 12 h at room temperature.
4. Open it cautiously. The use of sterile conditions from this point forward is essential. Remove the solid CPGs using syringe filter, and dry the solution using a SpeedVac evaporator.
5. Add 50 μL of anhydrous DMSO to dissolve the residue, and then 50 μL of trimethylamine trihydrofluoride (TEA·3HF) was then added. Shake it for 2 h at 65 °C.
6. Add 3 M sodium acetate (10 μL) and n-butyl alcohol (1 mL) to the reaction mixture, vortex several times, and store tubes at −80 °C for 0.5 h.
7. Centrifuge the tube for 10 min at 13,000 × *g*, 4 °C. Remove the supernatant and add 500 μL of ethanol to rinse the pellet. Centrifuge the tube for 5 min at 13,000 × *g*, 4 °C.
8. Discard the supernatant and redissolve them in DEPC-water for further reverse-Phase HPLC purification.

3.2.5 Purification of Caged RNA Oligonucleotides

1. Purify above modified oligoribonucleotide samples using Waters 2695 HPLC system with a Waters semi-preparative reverse-phase HPLC column. Run the purification process at 1.0 mL/min with a gradient of acetonitrile/0.05 M TEAB buffer. A, 0.05 M TEAB; B, acetonitrile. Unmodified oligonucleotides: A, 100–85% in 20 min; All the cRGD-modified RNAs: A, 100–75% in 25 min.

2. Monitor the analytic sample (5 μL) by UV absorption at 260 nm and record the retention time of each peak. Collect the peak fraction corresponding to the caged or native oligonucleotide products. Collect the preparative eluent with UV detector off according to retention time of the corresponding product peak.
3. Use the same method to collect all samples with the same retention time on HPLC trances, and combine these solutions to a 1.5 mL centrifuge tube. Remove the solvents by SpeedVac evaporator. As shown in Fig. 2, a series of modified or native RNAs strands are obtained.
4. Dissolve 100 μM oligonucleotides in 120 μL of DEPC-water. Then characterize all purified oligonucleotides by electrospray ionization mass spectrometry (ESI-Mass).
5. Measure the absorption at 260 nm of the obtained RNAs using NanoDrop spectrophotometer, and calculate the concentration of all synthetic RNAs.

3.2.6 Hybridization of Caged RNA Oligonucleotides with Their Complementary Strands

1. Mix the antisense oligonucleotides and its corresponding complementary sense RNAs in PBS buffer in a sterile 0.6 mL microcentrifuge tube. Heat the tube at 85 °C for 10 min. Cool the tube in the heating block to room temperature.
2. Confirm the formation of siRNA duplexes by 20% native polyacrylamide gel electrophoresis (PAGE).

*3.2.7 Photocleavage Ability Analysis of Caged siRNAs after Light Irradiation by Native PAGE (See **Note 4**)*

1. Clean the glass gel plates with water and 75% ethanol. Set gel plates for PAGE gel preparation.
2. Mix 20 mL of 20% acrylamide gel solution in a 100 mL conical flask. Add 200 μL of 10% (w/v) aqueous ammonium persulfate (APS) and 20 μL of *N*,*N*,*N*′,*N*′-tetramethylethylenediamine (TEMED).
3. Pour it quickly to the mode of plates. Insert a 0.75 mm gel comb immediately without introducing air bubbles. Place the plates on the bench until the gel polymerizes (about 0.5 h).
4. Set up the gel in a gel box using 1×TBE as the running buffer. Wash the wells using running buffer. Pre-run the gel at 200 V for 10 min.
5. Dissolve 10 mL hybridized siRNAs (10 μM) in 10 μL DEPC-water in 0.2 mL tube. Pipet 10 μL into another 0.2 mL tube.
6. Plate one of 0.2 mL tube on top of the LED lamp, and irradiate it 3 min at 7 mW/cm^2. Keep the other one in dark.
7. Add 2 μL of 6× RNA loading buffer (Bromophenol blue and Xylene Cyanol FF) to 10 μL of the corresponding solutions.

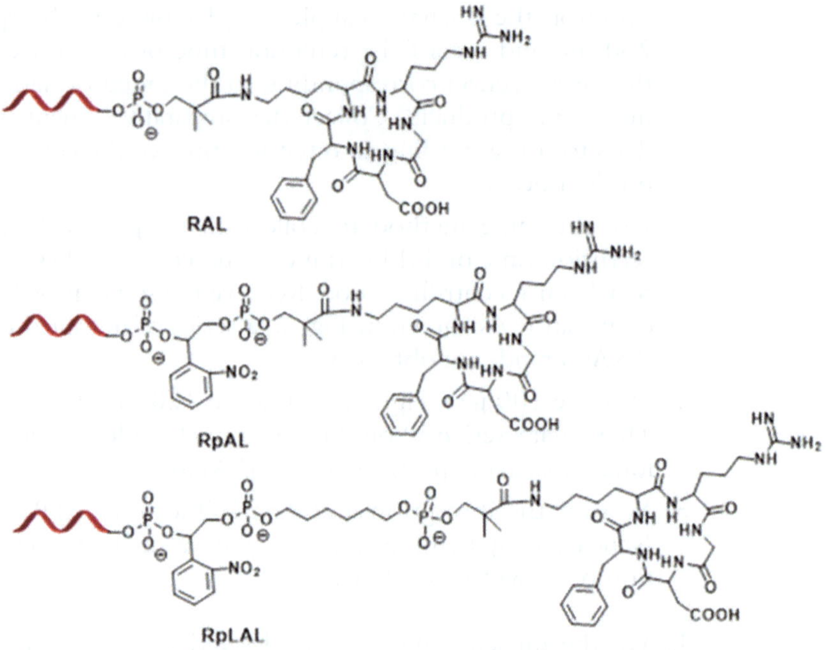

	Name	Sequence (5'-3')
GFP	AG	CUUGAAGAAGUCGUGCUGCTT
	SG	GCAGCACGACUUCUUCAAGTT
	RpAG	cRGD-linker1-PL-CUUGAAGAAGUCGUGCUGCTT
	RAG	cRGD-linker1-CUUGAAGAAGUCGUGCUGCTT
Firefly Luciferase	AL	GCGAAGAAGGAGAAUAGGGTT
	SL	CCCUAUUCUCCUUCUUCGCTT
	RAL	cRGD-linker1-PL-GCGAAGAAGGAGAAUAGG GTT
	RSL	cRGD-linker1-CCCUAUUCUCCUUCUUCGCTT
	RpAL	cRGD-linker1-PL- GCGAAGAAGGAGAAUAGGGTT
	RpSL	cRGD-linker1-PL-CCCUAUUCUCCUUCUUCGCTT
	RpLAL	cRGD-linker1-linker2-PL-GCGAAGAAGGAGAAUAGGGTT
Eg5	SE	CAACAA GGA UGAGUCUAUTT
	AE	AUAGACUUCAUCCUUGUUGTT
	RpAE	cRGD-linker1-PL-AUAGACUUCAUCCUUGUUGTT

Fig. 2 Synthesis of unmodified RNA and cRGD-modified RNA. Label abbreviation: *p or PL* photolabile linker, *R* cRGD, *AL* antisense strand of Firefly luciferase, *SL* sense strand of Firefly luciferase, *AG* antisense strand of GFP, *SG* sense strand of GFP, *SE* sense strand of Eg5, *AE* antisense strand of Eg5. (Figure adapted from Yu. et al. [11])

8. Load 6 μL samples into wells. Run the gel at 200 V until the dye front reached the bottom of the gel.
9. Disassemble the glass plates. Transfer the gel carefully to a container with 1‰ SYBR Gold nucleic acid gel stain using a plastic sheet. Incubate the gel at room temperature for 10 min. Wash the gel twice with DEPC-water. Image gel using a Chemiluminescence gel imaging system.

3.3 Photoregulation of Gene Silencing of Exogenous and Endogenous Gene Expression in Cell

3.3.1 Photoregulation of Gene Silencing of Firefly Luciferase and EGFP

Culture Cells

1. Culture human glioblastoma U87MG cells and U87MG-GFP cells using DEME supplemented with 10% v/v FBS and 1% v/v penicillin-streptomycin at 37 °C with 5% CO_2 humidified incubator in a 75 cm^2 cell culture flask.
2. Culture cells until over 90% confluency, and gently wash cells with 3 mL PBS. Replace PBS buffer 3 mL trypsin-EDTA solution. Detach the cells for 1–3 min, and add 9 mL DMEM medium to stop the digestion. Transfer cell solution to a 15 mL centrifuge tube and centrifuge at 1000 × *g* for 5 min.
3. Remove supernatant and resuspend cells with 3 mL DMEM medium. Add 15 mL DMEM medium to a new 75 cm^2 cell culture flask. Transfer 1 mL cell solution to cell culture flask for future experiments.
4. Use a cell counting chamber to estimate the amount of cells in left solution. Calculate and prepare 1×10^5 cells/mL. Seed these cells in 6-well cell culture plates with final volume of 2 mL DMEM medium, in 12-well cell culture plates with final volume of 1000 μL DMEM medium, or in 24-well cell culture plates with final volume of 500 μL DMEM medium in each well.
5. The cells were used to transfect until cell confluency reached ~70–80%.

Transfection of Firefly/Renilla Luciferase Reporter Genes and Caged siRNAs with Lipofectamine2000

1. Prepare Lipofectamine2000 solution for nucleic acid transfection according to the manufacturer's transfection protocol. Dissolve 40 μL (0.5 μL per well) Lipofectamine2000 in 1000 μL OptiMEM medium in a 1.5 mL microcentrifuge tube, and then vortex for 15 s. Incubate mixture for 5 min at room temperature.
2. Prepare 11 1.5 mL microcentrifuge tubes. Dissolve 8800 ng Firefly plasmid and 4400 ng Renilla plasmid with 1100 μL OptiMEM medium in one 1.5 mL microcentrifuge tube. Vortex this tube gently for 10 s. Transfer plasmid solution to ten tubes with 100 μL solution for each tubes. Add native or modified siRNAs for each tube with 40 pmol AL/SL, 40 pmol RpAL/SL, 40 pmol RpAL/RSL, 40 pmol RpAL/RpSL, 40 pmol AL/RSL, 40 pmol AL/RpSL, 40 pmol RAL/RSL, 40 pmol RAL/RpSL, and 40 pmol RAL/SL, respectively. The left one tube is negative control. Add OptiMEM to a final volume 200 μL of each tube. Gently vortex tubes for 10 s.
3. Add above 100 μL Lipofectamine2000 solution (Subheading 3.3.1, **step 6**) to each tube (Subheading 3.3.1, **step 7**), respectively. Add OptiMEM to a final volume of 400 μL of each tube. Gently vortex for 10 s. Incubate for 15–20 min at room temperature.

4. Remove DMEM medium of cultured cells. Add 450 μL Opti-MEM medium to each well. Add 50 μL transfection complexes (Subheading 3.3.1, **step 8**) to each well. The final volume is 500 μL. Cells transfected with each sample were divided into two groups, and each group contained four paralleled experiments.
5. Incubate transfection complex in cell culture plate for 4 h in a 37 °C incubator. Keep one group in dark and irradiate the other group for 3 min at 7 mW/cm^2. Then, replace the medium with 500 μL fresh DMEM medium containing 10% FBS in each well.
6. Incubate for another 24 h in a 37 °C incubator. Evaluate gene silencing activities of cRGD-modified caged siRNAs using the Dual-Luciferase Reporter Assay System according to the manufacturer's introduction.
7. Remove DMEM medium from cell culture plate. Rinse cells in 1 × PBS.
8. Add 100 μL 1xpassive lysis buffer into each well. Shake cell culture plate for 20 min at room temperature. Transfer about cell lysate to a 96-well luminescence test plate.
9. Measure fluorescence intensity of each well using Fluoroskan Ascent FL (Fig. 3). Set the lag time for 2.0 s and integration time for 10 s. Add luciferase assay reagent and stop reagent 30 μL. Use 5–10 s read time and one to two times delay to finish measurements.

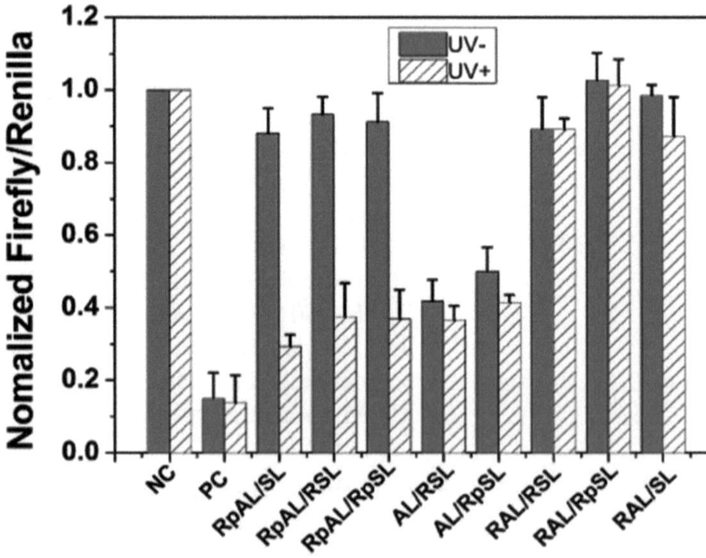

Fig. 3 Photomodulating firefly luciferase activity using cRGD caged siRNAs under different compositions of cRGD and linker modifications ($n = 3$). (Figure adapted from Yu. et al. [11])

Evaluate GFP Reporter Gene Expression with Caged siRNAs

1. Prepare Lipofectamine RNAiMAX for cell transfection according to the manufacturer's transfection protocol. Dissolve 48 μL (3 μL per well) Lipofectamine RNAiMAX in 820 μL Opti-MEM medium in a 1.5 mL microcentrifuge tube and then vortex for 15 s. Incubate the mixture for 5 min at room temperature.
2. Add 100 μL OptiMEM to four tubes. Add 40 pmol AG/SG, 40 pmol RpAG/SG, and RAG/SG to three tubes, respectively, and then add OptiMEM to above four tubes to make a final volume of 200 μL. Vortex tubes for 15 s.
3. Add 200 μL Lipofectamine RNAiMAX/OptiMEM (Subheading 3.3.1, **step 15**) to above four sample tubes (Subheading 3.3.1, **step 16**). Vortex for 15 s. Incubate for 15–20 min at room temperature.
4. Remove DMEM medium from cultured cells. Add 900 μL OptiMEM medium to each well. Then add 100 μL transfection complexes (Subheading 3.3.1, **step 17**) to each well. The final volume of each tube was 1000 μL. Cells transfected with each sample were divided into two groups. One group contains two paralleled experiments.
5. Incubate transfection complexes in cell culture plates for 4 h in a 37 °C incubator. Keep one group in dark and irradiate the other group for 3 min at 7 mW/cm^2. Then, replace medium with 1000 μL DMEM medium containing 10% FBS in each well.
6. Incubate the cells for another 48 h in a 37 °C incubator. Replace DMEM medium of each well with 0.5 mL of 1× PBS. Collect fluorescence images of cells in each well by fluorescence microscopy, using excitation and emission wavelengths of 488 nm and 509 nm for GFP (Fig. 4a).
7. Remove the PBS buffer and add 200 μL trypsin to each well. Detach the cells and then add 800 μL PBS buffer to stop the digestion. Transfer the cell solutions to 1.5 mL tubes. Centrifuge solutions for 5 min at 1000 × *g*. Resuspend the cells with 200 μL of 1× PBS and sieve solutions using a 400-mesh screen. Count fluorescent cells using flow cytometry (Fig. 4).

3.3.2 Evaluate Eg5 Gene Expression and Caged siRNAs in Cells

Real-Time qPCR

1. Prepare Lipofectamine RNAiMAX solution for transfection according to the manufacturer's transfection protocol. Dissolve 10 μL (0.5 μL per well) Lipofectamine RNAiMAX in 500 μL OptiMEM medium in a 1.5 mL microcentrifuge tube and then vortex for 15 s. Incubate the mixture for 5 min at room temperature.
2. Add 50 μL OptiMEM each to three tubes. Add 4 pmol AE/SE and 4 pmol RpAE/SE, respectively, to two tubes, and then add OptiMEM to the left third tube. Make a final volume of 100 μL of each tube. Vortex for 15 s.

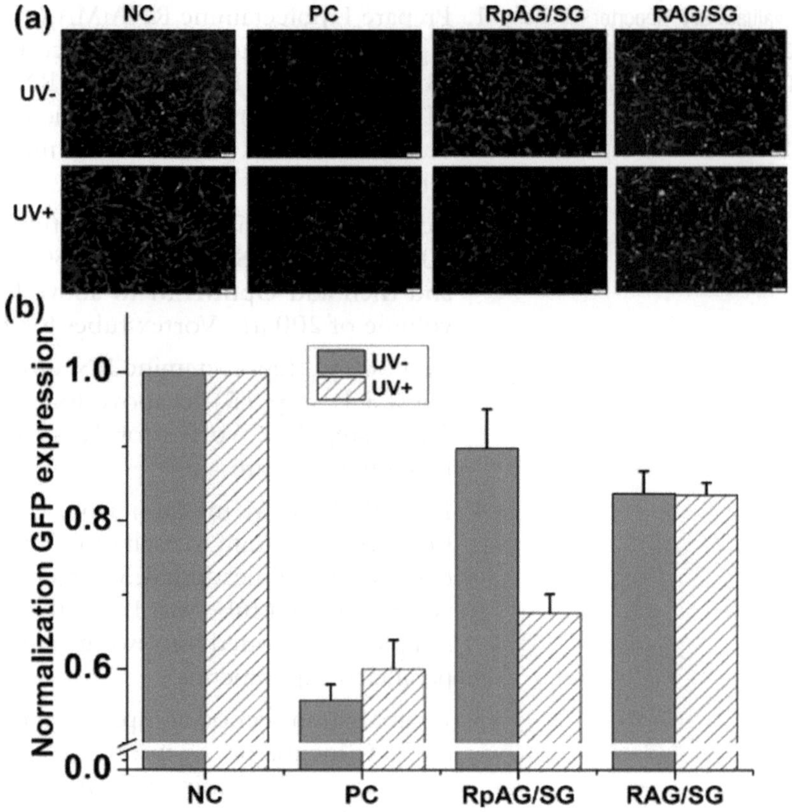

Fig. 4 Photomodulating GFP expression using cRGD caged siRNAs in U87MG-GFP cells. (**a**) GFP fluorescent images of cells treated with different conditions; the scale bars is 100 μm ($n = 3$). (**b**) Quantitative flow cytometry analysis of GFP fluorescence with/without light irradiation (365 nm, 3 min) using ($n = 3$). (Figure adapted from Yu. et al. [11])

3. Add 100 μL Lipofectamine RNAiMAX/OptiMEM each to three sample tubes. Vortex for 15 s. Incubate the solution for 15–20 min at room temperature.
4. Remove DMEM medium from cultured cells. Add 950 μL DMEM medium without FBS and penicillin-streptomycin to each well. Add 50 μL transfection complex to each well. The final volume is 1000 μL. Cells transfected with each sample were divided into two groups.
5. Incubate transfection complexes in cell culture plates for 4 h in a 37 °C incubator. Keep one group in dark and irradiate the other group for 3 min at 7 mW/cm^2. Then, replace medium with 1000 μL fresh DMEM medium containing 10% FBS in each well.
6. Incubate cells for another 48 h in a 37 °C incubator.

Isolation of mRNAs from Cell

1. Remove DMEM from the treated cells. Rinse the cells with 1 mL 1× PBS once. Add 1 mL TRIZOL reagent to cells. Incubate the cells for 2 min at room temperature. Transfer the solution to 1.5 mL microcentrifuge tube.
2. Add 0.2 mL chloroform to each tube. Vortex for 15 s. Centrifuge for 3 min at 12,000 × *g* at 4 °C. Transfer 0.35 mL liquid supernatant of each tube to a new 1.5 mL microcentrifuge tube. Add 0.35 mL isopropanol to each well. Vortex for 15 s. Centrifuge for 10 min at 12,000 × *g* at 4 °C.
3. Remove the supernatant from the tubes. Add 1 mL precool 75% aqueous ethanol. Vortex the tubes five times. Centrifuge for 3 min at 12,000 × *g* at 4 °C. Repeat this step twice.
4. Remove the supernatant from each tube. Dry each tube in the air at room temperature.

Reverse Transcription

1. Dissolve isolated mRNAs in 20–50 μL DEPC-water. Quantify the amount of mRNAs using NanoDrop2000. Calculate the amount of RNAs and prepare 300 ng RNA solutions.
2. Use HiScript Ⅱ Q RT SuperMix for qPCR kit for reverse transcription. Add 2 μL SuperMix and 300 ng RNA to 0.2 mL microcentrifuge tube. Add nuclease-free water to make a final volume of 10 μL. Centrifuge the solution for 15 s at 1000 × *g*. Carry out mRNA reverse transcription using PCR amplifier. Set up running program for 10 min at 25 °C, then 30 min at 50 °C, and last 5 min at 85 °C.
3. After the program was finished, add 50 μL nuclease-free water to each tube. Vortex each tube for 15 s. Centrifuge each sample solution for 20 s at 1000 × *g* at room temperature.

RT-PCR Testing

1. Use GoTaq@ qPCR Master Mix kit to test real-time fluorescence quantification (rt-qPCR). Each sample has six paralleled experiments.
2. Add Eg5 primers to three samples.
3. Add GAPDH primers to left three samples.
4. Add 10 μL Master Mix, 0.2 μM primers (2 μL), 4 μL nuclease-free water, and 4 μL primers (Eg5 forward primer 5′-CAGC TGAAAAGGAAACAGCC, Eg5 reverse primer 5′-ATGAACA ATCCACACCAGCA, GAPDH forward primer 5′-TGCAC CACCAACTGCTTAGC, and GAPDH reverse primer 5′-GGCATGGACTGTGGTCATGAG).
5. Vortex each tube for 15 s and then spin down for 20 s at 1000 × *g* at room temperature.
6. Set up RT-qPCR program as the following: hot-start activation 1 cycle for 2 min at 95 °C, denaturation 40 cycles for 15 s at 95 °C, annealing/extension for 60 s at 60 °C, and dissociation 1 cycle 60–95 °C (Fig. 5b).

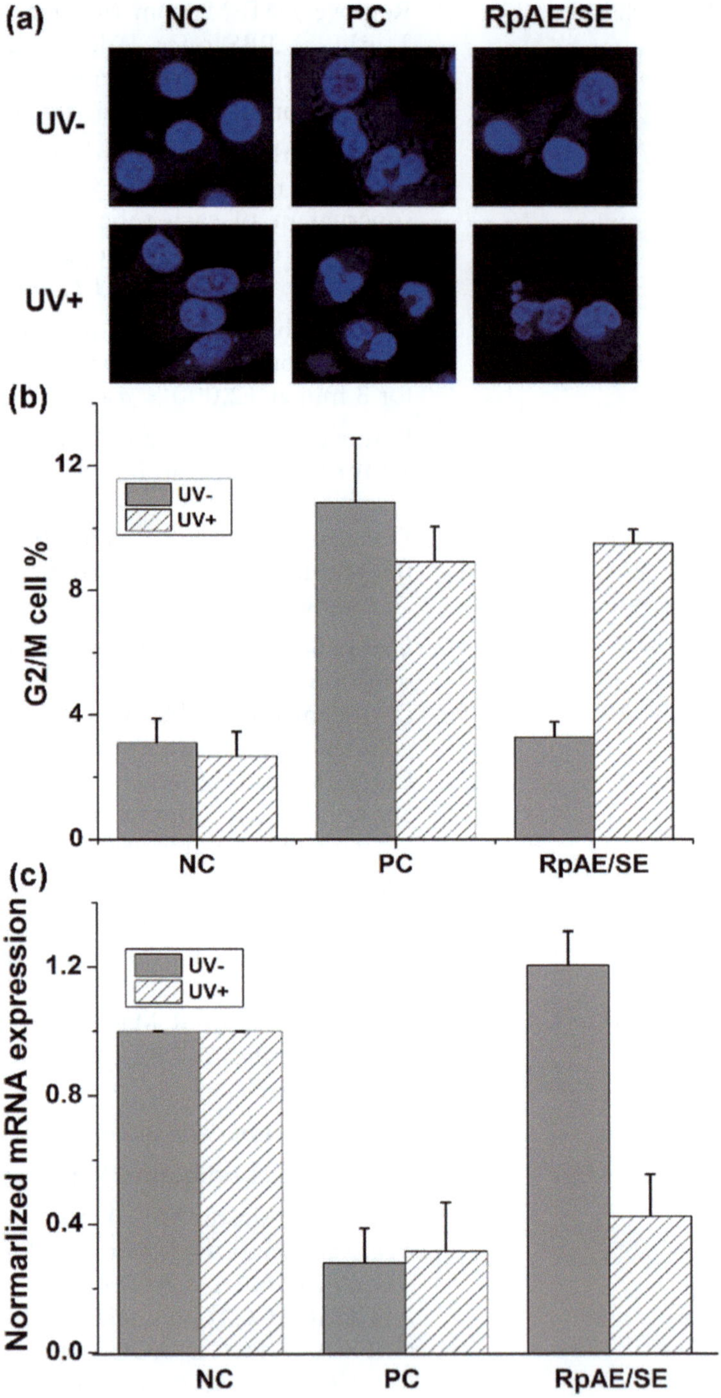

Fig. 5 (**a**) Photomodulating Eg5 gene silencing using cRGD caged siRNA (RpAE/SE) in U87MG cells. The cells were stained with Hoechst 33342 (blue); (**b**) cell phase analysis of G2/M of U87MG cells ($n = 3$) after cells were treated with 10 nM AS/SE or RpAE/SE with/without light activation; (**c**) relative level of Eg5 mRNA in U87MG cells ($n = 3$) after cells were treated with 10 nM AS/SE or RpAE/

Cell Cycle Analysis

1. Repeat Subheading 3.3.2, **steps 1–6**.
2. Remove DMEM medium from the culture plates, and rinse the cells in each well with 1 mL 1× PBS, and then add 0.3 mL trypsin-EDTA.
3. When the cells were detached from the plate, add 1 mL DMEM complete medium and transfer cells to 1.5 mL microcentrifuge tubes.
4. Centrifuge the samples for 5 min at 1000 × *g* at room temperature.
5. Remove supernatant and add 1 mL precool 1× PBS to each tube.
6. Vortex for 15 s and centrifuge for 5 min at 1000 × *g* at room temperature again.
7. Remove supernatant and resuspend cells in 50 μL PBS buffer.
8. Add 1 mL precool 70% aqueous ethyl alcohol to each tube and incubate for 12 h at 4 °C.
9. After, centrifuge for 5 min at 1000 × *g* at room temperature. Remove supernatant and leave around 50 μL.
10. Prepare propidium iodide staining solution. Add 576 μL propidium iodide (20×), 12 mL staining buffer, and 240 μL RNase A (50×) to a 15 mL tube.
11. Add 0.5 mL propidium iodide staining solution to each tube, and incubate cells for 30 min at 37 °C. Store these samples at 4 °C until use.
12. Centrifuge the solutions for 5 min at 1000 × *g* at room temperature again. Remove supernatant and add 300 μL of 1× PBS to resuspend cells. Sieve solution through a 400-mesh screen. Remove filtrate and count fluorescent cells using flow cytometry with an excitation wavelength of 488 nm. Use ModFit software to analyze the percentage of cells in each phase of the cell cycle (Fig. 5c).

Phenotypic Eg5 Inhibition Assay

1. Prepare six 35 mm laser confocal culture dishes for the confocal imaging. Each dish was seed 1 × 10^5 U87MG cells with 2 mL DMEM medium. Incubate the cells for 24 h in a 37 °C, 5% CO_2 incubator.
2. Dissolve 7 μL (1 μL per well) Lipofectamine2000 in 700 μL OptiMEM medium in a 1.5 mL microcentrifuge tube and then vortex for 15 s. Incubate the solution for 5 min for later use.

Fig. 5 (continued) SE with/without light activation. GAPDH was used as internal control. Error bars represented standard deviations from three independent experiments. (Figure adapted from Yu. et al. [11])

3. Prepare three 1.5 mL microcentrifuge tubes. Add 2 pmol AE/SE and 2 pmol RpAE/SE to two tubes, respectively. Add OptiMEM to a final volume 200 μL for all three tubes. Vortex for 10 s.
4. Add 200 μL Lipofectamine2000/OptiMEM each to above three tubes, respectively. Gently vortex for 10 s and incubate for another 15–20 min at room temperature.
5. Remove cell DMEM medium. Add 800 μL DMEM medium without FBS and penicillin-streptomycin to each well. Add 200 μL transfection complex to each dish. The final volume was 1000 μL. Cell dishes for each sample was divided into two groups.
6. Incubate cells with above transfection complex in culture dishes for 4 h in a 37 °C incubator. Keep one group in dark and irradiate the other group for 3 min at 7 mW/cm^2. Then, replace medium with 1000 μL fresh DMEM medium containing 10% FBS in each well. Incubate the cells for another 48 h in a 37 °C incubator.
7. Remove the medium and add 2 mL 4% formaldehyde to fix cells. Incubate cells for 30 min at room temperature. Rinse cells with 1× PBS three times.
8. Stain cells using 2 mL PBS buffer containing Hoechst 33342 (1 μL/mL), rhodamine-cyclopeptide (dilute it 10,000 times), and 1% BSA. Incubate the cells for 1 h and rinse the dishes with PBS buffer three times.
9. Image the cells using laser scanning confocal microscope with the excitation wavelength at 405 nm for Hoechst 33342 and 560 nm for rhodamine-cyclopeptide (Fig. 5a).

3.4 Intracellular Distribution and Cellular Uptake Pathways

3.4.1 Confocal Laser Scanning Microscopy

1. Prepare 7 35 mm laser confocal culture dishes. Seed each dish with 2 × 10^5 U87MG cells in 2 mL DMEM medium, and incubate cells for 24 h in a 37 °C, 5% CO_2 incubator.
2. Place 15 μL (2 μL per well) Lipofectamine2000 in 750 μL OptiMEM medium in a 1.5 mL microcentrifuge tube and then vortex for 15 s. Incubate the mixture for 5 min at room temperature.
3. Add 1400 ng Firefly plasmid and 700 ng Renilla plasmid to a 1.5 mL tube, and then adjust the volume to 700 μL with OptiMEM.
4. Transfer 500 μL mix solution to a new 1.5 mL microcentrifuge tube, and add 100 pmol Cy3-SL/RpAL to make a final volume of 1000 μL with OptiMEM. Transfer another 200 μL of the above mixture solution to a new 1.5 mL microcentrifuge tube, and add 40 pmol Cy3-SL/AL to make a final volume of 400 μL with OptiMEM.

5. Add 500 μL Lipofectamine2000 solution (prepared Subheading 3.4.1, **step 2**) to Cy3-SL/RpAL/plasmid solution. Add 200 μL Lipofectamine2000 solution (prepared Subheading 3.4.1, **step 2**) to Cy3-SL/AL/plasmid solution, and incubate the mixture for 15–20 min at room temperature.
6. Replace the medium in U87MG cell dish with 800 μL fresh OptiMEM per dish.
7. Add 200 μL transfection complex (Subheading 3.4.1, **step 4**) to each dish, and incubate the cells in a 37 °C, 5% CO_2 incubator.
8. Incubate cells with transfection complexes in culture dish for 4 h in a 37 °C incubator. Irradiate two dishes of RpAL/Cy3-SL for 3 min at 7 mW/cm^2. Then, replace the medium with 2000 μL fresh DMEM medium containing 10% FBS in each well. Incubate the cells for 2 h, 12 h, or 24 h in a 37 °C incubator.
9. Remove DMEM medium and rinse cultured cells with 2 mL of 1× PBS twice. Add 500 μL Hoechst33342 (10 μg/mL) and Lysotracker green (1 μL/mL) to each dish, and stain cells for 15 min at room temperature.
10. Remove Hoechst33342 and Lysotracker green solutions. Rinse cells with 2 mL of 1× PBS three times. Add 500 μL of 1× PBS to each dish, and image the cells by laser scanning confocal microscope with excitation wavelength at 561 for Cy3, 405 nm for Hoechst33342, and 488 nm for Lysotracker green.
11. Track the distribution of cRGD-modified siRNAs after different incubation times.

3.4.2 Uptake Pathway Experiments

1. Seed U87MG cells in 12-well plates with 1 mL of cell solution with 2×10^5 cells in each well. Culture the cells for 24 h in a 37 °C, 5% CO_2 incubator.
2. Prepare 100 μL of stock DMSO solutions of chlorpromazine (0.3 mg/μL), amiloride (0.08 mg/μL), methyl-β-cyclodextrin (1.04 mg/μL), and genistein (0.204 mg/μL), respectively. Dilute for 100 times of chlorpromazine, amiloride, and genistein and 10 times of methyl-β-cyclodextrin using DMSO to prepare working solutions.
3. Dilute the corresponding working solutions (4 μL) using OptiMEM (1600 μL), respectively, and gently vortex each tube. Remove cell medium and add 1600 μL of above solutions to cells in each well, respectively. Incubate cells for another 2 h. For the low-temperature experiment, replace DMEM medium with 1600 μL OptiMEM and incubate cells at 4 °C for 1 h.

4. Add both 8 μL Lipofectamine2000 and 810 μL OptiMEM to two 1.5 mL microcentrifuge tubes and vortex 15 s.
5. Add 120 pmol Cy3-SL/RpAL to a 1.5 L microcentrifuge tube and dilute to 810 μL with OptiMEM.
6. Prepare four 1.5 mL microcentrifuge tubes, and add 200 μL siRNA (Subheading 3.4.2, **step 5**) to each tube. Then add 400 μL prepared Lipofectamine 2000 solution. Mix solution thoroughly and incubate the mixture for 15 min at room temperature.
7. Add 400 μL transfection complexes (Subheading 3.4.2, **step 6**) to each well. For the negative control experiment, add 400 μL fresh OptiMEM medium. For each condition, prepare at least three paralleled wells.
8. Remove DMEM medium from the culture plates. Rinse the cells with 1 mL 1× PBS. Add 0.3 mL trypsin-EDTA to each well. Detach cells from the plate, and add 1 mL DMEM to stop digestion. Transfer the solution to 1.5 mL microcentrifuge tube. Centrifuge cells for 5 min at 1000 × g at room temperature. Remove the supernatant and add 1 mL precool 1× PBS to each tube. Sieve the cell solutions using a 400-mesh screen. Count fluorescent cells using flow cytometry with an excitation wavelength of Cy3 at 561 nm.
9. Quantify cell uptake of Cy3-SL/AL in comparison with the negative control and the positive control experiment with no addition of inhibitors at 37 °C.

3.5 Photoregulation of GFP Gene Silencing In Vivo

1. Repeat Subheading 3.4.1, **steps 1–3**.
2. Use a cell counting chamber to estimate cell number and prepare 6×10^5 cells/mL solution.
3. Inject 60 μL of cell solution to both inner thighs of each BALB/c nude mice ($n = 6$).
4. On day 4, randomize the mice into three groups.
5. Add 86.67 μL transfection regents (Entranster-in vivo) to three Eppendorf tubes. Then add 13 nmol AG/SG and 13 nmol RpAG/SG to two of the tubes. Make a final volume of 260 μL by adding PBS buffer. Vortex the mixture for 15 s and incubate for 15–20 min at room temperature.
6. For the first group ($n = 2$), inject 60 μL transfection regents/AG/SG solution in the left thigh tumor and 60 μL transfection regents/PBS buffer in the right thigh tumor. Treat the mice in the second group the same way as the first group. For the third group, inject 60 μL transfection regents/RpAG/SG solution in both the left and right thigh tumor.

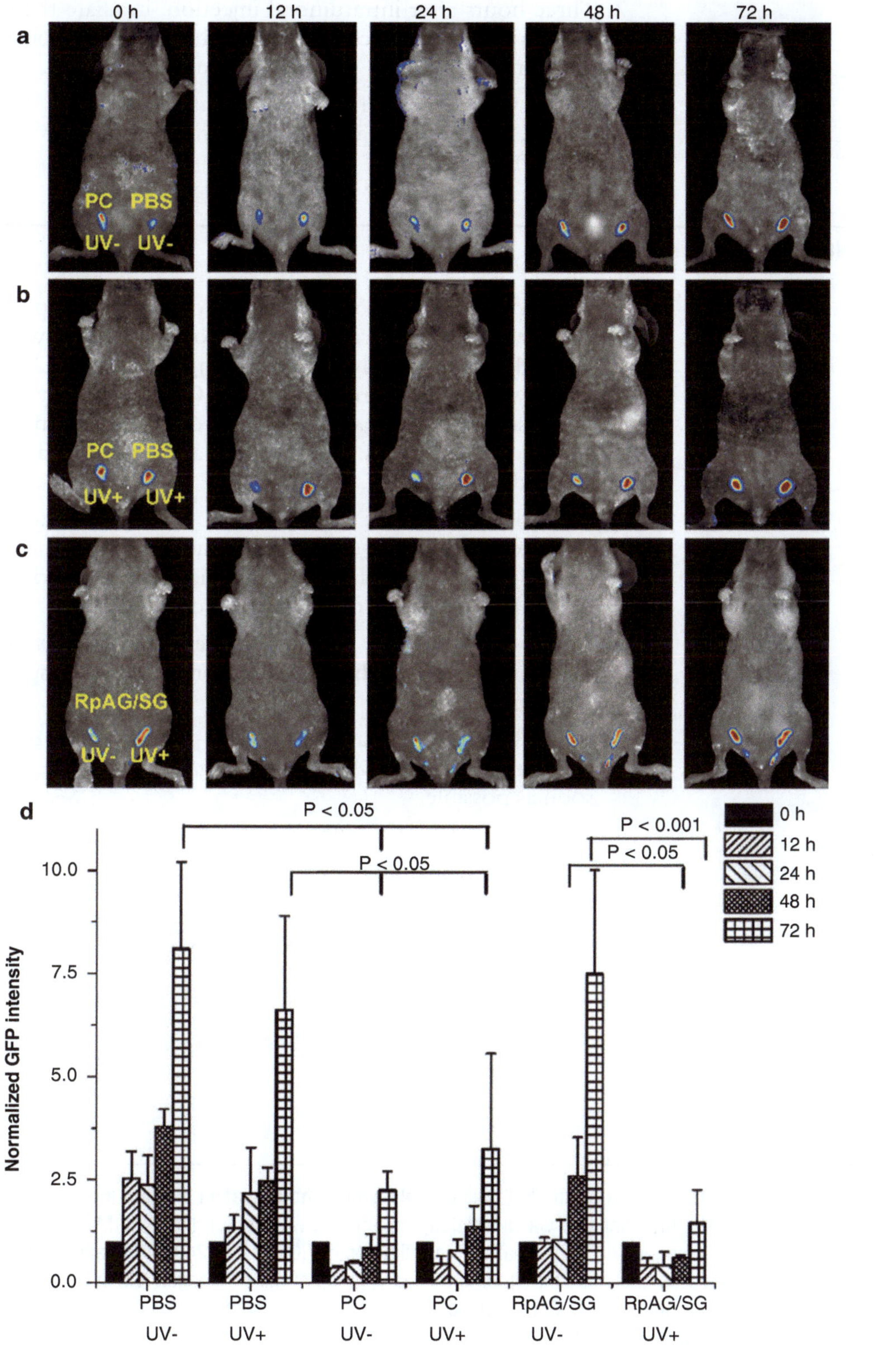

Fig. 6 Photoregulation GFP expression in tumors in vivo using cRGD caged siRNA. (**a**) Injection of PBS on the right and 3 nM linear positive control siRNAs on the left, respectively. (**b**) Injection of PBS on the right and 3 nM

7. Three hours after intratumoral injection, irradiate the tumors at both sides in the second group and the right side only in the third group for 5 min (365 nm, 15 min).
8. Image GFP fluorescence intensity in tumor tissues at 0, 12, 24, 48, and 72 h using the Mastro Antomated in vivo imaging (Fig. 6).

4 Notes

1. All reactions related to the synthesis of photolabile phosphoramidite (PL) were carried out in a room with yellow light to avoid the photocleavage of the caging group. $CDCl_3$ for ^{31}P NMR was filtrated through basic Al_2O_3 before every use. The synthesis of phosphoramidites required strict oxygen-free and anhydrous conditions. Make sure that all solvents are freshly distilled over CaH_2.
2. For the synthesis of terminal DMTr protection of 1,6-hexanediol, 4,4′-dimethoxytriphenylmethyl chloride in CH_2Cl_2 must be slowly added to avoid by-product with terminal DMTr protection of both two hydroxyl groups.
3. All the synthesis and evaluation of caged siRNAs should be used under yellow light or in dark condition unless specifically mentioned.
4. Store RNA amidites at −30 °C or −80 °C before dissolving them in acetonitrile. After dissolving in acetonitrile, use it as soon as possible.

Fig. 6 (continued) linear siRNAs on the left, respectively, with light irradiation (5 min). (**c**) Injection of 3 nM RpAG/SG on both sides, the right side was irradiated, the left side was kept in the dark. Mice from each test group were imaged (Xenogen IVIS system) at different time points (0 h, 12 h, 24 h, 48 h, and 72 h), and GFP fluorescence intensity was quantified for all tumors in each group of mice (Xenogen IVIS system) on different days (normalized to the intensity of GFP fluorescence at 0 h for the same tumor of each mice) ($n = 3$). (Figure adapted from Yu. et al. [11])

References

1. Elbashir SM, Harborth J, Lendeckel W, Yalcin A, Weber K, Tuschl T (2001) Duplexes of 21-nucleotide RNAs mediate RNA interference in cultured mammalian cells. Nature 411:494–498
2. Roberts TC, Ezzat K, EL Andaloussi S, Weinberg MS (2016) Synthetic SiRNA delivery: progress and prospects. In: Shum K, Rossi J (eds) SiRNA delivery methods: methods and protocols. Springer New York, New York, NY, pp 291–310. https://doi.org/10.1007/978-1-4939-3112-5_23
3. Broering R, Real CI, John MJ, Jahn-Hofmann-K, Ickenstein LM, Kleinehr K, Paul A, Gibbert K, Dittmer U, Gerken G, Schlaak JF (2013) Chemical modifications on siRNAs avoid toll-like-receptor-mediated activation of the hepatic immune system in vivo and in vitro. Int Immunol 26:35–46
4. Tluk S, Jurk M, Forsbach A, Weeratna R, Samulowitz U, Krieg AM, Bauer S, Vollmer J (2009) Sequences derived from self-RNA containing certain natural modifications act as suppressors of RNA-mediated inflammatory immune responses. Int Immunol 21:607–619
5. Chen C, Yang Z, Tang X (2018) Chemical modifications of nucleic acid drugs and their delivery systems for gene-based therapy. Med Res Rev 38:829–869
6. Ku SH, Jo SD, Lee YK, Kim K, Kim SH (2016) Chemical and structural modifications of RNAi therapeutics. Adv Drug Deliv Rev 104:16–28
7. Debart F, Dupouy C, Vasseur J-J (2018) Stimuli-responsive oligonucleotides in prodrug-based approaches for gene silencing. Beilstein J Org Chem 14:436–469
8. Casey JP, Blidner RA, Monroe WT (2009) Caged siRNAs for spatiotemporal control of gene silencing. Mol Pharm 6:669–685
9. Shah S, Rangarajan S, Friedman SH (2005) Light-activated RNA interference. Angew Chem Int Ed 44:1328–1332
10. Jain PK, Shah S, Friedman SH (2011) Patterning of gene expression using new photolabile groups applied to light activated RNAi. J Am Chem Soc 133:440–446
11. Yu L, Liang D, Chen C, Tang X (2018) Caged siRNAs with single cRGD modification for photoregulation of exogenous and endogenous gene expression in cells and mice. Biomacromolecules 19:2526–2534
12. Ji Y, Yang J, Wu L, Yu L, Tang X (2016) Photochemical regulation of gene expression using caged siRNAs with single terminal vitamin E modification. Angew Chem Int Ed 55:2152–2156
13. Yu L, Jing N, Yang Z, Zhang L, Tang X (2018) Caged siRNAs with single folic acid modification of antisense RNA for photomodulation of exogenous and endogenous gene expression in cells. Org Biomol Chem 16:7029–7035
14. Zhang L, Chen C, Fan X, Tang X (2018) Photomodulating gene expression by using caged siRNAs with single-aptamer modification. Chembiochem 19:1259–1263

Chapter 9

Exploring 5′-Biotinylation of the Sense Strand to Improve siRNA Specificity and Potency

Anne Mobergslien and Mouldy Sioud

Abstract

Although RNA interference is widely used for gene silencing, unintended gene modulation generated by off-target effects represents a major barrier to its applications in biology and medicine. Off-targeting can be induced by both the sense and antisense siRNA strands. An approach to minimizing off-target gene silencing by the sense strand would be the blockade of the 5′-end phosphorylation, thereby impeding its entry into the RNA-induced silencing complex (RISC). In this chapter, a biotin group at the 5′-end of the sense strand was used to inhibit its incorporation into RISC, thereby facilitating the antisense strand selection and enhancing siRNA cleavage potency. Biotin is a naturally occurring compound, and its presence in siRNA sequences will not induce additional side effects.

Key words Small interfering RNA, Off-target effects, Design tools, Antisense strand, Thermodynamics, Argonaute

1 Introduction

RNA interference (RNAi) is a naturally occurring mechanism of gene silencing that is mediated by double-stranded RNA and has been implicated in the control of endogenous and exogenous gene expression [1], *see* Chapter 1. Since its description by Tuschl and colleagues [2], small interfering RNAs (siRNAs) are used by most, if not all, investigators as basic method for knocking down gene expression in vitro and in vivo [3, 4]. Although this technique was initially believed to be highly specific, several studies have shown that siRNA can silence several genes in addition to the gene intended [5, 6]. Because siRNA can function as miRNAs, it is more likely that some nontarget genes will be inhibited by any given siRNA. Off-target signatures can be identified for both the sense and the antisense siRNA strands. In addition to gene silencing, siRNAs activated innate immunity in a sequence-dependent manner [7, 8]. These findings demonstrated a clear shortcoming of

Mouldy Sioud (ed.), *RNA Interference and CRISPR Technologies: Technical Advances and New Therapeutic Opportunities*, Methods in Molecular Biology, vol. 2115, https://doi.org/10.1007/978-1-0716-0290-4_9,

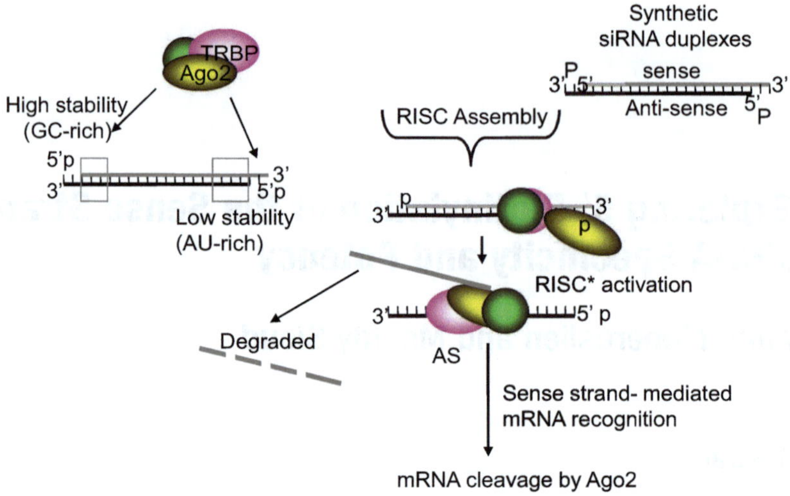

Fig. 1 Schematic representation of gene silencing by siRNAs. siRNAs are directly loaded into a multi-protein complex termed RNA-induced silencing complex (RISC) where the sense strand with high 5′-stability is cleaved by the nuclease AGO2 resulting in strand separation. Subsequently, the RISC containing the antisense strand (guide strand) seeks out and binds to complementary mRNA sequences. Bound mRNA molecules are then cleaved by AGO2, and cleaved mRNA fragments are rapidly degraded by cellular nucleases. Following dissociation, the active RISC is able to recycle and cleave additional mRNA molecules

the technology. Hence, identifying measures that can enhance siRNA specificity is critical for developing RNA into a reliable research and therapeutic tool.

Several strategies were proposed to control the activation of innate immunity and off-target gene silencing. These approaches include reducing siRNA concentration, chemically modifying siRNA, and applying more rigorous bioinformatics to siRNA design. For example, the incorporation of 2′-*O*-methyl group at position 1 and 2 of the passenger strand was used to reduce its silencing activity [9–11], *see* Chapter 2. Asymmetric siRNA design was also used to limit passenger (sense) strand entry into RISC [10] (Fig. 1). Also, fragmented passenger strand was used in order to eliminate its loading into RISC [11]. With respect to target specificity of the guide strand, incorporation of a 2′-*O*-methyl group at position 2 in the seed region reduced seed region-specific off-target silencing [9]. Similarly, the replacement of 2′-hydroxyl of uridine residues with either 2′-fluoro, or 2′-deoxy, or 2-*O*-methyl groups abrogated the activation of TLR7 and TLR8 by siRNAs [12]. Although some chemical modifications enhanced siRNA specificity, minimizing the degree of chemical modifications needed either to separate gene silencing from immune activation or to control the loading of siRNA strands into the RISC may be beneficial for therapeutic use. Here we used biotin, a natural compound found in a wide range of food sources, to modulate siRNA unwanted effects. We demonstrate its effectiveness in inhibiting gene silencing. Hence, such single modification can be added to the sense strand to impede its entry into RISC, thereby reducing siRNA off-target effects.

2 Materials

2.1 Cell Lines

Breast SKBR3 and colon SW480 cancer cell lines were purchased from American Type Culture Collection (ATCC). The cells were cultured in RPMI medium supplemented with antibiotics and 10% FCS. Human peripheral mononuclear cells (PBMCs) were prepared from buffy coats obtained from normal volunteers.

2.2 siRNA Sequences

Unmodified siRNAs and 5′-biotin modified siRNAs were purchased from Eurofins MWG (Ebersberg, Germany). The sequences of siRNA target sequences are the following:

Survivin: 5′-GAGCCAAGAACAAAAUUGC-3′.

Galectin-3: 5′-GUCUGGGCAUUCUGAUGUU-3′.

HER-2: siRNA-1: GGGCUGGCUCCGAUGUAUU; HER-2 iRNA-2.

5′-GGGACGAAUUC-UGCACAAU-3′.

2.3 siRNA Electroporation

1. Electroporation apparatus BTX ECM 830.
2. 4 mm gap cuvettes.
3. Chemically synthesized siRNAs.
4. Tissue culture flaks (25 and 75 cm^2).
5. RPMI 1640 medium.
6. RPMI 1640 medium supplemented with 10% FCS and antibiotics.
7. Trypsin/EDA.
8. Breast cancer cell line SKBR3.
9. Ovarian cancer cell line SW480.
10. Human peripheral blood mononuclear cells.

2.4 Preparation of Protein Extracts

1. Phosphate-buffered saline (PBS): Dissolve 8 g NaCl, 1.16 g Na_2HPO_4, 0.2 g KH_2PO_4, and 0.2 g KCl into 1 l distilled water. Adjust the pH to 7.0.
2. Lysis buffer: 10% NP40 in PBS supplemented with protease inhibitors (aprotinin, pepstatin, and leupeptin, 10 μg/ml each).
3. Eppendorf centrifuge.
4. Protein assay kit (Bio-Rad Laboratories).

3 Methods

3.1 Trypsinizing and Culturing SKBR3 and SW40 Cancer Cell Lines

As cells reach confluency, they must be subcultured. The first step in subculturing adherent cells is to detach cells from the surface of tissue culture flasks. We normally culture cell lines in 25 cm^2 or 75 cm^2 tissue culture flasks.

1. Examine the flask of cultured cells to check for cell density (70–80% confluent), and then remove all medium from primary culture with a sterile 10 ml pipette.
2. Wash the monolayer with PBS or serum-free medium.
3. Remove the PBS and add enough trypsin/EDTA solution to cover adhering cell layer.
4. Place the flask at 37 °C for 5 min, and then check flask under inverted microscope to make sure that the cells are rounded up and detached from the surface.
5. When a single cell suspension has been obtained, add 10 ml complete medium to 75 cm^2 tissue culture flask to inhibit further trypsin activity that might damage cells.
6. Transfer the cell suspension to a sterile 15 ml conical centrifuge tube.
7. Spin the cells down at 700 × *g* for 5 min in the table top centrifuge.
8. Discard the supernatant carefully.
9. Resuspend the cells in 5 ml of fresh complete media and count cell number.
10. Once the cell number has been determined, seed cells into a fresh 75 cm^2 tissue culture flask to get a final concentration of around 2×10^5 cells/ml, and incubate at 37 °C for 2–4 days and continue to passage as necessary (*see* **Note 1**).
11. If the cells are to be transfected by electroporation, transfer the desired number of cell c to 50 ml Falcon tube, add 10 ml complete medium, and incubate at 37 °C for 1–2 h in order for the cells to recover from trypsinization.

3.2 Electroporation of siRNAs into Cell Lines

Here we describe a protocol for electroporating 3×10^6 cells by square wave electroporation (*see* **Note 2**).

1. Spin down the cells (Subheading 3.1, **step 12**) and wash once with serum-free medium. Resuspend the cells in serum-free medium at 6×10^6 cells/ml.
2. Dispense RPMI complete medium into 25 cm^2 tissue culture flasks (5 ml/flask) and incubate at 37 °C.

3. Mix the cells (500 μl) with the desired siRNA concentration (0.5–6 μg/transfection), transfer the mixtures into the electroporation cuvette, and then incubate in ice for 2–5 min.
4. During the incubation time, set up the following electroporation parameters:
   - Volt: 500.
   - Pulse length: 2 ms.
   - Pulse: 01.
   - Interval: 01.0 s.
   - Polarity: unipolar.
5. Subsequently, place the metal walls in the cuvette against the metal walls in the electroporator, and then start pulsing.
6. Transfer the electroporated cells to 5 ml pre-warmed RPMI complete medium, and incubate at 37 °C prior to analyzing siRNA effects by Western and/or RT-PCR (*see* **Note 3**).

3.3 Preparation of Protein Extracts

1. Subsequent to transfection, scarp the cells and transfer cell suspension into 10 ml conical centrifuge tube.
2. Centrifuge for 5 min at 700 × *g*.
3. Remove supernatant, add 1 ml PBS buffer, and transfer cell suspension into 1.5 ml centrifuge tube.
4. Centrifuge 5 min at 2300 × *g*, remove supernatant, add 1 ml PBS to cell pellet and centrifuge 5 min at 2300 × *g*.
5. Add to cell pellet lysing buffer (50 μl), mix well, and place cell suspension in ice for 30 min. Vortex twice during the incubation.
6. Centrifuge cells for 10 min at 12,000 × *g*.
7. Save the supernatant which contains the cytoplasmic proteins. Measure protein concentration using Bio-Rad protein assay kit or other methods.
8. The effects of siRNA on gene expression can be analyzed by standard techniques such as Western blots and RT-PCR (*see* Figs. 2 and 3).

3.4 Experimental Evaluation of the Method

To assess the effect of 5′-end modification on siRNA gene silencing, we evaluated the effect of a biotin group at the 5′-end of the antisense (guide) strand to suppress gene silencing. Notably, certain structural features of siRNA, including 5′ phosphate group, are critical for loading into the RISC (Fig. 1).

As illustrated in Fig. 2a–c, 5′-biothylation of the antisense strand of three siRNA sequences directed against survivin, galectin-3, and HER-2 inhibited gene silencing more likely by inhibiting the selection of the guide strand by Ago-2 protein

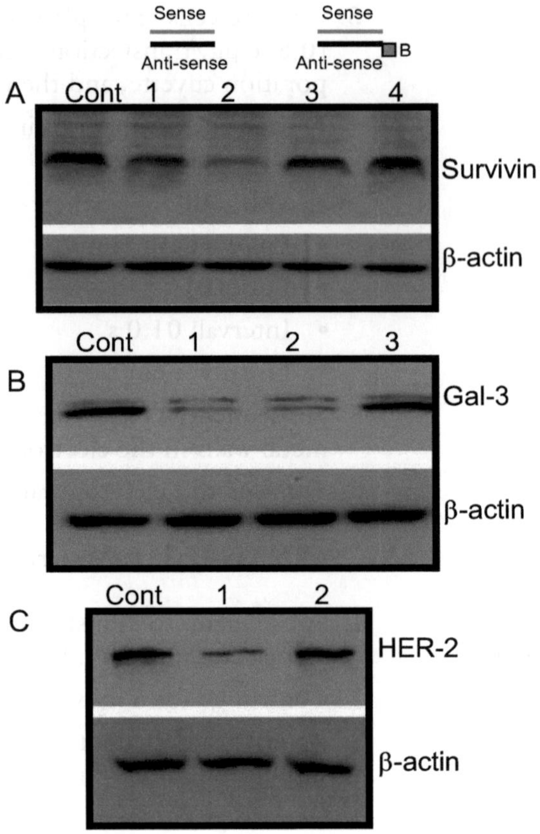

Fig. 2 Effect of 5′-biothylation of the guide strand on gene silencing. (**a**) SW40 colon cancer cells were transfected with either unmodified (lanes 1 and 2) or 5′-end biotin modified (lanes 3 and 4) siRNA against survivin for 18 h. Subsequently, protein extracts were prepared and the expression of survivin was determined by Western blots. Actin was used as an internal control for protein loading. Two siRNA concentrations (2 and 4 μg/transfection) were used. (**b**) SW40 cells were transfected with either unmodified (lanes 1 and 2) or 5′-end modified (lane 2) siRNA against galectin-3 for 48 h. As in A, gene silencing was determined by Western blots. (**c**) SKBR3 were transfected with either unmodified (lane 1) or 5′-end modified siRNA against HER-2 for 48 h, and then gene silencing was analyzed by Western blots. In all panels, the biotin was conjugated to the 5′-end of the antisense strand

[13]. These results are in agreement with early report suggesting the requirement of 5′ phosphate for siRNA loading into the RISC complex [2]. Overall, the data indicate that a siRNA duplex with biotin at 5′-end of the antisense strand is not competent in RNA interference. Therefore, the same modification can be used to block the incorporation of the sense (passenger) strand into the RISC complex.

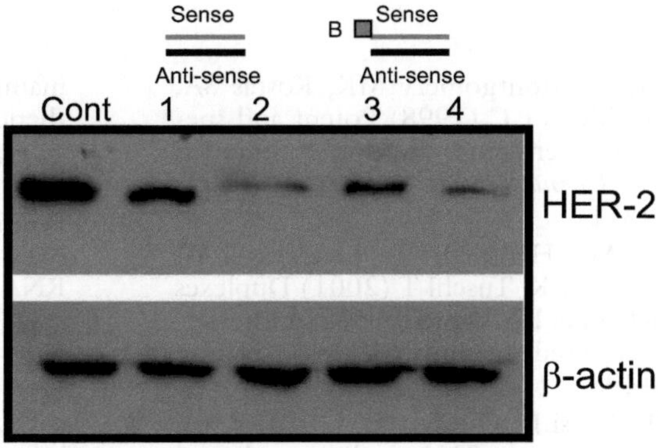

Fig. 3 Effect of 5′-biotinylation of the passenger (sense) strand on gene silencing. As in Fig. 2c, SKBR3 cells were transfected for 48 h with either unmodified or 5′-end modified siRNA against HER-2. Gene silencing was measured by Western blots. Two siRNA concentrations were tested (1 and 2 μ/transfection)

In the next experiments, we tested the effect of 5′-bionylation of the passenger strand on RNA gene silencing. As shown in Fig. 3, inhibition of phosphorylation of the 5′-end of the siRNA passenger strand through biotinylation did not inhibit gene silencing. In contrast, there is a clear enhancement of gene silencing by 5′-biotin modification of the passenger strand when compared to unmodified siRNAs (compare lanes 1 and 3) . This enhancement is more likely due to the more efficient loading of the guide strand into the RISC, thus supporting the notion that interfering with passenger strand loading into RISC may enhance RNAi (*see* Chapter 1). Notably, HER-2 siRNA is asymmetric siRNA [13], suggesting that 5′-biotinylation of the passenger strand can facilitate the loading of the sense strand into RISC, even in the context of a thermodynamically asymmetric siRNAs (*see* **Note 3**).

4 Notes

1. A large number of cells (4×10^7/800 μl) can be transfected in a single 4 mm gap disposable cuvette by square wave electroporation.
2. All culture incubations should be performed in a humidified 37 °C, 5% CO_2 incubator.
3. Reynolds and colleagues [14] published a set rules based on a systematic analysis of 180 siRNAs that underlined the importance of duplex thermodynamics, as determined by overall GC duplex content, lower stability at 3′-end of the sense strand, controlling for potential internal hairpins, and presence of a uridine at position 10 of the sense strand (*see* Chapter 1).

References

1. Fire A, Xu S, Montgomery MK, Kostas SA, Driver SE, Mello CC (1998) Potent and specific genetic interference by double-stranded RNA in *Caenorhabditis elegans*. Nature 391:806–811
2. Elbashir SM, Harborth J, Lendeckel W, Yalcin A, Weber K, Tuschl T (2001) Duplexes of 21-nucleotide RNAs mediate RNA interference in cultured mammalian cells. Nature 411:494–498
3. Lares MR, Rossi JJ, Ouellet DL (2010) RNAi and small interfering RNAs in human disease therapeutic applications. Trends Biotechnol 28:129–138
4. Behlke MA (2006) Progress towards in vivo use of siRNAs. Mol Ther 13:644–670
5. Jackson AL, Bartz SR, Schelter J, Kobayashi SV, Burchard J, Mao M et al (2003) Expression profiling reveals off-target gene regulation by RNAi. Nat Biotechnol 21:635–637
6. Clark PR, Pober JS, Kluger MS (2008) Knockdown of TNFR1 by the sense strand of an ICAM-1 siRNA: dissection of an off-target effect. Nucleic Acids Res 36:1081–1097
7. Sioud M (2005) Induction of inflammatory cytokines and interferon responses by double-stranded and single-stranded siRNAs is sequence-dependent and requires endosomal localization. J Mol Biol 348:1079–1090
8. Judge AD, Sood V, Shaw JR, Fang D, McClintock K, MacLachlan I (2005) Sequence-dependent stimulation of the mammalian innate immune response by synthetic siRNA. Nat Biotechnol 23:457–462
9. Jackson AL, Burchard J, Leake D, Reynolds A, Schelter J, Guo J, Johnson JM et al (2006) Position-specific chemical modification of siRNAs reduces off-target transcript silencing. RNA 12:1197–1205
10. Chang CI, Yoo JW, Hong SW, Lee SE, Kang HS, Sun X, Rogoff HA, Ban C, Kim S, Li CJ, Lee D (2009) Asymmetric shorter-duplex siRNA structures trigger efficient gene silencing with reduced non-specific effects. Mol Therapy 17:725–732
11. Fluiter K, Mook ORF, Baas F (2009) The therapeutic potential of LNA siRNAs: reduction of off-target effects by chemical modification of the siRNAs. Methods Mol Biol 487:189–203
12. Sioud M (2006) Single-stranded small interfering RNA are more immunostimulatory than their double-stranded counterparts: a central role for 2′-hydroxyl uridines in immune responses. Eur J Immunol 36:1222–1230
13. Schwarz DS, Hutvagner G, Du T, Xu Z, Aronin N, Zamore PD (2003) Asymmetry in the assembly of the RNAi enzyme complex. Cell 115:199–208
14. Reynolds A, Leake D, Boese Q, Scaringe S, Marshall WS et al (2004) Rational siRNA design for RNA interference. Nat Biotechnol 22:326–330

Chapter 10

In Vivo Delivery of Cassettes Encoding Anti-HBV Primary MicroRNAs Using an Ancestral Adeno-Associated Viral Vector

Njabulo Mnyandu, Patrick Arbuthnot, and Mohube Betty Maepa

Abstract

Chronic hepatitis B, a liver disease resulting from persisting hepatitis B virus (HBV) infection, remains a global health challenge despite the availability of an effective vaccine. Various preclinical studies using adeno-associated viruses (AAVs) to deliver anti-HBV RNA interference (RNAi) activators to mediate long-lasting HBV silencing show promise. Recent positive outcomes observed in clinical trials and the FDA approval of AAV-based drugs further demonstrate the potential of AAVs in antiviral therapeutic development. However, the prevalence of neutralizing antibodies against vectors based on extant AVV capsids limits the application of these vectors in human. The exciting reports on in silico designed and in vitro synthesized ancestral AAV (Anc80L65) with a potential to evade prevailing AAV neutralizing antibodies will significantly contribute to the success of these vectors in humans. Here, we describe methods for production and in vivo characterization of Anc80L65 expressing anti-HBV RNAi activators.

Key words Hepatitis B virus, MicroRNAs, Ancestral AAV, Anc80L65, Bioluminescence imaging, Q-PCR

1 Introduction

Although existence of an effective vaccine has alleviated the burden of hepatitis B (Hep B), high levels of chronic infections persist worldwide [1]. The current nucleosides or nucleotides analogues and immunomodulatory-based anti-HBV treatments have low efficacy [2, 3], prompting the search for more effective alternative therapies. In this regard, the delivery of effective RNA interference (RNAi) activators to inhibit HBV gene expression has been extensively explored. The microRNA (miR) RNAi pathway is the most studied and is being harnessed for therapeutic purposes [4–9]. This pathway can be reprogrammed by exogenous synthetic short interfering RNAs (siRs) [10] or expressed RNAi activators such as artificial miRs (amiRs) [11]. Effects of amiRs may be sustained,

Mouldy Sioud (ed.), *RNA Interference and CRISPR Technologies: Technical Advances and New Therapeutic Opportunities*, Methods in Molecular Biology, vol. 2115, https://doi.org/10.1007/978-1-0716-0290-4_10,

which is useful for therapeutic application, and can be efficiently expressed from tissue-specific RNA polymerase II promoters [12, 13]. Another advantage of using expressed RNAi activators is their compatibility with highly efficient viral vectors. While both adenoviral and lentiviral vectors can provide efficient delivery of anti-HBV pri-miRs to the liver [12, 14], their high immune stimulation [15] or integrative nature [16] raises safety concerns.

The biosafety profile of adeno-associated viral vectors (AAVs) is well established [17–19]. Wild-type AAVs are classified in different serotypes, named AAV1–AAV12, which have specific tissue tropisms [20]. These icosahedral viruses are about 20 nm in diameter, carry a single-stranded 4.7 kb genome, are non-enveloped, and belong to the *Parvoviridae* family of the *Dependovirus* genus. The AAV genome consists of inverted terminal repeats (ITRs) flanking *cap* and *rep* genes [21]. The *rep* gene expresses a quartet of polypeptides named by size, Rep78, Rep68, Rep52, and Rep40, which are needed for viral genome replication [reviewed in [22]]. On the other hand, the *cap* gene expresses viral structural proteins (VP1, VP2, and VP3) that make up the ~22 nm, $T = 1$ icosahedral capsid that comprises 60 VP subunits. VP1, VP2, and VP3 are arranged in a stoichiometric ratio of 1:1:10. The ITRs' *cis* acting elements bear Rep protein binding elements (RBEs) together with an origin of replication, which are essential for viral gene packaging, replication, and encapsidation [23–25]. The VP subunits in most AAVs, except for AAV4, AAV5, and AAV11, are assembled with the aid of assembly activation protein (AAP) expressed from the *cap* open reading frame (ORF) [26].

Wild-type AAVs naturally are replication incompetent, and productive infection is enabled by expression of adenovirus (Ad) helper genes, *E1A*, *E1B*, *E2A*, *E4*, and *ORF6*, together with viral-associated (*VA)* RNAs [27]. The wild-type AAV genes as well as AAV replication-enabling Ad genes are provided for production of RNAi-activating AAVs used in therapeutic application. A cocktail of plasmids carrying various genes is used to generate recombinant AAVs (rAAVs) (Fig. 1). This consists of sequences bearing ITR *cis* acting elements, which flank transgenes such as anti-HBV RNAi activator cassettes, Ad helper genes, *cap* and *rep* AAV genes (reviewed in [28]), and in some cases sequences encoding the AAP [29]. To generate rAAV genomes encoding HBV-targeting gene silencers, wild-type *cap* and *rep* genes are replaced by anti-HBV RNAi-activating sequences. Exclusion of *rep* and *cap* in the rAAVs renders the vectors replication-defective and also prevents unintended toxicity of Rep expression [30, 31]. Despite *rep* gene expression's harmful effects, its activity, as well as that of the *cap* gene, is vital for rAAV production. Hence *rep* and *cap* sequences are supplied *in trans*, along with other helper genes (AAP and/or Ad-helper). The rAAV genome may be single stranded (ssAAV) or self-complimentary (scAAV), with the latter conformation

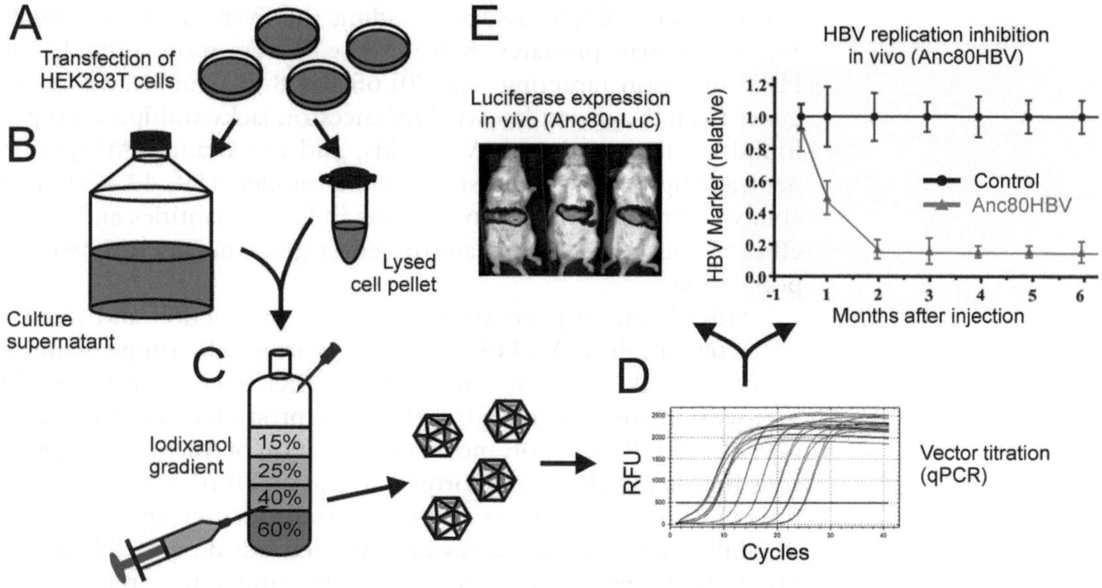

Fig. 1 Schematic representation of Anc80L65 vector production and characterization in vivo. (**a**) Four plasmids are transfected into HEK293T cells to propagate Anc80L65 vector. The cocktail of plasmids includes sequences encoding (1) ITRs with amiR/NLuc, (2) AAP, (3) Ad helper elements, and (4) Rep and Cap. (**b**) Spent medium is collected at 48, 72, 96, and 120 h after transfection. Cells are harvested at 120 h and lysed to release the viral particles. (**c**) After the virus in the spent media is precipitated and then resuspended with clarified cell lysate, the vector is purified following centrifugation through an iodixanol gradient. Anc80L65 are titrated using quantitative polymerase chain reaction (qPCR) (**d**) and then characterized in vivo (**e**) by analysis of nanoluciferase expression or HBV replication inhibition

achieving faster transgene expression kinetics following transduction of cells. The double-stranded conformation of self-complementary AAVs is achieved by mutating the terminal resolution sites (trs) located adjacent to one of the ITRs [32–34]. The optimal AAV genome packaging size of ssAAVs is between 4.1 and 4.9 Kb [35]. scAAVs have half the genome size and should ideally not be larger than 2.35 Kb. However, this is not a limitation for delivery of RNAi activators because RNAi expression cassettes are usually small enough to be accommodated by scAAVs. Studies using AAVs to deliver RNAi activators against HBV infection have shown efficient liver transduction and a sustained therapeutic effect without obvious adverse effects [13, 36].

Despite the attractive properties of AAVs, capsid-specific neutralizing antibodies that are prevalent in human populations form a barrier to effective therapeutic gene transfer [37–41]. Recently, synthetic vector variants bearing AAV ancestral capsids were designed in silico and synthesized in vitro [29, 42]. These vectors showed superior transduction efficiencies when compared to AAV2, AAV4, AAV5, AAV8, and AAV9 [42]. The development of the ancestral vector Anc80L65, with excellent transduction efficiency in multiple tissues including the livers of small animals and nonhuman primates, makes these vectors more suitable for HBV infection targeting. Anc80L65 has 8.6% amino acid divergence from the closest known AAV ancestor, lacks multiple epitopes found on the existing AAV capsids, and can evade AAV-specific neutralizing antibodies in small animal models [29, 43–45]. The ability to produce Anc80L65 vectors in large quantities and prove efficacy will provide the means to deliver gene therapy to a broader population.

This chapter reports on methods for production and in vivo characterization of Anc80L65 vectors bearing self-complementary genomes and expressing anti-HBV amiRs from a liver-specific mouse transthyretin receptor (mTTR) promoter (Anc80-HBV) [29, 46, 47]. Production and characterization of Anc80L65 expressing the nanoluciferase reporter (nLuc, Anc80-nLuc) is included for convenient evaluation of in vivo transduction efficiency and durability of transgene expression. All methods described here use HBV transgenic mice containing an integrated replication-competent greater-than-genome-length viral sequence [48]. Methods for vector production, in vivo transduction, and efficacy assessment are schematically depicted in Fig. 1 and described in detail below.

2 Materials

2.1 Anc80L65 Vector Production and Purification

2.1.1 Construction of scAAV ITR and Transgene Bearing Plasmids and Transfection of HEK293T Cells

1. Plasmid carrying scAAV ITRs (e.g., pBS-H1-RSV-GFP) [49].
2. Plasmid carrying anti-HBV *pri-miR* (e.g., pMTTR pri-miR-31/5-8-9) [12] or *nLuc* (pNL1.1NLuc, Promega, WI, USA) sequence.
3. Plasmid carrying Anc80L65 *cap* and AAV2 *rep* genes (e.g., pAnc80L65, Addgene number plasmid #68837) [29].
4. Plasmid carrying *AAP* gene (e.g., pAAP2, Addgene number#73211) [29].
5. Plasmid carrying adenoviral helper genes (e.g., pACG) [49].
6. Human embryonic kidney cells (HEK293T).
7. Complete Dulbecco's modified Eagle's medium (DMEM). Add fetal bovine to 10%, penicillin to 100 U/mL, and streptomycin to 100 μg/mL.
8. Tissue culture plates (15 cm^2).
9. Polyethylenimine Max. Dissolve 0.25 g PEImax in 225 mL sterile distilled water. Adjust pH to 7.0 with 1 M NaOH. Add sterile distilled water up to 250 mL and filter sterilize using 0.22 μm filter membrane. PEImax solution should not be frozen and is stable for 9 months at 4 °C.

10. OPTIMEM.
11. 15 mL centrifuge tubes.

2.1.2 Harvesting of Vector Particles from Spent Media and HEK293 Cell Pellet

1. Complete Dulbecco's modified Eagle's medium as described above.
2. 1× phosphate buffered saline (1× PBS).
3. Cell scrapers (23 cm).
4. 50 mL centrifuge tubes.
5. Virus lysate buffer: To 50 mM Tris–Cl, add NaCl to 150 mM and $MgCl_2$ to 5 mM. Adjust pH to 8.5.
6. Molecular biology grade polyethylene glycol 8000 (PEG 8000): Prepare 5× PEG by adding 500 g PEG 8000 and 115 g NaCl in up to 1 L of distilled water. Autoclave at 121 °C, 100 kPa for 30 min. Cool at room temperature and then store at 37 °C.
7. Benzonase.
8. A-4-38 rotor.
9. VacuCap 90 PF filters.
10. Sterile Schott bottles (500 mL).
11. JA-10 rotor.

2.1.3 Purification of Vectors by Iodixanol Gradient Centrifugation

1. 1× PBS.
2. PBS-MK: Add $MgCl_2$ to 1 mM and KCl to 2.5 mM in 1× PBS and filter sterilize.
3. PBS-MK-NaCl: Add NaCl to 1 M in PBS-MK and filter sterilize.
4. Optiprep.
5. Phenol red.
6. 60% iodixanol: Add 125 μL of phenol red to 50 mL Optiprep.
7. 40% iodixanol: Mix 30 mL Optiprep with 15 mL PBS-MK.
8. 25% iodixanol: Mix 20 mL Optiprep with 28 mL PBS-MK and 120 μL phenol red.
9. 15% iodixanol: Mix 12 mL Optiprep with 36 mL PBS-MK-NaCl.
10. Quickseal ultracentrifuge tubes.
11. 20 × 1.5″ gauge needle.
12. 3 mL and 10 mL syringes.
13. Pasteur pipettes (230 mm).
14. Fixed-angle 70.1. Ti rotor.

2.2 Quantification of Genome Copy Numbers in Purified Vector Stocks

1. Mini DNA extraction kit.
2. SYBR green master mix.
3. Vector-specific primers (e.g., mTTR forward, 5′ GCACTGG-GAGGATGTTGAGT 3′; mTTR reverse, 5′CCCCTGTTCAAACATGTCCT3′) [13].
4. DNA standards (e.g., serial dilutions of linearized plasmid with target sequence).
5. Real-time PCR instrument.

2.3 Determination of Anc80-nLuc Liver Transduction Using Bioluminescence Imaging

1. Nano-Glo® luciferase substrate.
2. 1× PBS.
3. Isoflurane.
4. Bioluminescence imager.

2.4 Determination of Liver Transduction by Anc80-HBV

1. Saline (SABAX, 0.9% sodium chloride).
2. Handheld homogenizer.
3. Spectrophotometer instrument.
4. Mini DNA extraction kit.
5. SYBR green master mix.
6. Vector-specific primers (e.g., mTTR forward, 5′ GCACTGG-GAGGATGTTGAGT 3′; mTTR reverse, 5′CCCCTGTTCAAACATGTCCT3′) [13].
7. DNA standards (e.g., serial dilutions of linearized plasmid with target sequence).
8. Real-time PCR instrument.

2.5 Assessment of Anc80-HBV Activity and Safety

1. HBsAg immunoassay kit.
2. Mini DNA extraction kit.
3. Ultra-Sensitive ABC Peroxidase Staining Kit.
4. Unlabeled mouse anti HBcAg primary mouse antibody (Abcam, MA, USA): Dilute 1:1000.
5. HBV DNA standards.
6. HBV-specific primers (e.g., HBVs forward, 5′ TGC ACC TGT ATT CCC ATC 3′; HBVs reverse, 5′ CTG AAA GCC AAA CAG TGG 3′) [4].
7. SYBR green master mix.
8. Real-time PCR instrument.
9. Polyinosinic-polycytidylic acid (Poly:IC).
10. Aspartate transaminase (AST) and alanine transaminase (ALT) assay kit.
11. Reverse Transcription Kit.

12. Interferon-β- (IFN-β)-specific primers (e.g., IFN-β forward, 5′-TCCAAATTGCTCTCCTGTTGTGCT-3′; IFN-β reverse, 5′ CCACAGGAGCTTCTGACACTGAAAA 3′) [4].
13. 2′ 5′ oligoadenylate synthase-*1 (OAS-1)*-specific primers (e.g., OAS1 forward, 5′ CGAGGGAGCATGAAAACACATTT 3′; OAS1 reverse, 5′ GCAGAGTTGCTGGTAGTTTATGAC 3′) [4].
14. Myxovirus resistance protein A (MxA)-specific primers (e.g., MxA forward, 5′ CTGGTGCTGAAACTGAAGAAAC 3′; MxA reverse, 5′ ATCTCAATCTCGTAGTCCTGGTA 3′) [4].
15. Glyceraldehyde-3-phosphate dehydrogenase (GAPDH)-specific primers (e.g., GAPDH forward, 5′ AGGGGTCATT-GATGGCAACAATATCCA 3′; GAPDH reverse, 5′ TTTAC-CAGAGTTAAAAGCAGCCCTGGTG 3′) [4].
16. Cytometric bead array (CBA) mouse inflammation kit.
17. Flow cytometer instrument.
18. Flow cytometer data analysis software.

3 Methods

3.1 Anc80L65 Production and Purification

3.1.1 Construction of scAAV ITR and Transgene Bearing Plasmids and Transfection of HEK293T Cells

1. In a *scAAV* ITR bearing plasmid, clone the liver-specific promoter-driven *pri-miR* or *nLuc* sequences in between the *ITRS* using standard cloning methods (*see* **Note 1**).
2. Grow four 15 cm^2 plates of HEK293 cells to 100% confluency in complete DMEM with supplements.
3. In a 15 mL centrifuge tube, prepare 515 μg of plasmid DNA mix consisting of a plasmid carrying (1) AAV *ITRs* and anti-HBV *pri-miR* or *nLuc* sequences, (2) adeno-viral helper genes, (3) *AAP* gene, and (4) Anc80L65 *cap* and AAV2 *rep* genes in a mass ratio of 0.25:1:2:2, respectively.
4. Add 4 mL OPTIMEM to the DNA.
5. In a separate tube, add PEImax at a mass ratio of three to five times the total DNA mass and add OPTIMEM to 4 mL.
6. Incubate tubes containing DNA and PEImax at room temperature for 5 min.
7. Combine both the DNA and PEImax mixes in one tube, vortex briefly, and incubate at room temperature for 15 min before adding 2 mL of the mixture per plate in a dropwise manner.
8. Incubate the cells at 37 ° C under 5% CO_2 culture conditions for 120 h.

3.1.2 Harvesting of Vector Particles from Spent Medium and HEK293T Cell Pellet

1. At time points 48, 72, 96, and 120 h post-transfection, collect spent media and store at 4 ° C. Replace spent media with 20 mL of fresh complete DMEM growth media at all time points, except the last time point.
2. After supernatant collection at 120 h, distribute 40 mL 1× PBS in equal volumes to four plates. Detach cells with a cell scraper and transfer to a 50 mL centrifuge tube.
3. Pellet harvested cells by centrifugation in A-4-38 rotor at 400 × *g* for 8 min.
4. Use 5 mL virus lysate buffer to resuspend, the pellet and keep the suspension at −80 °C until cell lysis is performed.
5. Release vector particles by 5 rounds of freeze-thaw lysing cycles (5 × 37 °C, 5 × −196 °C) using a warm water bath and liquid nitrogen.
6. Digest nucleic acids by adding benzonase to 50 units/mL and incubating the lysate in a 37 °C water bath for an hour. Vortex mix for 15 s at 10 min intervals.
7. Pellet the cell debris by centrifuging the suspension in an A-4-38 rotor at 2500 × *g* for 15 min, and transfer supernatant to a fresh 50 mL tube.
8. Repeat the centrifugation at the same conditions and transfer the supernatant to a fresh tube.
9. To precipitate Anc80L65 from the spent media collected at all the time points in **step 1**, the following steps must be done:
   (a) Filter the supernatants through VacuCap filters into sterile 500 mL Schott bottles.
   (b) Add 12.5 mL of 5 × PEG 8000 per 50 mL of filtrate and leave overnight at 4 °C (*see* **Note 2**).
   (c) Pellet the vectors by centrifugation of the filtrate at 2744 × *g* in a JA-10 rotor for 30 min and discard the supernatant.
10. Resuspend the PEG8000 pellet from **step 9(c)** with cell lysate supernatant collected in **step 8** above.

3.1.3 Purification of Vectors by Iodixanol Gradient Centrifugation

1. With a 10 mL syringe attached to a 20 × 1.5″ gauge needle, transfer the viral vector suspensions, each of up to 7 mL, through a Pasteur pipette into Quickseal ultracentrifuge tubes.
2. To the centrifuge tubes, slowly add 1.5 mL of each 15%, 25%, 40%, and 60% iodixanol solutions consecutively (*see* **Note 3**).
3. Balance the centrifuge tubes and seal them by melting the tips before ultracentrifugation.
4. Ultracentrifuge the gradients at 171,500 × *g* in a 70.1.Ti rotor at 4 °C for 2 h.

5. Fix the gradient tubes onto a stand bracket, and release the pressure by punching the top of the centrifuge tube with a 20 × 1.5″ gauge needle.
6. Extract virus by punching with another 20 × 1.5″ gauge needle attached to a 2--3 mL syringe at the base of the 40% iodixanol vector containing layer (*see* **Note 4**).

3.2 Anc80L65 Handling and Storage

Purified Anc80L65 particles should be aliquoted in smaller volumes to avoid repeated freeze-thaw cycles and stored at −80 °C. Viral stock aliquots can be thawed at room temperature when required. Any sample containing Anc80L65 should be considered infectious, and approved biosafety procedures should be followed. Administration to animals should be carried out according to protocols approved by institutional ethics committees. Purified Anc80L65 can be diluted in saline before administering 1×10^{11} viral particle equivalents (VPEs) per 30 g mouse in 300 μL. Systemic injection of mice is carried out via a tail vein. Blood samples can be collected by retro-orbital puncture from isoflurane-anesthetized mice.

3.3 Quantification of Genome Copy Numbers in Purified Vector Stocks

1. Isolate and purify DNA from 50 μL of the purified virus using a DNA Mini extraction kit according to the manufacturer's instructions, and elute DNA in 200 μL elution buffer or sterile distilled water.
2. Use 1 μL of DNA to perform absolute quantification of genome copies by q-PCR using a SYBR green master mix according to manufacturer's instructions in a real-time PCR instrument. Use vector-specific primer sets and a standard curve generated from known copy numbers of the target-containing fragment.

3.4 Determination of Anc80L65 liver transduction, activity and safety

1. Inject mice with 1×10^{11} VPEs of Anc80L65.
2. To determine liver transduction by An80-nLuc, perform bioluminescence imaging at one week after injection (*see* **Note 5**) using bioluminescence imager as follows:
   (a) Dilute Nano-Glo® Luciferase substrate 1:50 with 1× PBS.
   (b) Inject diluted Nano-Glo® luciferase substrate intraperitoneally (300 μL per 30 g mouse).
   (c) Anesthetize mice using isoflurane, perform live imaging, and quantify photon counts in a bioluminescence imaging system.
   (d) Repeat **steps** (**a**)–(**c**) weekly to determine luciferase expression over a 3-month period (*see* **Note 5**).

3. Measure transduction, activity, and safety of Anc80-HBV using the following standard procedures and/or specialized commercial kits:
   (a) To assess Anc80-HBV transduction, harvest and homogenize one lope of the liver starting from 2 weeks post infection (*see* **Note** 5). Isolate DNA and use 100 ng to perform qPCR using SYBR green master mix, vector specific primers, and a q-PCR machine.
   (b) To measure HBV gene expression silencing, harvest serum and liver tissue starting from 2 weeks post infection (*see* **Note** 5). Use 50 times diluted serum to perform HBV surface antigen ELISA using HBsAg immunoassay kit. Use liver sections to perform HBV core antigen immunohistochemical staining using ultrasensitive ABC peroxidase staining kit and anti-HBcAg primary antibody.
   (c) To quantify circulating HBV viral particles, isolate DNA from 10 μL of serum collected in **step** (**b**) above and elute in 50 μL of elution buffer using mini DNA extraction kit. Use 5 μL to performe q-PCR using SYBR green master-mix, HBV specific primers, q-PCR machine, and commercial HBV DNA standards.
   (d) Assess hepatic inflammatory marker expression using reverse transcriptase q-PCR (RT q-PCR) on RNA isolated from livers harvested at 6 h postinfection (*see* **Note 5**). Use a reverse transcription kit and SYBR green master mix according to manufacturer's instructions. Use RNA from livers injected with polyinosinic-polycytidylic acid (Poly: IC, 100 μg/mouse) as a positive control for induction of inflammation (*see* **Note 6**). Determine *IFN-β*, *OAS-1*, and *MxA* mRNA levels relative to *GAPDH* mRNA using gene-specific primer sets.
   (e) Use 50 μL of undiluted serum to quantify cytokine concentrations in serum collected at 6 h postinjection (*see* **Note 5**). Assay with a cytometric bead array kit according to the manufacturer's instructions. Use serum from mice injected with Poly:IC (100 μg/mouse) as a positive control for induction of inflammation (*see* **Note 5**). Quantify concentrations of IL-6, IL-10, MCP-1, IFN-γ, TNF-α, and IL-12p70 by running the samples on a flow cytometer. Perform data analysis using commercially available software.
   (f) Measure ALT and AST activity in serum harvested weekly for 8 weeks (*see* **Note 5**) using AST and ALT assays run in a routine clinical laboratory. Use 200 μL serum diluted 1:1 with saline (*see* **Note** 7).

(g) Detect hepatic inflammation using hematoxylin and eosin staining. Fibrosis may be detected by Sirius red staining according to standard procedures in a routine laboratory.

4 Notes

1. Make sure the total sequence does not exceed the maximum packaging capacity of 2.35 kb for scAAVs.
2. Be sure to use molecular-grade PEG8000 for easy resuspension of the pellet.
3. If unavoidable, remove bubbles that may form during preparation of the iodixanol gradient by tapping on the bubbles located in the middle of the tube. Bubbles located at the top of the tubes may be removed by careful aspiration.
4. Careful extraction of the rAAV layer is important to avoid contaminants, such as empty capsids that are typically located within the 25% layer.
5. Time points recommended represent the earliest or latest times which are most likely to show a significant effect. Time points in between this range may also be used.
6. Poly:IC must be injected into mice by hydrodynamic injection.
7. If the serum is limited, a 1:4 dilution can be used.

Acknowledgments

The work in the Wits/SAMRC Antiviral Gene Therapy Research Unit is funded by the South African Medical Research Council (SAMRC), South African National Research Foundation (NRF), and Poliomyelitis Research Foundation (PRF).

References

1. Naghavi M, Wang H, Lozano R, Davis A, Liang X, Zhou M (2015) Global, regional, and national age-sex specific all-cause and cause-specific mortality for 240 causes of death, 1990-2013: a systematic analysis for the global burden of disease study 2013. Lancet 385(9963):117–171
2. Billioud G, Pichoud C, Puerstinger G, Neyts J, Zoulim F (2011) The main hepatitis B virus (HBV) mutants resistant to nucleoside analogs are susceptible in vitro to non-nucleoside inhibitors of HBV replication. Antivir Res 92 (2):271–276
3. Sypsa VA, Mimidis K, Tassopoulos NC, Chrysagis D, Vassiliadis T, Moulakakis A et al (2005) A viral kinetic study using pegylated interferon alfa-2b and/or lamivudine in patients with chronic hepatitis B/HBeAg negative. Hepatology 42(1):77–85
4. Carmona S, Ely A, Crowther C, Moolla N, Salazar FH, Marion PL et al (2006) Effective inhibition of HBV replication in vivo by anti-HBx short hairpin RNAs. Mol Ther 13 (2):411–421
5. Giering JC, Grimm D, Storm TA, Kay MA (2008) Expression of shRNA from a tissue-specific pol II promoter is an effective and safe RNAi therapeutic. Mol Ther 16 (9):1630–1636
6. Grimm D, Streetz KL, Jopling CL, Storm TA, Pandey K, Davis CR et al (2006) Fatality in

mice due to oversaturation of cellular microRNA/short hairpin RNA pathways. Nature 441(7092):537–541

7. Montgomery MK, Xu S, Fire A (1998) RNA as a target of double-stranded RNA-mediated genetic interference in Caenorhabditis elegans. Proc Natl Acad Sci U S A 95 (26):15502–15507
8. Elbashir SM, Lendeckel W (2001) Tuschl T. RNA interference is mediated by 21- and 22-nucleotide RNAs. Genes Dev 15 (2):188–200
9. McCaffrey AP, Nakai H, Pandey K, Huang Z, Salazar FH, Xu H et al (2003) Inhibition of hepatitis B virus in mice by RNA interference. Nat Biotechnol 21(6):639–644
10. Marimani MD, Ely A, Buff MC, Bernhardt S, Engels JW, Scherman D et al (2015) Inhibition of replication of hepatitis B virus in transgenic mice following administration of hepatotropic lipoplexes containing guanidinopropyl-modified siRNAs. J Control Release 209:198–206
11. Ely A, Naidoo T, Arbuthnot P (2009) Efficient silencing of gene expression with modular trimeric pol II expression cassettes comprising microRNA shuttles. Nucleic Acids Res 37 (13):e91
12. Mowa MB, Crowther C, Ely A, Arbuthnot P (2014) Inhibition of hepatitis B virus replication by helper dependent adenoviral vectors expressing artificial anti-HBV pri-miRs from a liver-specific promoter. Biomed Res Int 2014:718743
13. Maepa MB, Ely A, Grayson W, Arbuthnot P (2017) Sustained inhibition of HBV replication in vivo after systemic injection of AAVs encoding artificial antiviral primary MicroRNAs. Mol Ther Nucleic Acids 7:190–199
14. Ivacik D, Ely A, Ferry N, Arbuthnot P (2015) Sustained inhibition of hepatitis B virus replication in vivo using RNAi-activating lentiviruses. Gene Ther 22(2):163–171
15. Crowther C, Ely A, Hornby J, Mufamadi S, Salazar F, Marion P et al (2008) Efficient inhibition of hepatitis B virus replication in vivo, using polyethylene glycol-modified adenovirus vectors. Hum Gene Ther 19(11):1325–1331
16. Qian W, Wang Y, Li RF, Zhou X, Liu J, Peng DZ (2017) Prolonged integration site selection of a lentiviral vector in the genome of human keratinocytes. Med Sci Monit 23:1116–1122
17. Kassner U, Hollstein T, Grenkowitz T, Wuhle-Demuth M, Salewsky B, Demuth I et al (2018) Gene therapy in lipoprotein lipase deficiency: case report on the first patient treated with alipogene tiparvovec under daily practice conditions. Hum Gene Ther 29(4):520–527
18. Maguire AM, Simonelli F, Pierce EA, Pugh EN Jr, Mingozzi F, Bennicelli J et al (2008) Safety and efficacy of gene transfer for Leber's congenital amaurosis. N Engl J Med 358 (21):2240–2248
19. Rafii MS, Baumann TL, Bakay RA, Ostrove JM, Siffert J, Fleisher AS et al (2014) A phase1 study of stereotactic gene delivery of AAV2-NGF for Alzheimer's disease. Alzheimers Dement 10(5):571–581
20. Zincarelli C, Soltys S, Rengo G, Rabinowitz JE (2008) Analysis of AAV serotypes 1-9 mediated gene expression and tropism in mice after systemic injection. Mol Ther 16(6):1073–1080
21. Srivastava A, Lusby EW, Berns KI (1983) Nucleotide sequence and organization of the adeno-associated virus 2 genome. J Virol 45 (2):555–564
22. Samulski RJ, Muzyczka N (2014) AAV-mediated gene therapy for research and therapeutic purposes. Annu Rev Virol 1 (1):427–451
23. Buller RM, Rose JA (1978) Characterization of adenovirus-associated virus-induced polypeptides in KB cells. J Virol 25(1):331–338
24. Kronenberg S, Kleinschmidt JA, Bottcher B (2001) Electron cryo-microscopy and image reconstruction of adeno-associated virus type 2 empty capsids. EMBO Rep 2(11):997–1002
25. Xie Q, Bu W, Bhatia S, Hare J, Somasundaram T, Azzi A et al (2002) The atomic structure of adeno-associated virus (AAV-2), a vector for human gene therapy. Proc Natl Acad Sci U S A 99 (16):10405–10410
26. Earley LF, Powers JM, Adachi K, Baumgart JT, Meyer NL, Xie Q et al (2017) Adeno-associated virus (AAV) assembly-activating protein is not an essential requirement for capsid assembly of AAV serotypes 4, 5, and 11. J Virol 91(3):epub
27. Myers MW, Laughlin CA, Jay FT, Carter BJ (1980) Adenovirus helper function for growth of adeno-associated virus: effect of temperature-sensitive mutations in adenovirus early gene region 2. J Virol 35(1):65–75
28. Goncalves MA (2005) Adeno-associated virus: from defective virus to effective vector. Virol J 2:43
29. Zinn E, Pacouret S, Khaychuk V, Turunen HT, Carvalho LS, Andres-Mateos E et al (2015) In silico reconstruction of the viral evolutionary lineage yields a potent gene therapy vector. Cell Rep 12(6):1056–1068

30. Hermanns J, Schulze A, Jansen-Dburr P, Kleinschmidt JA, Schmidt R, zur Hausen H (1997) Infection of primary cells by adeno-associated virus type 2 results in a modulation of cell cycle-regulating proteins. J Virol 71 (8):6020–6027
31. Schmidt M, Afione S, Kotin RM (2000) Adeno-associated virus type 2 Rep78 induces apoptosis through caspase activation independently of p53. J Virol 74(20):9441–9450
32. Ferrari FK, Samulski T, Shenk T, Samulski RJ (1996) Second-strand synthesis is a rate-limiting step for efficient transduction by recombinant adeno-associated virus vectors. J Virol 70(5):3227–3234
33. McCarty DM, Monahan PE, Samulski RJ (2001) Self-complementary recombinant adeno-associated virus (scAAV) vectors promote efficient transduction independently of DNA synthesis. Gene Ther 8(16):1248–1254
34. Ling C, Wang Y, Lu Y, Wang L, Jayandharan GR, Aslanidi GV et al (2015) Enhanced transgene expression from recombinant single-stranded D-sequence-substituted adeno-associated virus vectors in human cell lines in vitro and in murine hepatocytes in vivo. J Virol 89(2):952–961
35. Dong JY, Fan PD, Frizzell RA (1996) Quantitative analysis of the packaging capacity of recombinant adeno-associated virus. Hum Gene Ther 7(17):2101–2112
36. Chen CC, Sun CP, Ma HI, Fang CC, Wu PY, Xiao X et al (2009) Comparative study of anti-hepatitis B virus RNA interference by double-stranded adeno-associated virus serotypes 7, 8, and 9. Mol Ther 17(2):352–359
37. Calcedo R, Vandenberghe LH, Gao G, Lin J, Wilson JM (2009) Worldwide epidemiology of neutralizing antibodies to adeno-associated viruses. J Infect Dis 199(3):381–390
38. Boutin S, Monteilhet V, Veron P, Leborgne C, Benveniste O, Montus MF et al (2010) Prevalence of serum IgG and neutralizing factors against adeno-associated virus (AAV) types 1, 2, 5, 6, 8, and 9 in the healthy population: implications for gene therapy using AAV vectors. Hum Gene Ther 21(6):704–712
39. Calcedo R, Morizono H, Wang L, McCarter R, He J, Jones D et al (2011) Adeno-associated virus antibody profiles in newborns, children, and adolescents. Clin Vaccine Immunol 18 (9):1586–1588
40. Nathwani AC, Tuddenham EG, Rangarajan S, Rosales C, McIntosh J, Linch DC et al (2011) Adenovirus-associated virus vector-mediated gene transfer in hemophilia B. N Engl J Med 365(25):2357–2365
41. Fitzpatrick Z, Leborgne C, Barbon E, Masat E, Ronziti G, van Wittenberghe L et al (2018) Influence of pre-existing anti-capsid neutralizing and binding antibodies on AAV vector transduction. Mol Ther 9:119–129
42. Santiago-Ortiz J, Ojala DS, Westesson O, Weinstein JR, Wong SY, Steinsapir A et al (2015) AAV ancestral reconstruction library enables selection of broadly infectious viral variants. Gene Ther 22(12):934–946
43. Carvalho LS, Xiao R, Wassmer SJ, Langsdorf A, Zinn E, Pacouret S et al (2018) Synthetic adeno-associated viral vector efficiently targets mouse and nonhuman primate retina in vivo. Hum Gene Ther 29(7):771–784
44. Suzuki J, Hashimoto K, Xiao R, Vandenberghe LH, Liberman MC (2017) Cochlear gene therapy with ancestral AAV in adult mice: complete transduction of inner hair cells without cochlear dysfunction. Sci Rep 7:45524
45. Gu X, Chai R, Guo L, Dong B, Li W, Shu Y et al (2019) Transduction of adeno-associated virus vectors targeting hair cells and supporting cells in the neonatal mouse cochlea. Front Cell Neurosci 13:8
46. Jungmann A, Leuchs B, Rommelaere J, Katus HA, Muller OJ (2017) Protocol for efficient generation and characterization of adeno-associated viral vectors. Hum Gene Ther Methods 28(5):235–246
47. Grimm D (2002) Production methods for gene transfer vectors based on adeno-associated virus serotypes. Methods 28 (2):146–157
48. Marion PL, Salazar FH, Liittschwager K, Bordier BB, Seegers C, Winters MA et al (2003) A transgenic mouse lineage useful for testing antivirals targeting hepatitis B virus. In: Schinazi RF, Sommadossi J-P, Rice CM (eds) Frontiers in viral hepatitis. Elsevier Science, Amsterdam, pp 197–202
49. Grimm D, Pandey K, Nakai H, Storm TA, Kay MA (2006) Liver transduction with recombinant adeno-associated virus is primarily restricted by capsid serotype not vector genotype. J Virol 80(1):426–439

Chapter 11

Generating DNA Expression Cassettes Encoding Multimeric Artificial MicroRNA Precursors

Fiona T. van den Berg, Abdullah Ely, and Patrick Arbuthnot

Abstract

RNA interference (RNAi) is a promising tool for the treatment of chronic viral infection, such as that caused by the hepatitis B virus (HBV). RNAi activators, including expressed primary microRNA (pri-miRNA) mimics, can effectively silence viral gene expression and thereby inhibit viral replication. Here we describe a protocol for the design, generation and functional assessment of cassettes encoding effective single and multimeric pri-miRNA mimics. Artificial miRNAs targeting viral genes can be identified in silico and used to design corresponding pri-miRNA mimics. A two-step generation and TA cloning protocol can be used to produce single mimics, while the strategic use of restriction sites enables concatenation of mimics in a sub-cloning protocol. Basic gene silencing function of pri-miRNA mimics in cell culture can then be assessed using a dual luciferase assay and appropriate minimal targets. The methods described here for the generation of effective pri-miRNA mimics targeting HBV can be applied in the silencing of other viral or endogenous genes.

Key words Primary microRNA, Artificial microRNA, RNAi, Gene therapy, Hepatitis B virus

1 Introduction

RNA interference (RNAi)-based gene therapy has gained attention for its use to control viral infections (*see* Chapters 1 and 2). The development of this approach is especially important for chronic viral infection, such as that caused by the hepatitis B virus (HBV). Upon infection of a hepatocyte by HBV, the viral genome is released into the nucleus where it forms covalently closed circular DNA (cccDNA) [1]. This replication intermediate acts as the template for the transcription of pre-genomic RNA (pgRNA) and protein-coding messenger RNAs (mRNAs), resulting in the production of new infectious virions. cccDNA is a stable replication intermediate, and its persistence in carriers of the virus remains an obstacle to curing infection [2]. In an alternative approach, RNAi activators can be used to silence viral gene expression from cccDNA and inhibit viral replication. Sustained expression of artificial

Mouldy Sioud (ed.), *RNA Interference and CRISPR Technologies: Technical Advances and New Therapeutic Opportunities*, Methods in Molecular Biology, vol. 2115, https://doi.org/10.1007/978-1-0716-0290-4_11,

antiviral RNAi activators may therefore provide durable suppression of viral replication as part of a gene therapy treatment [3].

Various RNAi activators have been widely investigated for therapeutic applications, and small interfering RNAs (siRNAs) and short hairpin RNAs (shRNAs) are now common tools in molecular biology. However, artificial miRNAs derived from larger precursor microRNA (pre-miRNA) or primary microRNA (pri-miRNA) mimics have emerged as effective and safe RNA activators in applications requiring sustained gene silencing. RNAi activators rely on components of the endogenous mammalian miRNA biogenesis pathway [4, 5] to mediate potent and specific posttranscriptional silencing of target genes. Long-term, high-level expression of shRNAs expressed from a polymerase III (pol III) promoter led to dose-dependent liver injury and death in mice, which was attributed to saturation of RNAi pathway components [6, 7]. This effect can be avoided through the controlled expression of artificial miRNAs derived from pri-miRNA mimics [8–10]. The use of a pol II promoter with inducible or tissue-specific expression also provides more versatile expression of pri-miRNA mimics [11–14]. Several pri-miRNA mimics may also be expressed simultaneously as a polycistronic construct to enable simultaneous silencing of multiple viral targets in a combinatorial approach (Fig. 1) [15–18].

In this chapter, we describe a protocol for the design, cloning and testing of single and multimeric pri-miRNA mimics using HBV as model for gene silencing [14, 18–20]. Artificial miRNAs can be

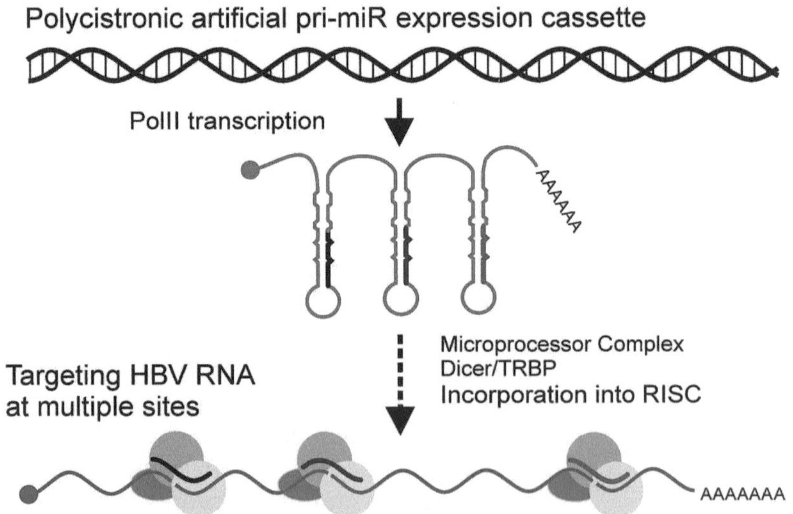

Fig. 1 Multimeric pri-miRNA mimics expressed from a polycistronic pol II expression cassette. Multimeric pri-miRNA mimics are processed in the endogenous mammalian miRNA biogenesis pathway to produce several artificial miRNAs. MiRNAs then direct silencing machinery to the respective viral targets, thus silencing multiple viral genes simultaneously

selected using freely available online design tools and used to replace the miRNA of a naturally occurring pri-miRNA sequence to create a mimic or artificial pri-miRNA (apri-miRNA). miRNA precursors of miR-30a, miR-31 and miR-122 have been well-characterized for this purpose [18, 19, 21, 22]. Pri-miRNA mimic constructs can be easily generated using a two-step approach and inserted into a TA cloning vector. Multimeric mimics may be generated using restriction sites that allow for mimic concatenation [18] and then inserted into mammalian expression vectors for functional assessment. Basic RNAi function can then be assessed using a dual luciferase reporter assay with a minimal viral target. Further experiments may also be required to characterize the function of expression cassettes. Such investigations may include Northern blot hybridization to verify processing of the artificial pri-miRNAs or an antigen assay to determine inhibition of viral replication. The methods described here can be applied to the generation of pri-miRNA mimics targeting other viral or endogenous genes.

2 Materials

2.1 Artificial miRNAs and pri-miRNAs

1. Computer with Internet connection and web browser.
2. Word processing software.
3. Graphic design software (optional).

2.2 Primer Extension and PCR

1. DNA oligonucleotides.
2. *Taq* DNA polymerase Master Mix.
3. PCR-grade water.
4. Thermal Cycler.
5. PCR tubes (0.2 ml).

2.3 Plasmid Ligation, Purification, and Restriction Digest

1. TA cloning vector.
2. Mammalian expression plasmid.
3. psiCheck™-2 (Promega).
4. T4 DNA ligase and ligation buffer.
5. Chemically competent *E. coli* cells.
6. LB growth medium: 1% w/v tryptone, 0.5% w/v yeast extract, and 0.5% w/v sodium chloride.
7. LB agar: LB growth medium with 1.5% w/v agarose.
8. Ampicillin 1000× stock: 0.1 g/ml ampicillin sodium salt, 50% v/v ethanol, and 0.2 μm filter.
9. Bacterial Petri dishes (10 cm).

10. Bacterial culture tubes (10 ml).
11. Incubator with orbital shaker (37 °C).
12. Incubator (37 °C).
13. Plasmid Mini Kit.
14. Reaction tubes (1.7 ml).
15. Centrifuges (Eppendorf 5415R, rotor F-45-24-11; 5810R, rotor A-4-81).
16. Restriction enzymes and reaction buffers.
17. DNA sequencing facility.
18. Nano-volume spectrophotometer.

2.4 Agarose Gel Electrophoresis

1. Agarose.
2. 1× TAE buffer: 40 mM Tris base, 20 mM glacial acetic acid, 1 mM EDTA, pH 8.5.
3. Ethidium bromide (10 mg/ml).
4. Molecular weight marker.
5. 6× DNA loading dye.
6. Gel electrophoresis equipment (gel chambers, casting trays, combs).
7. Power supply.
8. Gel imaging system.
9. UV transilluminator.
10. Gel Extraction Kit.

2.5 Cell Culture

1. Adherent cell lines (HEK293, HeLa, Huh-7).
2. Complete cell culture medium: Dulbecco's modified Eagle medium (DMEM), 10% fetal bovine serum (FBS), 100 U/ml penicillin, and 0.1 mg/ml streptomycin.
3. Antibiotic 1000× stock: 100,000 U/ml penicillin and 100 mg/ml streptomycin.
4. Phosphate buffered saline (PBS).
5. TrypLE Express dissociation reagent.
6. CO_2 incubator (37 °C, 5% CO_2).

2.6 Dual Luciferase Gene Silencing Assay

1. Plasmid Maxi Kit.
2. GFP reporter plasmid (pCI-eGFP [23]).
3. Transfection Reagent, Lipofectamine® 2000.
4. Opti-MEM® I reduced serum medium.
5. Fluorescence microscope.

6. Dual-Luciferase® Reporter Assay System: passive lysis buffer, LAR II reagent, and Stop & Glo® Reagent.
7. Dual injection Luminometer (Veritas™).

3 Methods

3.1 Selection of Artificial miRNAs

1. Identify the desired target region in the HBV genome. The target sequence should be present in the infectious molecular clone and/or relevant viral isolates.
2. Use a free online siRNA design tool to identify siRNA candidate sequences complementary to the target. Copy and paste the target sequence into the design tool, such as the Oligowalk Web Server for siRNA Design [24] or siDirect [25] (*see* **Note 1**). Set the siRNA sequence length to 21 nt.
3. Identify the best siRNA candidates for use as artificial miRNAs by refining the list of siRNA output sequences. Use the following selection criteria to exclude sequences with (1) no 5′ U at position 1, (2) a G or C at position 2, (3) a U or A in positions 20–21, and (4) a run ≥4 for any single base (*see* **Note 2**).

3.2 Design of Artificial pri-miRNAs (apri-miRNAs)

1. Select a naturally occurring miRNA precursor (e.g. pre miR-30a, pre-miR-31 and pre-miR-122) (*see* **Note 3**).
2. Use sequence annotations for the pre-miRNA and mature miRNA available on the miRBase database [26] (e.g. hsa-mir-31) (*see* **Note 1**). Download the stem loop RNA sequence.
3. Extend the sequence by 50 nt on either side of the expected Drosha-DGCR8 cleavage site to create a longer pri-miRNA sequence. Use the genomic coordinates provided on miRbase and a genome browser such as Ensembl [27].
4. Replace the predominant miRNA sequence of the pre-miRNA stem loop with the artificial miRNA. The complementary sequence of the stem must also be replaced to preserve base pairs and mismatches as far as possible (*see* **Note 4**).
5. Predict the RNA structure of the new apri-miRNA sequences using the mfold web server [28]. Copy and paste the sequences into the RNA Folding Form and use default parameters. Check that folding features of the original and artificial pri-miRNA are preserved.

3.3 Two-Step Generation of Sequence Encoding an apri-miRNA

1. For the first reaction (Fig. 2), design and order a pair of partially complementary ~60 b DNA oligonucleotides (one sense, one antisense) to generate the central ~100 bp region of the DNA sequence for the apri-miRNA. The oligonucleotides should have an overlap of about 20 bp.

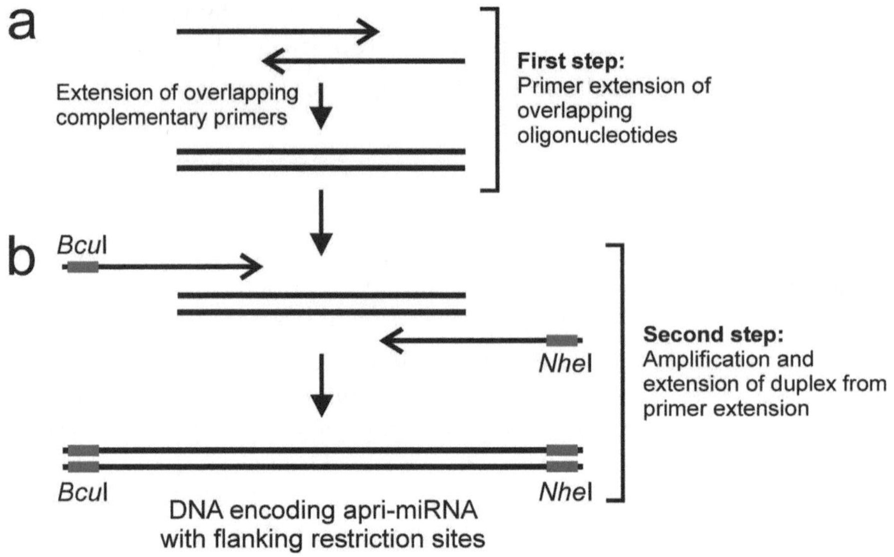

Fig. 2 Two-step reaction for the generation of pri-miRNA mimic constructs. (**a**) In the first step, overlapping complementary primers are used to generate the central ~100 bp of the DNA sequence. (**b**) In the second step, the product of the first reaction is used as a PCR template and extended using a second pair of primers with 5′ restriction sites. The final product is the complete DNA sequence encoding the pri-miRNA mimic with flanking restriction sites

2. Use 1 μl of each oligonucleotide (10 μM) in a 20 μl reaction using a *Taq* DNA polymerase Master Mix. Set the PCR program for primer extension as follows: denaturation at 94 °C for 5 min; 30 cycles: 94 °C for 10 s, annealing temperature for 30 s; 72 °C for 40 s; final extension at 72 °C for 7 min.
3. For the second step, a PCR, design and order another pair of oligonucleotides (one sense, one antisense) partially complementary to the product of the first reaction. These oligonucleotides should extend the ~100 bp first-round product to the full DNA sequence. Unique, but compatible, 5′ and 3′ restriction sites (e.g., 5′ *Nhe*I site and a 3′ *Bcu*I site) must be included for cloning (*see* **Note 5**).
4. Dilute the primer extension template from the first reaction 100-fold, and use 1 μl as the template in the second PCR. Use the same reaction volumes and PCR program as in **step 2** above.

3.4 Agarose Gel Electrophoresis

1. Analyze the size of PCR products using agarose gel electrophoresis. Prepare DNA samples: 5 μl of each PCR, 10 μl sterile water, 3 μl 6× DNA loading dye.
2. Prepare a gel (1.5% w/v agarose, 1× TAE buffer, 35 μg/100 ml ethidium bromide) (*see* **Note 6**).

3. Load DNA samples into the wells. Also load a DNA ladder (800 ng) in a separate well.
4. Subject the gel to electrophoresis (80 V, 45 min) in a 1× TAE buffer.
5. View the agarose gel using UV light and a gel imaging system.

3.5 Clone Single pri-miRNA Construct

1. Insert the PCR product containing the full pri-miRNA mimic sequence into an appropriate TA cloning plasmid, such as pTZ57R/T (Thermo Fisher Scientific). Perform a ligation reaction with T4 DNA ligase according to the manufacturer's instructions using 2 μl of a 1–20× dilution of the amplicon-containing PCR mix.
2. Use 10 μl of the ligation reaction to transform chemically competent *E. coli* cells, and incubate overnight on LB-agar growth medium with an appropriate antibiotic (e.g. 0.1 mg/ml ampicillin).
3. Select a single colony, inoculate 5 ml LB growth medium (0.1 mg/ml ampicillin), incubate overnight at 37 °C, and then purify the plasmid (Plasmid Mini Kit).
4. Verify the size of the DNA insert by restriction enzyme digestion (1 μg plasmid DNA), followed by agarose gel electrophoresis (1% w/v agarose gel).
5. Determine the DNA sequence of the insert by Sanger sequencing.

3.6 Clone Multimeric pri-miRNA Construct

1. To create a multimeric construct, perform a restriction digest using the plasmid with the apri-miRNA sequence to be in the first position. Use the 3′ restriction site of the insert (3′ *Bcu*I) and a nearby 3′ site (e.g. *Eco*RI) in the plasmid multiple cloning site (MCS). This will produce a linearized plasmid with the first apri-miRNA sequence (Fig. 3).
2. Perform a restriction digest on the plasmid with the apri-miRNA sequence to be in the second position. Use the 5′ site of the insert (*Nhe*I) and the same 3′ site of the plasmid MCS (*Eco*RI) as in step 1. This will produce the second apri-miRNA sequence.
3. Purify both digest fragments using agarose gel electrophoresis followed by gel extraction. View and excise the relevant DNA bands using a UV transilluminator.
4. Determine the concentration of the purified DNA by measuring absorbance at 260 nm using a nano-volume spectrophotometer.
5. To create a plasmid with the multimeric construct, perform a ligation reaction of the purified fragments using T4 DNA ligase according to the manufacturer's instructions.

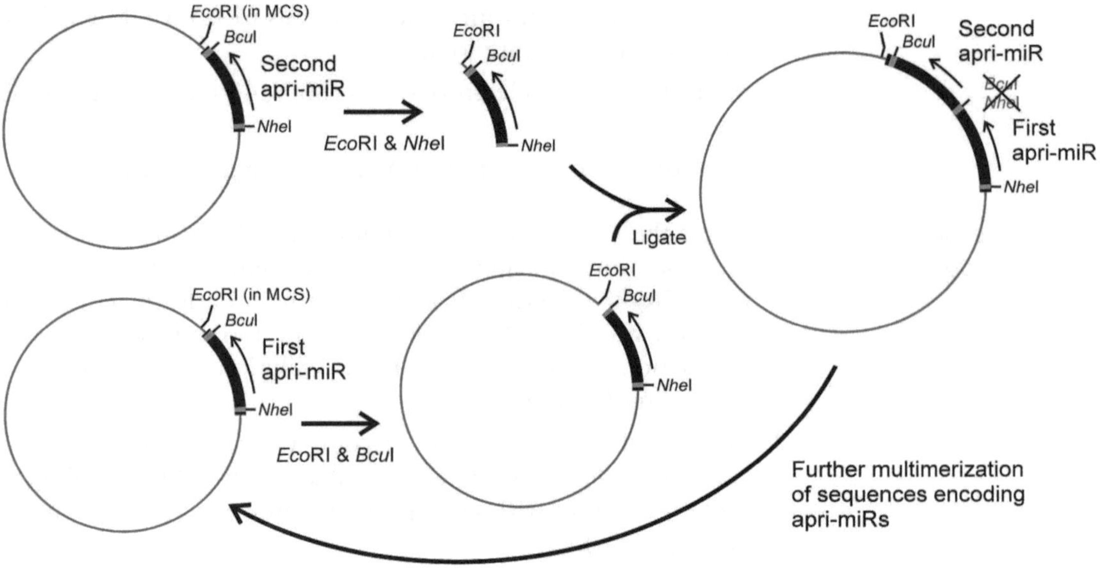

Fig. 3 Cloning strategy to generate multimeric pri-miRNA mimic expression cassettes. A restriction digest (*Eco*RI & *Bcu*I) is used to linearize the plasmid encoding the first apri-miRNA. Another restriction digest (*Eco*RI & *Nhe*I) is used to excise the second apri-miRNA. Ligation of these digest products produces a plasmid encoding two apri-miRNAs. The unique but complementary restriction sites flanking the apri-miRNAs (3′ *Bcu*I of first apri-miRNA; 5′ *Nhe*I of second apri-miRNA) enable ligation of the two pri-miRNAs with destruction of the original sites. This enables further a pri-miRNA multimerization using the same cloning strategy

6. Transform chemically competent *E. coli* cells, culture overnight in 5 ml of LB (0.1 mg/ml ampicillin), purify the plasmid, and verify the construct as in Subheading 3.5, **steps 2–4**.
7. To add a third apri-miRNA sequence, repeat **steps 1–6** using the plasmid with the dimeric apri-miRNA construct for **step 1** (Fig. 3).

3.7 Insert Constructs into Mammalian Expression Plasmid

1. Select an appropriate mammalian expression plasmid with a pol II promoter, for example, pCI-neo (Promega). Ensure that the expression plasmid has restriction sites in the MCS to enable directional cloning.
2. Perform a restriction digest using compatible enzymes for the insert (TA cloning plasmid construct) and the expression plasmid.
3. Purify the digest fragments by gel extraction and quantify using nano-volume spectrophotometer.
4. Set up a ligation reaction using a ratio of 3 pmol of insert to 1 pmol of plasmid (*see* **Note 7**).
5. Transform chemically competent *E. coli* cells, culture overnight in 5 ml of LB with appropriate antibiotic, purify the plasmid, and verify the construct as in Subheading 3.5, **steps 2–4**.

3.8 Insert Target Sequence into Dual Luciferase Target Plasmid

1. Obtain the minimal target sequence complementary to the artificial miRNA.
2. Order single-stranded DNA oligonucleotides constituting each strand of the minimal target duplex. Include terminal 5′ and 3′ restriction site overhangs (e.g. *Xho*I and *Not*I) to enable direct cloning into the MCS of psiCheck™-2 (Promega) (*see* **Note 8**). Also include a restriction site (e.g. *Eco*RV) just before the 5′ or 3′ terminal restriction site overhang to assist in the verification of the cloned target plasmid (**step** 7) (*see* **Note 9**).
3. To anneal the two oligonucleotides of each target duplex, add 2 μl of a 100 μM stock of each oligonucleotide in 20 μl of sterile water. Incubate at 80 °C for 5 min in a heating block. Disconnect heating block from the power source and allow to cool to room temperature gradually.
4. Perform restriction digestion with appropriate enzymes to linearize psiCheck™-2.
5. Purify the linearized plasmid by agarose gel electrophoresis followed by gel extraction. Quantify the purified DNA using a nano-volume spectrophotometer.
6. Insert the target duplex into the linearized psiCheck™-2 using a T4 DNA ligase reaction and a ratio of 3 pmol of insert to 1 pmol of plasmid (*see* **Note** 7).
7. Transform chemically competent *E. coli* cells, culture overnight in 5 ml of LB with appropriate antibiotic, purify the plasmid, and verify the construct using a restriction enzyme digest (*Eco*RV) as in Subheading 3.5, **steps 2–4**.

3.9 Dual Luciferase Gene Silencing Assay

1. Purify all required plasmids using a large-scale purification protocol (Plasmid Maxi Kit). Include the original mammalian expression plasmid without a pri-miRNA expression cassette as a mock control. Also, include a reporter plasmid as a marker of transfection efficiency (e.g. pCI-eGFP) (*see* **Note 10**).
2. Culture a common human monolayer cell line suitable for transfection such as HEK293.
3. At 24–48 h before transfection, seed cells at 50% confluency (~0.95×10^5 cells) in a 24-well tissue culture dish in 1 ml of antibiotic-free medium (*see* **Note 11**).
4. Reduce the well medium to 500 μl before transfection. Cells should be ~80% confluent and evenly dispersed at the time of transfection.
5. Transfect cells with 1 μg total plasmid DNA (750 ng apri-miRNA expression plasmid, 150 ng dual luciferase target plasmid, 100 ng GFP reporter) per well in triplicate. Use 1 μl Lipofectamine® 2000 Reagent (Thermo Fisher Scientific) and 100 μl Opti-MEM® per 1 μg DNA according to the

manufacturer's instructions. Include three control wells each transfected with 1 μg pCI-eGFP as an indicator of total transfection efficiency.

6. At 16–24 h post-transfection, if required, gently replace the medium in each well with 1 ml of fresh medium using a micropipette.
7. At 48 h post-transfection, check GFP expression using fluorescence microscopy and save the images. Only proceed with the assay if reporter expression is satisfactory and cells are adherent (*see* **Note 12**).
8. Perform the Dual-Luciferase® Reporter Assay (Promega) 48 h after transfection according to the manufacturer's protocol. Thaw and dilute the assay reagents prior to cell lysis (*see* **Notes 13** and **14**).
9. Remove media from the cells and add 100 μl passive lysis buffer per well. Incubate at room temperature for 15 min, mix sample by pipetting, and transfer 10 μl of each sample to a 96-well white or black polystyrene assay plate.
10. Perform the assay measurements using a dual injection Veritas™ Luminometer (Promega) (*see* **Note 15**).
11. Calculate the average ratio of *Renilla* luciferase signal to Firefly luciferase signal (h*R*luc:hFluc). Results can be normalized relative to the mock control sample (*see* **Note 16**).

4 Notes

1. Current website addresses: OligoWalk Web Server (http://rna.urmc.rochester.edu/cgi-bin/server_exe/oligowalk/oligowalk_form.cgi); siDirect version 2.0 (http://sidirect2.rnai.jp/); miRbase (http://www.mirbase.org/); Ensembl genome browser (https://www.ensembl.org/index.html); mfold web server (http://unafold.rna.albany.edu/?q=mfold).
2. These criteria are based on generally favorable features for effective miRNAs. The strand of the miRNA duplex with lower 5′ stability is typically selected as the mature miRNA [29, 30], while the RNAi silencing complex has a preference for miRNAs with a 5′ U or A [31]. miRNA sequences may still be effective without adhering to all the criteria listed.
3. Other naturally occurring miRNA precursors can be used [32], but only one mature miRNA should be predominant. Fully artificial pri-miRNA sequences and processing motifs may also be used, but function must be determined empirically [33, 34].
4. Word processing or presentation software may be used for this step, but graphic design software is especially useful.

5. Restriction sites of the apri-miRNA construct should be chosen with care for the cloning strategy. Terminal 5′ and 3′ restriction sites should be unique, but with compatible cohesive ends to allow for ligation, with destruction of the original restriction sites, in the creation of multimeric constructs. Restriction sites of the PCR product should not be present in the TA cloning plasmid. Restriction sites of the PCR product and TA cloning plasmid should also allow for directional cloning into the final mammalian expression plasmid.
6. Ethidium bromide is an intercalating agent and known mutagen. Wear nitrile gloves when handling all gels and equipment. Add ethidium bromide to the gel in the fume hood. Dispose of gels and buffer in accordance with institutional regulations.
7. If ligation is unsuccessful at a ratio of 3 pmol of insert to 1 pmol of plasmid, change the ratio in the range from 1 to 1 up to 7 to 1.
8. psiCheck™-2 contains genes for both Firefly luciferase and *Renilla* luciferase. The MCS is located within the 3′ UTR of the *Renilla* luciferase gene.
9. Choose a restriction site to include in the target insert that is also present in psiCheck™-2 (e.g. *Eco*RV). Together, these sites should give digest products suitable for verification of a plasmid containing an insert.
10. Final plasmid DNA concentrations following purification should be at ~1 μg/μl. Prepare a further ~400 ng/μl solution for dual luciferase target plasmids and a ~200 ng/μl solution for the GFP reporter plasmid.
11. Resuspend cells in a single final volume before seeding 24-well plates. Gently rock the cell suspension after adding cells to each row of a plate. Gently rock the plate four times back and forth and four times left to right, and repeat before placing in the incubator for even cell coverage.
12. Transfection efficiency will vary for different cell lines and should be established prior to the assay using a reporter plasmid.
13. The Dual-Luciferase® Reporter Assay contains components for three reagents: passive lysis buffer (cell lysis), LAR II reagent (beetle luciferin substrate for Firefly luciferase), and Stop & Glo® reagent (coelenterazine substrate for *Renilla* luciferase).
14. To ensure sufficient volumes for the LAR II and Stop & Glo® reagents, make a mastermix for (number of assay samples +3) and add 500 μl for priming of the luminometer.
15. The luminometer assay program must be set to first add the LARII reagent and measure the Firefly luciferase signal and then, second, to add the Stop & Glo® reagent and measure the

Renilla luciferase signal. The program can be adjusted to utilize 50 μl of both the LAR II and Stop & Glo® reagents.

16. Efficient pri-miRNA mimics have target silencing of 80% or more relative to the mock sample. Lower efficiency may be observed for certain mimics in a multimeric construct.

References

1. Seeger C, Mason WS (2015) Molecular biology of hepatitis B virus infection. Virology 479-480:672–686
2. Wang YJ, Yang L, Zuo JP (2016) Recent developments in antivirals against hepatitis B virus. Virus Res 213:205–213
3. Valdmanis PN, Kay MA (2017) Future of rAAV gene therapy: platform for RNAi, gene editing, and beyond. Hum Gene Ther 28:361–372
4. Ha M, Kim VN (2014) Regulation of microRNA biogenesis. Nat Rev Mol Cell Biol 15:509–524
5. Bartel DP (2018) Metazoan MicroRNAs. Cell 173:20–51
6. Grimm D, Streetz KL, Jopling CL, Storm TA, Pandey K, Davis CR, Marion P, Salazar F, Kay MA (2006) Fatality in mice due to oversaturation of cellular microRNA/short hairpin RNA pathways. Nature 441:537–541
7. Grimm D, Wang L, Lee JS, Schurmann N, Gu S, Borner K, Storm TA, Kay MA (2010) Argonaute proteins are key determinants of RNAi efficacy, toxicity, and persistence in the adult mouse liver. J Clin Invest 120:3106–3119
8. Castanotto D, Sakurai K, Lingeman R, Li H, Shively L, Aagaard L, Soifer H, Gatignol A, Riggs A, Rossi JJ (2007) Combinatorial delivery of small interfering RNAs reduces RNAi efficacy by selective incorporation into RISC. Nucleic Acids Res 35:5154–5164
9. McBride JL, Boudreau RL, Harper SQ, Staber PD, Monteys AM, Martins I, Gilmore BL, Burstein H, Peluso RW, Polisky B et al (2008) Artificial miRNAs mitigate shRNA-mediated toxicity in the brain: implications for the therapeutic development of RNAi. Proc Natl Acad Sci U S A 105:5868–5873
10. Boudreau RL, Martins I, Davidson BL (2009) Artificial MicroRNAs as siRNA shuttles: improved safety as compared to shRNAs in vitro and in vivo. Mol Ther 17:169–175
11. Zeng Y, Cai X, Cullen BR (2005) Use of RNA polymerase II to transcribe artificial microRNAs. Methods Enzymol 392:371–380
12. Zhou H, Xia XG, Xu Z (2005) An RNA polymerase II construct synthesizes short-hairpin RNA with a quantitative indicator and mediates highly efficient RNAi. Nucleic Acids Res 33:e62
13. Stegmeier F, Hu G, Rickles RJ, Hannon GJ, Elledge SJ (2005) A lentiviral microRNA-based system for single-copy polymerase II-regulated RNA interference in mammalian cells. Proc Natl Acad Sci U S A 102:13212–13217
14. Mowa MB, Crowther C, Ely A, Arbuthnot P (2014) Inhibition of hepatitis B virus replication by helper dependent adenoviral vectors expressing artificial anti-HBV pri-miRs from a liver-specific promoter. Biomed Res Int 2014:718743
15. Chung KH, Hart CC, Al-Bassam S, Avery A, Taylor J, Patel PD, Vojtek AB, Turner DL (2006) Polycistronic RNA polymerase II expression vectors for RNA interference based on BIC/miR-155. Nucleic Acids Res 34:e53
16. Liu YP, Haasnoot J, ter Brake O, Berkhout B, Konstantinova P (2008) Inhibition of HIV-1 by multiple siRNAs expressed from a single microRNA polycistron. Nulceic Acids Res 36:2811–2824
17. Aagaard LA, Zhang J, von Eije KJ, Li H, Sætrom P, Amarzguioui M, Rossi JJ (2008) Engineering and optimization of the miR-106b cluster for ectopic expression of multiplexed anti-HIV RNAs. Gene Ther 15:1–14
18. Ely A, Naidoo T, Arbuthnot P (2009) Efficient silencing of gene expression with modular trimeric Pol II expression cassettes comprising microRNA shuttles. Nucleic Acids Res 37:e91
19. Ely A, Naidoo T, Mufamadi S, Crowther C, Arbuthnot P (2008) Expressed anti-HBV primary microRNA shuttles inhibit viral replication efficiently in vitro and in vivo. Mol Ther 16:1105–1112
20. Maepa MB, Ely A, Grayson W, Arbuthnot P (2017) Sustained inhibition of HBV replication in vivo after systemic injection of AAVs encoding artificial antiviral primary MicroRNAs. Mol Ther Nucleic Acids 7:190–199
21. Zeng Y, Wagner EJ, Cullen BR (2002) Both natural and designed micro RNAs can inhibit

the expression of cognate mRNAs when expressed in human cells. Mol Cell 9:1327–1333

22. Zeng Y, Cullen BR (2003) Sequence requirements for micro RNA processing and function in human cells. RNA 9:112–123
23. Passman M, Weinberg M, Kew M, Arbuthnot P (2000) In situ demonstration of inhibitory effects of hammerhead ribozymes that are targeted to the hepatitis Bx sequence in cultured cells. Biochem Biophys Res Commun 268:728–733
24. Lu ZJ, Mathews DH (2008) Efficient siRNA selection using hybridization thermodynamics. Nucleic Acids Res 36:640–647
25. Naito Y, Yoshimura J, Morishita S, Ui-Tei K (2009) siDirect 2.0: updated software for designing functional siRNA with reduced seed-dependent off-target effect. BMC Bioinformatics 10:392
26. Kozomara A, Birgaoanu M, Griffiths-Jones S (2019) miRBase: from microRNA sequences to function. Nucleic Acids Res 47:D155–D162
27. Zerbino DR, Achuthan P, Akanni W, Amode MR, Barrell D, Bhai J, Billis K, Cummins C, Gall A, Giron CG et al (2018) Ensembl 2018. Nucleic Acids Res 46:D754–D761
28. Zuker M (2003) Mfold web server for nucleic acid folding and hybridization prediction. Nucleic Acids Res 31:3406–3415
29. Khvorova A, Reynolds A, Jayasena SD (2003) Functional siRNAs and miRNAs exhibit strand bias. Cell 115:209–216
30. Schwarz DS, Hutvagner G, Du T, Xu Z, Aronin N, Zamore PD (2003) Asymmetry in the assembly of the RNAi enzyme complex. Cell 115:199–208
31. Frank F, Sonenberg N, Nagar B (2010) Structural basis for 5′-nucleotide base-specific recognition of guide RNA by human AGO2. Nature 465:818–822
32. van den Berg FT, Rossi JJ, Arbuthnot P, Weinberg MS (2016) Design of effective primary MicroRNA mimics with different basal stem conformations. Mol Ther Nucleic Acids 5: e278
33. Auyeung VC, Ulitsky I, McGeary SE, Bartel DP (2013) Beyond secondary structure: primary-sequence determinants license pri-miRNA hairpins for processing. Cell 152:844–858
34. Fang W, Bartel DP (2015) The menu of features that define primary MicroRNAs and enable de novo design of MicroRNA genes. Mol Cell 60:131–145

Chapter 12

MiR-302-Mediated Somatic Cell Reprogramming and Method for Generating Tumor-Free iPS Cells Using miR-302

Shi-Lung Lin, Jack S. Chen, and Shao-Yao Ying

Abstract

Human induced pluripotent stem cells (iPSCs) by four factors have the risks of teratoma formation and potential tumorigenicity. To overcome this major hurdle, we examined the mechanism(s) by which the cell cycle genes of embryonic cells were regulated. Naturally occurring embryonic stem cells (ESCs) possess two unique stemness properties: pluripotent differentiation into all cell types and self-renewal with no risk of tumor formation. Despite overwhelming reports describing iPSC pluripotency, there have been no observations of tumor prevention mechanism that suppresses tumor formation similar to that in naturally occurring ESCs. The ESC-specific microRNA (miRNA), miR-302, regulates human iPSC tumorigenicity through co-suppression of both cyclin E-CDK2 and cyclin D-CDK4/6 cell cycle pathways during G1-S phase transition. MiR-302 also silenced BMI-1, a cancer stem cell marker gene, to promote the expression of two senescence-associated tumor suppressor genes, p16Ink4a and p14/p19Arf. Together, the combinatory effect of reducing G1-S cell cycle transition and increasing p16/p14(p19) expression resulted in a relatively attenuated cell cycle rate similar to that of 2-to-8-cell-stage embryonic cells in early mammalian zygotes (20–24 h/cycle), as compared to the fast proliferation rate of iPSCs induced by four defined factors Oct4-Sox2-Klf4-c-Myc (12–16 h/cycle). In addition to the prevention of stem cell tumorigenicity, the mechanism underlying miR-302-mediated iPSCs also includes the initiation of global genomic DNA methylation, activation of ESC-specific gene expression, and inhibition of developmental signaling. Overall, we have established an effective protocol to express the intronic miR-302 cluster, according to its own natural biogenesis mechanism to generate tumor-free iPSCs for use in biology and therapy.

Key words miR-302, microRNA (miRNA), RNAi, iPSC, ESC, Stem cell, Cell cycle, CDK2, CDK4/6, Cyclin D, BMI-1, p16Ink4a, p14/p19Arf, p21Cip1/Waf1, CDKN1A, RNAi, DND demethylation

1 Introduction

The induction of human pluripotent stem cells (iPSCs)[1] has advanced rapidly, and stem cell products are now developing for therapeutic applications. However, questions remain regarding the tumorigenic potential of such cells. MicroRNAs (miRNAs) are short single-stranded noncoding RNAs, approximately 20–25 nucleotide long, which represent a class of small regulatory RNAs. By inhibiting the translation of target mRNAs via RNA

Mouldy Sioud (ed.), *RNA Interference and CRISPR Technologies: Technical Advances and New Therapeutic Opportunities*, Methods in Molecular Biology, vol. 2115, https://doi.org/10.1007/978-1-0716-0290-4_12,

interference (RNAi), miRNAs regulate gene expression posttranscriptionally and therefore play an important role in a wide range of cellular processes [2]. The embryonic stem (ES) cell-specific intronic miRNA, miR-302, has been used to reprogram both human normal and skin cancer cells into a hESC-like state [3].

The biogenesis of an intergenic miRNA starts with the synthesis of a primary miRNA transcript (pri-miRNA), either catalyzed by type II or III RNA polymerase (Pol-II/III) in the case of intergenic miRNAs or deriving from introns that are removed from premature messenger RNAs following DNA transcription in the case of intronic miRNAs. Pri-miRNAs are processed in the nucleus by the ribonuclease Drosha into a miRNA precursor (pre-miRNA) approximately 60-nt in length (*see* Chapter 1). After being transported into the cytoplasm, these pre-miRNAs are further processed into mature and functional miRNAs by the cytoplasmic ribonuclease Dicer [4]. Mature and functional miRNAs subsequently associate with a number of proteins to form the RNA-induced silencing complex (RISC) that bind with target mRNAs having total or partial complementary sequences to the miRNAs and initiate the inhibition of subsequent protein translation via RNA interference (RNAi) [5]. Thus far, miRNAs have been shown to play an instrumental role in a wide variety of biological processes, including cell cycle regulation, apoptosis, control of metabolic pathways, and induction of somatic cell reprogramming [6].

Stem cells possess a unique ability to differentiate into nearly all types of cells in the body, and therefore they provide significant potentials for cell-based therapies to restore tissues or organs damaged due to injuries, degenerative diseases, cancer, and aging. Somatic cell reprogramming (SCR) is an epigenetic process essential for generating induced pluripotent stem cells (iPSCs) [7]. Deciphering the underlying mechanisms of SCR and developing methodologies to reprogram adult differentiated cells to a pluripotent state not only circumvent certain ethical and political controversies surrounding the use of human embryonic stem cells (hESCs) but also offer a possible solution for generating patient-specific pluripotent stem cells.

In 2006, Takahashi and Yamanaka developed a method for generating iPSCs, using retroviral delivery of four defined transcription factors (Oct4, Sox2, Klf4, and c-Myc; also known as OSKM) in vitro [8]. Subsequently, Yu and coworkers developed another SCR method using a different set of four defined factors (Oct4, Sox2, Nanog, and LIN28; also known as OSNL) [9]. Then, iPSC-based SCNT therapy has been used to treat sickle cell anemia in a transgenic mouse model [10]. Despite these breakthroughs, the mechanisms underlying the four factor-induced iPSC transformation remain unclear, and a number of concerns remain. First, oncogenic c-Myc is required to boost the success rate of iPSC formation, and the success rate is less than 0.002% when c-Myc is omitted. Second, the iPSC population is heterogeneous due to the variable delivery efficiency of the four factors. As a matter of fact, the best stoichiometric combination of the four factors for reprogramming is still unidentified. Lastly, the combined effect of the four factors seems to maintain only the intrinsic network of embryonic gene expression [11, 12], whereas the effect that leads to a global cancellation of somatic gene expression patterns remains largely unknown. Taken together, the efficiency and safety of the four factor-induced iPSCs is questionable.

The process of SCR is initiated by global DNA demethylation, which erases most of the genomic methylation sites and resets the cellular gene expression pattern to an embryonic stem cell (ESC)-like profile. Our previous studies have shown, for the first time, that a group of noncoding RNAs (ncRNAs) rather than proteins are responsible for initiating global DNA demethylation in human iPSCs [3, 13]. These ncRNAs belong to the ESC-specific miRNA family, miR-302, that is most abundantly expressed in human ESCs and iPSCs but not in differentiated somatic cells [14–16]. Native miR-302 consists of four sense homologues (miR-302b, c, a, and d) and three antisense members (miR-302b*, c*, and a*), all of which are concurrently transcribed as a polycistronic RNA cluster along with another miRNA, miR-367 [14]. As shown in Fig. 1, miR-302 suppresses the translation of its target genes through complementary binding and formation of RISCs in the 3′-untranslated regions (3′-UTRs) on its target gene transcripts [17, 18]. Due to this unique function in gene regulation, miR-302, by replacing all four of the previously defined reprogramming factors (either OSKM or OSNL), plays a critical role in silencing lysine-specific histone demethylase 1/2 (LSD1/2, AOF2/1, or KDM1/1B) and DNA (cytosine-5-)-methyltransferase 1 (DNMT1) to induce global DNA demethylation and stimulating cellular Oct4-Sox2-Nanog expression to promote the reprogramming of somatic cells to ESC-like iPSCs [13].

In addition to miR-302-mediated SCR mechanism, we further identified a parallel tumor suppression mechanism, in which miR-302 attenuated both cyclin E-cyclin-dependent kinase 2 (CDK2) and cyclin D-CDK4/6 cell cycle pathways to inhibit iPSC tumorigenicity [19]. This novel property provides a breakthrough advantage in the preparation of tumor-free iPSCs for clinical trials and therapies because tumorigenicity is still one of the major obstacles hindering stem cell applications. Based on the dual function of miR-302 in SCR induction and tumor suppression, we have successfully developed and tested an inducible miR-302 expression system for generating tumor-free iPSCs derived from various human normal and cancerous cells [3, 13, 19]. To distinguish between the miRNA-mediated induced pluripotent stem cells from the previously described iPSCs induced by four defined protein factors (Oct4, Sox2, Klf4, and c-Myc), we referred to miR-302-induced tumor-free iPSCs as mirPSCs.

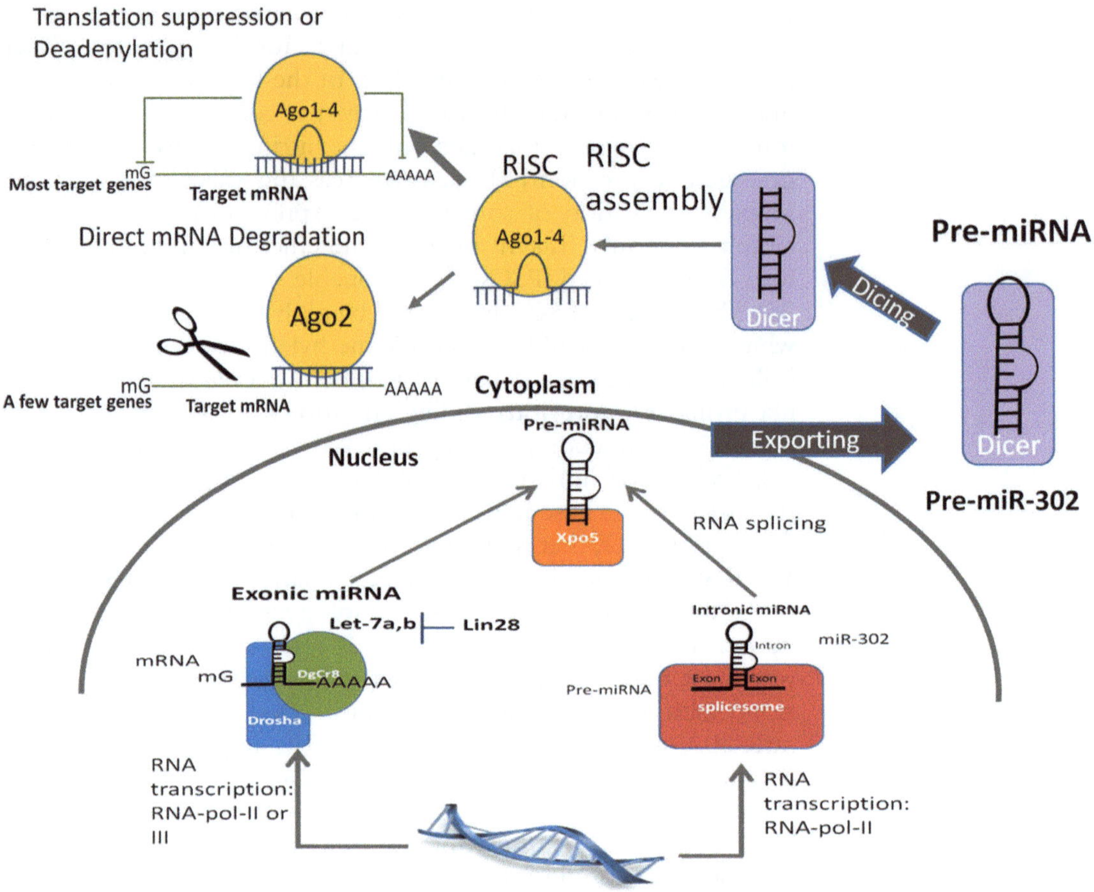

Fig. 1 The natural biogenesis mechanism of the intronic microRNA miR-302. MiR-302 is a natural intronic microRNA (miRNA), encoded in the intronic region of the La ribonucleoprotein domain family member 7 (LARP7 or PIP7S) gene. During miR-302 biogenesis, primary miRNA precursors (pri-miRNA) are first transcribed by type II RNA polymerases (Pol-II) along with the LARP7 gene transcripts and spliced by cellular spliceosomes to form hairpin-like miRNA precursors (pre-miRNA), which are then exported out of the nucleus by Exportin-5 (Xpo5) and further processed by Dicer-like RNaseIII endoribonucleases in cytoplasm to form mature miRNAs. Consequently, mature miR-302 molecules are assembled into RNA-induced silencing complexes (RISC) with argonaute proteins (Ago1-4) that elicit gene-specific silencing

Each individual mirPSC could grow into a colony or embryoid body with a relatively slow cell cycle rate (20–24 h/cycle). Due to miR-302-mediated cell cycle attenuation, mirPSCs often presented a quiescent cell-like morphology and form three-dimensional sphere colonies similar to that of an early zygote before the morula stage (Fig. 2a). The mirPSCs strongly and homogeneously expressed Oct4, Sox2, Nanog, Lin28, and many other major hESC markers (Fig. 2b, c), resembling the gene expression pattern of hESCs, as determined by immunocytochemical staining and Western blotting analyses. The transcriptome of mirPSCs contained an average of >92% similarity to hESCs as determined

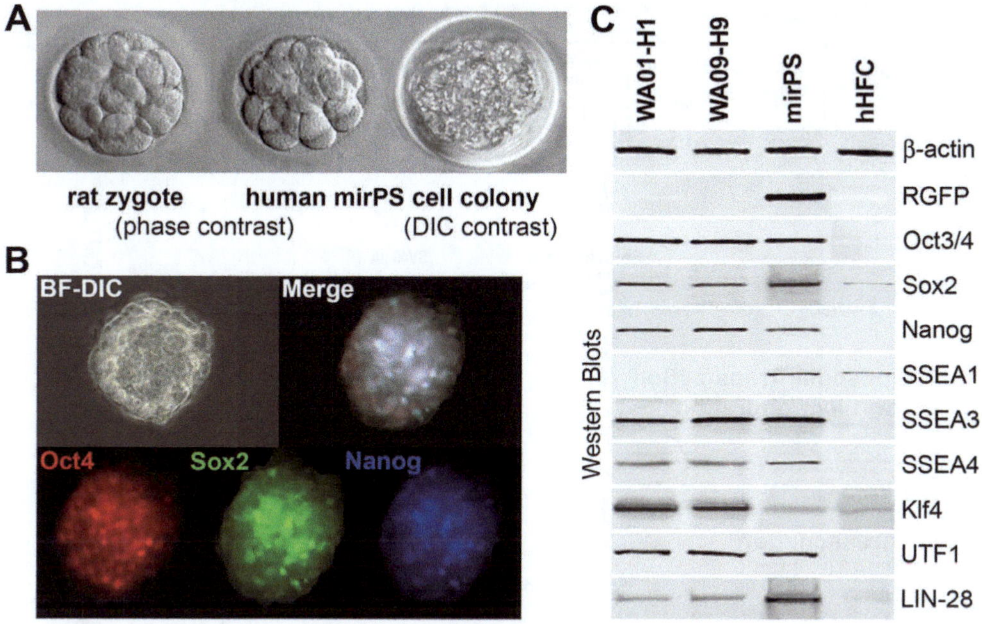

Fig. 2 Morphological and genetic properties of mirPSCs. (**a**) A morphological comparison between a morula-staged rat embryo and a mirPSC colony at 16–32-cell stage. BF-DIC, bright field with differential interference contrast. (**b**) Fluorescence microscope examination showing the homogeneous expression of the core reprogramming factors Oct3/4, Sox2, and Nanog in a mirPSC-derived embryoid body. (**c**) Western blots confirming the expression patterns of major hESC-specific markers in mirPSCs compared to those found in hESCs H1 and H9 ($n = 4$, $p < 0.01$)

by microarray analysis of global gene expression [3, 13]. Furthermore, genomic DNA demethylation, the first sign of SCR, was detected in the mirPSCs [3, 13]. Particularly, we noted that mirPSCs were pluripotent but not tumorigenetic, for they preferentially formed teratoma-like tissue cysts in the uterus or peritoneal cavity of pseudopregnant immunocompromised SCID-beige mice [3, 13]. These tissue cysts, not similar to teratomas, contained various but relatively organized tissue regions derived from all three embryonic germ layers. Furthermore, the presence of partial homeobox (HOX) gradient expression was frequently observed in these tissue cysts (Lin et al. unpublished). When xenografted into normal male mice, the mirPSCs were differentiated and assimilated by the surrounding tissues, presenting a potential application for regenerative medicine [19]. Taken together, our findings have established that miR-302 reprogrammed human somatic cells to form hESC-like iPSCs with a high level of pluripotency but less or no tumorigenicity.

MiR-302-mediated gene silencing is dose-dependent. The tumor suppression by miR-302 only occurred when its cellular concentration reached 1.1- to 1.3-folds of that in hESC H1 or H9 cells, which was corresponding to an optimal level of

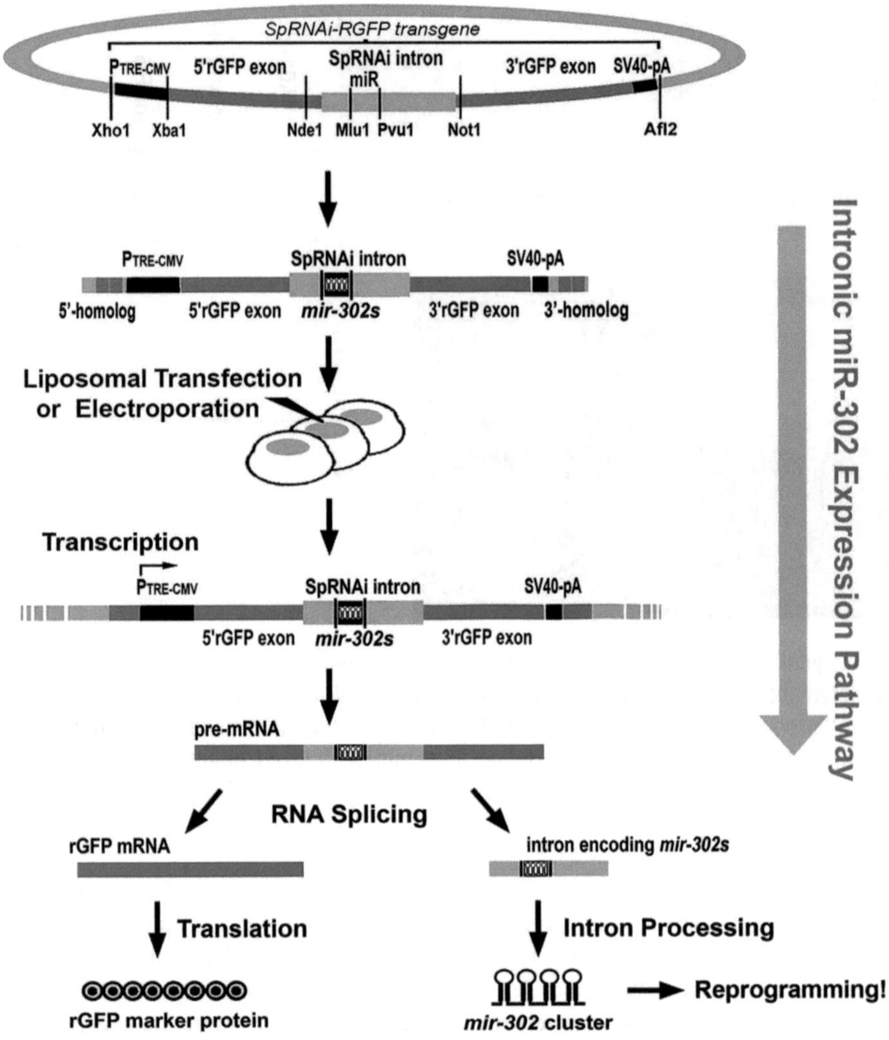

Fig. 3 Schematic procedure for generating iPSCs using intronic miR-302 expression. The miR-302-expressing *SpRNAi-RGFP* vector is introduced into human somatic cells by either electroporation or liposomal transfection. The expression of miR-302 relies on the natural intronic miRNA biogenesis pathway, in which the miR-302 cluster is co-expressed with the encoded *RGFP* gene transcripts and further processed into individual mature miR-302 molecules by RNA splicing enzymes (spliceosomal components) and RNaseIII Dicers

miR-302 for iPSC reprogramming [13, 19]. When miR-302 levels, introduced by the miR-302 expression system described in Fig. 3, were equal to or lower than those in hESCs, only large tumor suppressor homolog 2 (LATS2), but not CDK2, was silenced. When miR-302 levels were higher than those in hESCs, both CDK2 and LATS2 were silenced and hence cell cycle was attenuated at the G1-phase checkpoint [19]. Given that LATS2 inhibited the cyclin E-CDK2 pathway to block G1-S cell cycle transition [20], the direct silencing of CDK2 by miR-302 could enforce this LATS2 function even in the absence of LATS2. Moreover, the

CDK2 silencing effect also counteracted the suppressive effect of miR-367 on CDKN1C (p57, Kip2), a cell cycle inhibitor against both CDK2 and CDK4, subsequently leading to a reduced cell cycle rate (Lin et al. unpublished). Through such a dose-dependent gene silencing mechanism, miR-302 fine-tuned or even altered the function of its target genes at different rates during different stages of development [13, 17]. Conceivably, early human embryonic cells prior to the morula stage may express a higher miR-302 level than that in blastocyst-derived hESCs. These findings may account for the fact that 2-to-8-cell-stage embryonic cells have a relatively slower cell cycle rate (20–24 h/cycle) than that of hESCs (15–16 h/cycle) [21]. Even though both early zygotes and hESCs were known to display a short G1 phase, the G1 phase of 2-to-8-cell-stage zygotic cells is still significantly longer (4 ± 1 h) than that of hESCs. On the other hand, iPSC generation was reported to be involved either a cell cycle-dependent (with Klf4) or cell cycle-independent (with Nanog) reprogramming process [22], in which the former cannot be explained by miR-302's function. In fact, this hypothesis overlooked the fact that Klf4 is an upstream transcription factor of Nanog [23, 24]. Klf4 induced not only Nanog but also many other oncogenes, which promoted fast cell proliferation unrelated to the function of Nanog in reprogramming [11, 23–25]. Therefore, fast cell cycle, although can increase iPSC number, is not essential for SCR initiation to form iPSCs.

MiR-302 functions differently in normal cells than from cancer cells. We previously showed that miR-302 triggered a massive apoptosis (programmed cell death) in fast proliferative tumor/cancer cells whereas relatively slow growing normal cells could tolerate the inhibitory effect on cell proliferation [3, 19]. It is conceivable that tumor/cancer cells may not survive in such a relatively slow cell cycle state due to their high metabolism and rapid consumption rates, providing a beneficial advantage in preventing tumor formation.

Based on the scheme depicting the mechanisms underlying miR-302's tumor suppressing function (Fig. 4), miR-302 not only inhibits both cyclin E-CDK2 and cyclin D-CDK4/6 pathways to block >70% of the G1-S cell cycle transition but it also silences polycomb ring finger oncogene BMI-1 to increase the expression of cell cycle inhibitors p16Ink4a and p14/p19Arf [19]. As a result, 16Ink4a inhibits cyclin D-dependent CDK4/6 activity by phosphorylating the retinoblastoma protein Rb and subsequently preventing it from releasing E2F-dependent transcription required for S phase entry [26, 27]. Furthermore, p14/p19Arf prevents HDM2, a Mdm2 p53 binding protein homolog, from binding to p53 and permits the p53-dependent transcription responsible for G1 arrest or apoptosis [28]. By simultaneously activating multiple cell cycle attenuation and tumor suppression pathways, miR-302 induces iPSC generation while preventing stem cell tumorigenicity.

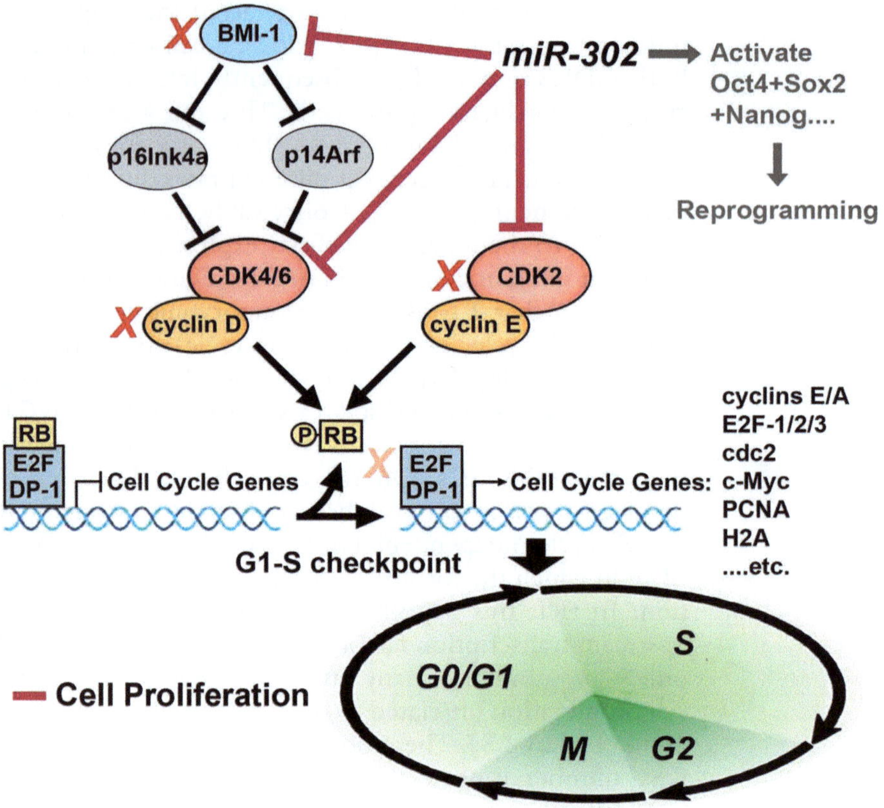

Fig. 4 Mechanism of miR-302-mediated tumor suppression in human iPSCs. MiR-302 not only concurrently suppresses G1-phase checkpoint regulators CDK2, cyclin D, and BMI-1 but also indirectly activates p16Ink4a and p14/p19Arf to suppress most (>70%) of the cell cycle activities during SCR. E2F is also a predicted target of miR-302. Relative quiescence at the G0/G1 state may prevent possible random growth and/or tumor-like transformation of the reprogrammed iPSCs, leading to a more accurate and safer reprogramming process, by which premature cell differentiation and tumorigenicity are both inhibited

In addition, since DNMT1-deficient ESCs have been observed to present a significant reduction of gene loss and mutation rates [29], miR-302 may help enhancing the genomic stability of both ESCs and iPSCs via LSD1/AOF2/KDM1-associated DNMT1 degradation [13, 19]. In fact, the online miRNA-target prediction program (http://www.ebi.ac.uk/enright-srv/microcosm/cgi-bin/targets/v5/detail_view.pl?transcript_id=ENST00000359526) provided by the European Bioinformatics Institute has also shown that DNMT1 is a direct target of miR-302a–d, further indicating that miR-302 can silence DNMT1 to induce global demethylation and genome stabilization. These currently identified miR-302 functions suggest that the anti-tumorigenicity and genomic stability of iPSCs and ESCs may be improved by elevating their miR-302 expression.

On the other hand, overexpression of miR-302 is not always beneficial. Human iPSCs have been reported to exhibit problems including early senescence and limited expansion [30, 31]. When miR-302 levels increased beyond 1.5- to 1.7-folds of that in hESCs, the degree of iPSC senescence was increased in proportion to the concentration of miR-302. Normal adult cells also undergo a limited number of divisions and then reach a quiescent state called replicative senescence. Cells that escape replicative senescence often become immortal cells; thus, replicative senescence is a cellular defense mechanism against tumor/cancer formation. Because miR-302 increased p16Ink4a and p14/p19Arf expression, two of the major regulators in replicative senescence, the overexpression of miR-302 may exacerbate early senescence in iPSCs. Previous studies have identified several genes encoded in the Ink4a/Arf locus, such as p16Ink4a and p14/p19Arf, which pose a barrier to the success of iPSC reprogramming [30, 32–34]. In particular, both of the p16Ink4a-RB and p14/19Arf-p53 activities are gradually enhanced and finally commit to the onset of early senescence after serial passaging of iPSCs, [30, 33]. In murine iPSCs, p19Arf rather than p16Ink4a is the main factor activating senescence, whereas in humans p16Inka plays a more important role than p14Arf [30, 32]. Our recent observations have further ruled out the effect of telomerase in mirPSC senescence (Fig. 5).

Due to forced miR-302 elevation, mirPSCs often display a higher degree of senescence than the iPSCs induced by the four defined transcription factors. Under the same feeder-free conditions, the four factor-induced iPSCs could be cultivated over 30~50 passages, while mirPSCs only reach a maximum of 26 passages. However, this early senescence was not due to the shortening of telomere length. It is known that LSD1/AOF2/KDM1 and histone deacetylase 2 (HDAC2) suppress the transcription of human telomerase reverse transcriptase (hTERT), and the deficiency of both epigenetic regulators has been shown to increase hTERT expression [35, 36]. Both LSD1/AOF2/KDM1 and HDAC2 are also targeted by miR-302 [13]. Using telomeric repeat amplification protocol (TRAP) and telomerase activity ELISA assays, we have confirmed that miR-302 silenced both LSD1/AOF2/KDM1 and HDAC2 to enhance the hTERT activity and subsequently maintain normal or longer telomere length in mirPSCs (Fig. 5a, b). In view of miR-302's ability to simultaneously regulate p16Ink4a, p14/p19Arf, and hTERT activities, mirPSCs may serve as a model for investigating the early senescence of iPSCs.

The balance between the mechanisms of tumorigenicity and senescence determines the safety and efficiency of iPSC generation. Our studies have shown that both mechanisms are regulated by the p16Ink4a-RB and p14/19Arf-p53 pathways [19]. In mirPSCs, miR-302 and its homologues modulated the levels of p16Ink4a and p14/19Arf via fine-tuning the BMI-1 expression [19], providing important insight into the mechanism underlying stem cell self-renewal without the formation of tumors. In four factor-

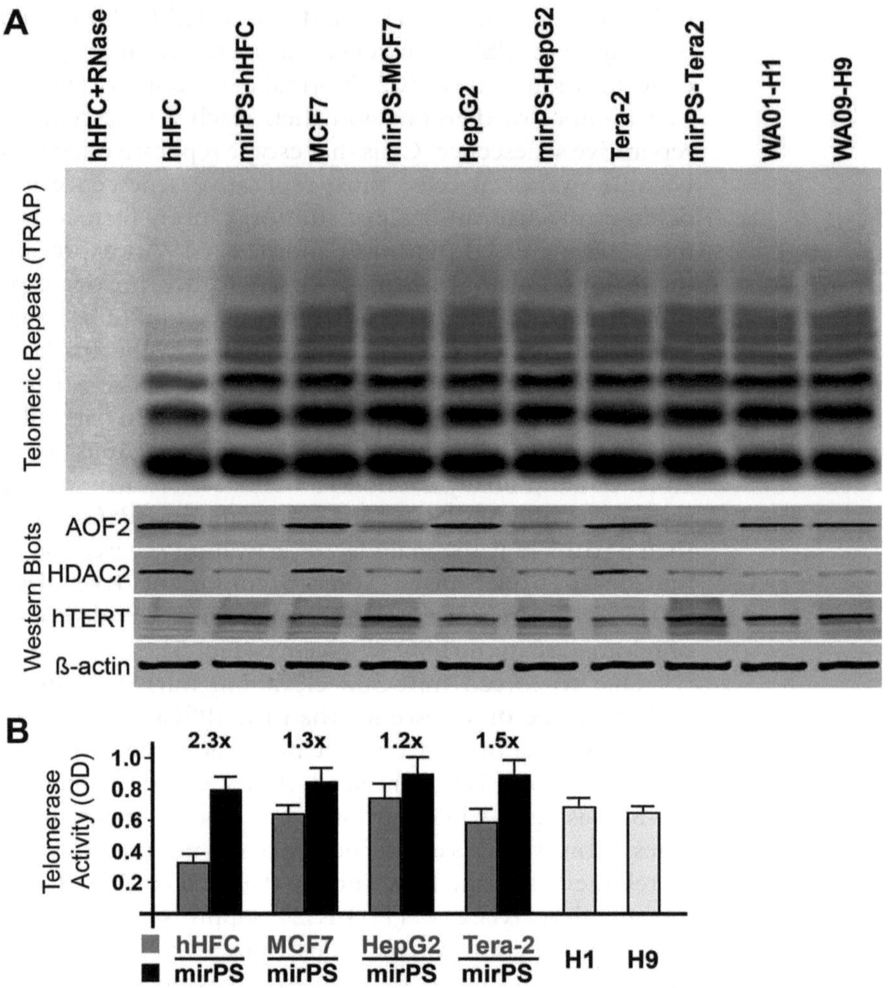

Fig. 5 Analysis of telomerase activity in various mirPSC lines in response to the miR-302 level elevated approximately 1.3-folds of the level in hESCs. (**a**) Telomerase activities shown by TRAP assays ($n = 5$, $p < 0.01$) in the upper panel. Telomerase activity was sensitive to RNase treatment (hHFC+RNase). Western blotting (lower panel) confirming the increase of hTERT and decrease of LSD1/AOF2/KDM1 and HDAC2 expression in various mirPSC lines ($n = 5$, $p < 0.01$). (**b**) Telomerase activities measured by telomerase PCR ELISA assays (OD470–OD680; $n = 3$, $p < 0.01$)

induced iPSCs, Oct4, Sox2, and Klf4 were also found to repress the expression of p16Ink4a and p14/19Arf, leading to fast iPSC proliferation [32]. Conceivably, the counteracting effect between miR-302 and Oct4-Sox2-Klf4 on the p16Ink4a and p14/19Arf expression may be the key in reaching the balance between tumorigenicity and senescence in iPSCs. On the other hand, we also found that miR-302 and Oct4-Sox2-Nanog formed a positive feedback cycle to fulfill the completion of iPSC reprogramming [13]. In this reprogramming cycle, miR-302 and Oct4-Sox2-Nanog are reciprocally induced to constantly maintain the ESC-like pluripotent state of iPSCs. Therefore, at a certain point in the cycle, the interaction between miR-302 and Oct4-Sox2-Nanog/Klf4 may result in a stable iPSC state without the problem of either tumorigenicity or senescence. Based on this understanding, the adequate combination of miR-302 and Oct4-Sox2-Nanog/Klf4 for generating tumor-free and senescence-free iPSCs presents a future challenge.

Through the natural mechanism of intronic miRNA biogenesis, we have identified the optimal miR-302 concentration for inducing tumor-free iPSCs [18, 19, 37]. MiR-302 is encoded in the intron region of the *La ribonucleoprotein domain family member 7* (*LARP7*, *PIP7S*) gene and expressed via an intronic miRNA biogenesis mechanism [3, 15, 37]. As described earlier, native miR-302 consists of four familial sense homologues (miR-302b, c, a, and d) and three distinct antisense members (miR-302b*, c*, and a*), all of which were simultaneously transcribed as a polycistronic RNA cluster along with miR367 [14]. Numerous attempts were made to investigate the "miR-302" function, but no complete expression of all miR-302a-d familial members was achieved. Also, no studies were focused on the dose-dependent effect of miR-302, which determined the real miR-302 function. Therefore, our current approach using intronic miR-302 expression was the first successful protocol to display full expression of all native miR-302a-d familial members in human iPSCs as determined by microarray and Northern blot analyses [3, 13]. In this chapter, we show the step-by-step procedure of this approach.

We have first developed a *pSpRNAi-RGFP* vector mimicking the process of miRNA biogenesis from mammalian introns [18, 33, 38, 39]. As shown in Fig. 3, primary intronic miRNA precursors (pri-miRNAs) are transcribed by type II RNA polymerases (Pol-II) and excised by spliceosomal components and/or Drosha to form miRNA precursors (pre-miRNAs), which are then exported out of the nucleus by Ran-GTP and Exportin-5 and further processed by a Dicer-like RNaseIII endoribonuclease in cytoplasm to form mature miRNAs [18, 37, 39–42]. However, the role of Drosha may be important but not required in this mechanism because the depletion of >85% Drosha with siRNA reduces less than a half of the miR-302 expression (Lin et al. unpublished), indicating that some other nuclear RNaseIII-like enzymes may replace Drosha for processing intronic pre-miRNAs [40]. Also, since mammalian introns often contain nonsense (i.e., translational stop) codons recognized by the nonsense-mediated decay (NMD) system, a cellular RNA surveillance mechanism [43, 44], the non-hairpin structures of an intron can be quickly degraded by NMD to prevent excessive RNA accumulation. RNA oversaturation has been reported to be a major problem for the direct (exonic) expression of short hairpin RNAs (shRNAs) and miRNAs in mammalian cells [45]. Under this strictly coordinated regulation by Pol-II transcription, RNA splicing and

NMD mechanisms, intronic miRNA biogenesis has a great evolutional advantage in safeguarding cellular miRNA concentration. In the present chapter, we describe the experimental procedures that led to the generation of tumor-free iPSCs.

2 Materials

2.1 Synthetic Oligonucleotides

1. All synthetic oligonucleotides required for the construction of the miR-302 familial cluster are listed in Table 1 (*see* **Note 1**). Each synthetic oligonucleotide was prepared in a final concentration of 200 pmole/μL.
2. 2× hybridization buffer: 200 mM KOAc, 60 mM HEPES KOH, 4 mM MgOAc (pH 7.4 at 25 °C).
3. Incubation chambers: 94 °C to 65 °C and 4 °C.

2.2 Construction of the miR-302 Cluster

1. 10× digestion buffers for each individual restriction enzyme.
2. Restriction enzymes, such as *XhoI*, *MluI*, *BglII*, *NdeI*, *XbaI*, *PvuI*, and *BamHI*, were prepared with a stock concentration of 10 units/μL.
3. Digestion reaction mix: 10 μL of autoclaved ddH_2O, 2 μL of 10× digestion buffer, 4 μL of restriction enzyme(s), and 4 μL of each individual hybridized oligonucleotide (e.g., the hybrid of miR-302a-sense and miR-302a-antisense); the reaction mix was prepared right before use.
4. 10× ligation buffer: 660 mM Tris–HCl (pH 7.5 at 20 °C), 50 mM $MgCl_2$, 50 mM dithioerythritol, and 10 mM ATP.
5. T4 DNA ligase (5 U/μL).
6. Ligation reaction mix: 6 μL of each of the digested hybridized oligonucleotides (four kinds of hybridized oligonucleotides in total 24 μL), 3 μL of 10× ligation buffer, and 3 μL of T4 ligase; the reaction mix was prepared right before use.
7. Incubation chambers: 37 °C, 75 °C, 10 °C, and 4 °C.
8. 70-base-pair (bp) cutoff purification filters, such as, but not limited to, QIAquick Gel Extraction Kit (Qiagen, Valencia, CA).
9. 100-bp cutoff purification filters, such as, but not limited to, QIAquick PCR Purification Kit (Qiagen).
10. Microcentrifuge ≥10,000 × *g*.

2.3 Insertion of the miR-302 Cluster into an Expression Vector

1. Expression vector (0.5 μg/μL), such as *pSpRNAi-RGFP* (*see* **Note 2**; [23, 26]) and *pLVX-AcGFP-N1*. Any expression vector containing an either *MluI/PvuI* or *XhoI/BamHI* multiple cloning site for insertion of the miR-302 cluster in the 5-′-untranslated region (5′-UTR) or in-frame intron of an reporter gene (i.e., EGFP or RGFP) can be used here.

Table 1
Synthetic sequences used for constructing the miR-302 familial cluster

Name	Sequence[a]
miR-302a-sense	5′-GATATCTCGA GTCACGCGTT **CCCACCACTT AAACGTGGAT GTACTTGCTT TGAAACTAAA GAAGTAAGTG CTTCCATGTT TTGGTGATGG** ATAGATCTCT C-3′
miR-302a-antisense	5′-GAGAGATCTA TCCATCACCA AAACATGGAA GCACTTACTT CTTTAGTTTC AAAGCAAGTA CATCCACGTT TAAGTGGTGG GAACGCGTGA CTCGAGATAT C-3′
miR-302b-sense	5′-ATAGATCTCT **CGCTCCCTTC AACTTTAACA TGGAAGTGCT TTCTGTGACT TTGAAAGTAA GTGCTTCCAT GTTTTAGTAG GAGT**CGTCAT ATGAC-3′
miR-302b-antisense	5′-GTCATATGAC GACTCCTACT AAAACATGGA AGCACTTACT TTCAAAGTCA CAGAAAGCAC TTCCATGTTA AAGTTGAAGG GAGCGAGAGA TCTAT-3′
miR-302c-sense	5′-GCCATATGCT **ACCTTTGCTT TAACATGGAG GTACCTGCTG TGTGAAACAG AAGTAAGTGC TTCCATGTTT CAGTGGAGG**C GTCTAGACAT-3′
miR-302c-antisense	5′-ATGTCTAGAC GCCTCCACTG AAACATGGAA GCACTTACTT CTGTTTCACA CAGCAGGTAC CTCCATGTTA AAGCAAAGGT AGCATATGGC-3′
miR-302d-sense	5′-CGTCTAGACA **TAACACTCAA ACATGGAAGC ACTTAGCTAA GCCAGGCTAA GTGCTTCCAT GTTTGAGTGT T**CGCGATCGG ATCCAC-3′
miR-302d-antisense	5′-GTGGATCCGA TCGCGAACAC TCAAACATGG AAGCACTTAG CCTGGCTTAG CTAAGTGCTT CCATGTTTGA GTGTTATGTC TAGACG-3′

[a]Bold letters indicate the sequences of each individual miR-302 precursor (pre-miR302)

2. 10× restriction buffer, such as *MluI* and *PvuI*, for digesting *pSpRNAi-RGFP* or *XhoI* and *BamH1* for digesting *pLVX-AcGFP-N1*.
3. Restriction enzymes, such as *MluI* and *PvuI* (for *pSpRNAi-RGFP*) or *XhoI* and *BamH1* (for *pLVX-AcGFP-N1*) in a stock concentration of 10 units/μL.
4. Digestion reaction mix: 13 μL of the miR-302 cluster, 2 μL of 10× digestion buffer, 4 μL of restriction enzymes (*MluI/PvuI* or *XhoI/BamHI*), and 1 μL of the expression vector; prepare the reaction mix right before use.
5. 100-bp cutoff purification filters.
6. Microcentrifuge ≥10,000 × *g*.
7. 10× ligation buffer: 660 mM Tris–HCl (pH 7.5 at 20 °C), 50 mM $MgCl_2$, 50 mM dithioerythritol, and 10 mM ATP.
8. T4 DNA ligase (5 U/μL).

9. Ligation reaction mix: 25 μL of the digested miR-302 cluster and expression vector mix, 3 μL of 10× ligation buffer, and 2 μL of T4 ligase; prepare the reaction mix just before use.
10. Incubation chambers: 37 °C, 75 °C, 10 °C, and 4 °C.
11. Luria-Bertani (LB) agar plates containing 50 mg/mL kanamycin for *pSpRNAi-RGFP* or 100 mg/mL ampicillin for *pLVX-AcGFP-N1* clone selection.
12. Low-salt LB broth containing 50 mg/mL kanamycin for *pSpRNAi-RGFP* or 100 mg/mL ampicillin for *pLVX-AcGFP-N1* clone amplification.
13. Ready-to-use transformation competent *E. coli* cells, such as DH5α.
14. Incubation shaker: 37 °C, >180 rpm vortex.
15. Plasmid extraction kit.

2.4 Transfection or Electroporation

1. Electroporator, such as, but not limited to, Neon Transfection System (Invitrogen, Carlsbad, CA) or Multiporator Electroporation System (Eppendorf, Hamburg, Germany).
2. Electroporation buffer (*see* **Note 3**).
3. Polysomal or liposomal reagents, such as, but not limited to, FuGENE HD Transfection Reagent (Roche Diagnostics, Indianapolis, IN).
4. Target cells (>2000 cells).
5. Cell culture medium, depending on the cell type of interest.
6. 1× trypsin/EDTA solution.

2.5 Selection and Cultivation of miR-302-Positive mirPSCs

1. Feeder-free mirPSC medium: KnockOut DMEM/F-12 medium supplemented with 10% KnockOut serum replacement (Invitrogen), 1% MEM nonessential amino acids, 100 μM ß-mercaptoethanol, 1 mM GlutaMax, 1 mM sodium pyruvate, 0.1 μM A83-01, 0.1 μM valproic acid (Stemgent, San Diego, CA), 10 ng/mL bFGF, 10 ng/mL FGF-4, 5 ng/mL LIF, and a mix of 50 IU/mL penicillin and 50 μg/mL streptomycin (*see* **Note 4**).
2. G418 (a final concentration ranged from 100 to 300 μg/mL for culturing *pSpRNAi-RGFP*-transfected cells) or puromycin (a final concentration ranged from 15 to 100 μg/mL for culturing *pLVX-AcGFP-N1*-transfected cells).
3. Incubator for cell culture: 37 °C, 5% CO_2.

3 Methods

3.1 Construction of the miR-302 Cluster

The double-stranded DNA of each individual miR-302 precursor is formed by annealing the sense strand of each miR-302 familial member to its respective antisense strand; for example, miR-302a-sense to miR-302a-antisense, miR-302b-sense to miR-302b-antisense, miR-302c-sense to miR-302c-antisense, and miR-302d-sense to miR-302d-antisense (*see* Table 1). Then, the double-stranded DNAs of all four miR-302 (a, b, c, and d) precursors are separately cleaved by different restriction enzymes to generate various cohesive ends. Based on these cohesive ends, all four miR-302 precursors can be sequentially ligated into one familial cluster ready for co-expression. All synthetic strands of miR-302 members must be purified by polyacrylamide gel electrophoresis (PAGE) and stored at −20 °C.

1. Hybridization: Mix the synthetic sense and antisense strands of each miR-302 member (each 1 nmole) in 10 μL of autoclaved ddH_2O; add 10 μL of 2× hybridization buffer, mix and heat to 94 °C for 3 min, and then slowly cool to 65 °C in 30 min. Stop the reaction on ice.
2. Restriction enzyme digestion: Prepare one digestion reaction mix for each hybridized miR-302 member (a, b, c, and d, respectively), containing 4 μL of the hybridized DNA, 2 μL of 10× digestion buffer, 4 μL of restriction enzymes, and 10 μL of autoclaved ddH_2O. Use 4 μL of *BglII* for miR-302a cleavage, 2 μL of *BglII* and 2 μL of *NdeI* for miR-302b cleavage, 2 μL of *NdeI* and 2 μL of *XbaI* for miR-302c cleavage, and 4 μL of *XbaI* for miR-302d cleavage. Incubate the reaction at 37 °C for 4 h and then stop at 4 °C. Purify each of the digested miR-302 member (a, b, c, and d, respectively) using a 70-bp cutoff purification filter, following the manufacturer's protocol, and recover the DNA in 30 μL of autoclaved ddH_2O.
3. Cohesive end ligation: Combine all four miR-302 members and mix well. Prepare a ligation reaction mix, containing 24 μL of the miR-302 mixture, 3 μL of 10× ligation buffer, and 3 μL of T4 ligase, and incubate the reaction at 10 °C for 16 h and then stop at 4 °C. This forms the miR-302 familial cluster containing miR-302a, b, c, and d in a 5′ to 3′ sequential order. Purify the miR-302 cluster using a 100-bp cutoff purification filter and recover it in 30 μL of autoclaved ddH_2O.

3.2 Insertion of the miR-302 Cluster into an Expression Vector

Intronic miRNA can be expressed from the 5′-UTR, 3′-UTR, or in-frame intron region of a gene; hence, any expression vector containing an insertion site in these regions can be used for miR-302 expression. For example, the *pSpRNAi-RGFP* plasmid vector possesses an *MluI/PvuI* insertion site in its in-frame intron, while the retroviral *pLVX-AcGFP-N1* vector (Clontech) contains a *XhoI/BamHI* cloning site in the 5′-UTR of its *AcGFP* gene. Both vectors have been tested for expressing the miR-302 cluster. However, due to the low stability of the highly structured miR-302 cluster, we have noticed that during the processes of vector amplification and extraction, the transformed *E. coli* competent cells cannot be stored at 4 °C or some of the hairpin pre-miRNA structures may be lost. For storage, the vector containing the miR-302 cluster is stable at 4 °C for up to 4 months and at −80 °C for over 2 years.

1. Restriction enzyme digestion: Mix 1 μL of the expression vector to 13 μL of the miR-302 cluster, and prepare one digestion reaction mix containing the 14 μL mixture, 2 μL of 10× digestion buffer, and either 2 μL of *MluI* and 2 μL of *PvuI* (for *pSpRNAi-RGFP*) or 2 μL of *XhoI* and 2 μL of *BamH1* (for *pLVX-AcGFP-N1*). Incubate the reaction at 37 °C for 4 h and then 4 °C. Purify the digested reaction using a 100-bp cutoff purification filter, and recover the DNAs in 30 μL of autoclaved ddH_2O.
2. Cohesive end ligation: Prepare a ligation reaction mix, containing 25 μL of the cleaved vector and miR-302 cluster mixture, 3 μL of 10× ligation buffer, and 2 μL of T4 ligase. Incubate the reaction at 10 °C for 16 h and then 4 °C. This forms the miR-302-expressing vector.
3. Vector selection: Add 5 μL of the miR-302-expressing vector to the ready-to-use transformation competent *E. coli*-DH5α cells, mix, and incubate the mixture at 4 °C for 10 min, following the manufacturer's protocol. Next, pour and smear the mixture evenly onto an antibiotic-containing LB agar plate (50 mg/mL kanamycin for *pSpRNAi-RGFP* or 100 mg/mL ampicillin for *pLVX-AcGFP-N1*), and incubate the transformed *E. coli*-DH5α cells at 37 °C, overnight.
4. Vector amplification: Pick and transfer each single cell colony, with a sterilized platinum loop or pipet tip, from the LB agar plate into 30 mL of antibiotic-containing LB broth (50 mg/mL kanamycin for *pSpRNAi-RGFP* or 100 mg/mL ampicillin for *pLVX-AcGFP-N1*), respectively. Further incubate the cell-containing LB broth on a shaker (>180 rpm) at 37 °C, overnight.
5. Vector extraction: Isolate and recover the amplified vector in 30 μL of autoclaved ddH_2O using a plasmid extraction mini-prep filter, following the manufacture's protocol. To confirm the insertion of the miR-302 cluster, the isolated vector can be digested with either *MluI* and *PvuI* (for *pSpRNAi-RGFP*) or *XhoI* and *BamH1* (for *pLVX-AcGFP-N1*) to generate a ~350 base-pair DNA band on 2% agarose gel electrophoresis.

3.3 Transfection or Electroporation

To express the miR-302 cluster in the cells of interest, we recommend either electroporation or liposomal/polysomal transfection. Although the *pLVX-AcGFP-N1* vector can also be used for lentiviral production, this approach must be performed with extreme caution due to miR-302's unknown in vivo function(s). For miR-302-induced iPSC generation, the cell types currently tested are human skin- and hair-derived somatic cells, including melanocytes, keratinocytes, fibroblasts, and neural-like HEK-293 as well as several cancer cell lines, including human melanoma Colo-829, breast carcinoma MCF-7, prostate cancer PC3, hepatocellular carcinoma HepG2, and embryonal teratocarcinoma Tera-2 cells. Notably, the reprogramming efficiency may vary in different cell types; for example, using electroporation, the average iPSC formation rates are >70% for HEK-293 cells, 15–20% for keratinocytes, 7–10% for melanocytes, <1~3% for adult fibroblasts, and <1% for all tested tumor/cancerous cells, respectively. The reprogramming efficiency in other cell types remains to be determined.

1. Electroporation: Add 2000–200,000 cells and 15–40 μg of the miR-302-expressing vector in 250 μL of electroporation buffer, mix well, and transfer into a 400 μL cuvette with aluminum electrodes.
2. Perform electroporation tests following the manufacturer's protocol (*see* **Note 3**). After electroporation, grow the cells in feeder-free mirPSC medium at 37 °C under 5% CO_2.
3. Liposomal/polysomal transfection (*see* **Note 5**). Grow cells to 50% confluency in a 100 mm culture dish, and replace the cell culture medium by 9 mL of serum-free cell culture medium 4 h prior to transfection.
4. For transfection preparation, add 15 μg of the miR-302-expressing vector into 1 mL of serum-free cell culture medium and mix well.
5. Next, add 50 μL of FuGENE HD reagent into the center of the vector-medium mixture, and vortex for 10 seconds and place the mixture at room temperature for 15 min (*see* **Note 6**).
6. After incubation, add the mixture dropwise, covering the whole 100 mm culture dish, and gently shake the culture dish (back and forth and right and left) several times to evenly distribute the mixture (*see* **Note 7**).
7. Incubate the cells at 37 °C under 5% CO_2 for 12–18 h, and then replace the medium with feeder-free mirPSC medium. Continue to grow the transfected cells in feeder-free mirPSC medium at 37 °C under 5% CO_2.

3.4 Selection and Cultivation of miR-302-Positive mirPSCs

Since the *pSpRNAi-RGFP* and *pLVX-AcGFP-N1* vectors contain an antibiotic-resistant gene against G418 and puromycin, respectively, the positive miR-302-transfected cells can be selected by using either G418 (for *pSpRNAi-RGFP*) or puromycin (for *pLVX-AcGFP-N1*). The positive miR-302-transfected cells also express either red fluorescent RGFP (*pSpRNAi-RGFP*) or green AcGFP (*pLVX-AcGFP-N1*) for color identification under a fluorescent microscope or cell sorting by a FACS flow cytometry machine. Both antibiotic selection and color identification/sorting ensure the purity of the miR-302-transfected cell population.

3.4.1 Antibiotic Selection

When the transfected cells start to express red or green fluorescent GFP, add either G418 (100–300 μg/mL for *pSpRNAi-RGFP*-transfected cells) or puromycin (15–100 μg/mL for *pLVX-AcGFP-N1*-transfected cells) to the cell culture medium and mix well. The optimal antibiotic concentration for mirPSC selection may vary dependent on the original cell types. Incubate the cells at 37 °C under 5% CO_2 for 24–48 h, and then replace the medium with fresh feeder-free mirPSC medium. Continue to grow the cells in feeder-free mirPSC medium at 37 °C under 5% CO_2 for 3 more days, and observe the purity of the fluorescent cells. If there are still many non-transfected (nonfluorescent) cells, repeat the steps of this section using a higher antibiotic concentration until the fluorescent cell population is relatively pure.

3.4.2 mirPSC Culturing and Passaging

Under the above feeder-free culture condition, mirPSCs tend to form large embryoid body-like colonies. When a mirPSC colony contains more than 2000 cells, divide the colony into several small pieces with a scalpel and transfer the cells to a new collagen-coating culture dish in feeder-free mirPSC medium. Incubate the cells at 37 °C under 5% CO_2 (*see* **Note 8**).

4 Notes

1. All synthetic oligonucleotides are purified by polyacrylamide gel electrophoresis (PAGE) to ensure their highest purity and sequence integrity.
2. The detailed method for constructing the *SpRNAi-RGFP* expression vector has been described in our previous publications [37, 38].
3. Electroporation buffer varies depending on the electroporation machine used. Please follow the manufacturer's suggestions.
4. The addition of A83-01, a SMAD2 inhibitor, and valproic acid, a HDAC inhibitor, is optional, depending on the cell type of interest.

5. The protocol shown here is optimized for FuGENE HD (Roche). For other liposomal/polysomal reagents, a different protocol may be adopted. Please follow the manufacturer's protocol for this section if not using FuGENE HD. Also, the transfection volume is prepared for a 100 mm culture dish containing cells at 50% confluency (in 10 mL of cell culture medium). If a smaller volume is desired, please reduce the amount of all materials and reagents in proportion to the volume of cell culture medium.
6. Maximum 30 min at room temperature.
7. Do not swirl the cell culture dish to mix the mixture. This results in uneven distribution of the transfection mixture.
8. When the mirPSC colony grows too big (>5000 cells), the cells in the center of the colony may undergo apoptosis, which causes the whole colony to differentiate or die. The apoptotic mirPSC colony cannot be recovered for further passaging. Even under the best cell culture condition, the mirPSCs can be passaged up to 26 generations, based on the current studies [13, 19].

References

1. Takahashi K, Yamanaka S (2006) Induction of pluripotent stem cells from mouse embryonic and adult fibroblast cultures by defined factors. Cell 126(4):663–676
2. Ha M, Kim N (2014) Regulation of microRNA biogenesis. Nature Rev Mol Cell Biol 15:509–524
3. Lin SL, Chang DC, Chang-Lin S, Lin CH, Wu DTS, Chen DT, Yang SY (2008) Mir-302 reprograms human skin cancer cells into a pluripotent ES-cell like state. RNA 14 (10):2115–2124
4. Song MS, Rossi JJ (2017) Molecular mechanisms of Dicer: endonuclease and enzymatic activity. Biochem J 474:1603–1618
5. Kim Y, Kim VN (2012) MicroRNA factory: RISC assembly from precursor microRNAs. Mol Cell 46:384–386
6. Ambros V (2004) The functions of animal miRNAs. Nature 435:350–355
7. Ma T, Xie M, Laurent T, Ding S (2013) Progress in the reprogramming of somatic cells. Circ Res 112:562–574
8. Takahashi K, Tanabe K, Ohnuki M, Narita M, Ichisaka T, Tomoda K, Yamanaka S (2007) Induction of pluripotent stem cells from adult human fibroblasts by defined factors. Cell 131:861–872
9. Yu J, Vodyanik MA, Smuga-Otto K, Antosiewicz-Bourget J, Frane JL, Tian S, Nie J, Jonsdottir GA, Ruotti V, Stewart R, Slukvin II, Thomson JA (2007) Induced pluripotent stem cell lines derived from human somatic cells. Science 318:1917–1920
10. Hanna J, Wernig M, Markoulaki S, Sun CW, Meissner A, Cassady JP, Beard C, Brambrink T, Wu LC, Townes TM, Jaenisch R (2007) Treatment of sickle cell anemia mouse model with iPS cells generated from autologous skin. Science 318:1920–1923
11. Kim J, Chu J, Shen X, Wang J, Orkin SH (2008) An extended transcriptional network for pluripotency of embryonic stem cells. Cell 132:1049–1061
12. Young RA (2011) Control of the embryonic stem cell state. Cell 144:940–954
13. Lin SL, Chang D, Lin CH, Ying SY, Leu D, Wu DTS (2011) Regulation of somatic cell reprogramming through inducible mir-302 expression. Nucleic Acids Res 39:1054–1065
14. Suh MR, Lee Y, Kim JY, Kim SK, Moon SH, Lee JY, Cha KY, Chung HM, Yoon HS, Moon SY, Kim VN, Kim KS (2004) Human embryonic stem cells express a unique set of microRNAs. Dev Biol 270:488–498
15. Barroso-delJesus A, Romero-Lopez C, Lucena-Aguilar G, Melen GJ, Sanchez L, Ligero G, Berzal-Herranz A, Menendez P (2008) Embryonic stem cell-specific miR302-367 cluster: human gene structure and

functional characterization of its core promoter. Mol Cell Biol 28:6609–6619

16. Wilson KD, Venkatasubrahmanyam S, Jia F, Sun N, Butte AJ, Wu JC (2009) MicroRNA profiling of human-induced pluripotent stem cells. Stem Cells Dev 18:749–758
17. Ying SY, Lin SL (2004) Intron-derived microRNAs—fine tuning of gene functions. Gene 342:25–28
18. Lin SL, Kim H, Ying SY (2008) Intron-mediated RNA interference and microRNA (miRNA). Front Biosci 13:2216–2230
19. Lin SL, Chang D, Ying SY, Leu D, Wu DTS (2010) MicroRNA miR-302 inhibits the tumorigenecity of human pluripotent stem cells by coordinate suppression of CDK2 and CDK4/6 cell cycle pathways. Cancer Res 70:9473–9482
20. Li Y, Pei J, Xia H, Ke H, Wang H, Tao W (2003) Lats2, a putative tumor suppressor, inhibits G1/S transition. Oncogene 22:4398–4405
21. Becker KA, Ghule PN, Therrien JA, Lian JB, Stein JL, van Wijnen AJ, Stein GS (2006) Self-renewal of human embryonic stem cells is supported by a shortened G1 cell cycle phase. J Cell Physiol 209:883–893
22. Hanna J, Saha K, Pando B, van Zon J, Lengner CJ, Creyghton MP, van Oudenaarden A, Jaenisch R (2009) Direct cell reprogramming is a stochastic process amenable to acceleration. Nature 462:595–601
23. Rowland BD, Bernards R, Peeper DS (2005) The KLF4 tumour suppressor is a transcriptional repressor of p53 that acts as a context-dependent oncogene. Nat Cell Biol 7:1074–1082
24. Jiang J, Chan YS, Loh YH, Cai J, Tong GQ, Lim CA, Robson P, Zhong S, Ng HH (2008) A core Klf circuitry regulates self-renewal of embryonic stem cells. Nat Cell Biol 10:353–360
25. Nandan MO, Yang VW (2009) The role of Krüppel-like factors in the reprogramming of somatic cells to induced pluripotent stem cells. Histol Histopathol 24:1343–1355
26. Parry D, Bates S, Mann DJ, Peters G (1995) Lack of cyclin D–Cdk complexes in Rb-negative cells correlated with high levels of p16INK4/MTS1 tumor suppressor gene product. EMBO J 14:503–511
27. Quelle DE, Zindy F, Ashmun RA, Sherr CJ (1995) Alternative reading frames of the NK4a tumor suppressor gene encode two unrelated proteins capable of inducing cell cycle arrest. Cell 83:993–1000
28. Kamijo T, Zindy F, Roussel MF, Quelle DE, Downing JR, Ashmun RA, Grosveld G, Sherr CJ (1997) Tumor suppression at the mouse INK4a locus mediated by the alternative reading frame product p19ARF. Cell 91:649–659
29. Chan MF, van Amerongen R, Nijjar T, Cuppen E, Jones PA, Laird PW (2001) Reduced rates of gene loss, gene silencing, and gene mutation in Dnmt1-deficient embryonic stem cells. Mol Cell Biol 21:7587–7600
30. Banito A, Rashid ST, Acosta JC, Li S, Pereira CF, Geti I, Pinho S, Silva JC, Azuara V, Walsh M, Vallier L, Gil J (2009) Senescence impairs successful reprogramming to pluripotent stem cells. Genes Dev 23:2134–2139
31. Feng Q, Lu SJ, Klimanskaya I, Gomes I, Kim D, Chung Y, Honig GR, Kim KS, Lanza R (2010) Hemangioblastic derivatives from human induced pluripotent stem cells exhibit limited expansion and early senescence. Stem Cell 28:704–712
32. Li H, Collado M, Villasante A, Strati K, Ortega S, Cañamero M, Blasco MA, Serrano M (2009) The Ink4/Arf locus is a barrier for iPS cell reprogramming. Nature 460:1136–1139
33. Utikal J, Polo JM, Stadtfeld M, Maherali N, Kulalert W, Walsh RM, Khalil A, Rheinwald JG, Hochedlinger K (2009) Immortalization eliminates a roadblock during cellular reprogramming into iPS cells. Nature 460:1145–1148
34. Marión RM, Strati K, Li H, Murga M, Blanco R, Ortega S, Fernandez-Capetillo O, Serrano M, Blasco MA (2009) A p53-mediated DNA damage response limits reprogramming to ensure iPS cell genomic integrity. Nature 460:1149–1153
35. Won J, Yim J, Kim TK (2002) Sp1 and Sp3 recruit histone deacetylase to repress transcription of human telomerase reverse transcriptase (hTERT) promoter in normal human somstic cells. J Biol Chem 277:38230–38238
36. Zhu O, Liu C, Ge Z, Fang X, Zhang X, Straat K, Bjorkholm M, Xu D (2008) Lysine-specific demethylase 1 (LSD1) is required for the transcriptional repression of the telomerase reverse transcriptase (hTERT) gene. PLoS One 3:e1446
37. Lin SL, Chang D, Wu DY, Ying SY (2003) A novel RNA splicing-mediated gene silencing mechanism potential for genome evolution. Biochem Biophys Res Commun 310:754–760
38. Lin SL, Ying SY (2006) Gene silencing *in vitro* and *in vivo* using intronic microRNAs. In: Ying

SY (ed) *MicroRNA* protocols. Humana Press, Totowa, NJ, pp 295–312

39. Lin SL, Chang D, Ying SY (2005) Asymmetry of intronic pre-microRNA structures in functional RISC assembly. Gene 356:32–38
40. Danin-Kreiselman M, Lee CY, Chanfreau G (2003) RNAse III-mediated degradation of unspliced pre-mRNAs and lariat introns. Mol Cell 11:1279–1289
41. Yi R, Qin Y, Macara IG, Cullen BR (2003) Exportin-5 mediates the nuclear export of pre-microRNAs and short hairpin RNAs. Genes Dev 17:3011–3016
42. Lund E, Guttinger S, Calado A, Dahlberg JE, Kutay U (2004) Nuclear export of microRNA precursors. Science 303:95–98
43. Zhang G, Taneja KL, Singer RH, Green MR (1994) Localization of pre-mRNA splicing in mammalian nuclei. Nature 372:809–812
44. Lewis BP, Green RE, Brenner SE (2003) Evidence for the widespread coupling of alternative splicing and nonsense-mediated mRNA decay in humans. Proc Natl Acad Sci U S A 100:189–192
45. Grimm D, Streetz KL, Jopling CL, Storm TA, Pandey K, Davis CR, Marion P, Salazar F, Kay MA (2006) Fatality in mice due to oversaturation of cellular microRNA/short hairpin RNA pathways. Nature 441:537–541

Chapter 13

Urinary MicroRNAs as Emerging Class of Noninvasive Biomarkers

Jaroslav Juracek and Ondrej Slaby

Abstract

MicroRNAs (miRNAs) are endogenous noncoding RNAs, which regulate gene expression on the post-transcriptional level. Since miRNAs are involved in the regulation of apoptosis, cellular proliferation, differentiation, and other important cellular processes, their deregulation is important for the development of a wide range of diseases including cancer. Apart from tissue, specific disease-related miRNA signatures can be found in body fluids as well. Especially for urologic diseases or injuries, urine miRNAs represent a promising group of biomarkers. Despite a large number of studies describing the importance of urinary miRNAs, there is a lack of recommendations for urine management and subsequent miRNA analysis. Thus, in this chapter, we aim to describe the origin and functions of urinary miRNAs and discuss the technical aspects of their detection including the pre-analytical phase principles and new directions in quantification, which could forward urine miRNA into clinical practice.

Key words MicroRNA, Urine, Biomarkers, MiRNA detection, MiRNA quantification

1 Introduction

MicroRNAs (miRNAs) are endogenous, 19–25 nucleotides long, single-stranded RNAs, which regulate gene expression by binding to target messenger RNA (mRNA). The discovery of miRNAs is connected with Victor Ambros and Gary Ruvkun who described in 1993 a new mechanism of gene expression in *Caenorhabditis elegans* through small nonprotein-coding RNA lin-4. Since then origin, biogenesis and biological functioning of miRNAs were extensively studied. According to the miRBase database (release 22.1, http://www.mirbase.org/), there is currently 2656 human mature miRNA sequences which have the capacity to regulate more than two-thirds of human protein-coding genes. This makes them key molecules in developmental, physiological, and pathological processes.

Mouldy Sioud (ed.), *RNA Interference and CRISPR Technologies: Technical Advances and New Therapeutic Opportunities*, Methods in Molecular Biology, vol. 2115, https://doi.org/10.1007/978-1-0716-0290-4_13,

While regulatory functions on the cellular level were well described, in 2008 there was the first evidence that miRNAs can occur in circulation as well. In the following years, the potential of circulating miRNAs as a new class of biomarkers was explored considering their high stability, analytical performance, and excellent availability for semi- or noninvasive diagnostics. The key study evaluating the presence of miRNAs in body fluids was published by Weber et al., who verified the level and distribution of miRNAs in the 12 types of body fluids (amniotic fluid, cerebrospinal fluid, plasma, pleural fluid, peritoneal fluid, breast milk, colostrum, saliva, seminal fluid, tears, bronchial lavage and urine). The following studies highlighted miRNA detection in body fluids as a promising diagnostic utility.

Except for blood serum and plasma where miRNAs were firstly described, the presence of these molecules was confirmed also in urine. Urine is a less complex body fluid than blood, represents a more stable environment, and allows noninvasive sampling. Especially for urologic diseases or injuries, it can be thus a valuable source of miRNA-based biomarkers. Currently, there is a large number of studies describing the importance of urinary miRNAs in oncology, nephrology, and cardiology. However, miRNA research in urine lacks recommendations or guidelines for urine sampling and processing as well for miRNA isolation and quantification. In this chapter, we hence discuss mainly the technical aspects of urine miRNAs detection and show new directions in their quantification leading to their usage as clinical biomarkers.

2 Origin and Significance of Urinary miRNAs

2.1 Origin

MicroRNAs were discovered and are most frequently described in cells; nevertheless, the need for noninvasive diagnostics led to extensive interest in the detection of circulating nucleic acids also in body fluids. The first description of circulating microRNAs was published by Lawrie et al. who identified miR-21, miR-155, and miR-210 in serum from diffuse large B-cell lymphoma patients [1]. In urine, the detection of cell-free miRNAs was introduced in 2010 by Hanke et al. Her team presented a robust technology platform providing proof that miRNAs are available in sufficient quantity enabling noninvasive diagnostics of bladder cancer [2]. Despite confirmation of miRNA presence in urine and their good analytical characteristics, it was still unclear how miRNAs enter the circulation.

Although precise mechanisms of miRNA secretion are not fully known, several pathways have been proposed to explain miRNA transport outside the cells. The first main pathway is passive leakage which occurs due to apoptosis, necrosis, cell injury, or inflammation. Within this process, the cytoplasmic membrane is damaged and the cellular content is uncontrollably released into the extracellular space. Passive leakage is an energy-free mechanism and under normal circumstances is not a significant source of circulating miRNAs.

Specific miRNA patterns identified in body fluids however led to the conviction that cell injury is not the only way of miRNA extracellular release. Considering the general instability of free RNA, other important factor is remarkable stability of these molecules outside the cell. Body fluids have been reported to contain high levels of RNase activity, and while synthetic miRNAs degrade shortly after introduction, the level of endogenous miRNAs is not significantly altered [3]. This indicated that extracellular miRNAs must be protected against RNase activity, and it might be connected to the second pathway of miRNA secretion—active transport. The first described way is secretion via extracellular vesicles (EVs) which was reported even before the discovery of miRNAs in body fluids [4]. Extracellular vesicles represent cell-derived circular membrane fragments with the characteristics of the cell of origin and containing cytosol. They comprise of exosomes, shedding vesicles, and apoptotic bodies [5] which differ in size, content, and formation process. Exosomes are released to the extracellular environment after fusion of late multivesicular bodies with the plasma membrane. The other two types of vesicles are released directly from the plasma membrane in living or dying cells [6]. In a wider perspective, the extracellular vesicles in urine originate mainly from renal tubule epithelial cells; however, they can be derived from any epithelial cell type that faces the urinary space. Therefore, it is not surprising that the proteome of urinary exosomes includes proteins that are characteristic of every renal epithelial cell type, as well as podocytes and transitional epithelia from the urinary system [7]. Under physiological conditions, the extracellular vesicle release is connected with the detoxification of the organism or fetal development [8]. However, higher secretion of EVs was described within the stress environment—especially in pathological processes such as cancer. As mentioned by Aubertin et al., anticancer treatment can significantly increase the levels of circulating EVs [9]. Similarly, extracellular miRNAs were found enriched within apoptotic bodies formed by dying cells [10].

Interestingly, even after the isolation of EVs fraction from serum [11] or follicular fluid [12], a big number of cell-free miRNAs was still detectable in vesicle-free fraction. This suggested other forms of non-vesicle-associated miRNAs, and in accordance with this, later studies have confirmed the association of miRNAs with lipoproteins [13] and RNA-binding proteins such as Ago2 [14, 15] or nucleophosmin 1 [16].

2.2 Function of Urinary miRNAs

Together with a role in gene expression regulation, the discovery of extracellular miRNAs being transferred between cells via extracellular vesicles [4] was what triggered the assumption they may be involved in intercellular communication. Follow-up studies demonstrated that after this transfer miRNAs can alter gene expression and cause functional changes within recipient cells [17].

The selective release is supported by different miRNA content of extracellular vesicles and parental cells. Moreover, cells preferentially release and retain specific miRNA subpopulations as shown by Pigati et al. [18]. It is also convenient that EVs released from a given cell type express on their surface the adhesion molecules of the cell of origin, which makes easier to be captured through specific receptor-ligand interaction by target cells that specifically recognize them rather than just by any cell present in the microenvironment [19]. Although specific pathways that control exosome loading and release are not fully known, recent studies demonstrated that the protein heterogeneous nuclear ribonucleoprotein A2B1 and Rab guanosine triphosphatases are involved in these processes [20, 21].

Under physiological conditions, intercellular transport of miRNAs is connected primarily to degradation pathways or developmental processes. Gene ontology analysis of predicted and validated targets of miRNAs shuttled by EVs released by adult human mesenchymal stem cells showed their involvement in multi-organ development, cell survival, differentiation, and immune system regulation [22]. Pro-inflammatory miR-155 and the immune-suppressive miR-146a were found abundant in the extracellular exosomes isolated from dendritic cell cultures. Subsequent uptake by recipient dendritic cells led to miRNA-mediated target gene repression [23]. Exosome transfer also plays a role in T-cell mediated immune response where EVs from endothelial cells modulate T cells and prevent chronic inflammation in tissues through the transfer of anti-inflammatory miRNA [24]. In urine, however, extracellular vesicles originate from nephron segments, and therefore targets for the identified miRNAs are mainly genes encoding membrane transporters and their regulators. This highlights the possibility that urinary miRNAs could modulate key renal tubular functions [25]. An example of miRNA upregulated in urinary exosomes is miR-194-5p, which is involved in innate immunity regulation. Higher secretion of this miRNA may result from the elevated inflammation in glomeruli and podocytes during nephrotic syndrome [26].

The paracrine or even endocrine signaling is most frequently discussed within the process of cancer initiation and progression. Within the tumor microenvironment, cancer cells interact and modulate normal host cells and extracellular matrix as well as effector molecules produced by the surroundings, and to do that, they need to trigger intracellular signaling pathways. Currently, emerging evidence has revealed a variety of exosomal functions, including the induction of immunity [27], epithelial-mesenchymal transition [28], and apoptosis [29] and intercellular transfer of membrane receptors [30], DNA [31], and noncoding RNA [32, 33]. In urinary tract cancers, the role of EVs-wrapped miRNAs is less clear. One of the few reports showed the potential role of urinary exosomal miRNAs in cell communication in renal carcinoma. Based on functional analysis, Butz et al. revealed that the targets of identified miRNAs can play a role in the cell cycle, tumorigenesis, and angiogenesis [34].

3 Analytical Aspects of Urinary miRNAs

Urine represents an easily approachable compartment and a noninvasive source for extracellular miRNAs. The potential use of urinary miRNAs in clinical practice is supported by their exceptional stability and good analytical performance. Despite the large amount of studies focused on the importance of urinary miRNAs, there is a low consensus regarding basic experimental procedures such as RNA isolation, quality control of samples, optimal quantification methods, and data normalization [35]. For this reason, there is often a very low overlap in identified miRNAs, and comparisons between studies from different laboratories are almost impossible. Moreover, most studies differ in the pre-analytical phase which affects greatly the yield and purity of obtained RNA [36]. The only approach enabling consistency and transparency in urinary miRNAs detection is to provide sufficient details on all procedures and methods.

3.1 Stability of Urinary microRNAs

The superior stability of urinary miRNAs even under extreme conditions is considered as the main advantage making them ideal noninvasive biomarkers. The first study focused on laboratory performance of urinary miRNAs evaluated their stability under clinically relevant storage conditions. Mall et al. demonstrated a surprising degree of stability of miRNAs in the urine after storage at room temperature and ten freeze-thaw cycles at −80 °C [37]. Moreover, some reports show that EC miRNAs in body fluids remain stable even when they subjected conditions like boiling, high or low pH, or prolonged storage time [38]. This supports findings of Liu et al. who observed good stability of miR-126 in urine samples kept at room temperature for an extended period of time, 4 °C, −20 °C, and − 80 °C for up to 12 h or subjected to 10 freeze-and-thaw cycles [39]. The presence of proteins in the urine or pH variation did not affect the miRNA level in the urine neither. Similarly, Yun et al. presented that levels of the miRNAs were not significantly changed in any of the experimental samples even after exposer to seven cycles of freezing and thawing or incubation at room temperature for 72 h [40]. Even in the case of miRNAs encapsulated in exosomes, it was confirmed that this fraction was resistant to RNase digestion and stable at 4 °C 24 h, after long-term storage at −80 °C and 5 repeat freeze-thaw cycles [41].

The stability of urine miRNAs is generally explained by conjugation with RNA-binding proteins or by wrapping in extracellular vesicles as described before. However, Mall et al. showed that after trypsin digestion, there was no change in the levels of analyzed miRNAs suggesting that unusual stability of urinary miRNA is not due to its association with proteins in urine [37]. Contrary to this, some studies reported that EV-associated miRNAs represent the minority, whereas approximately 90% of all miRNAs in blood plasma after cell debris removal are protein-bound [14]. However, the latest research shows that miRNAs detected in exosomes represented about 43% of all miRNAs expressed in total urine [42]. This significant difference may be caused by improved isolation strategies enabling an increase in the yield of urine exosomes. The last aspect influencing extracellular miRNAs stability is the primary structure. Bail et al. demonstrated that mutation of 3′ terminal seven nucleotides region of miR-382 increased the stability of the RNA [43]. Moreover, miRNA half-lives appear to significantly correlate with their content of AU and UA dinucleotides [44].

3.2 Urine Management

To obtain high-quality miRNAs and reliable miRNA measurement, there are several specific features that need to be considered when performing miRNAs studies. The first step is urine collection and processing. Most studies use voided urine samples for miRNA analyses which can be further separated using centrifugation into urine sediment and urine supernatant. Urine sediment consists of a mixture of salts, proteins, metabolites, and different cell types which originate in the urogenital tract. For this reason, significantly higher levels of miRNAs were detected in this fraction when compared with supernatant [37]. Nevertheless, the molecular content of urine sediment may differ dramatically under various pathological conditions. Moreover, sediment reflects mainly intracellular miRNA expression level.

On the other hand, miRNAs in urine supernatant originate from shedding microvesicles, exosomes, and apoptotic bodies or from complexes with high-density lipoproteins (HDL) particles and RNA-binding proteins. The inclusion of apoptotic bodies remains, however, questionable because their size is comparable with cell debris and blood platelets, and protocols for urine processing commonly involve removal of these urine components [45]. Despite this, urine supernatant is presented as a more suitable source of extracellular miRNAs.

Another important aspect is urine preservation and storage. Previous studies demonstrated that miRNAs alone are relatively stable in the urinary environment [37, 41]. However, Armstrong et al. are suggesting that some pre-analytical handling conditions are better for small RNA recovery and microRNA profiling than others. Specifically, they recommend spinning out cells/debris and isolate microRNA from the cell-free biofluid immediately after collection. The less preferred option is to hold the sample at room temperature and process within 8 h or use urine one time frozen/thawed [36]. They also highlight precipitation of uromodulin at low temperatures which can entrap EVs. Apart from this, some studies use additives such as RNAlater buffer [46], EDTA [47], guanidinium thiocyanate [48], or commercial RNA stabilizers [49] to stabilize and preserve nucleic acids. An interesting method for urinary miRNAs preservation and storage presents Zhang et al. who introduced the nylon membrane, which exhibits strong binding affinities for nucleic acids via electrostatic interaction. MiRNAs preserved on nylon membranes can be extracted and subsequently analyzed using the traditional downstream analytical applications without any degradation detected [50].

Within EVs biomarker studies, there are additional requirements for sample characterization before proceeding with RNA analysis. Depending on the purification method, different size distribution, morphology, yield, presence of exosomal protein markers, or RNA, the content of EVs is obtained. In connection with the pre-analytical phase, there is also the influence of the Tamm-Horsfall protein (THP/uromodulin) complex, which forms in the urine at reduced temperatures. It is a major protein component of urine which entraps large amounts of EVs or naked miRNAs. Storage and processing of urine may therefore greatly reduce final EVs yield and affect the biomarker profile [51]. The solution could be, for instance, disruption of the interaction between THP filaments and EVs with the help of alkaline pH and lowered ionic concentration as proposed by Puhka et al. [52].

3.3 Purification of Urinary microRNAs

While miRNAs bound with proteins or lipoproteins are occasionally suggested to derive from the passive release of cytosolic components during cell death, encapsulation into EVs is considered as a highly selective process [53]. MiRNAs entrapped in EVs are isolated from the urine environment, and therefore their level should not be influenced by external influences. Suitability of urine EVs fraction for diagnostics is supported by different RNA profile of exosomes and parent cells [54]. On the one hand, despite that some of the miRNAs are enriched in specific fluids [55], some of them might not be detected in exosomes. On the other hand, their expression levels might be particularly high in the unfractionated urine, indicating that absence in exosomes is not due to technological failure [42].

Considering this fact, choosing a suitable fraction of urine as well as proper purification technique represents one of the crucial steps within miRNA analysis. However, currently, there is no standard method for urinary miRNA isolation. In the study of Martínez-Fernández et al., a systematic evaluation of purification methods using commercial extraction kits is presented [56]. Authors purified RNA from voided urine with four selected kits: Norgen Total RNA kit and Norgen miRNA kit (Norgen Biotek), miRCURY kit (Exiqon), and miRNeasy serum/plasma kit (Qiagen). Considering the similar functional principle, none of the methods, however, outperformed significantly other compared kits. Given the low miRNAs concentration in urine, their association with proteins can be used for improved collection and isolation as well. Xu et al. reported the use of 10-nm core-sized magnetic nanoparticles with a carboxylic acid coating that can adsorb low-concentration proteins and form protein corona. Aggregation of nanoparticles with miRNA-associated proteins and microvesicle-related proteins can subsequently simplify and enrich miRNA isolation from urine [57].

Overall, commercial kits validated for extracting the urinary miRNAs are often used for both free and vesicle-encapsulated miRNAs, which may change the representation of miRNAs in the purification product. For this reason, some studies focus only on an exosomal fraction of urine. Commonly, ultracentrifugation or differential centrifugation, immunoaffinity-based capture, and size exclusion chromatography have been used for EVs collection [58]. Alternatively, polymer precipitation often used in commercial kits [59] or microfluidic-based platforms [60] have emerged as a possible replacement. In regard to set the most suitable technique, Royo et al. compared five isolation methods (3 commercial isolation kits, ultracentrifugation, and lectin-based purification), and using a multiplex miRNA assay showed high correlation through all techniques [61]. Interestingly, in their previous work, they showed that different isolation techniques have a greater affinity for urine EVs harboring certain membrane markers, even though most markers were present in all preparations [62]. Similarly, Gheinani et al. compared the size distribution, morphology and yield within ultracentrifugation (UC), polyethylene glycol precipitation (PEG), protein concentration and size-exclusion chromatography (C-SEC), ultracentrifugation and SEC (UC-SEC), and polyethylene glycol precipitation and SEC (PEG-SEC). Although the UC method produced the highest particle yield, the EVs had high dispersion and contained protein aggregates and other particles. After considering all parameters, UC-SEC was evaluated as the best method yielding high purity exosomes [42]. However, Yasui et al. proclaim that none of the existing methodologies have satisfied the requirements for simple and non-invasive urine-based early disease diagnoses, and a lot of them require a large urine input volume [58]. For this reason, they constructed a device composed of nanowires anchored into a microfluidic substrate which enables high-efficiency EVs collections and in situ miRNA extractions. The method combines the mechanical stability of nanowires anchored into substrates during buffer flow and the electrostatic collection of EVs onto the nanowires. Moreover, expression analysis showed that the device extracted a larger variety of species of miRNAs than conventional methods and detected disease-related miRNAs from urine samples of just 1 ml [58]. Also, Ku et al. provided a protocol for urinary EVs enrichment by utilizing an automated microfluidic system, termed acoustic trap [63]. It is microfluidic technology that utilizes the scattering of ultrasound waves off-large pre-seeded polystyrene particles to capture EVs. See the pilot study for more information [64].

An important step for subsequent expression analysis is quality control of obtained RNA. RNA purified from body fluids including urine does not meet the standard quality criteria (evaluated based on A260:A230 and A260:A280 ratios measured using a spectrometer) due to a very low concentration. Despite the low quality, there is no negative effect on further analysis; therefore, new criteria or approach for quality and quantity assessment is needed [65]. Another aspect is contamination with miRNAs released from the circulating blood cells. Especially hemolysis occurring within the sample can enrich urine with red blood cell-specific miRNA, such as miR-451 or miR-16 [66]. This is highly disturbing considering the fact that numerous studies used miR-16 as a reference gene for data normalization in body fluid measurements. Recently the team of Mikael Kubista presented a quality control panel for optimization of miRNA experiments from biofluid samples. The panel uses the two-tailed RT-qPCR technology and is based on two sets of synthetic spike-in molecules and three endogenous miRNAs [67] in order to assess RNA quality, its degradation degree, and contamination. In conclusion, the adaptation of these requirements into MIQE guidelines could help standardize body fluid biomarker studies.

4 Quantification

The study by Weber et al. revealed that urine contains the least amount of detectable miRNA from all tested fluids, whether it is due to uptake by kidneys or destruction in urine environment. From this reason, sensitive and specific quantification method is important for miRNA expression analysis. The most common techniques for miRNA profiling are miRNA microarrays, qRT-PCR arrays, and next-generation sequencing. There is, however, a broad spectrum of methods for subsequent validation.

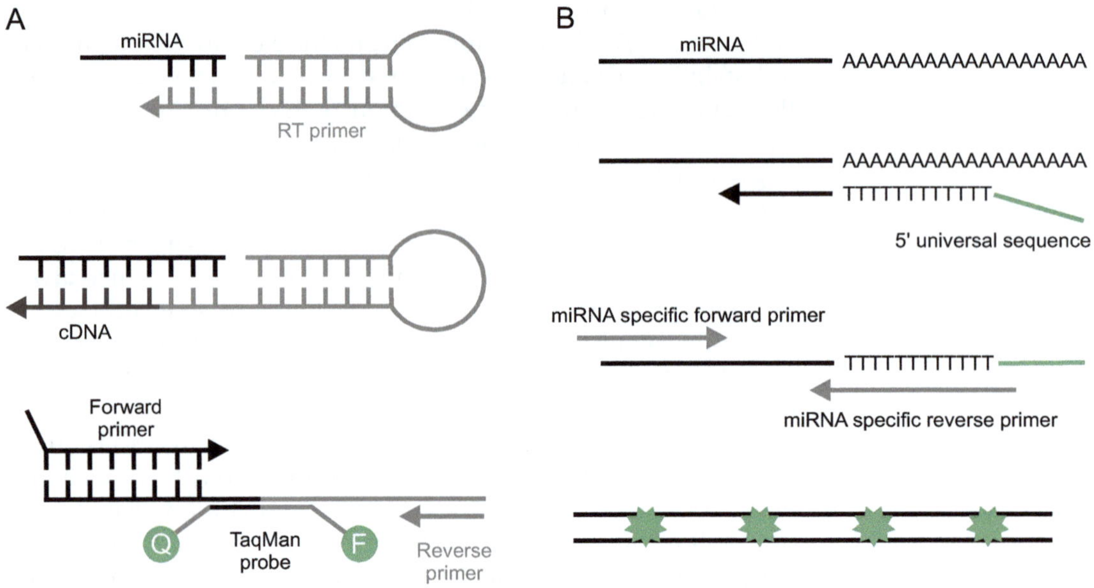

Fig. 1 Basic approaches commonly used for miRNA detection. (**a**) miRNA-specific stem-loop reverse transcription in combination with TaqMan qPCR primer/probe assay. The mature miRNA is extended using a stem-loop primer. For amplification, one miRNA-specific and one universal primer is used. (**b**) Poly-A tailing with locked-nucleic-acid (LNA)-enhanced primers. Mature miRNA is polyadenylated and transcribed into cDNA. Quantification is performed using miRNA-specific primer and LNA-modified primer with nonspecific intercalating dye SYBR Green as detection system

4.1 PCR-Based Methods

Currently, the most frequently used methodology is quantification using RT-qPCR (Fig. 1). Within this method, two basic approaches are commonly used for miRNA detection. The first approach combines poly-A tailing with amplification using microRNA-specific and locked-nucleic-acid (LNA)-enhanced forward and reverse primers and SYBR Green system [68]. Similarly, the universal reverse transcription was performed using a uniform extension of 3′ ends with a linker adapter [69] or a combination of tailing and adaptor ligation (TaqMan Advanced miRNA assay).

The second, originally provided by Life Technologies (now Thermo Fisher Scientific), is miRNA-specific stem-loop reverse transcription in combination with TaqMan qPCR primer/probe assay [70]. Even stem-loop primers represent probably the most frequently used approach, this method possesses some limitations. For instance, false-positive results can appear in the presence of genomic or plasmid DNA containing sequences of the corresponding pre-miRNA. Additionally, different 3′ isomiR species of a particular miRNA can be reverse transcribed and cross-detected [71].

For this reason, novel specific modification of RT-qPCR method are developed (Fig. 2). The two-tailed RT-qPCR uses specific structured RT primers composed of two hemiprobes

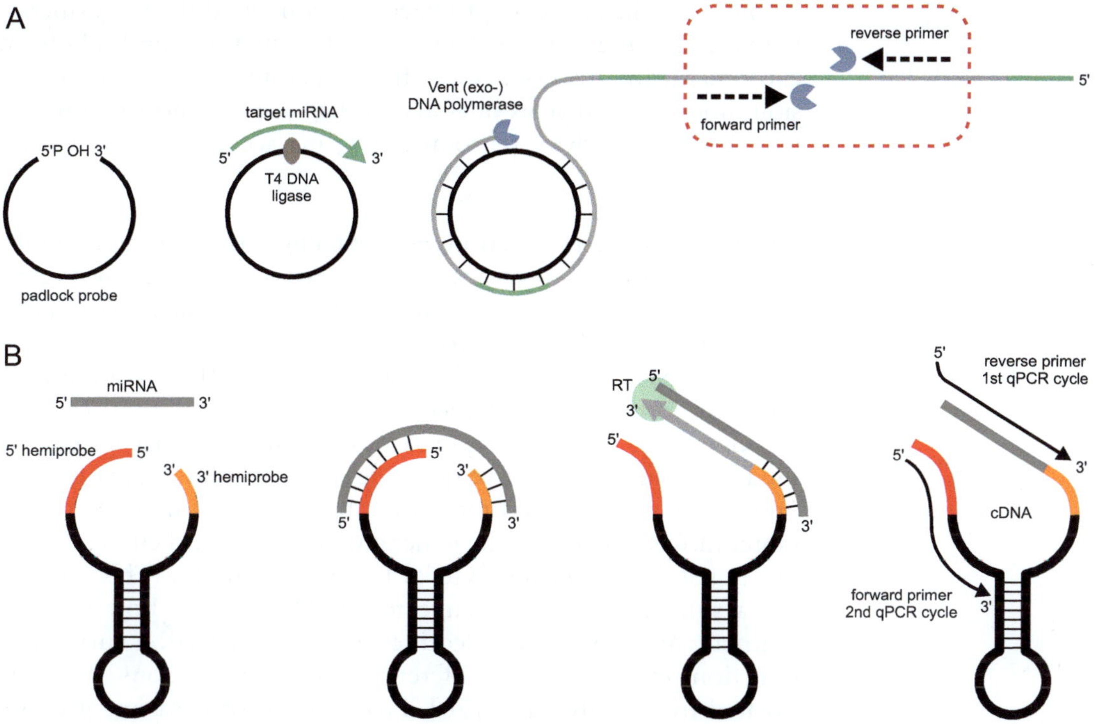

Fig. 2 Novel modification of RT-qPCR method. (**a**) One-step rolling circle quantitative PCR (RC-qPCR). The padlock probe hybridizes to a target miRNA; after ligation, they create a single-stranded circular DNA molecule, which is then a substrate for rolling circle amplification (RCA). The RCA concatemer product is detected by qPCR. (**b**) Two-tailed RT-qPCR. Two-tailed RT primer containing two hemiprobes binds target miRNA forming a stable complex. Reverse transcriptase (RT) binds the 3′ end of the hybridized two-tailed RT primer and elongates it to form tailed cDNA. The cDNA is amplified by qPCR using two target-specific primers

connected by a hairpin. Hemiprobes cooperatively bind miRNA sequence on both 5′ and 3′ ends and increase sensitivity and specificity [72]. Another novel method is rolling circle-quantitative PCR (RC-qPCR) which combines rolling circle amplification (RCA) and qPCR. The enzymes needed for both separate reactions are replaced by one Vent (exo-) DNA polymerase. Within the reaction, the 3′ and 5′ ends of the padlock probe are in the presence of target miRNA ligated to form a circular template. MiRNA initiates the rolling circle amplification in the presence of DNA polymerase, producing long concatemers for qPCR [73]. As a modification of the conventional PCR method, the digital PCR can be considered as well. Specifically, the droplet-based platform (ddPCR) which is applied within miRNA analysis in body fluids. This method enables absolute quantification by counting miRNA molecules encapsulated in discrete, volumetrically defined droplet partitions. With the eligible protocol, ddPCR provided higher precision and less day-to-day variation than RT-qPCR [74]. The same method was used for the quantification of urinary exosomal miRNAs by

Wang et al. The authors optimized and compared the performance between qPCR and ddPCR platforms for the three miRNAs from exosomal urine fraction. After the evaluation of sensitivity, precision, linearity, and reproducibility, it has been found that ddPCR has a higher sensitivity, reproducibility, and accuracy in comparison to qPCR [75].

4.1.1 Data Normalization

One of the biggest limitations and challenges of PCR based methods is data normalization using endogenous and exogenous reference controls for data correction. The main goal is to remove nonbiological variability across samples, so the biological differences can be easily identified. To achieve this, different studies use different normalization methods.

Generally, two main approaches for miRNA analysis can be used—absolute and relative quantification. Absolute quantification provides a means to estimate the absolute copy number of the target miRNA based on the generation of a standard curve, carried out in parallel with qRT-PCR of biological samples. This method has a sufficient dynamic range for miRNA assays; however, the synthetic miRNA oligonucleotides used for calibration curve construction may perform differently than endogenous miRNAs isolated from body fluids [76]. Probably for this reason, absolute quantification is not widely used in the published literature.

Relative quantitation compares a relative measure (expressed as quantification cycle (Cq), the cycle number at which a PCR amplification curve crosses a threshold line set within the exponential growth region) of biological sample to a reference gene. Currently, there is no universal endogenous reference gene or miRNA available for expression data normalization in urine (list of common normalizers in urine is summarized in Table 1). Often reference control is selected based on literature; nevertheless, the best approach is to analyze expression data using different algorithms including geNorm, NormFinder, or BestKeeper. In this case, the selected molecule needs further validation in samples to examine their steady expression across all analyzed groups. Among the most commonly used reference genes are small-nucleolar RNAs, noncoding RNAs involved in the site-specific modification of other RNAs. Despite numerous studies used RNU44, RNU48, RNU43, or RNU6B as normalizers, their level may be deregulated between malignant and nonmalignant tissue and associated with a prognosis of some type of cancer. Obviously, this can introduce bias in miRNA expression evaluation [77].

Biological and technical limitations could be hypothetically eliminated with the use of other miRNAs for data normalization. It is, however, impossible to use single miRNA as a universal normalizer, because miRNAs stable in tissue doesn't have to be stable in body fluids and vice versa. Additionally, the biological function of miRNAs must be considered as well—we assume that

Table 1
Normalization methods used in studies performing urine miRNA analysis published in the time period 2015–2019. Only studies using human urine samples and RT-qPCR method were involved. Studies with the incomplete or unclear experimental design were omitted

Author/year	Study subjects	Urine fraction	Quantification	Normalizer	Ref
Khalid/2019	Kidney transplant patients	Urine supernatant	Relative	Cel-miR-39	[89]
Fredsøe/2019	Prostate cancer, BHP	Urine supernatant	Relative	Mean of miR200b-3p, miR-27b-3p, and miR-30b-5p	[49]
Eissa/2019	Bladder cancer, benign urological conditions, healthy controls	Urine pellet	Relative	RNU6	[90]
Chen/2019	Idiopathic nephrotic syndrome patients	Urine exosomes	Absolute	Urine volume, calibration curves	[26]
Hofbauer/2018	Bladder cancer, controls with microscopic hematuria	Whole urine	Relative	RNU48, miR-103	[91]
Ghorbanmehr/2019	Bladder cancer, prostate cancer, BHP, healthy controls	Voided urine	Relative	5S rRNA	[92]
Prabu/2019	Subjects with normal glucose tolerance, type 2 diabetes patients	Extracellular vesicles	Relative	Cel-miR-39	[93]
Juracek/2018	Bladder cancer, renal cancer, healthy controls	Urine supernatant	Absolute	Calibration curve	[47]
Perez-Hernandez/2018	Hypertensive patients with and without albuminuria, healthy controls	Urine exosomes	Relative	Median normalization, cel-miR-39	[94]
Beltrami/2018	Diabetic kidney disease patients (DKD), diabetic patients without DKD	Urine supernatant	Relative	hsa-miR-191-5p	[95]
Wang/2018	Tuberculosis patients, healthy controls	Urine supernatant	Relative	U6	[96]
Catapano/2018	Duchenne muscular dystrophy patients, healthy controls	Urine exosomes	Relative	miR-16c-5p	[97]
Mytsyk/2018	Renal cancer patients	Whole urine	Relative	U6	[98]

(continued)

Table 1 (continued)

Author/year	Study subjects	Urine fraction	Quantification	Normalizer	Ref
Petrozza/2017	Clear cell RCC patients	Whole urine	Relative	SNORD61, SNORD68, RNU6–2	[99]
Xie/2017	Type 2 diabetes patients	Urine exosomes	Relative	U6	[100]
Kao/2017	Gastric cancer patients	Whole urine	Relative	U6	[101]
Millán/2017	Kidney transplant recipients	Urinary cell pellet	Relative	miR-103	[102]
Fedorko/2017	Clear cell RCC patients	Urine supernatant	Absolute	Calibration curve	[103]
Xu/2017	Prostate cancer patients, BHP, healthy controls	Urine exosomes	Relative	Cel-miR-39	[104]
Mearini/2017	Bladder cancer patients, healthy controls	Urine sediment	Relative	RNU6	[105]
Gan/2017	Preeclampsia pregnancy, healthy pregnancy	Urine cell pellet	Relative	U6	[106]
Domenico/2017	Kidney transplant recipients with graft dysfunction	Urine cell sediment	Relative	Cel-miR-39	[107]
Min/2017	Immunoglobulin A nephropathy patients, healthy controls	Urine exosomes	Relative	RNU6	[108]
Du/2017	Bladder cancer patients, healthy controls	Urine supernatant	Relative	miR-532-5p, let-7b-5p	[109]
Li/2017	Clear cell RCC patients, healthy controls	Urine supernatant	Relative	Cel-miR-39	[110]
Gracia/2017	Healthy volunteers	Urine exosomes	Relative	RNU6B	[25]
Liang/2017	Immunoglobulin A nephropathy patients, healthy controls	Urine cell sediment	Relative	U6	[111]
Ingelmo-Torres/2017	Non-muscle invasive bladder cancer patients	Urinary cell pellet	Relative	miR-103-3p, miR-30c-5p	[112]
	Urothelial carcinoma patients, healthy controls	Urinary exosomes	Relative	Creatinine concentration	[113]

Matsuzaki/ 2017					
Huang/2017	Primary focal segmental glomerulosclerosis, minimal change disease patients	Urine exosomes	Relative	RNU6	[114]
Urquidi/2016	Urothelial carcinoma patients, healthy controls	Urinary cell pellet	Relative	Mamm U6	[115]
Foj/2017	Prostate cancer patients, healthy controls	Urine cell pellet, urine exosomes	Relative	Cel-miR-39	[116]
Zou/2017	Patients who underwent cardiovascular surgery	Urine supernatant	Relative	Cel-miR-39	[117]
Andreu/2017	Bladder cancer patients, healthy controls	Urine exosomes	Relative	RNU1A1, RNU5G	[118]
Eissa/2017	Diabetic nephropathy patients, healthy controls	Urine exosomes	Relative	RNU6–2	[119]
Pospisilova/ 2016	Bladder cancer patients, healthy controls	Urine supernatant	Relative	miR-191, miR-28-3p and miR-200b	[120]
Khafaei/2015	Mustard gas-injured patients, healthy controls	Urine supernatant	Relative	5S rRNA	[121]
Zununi Vahed/ 2016	Kidney recipients, healthy controls	Urine cell pellet	Relative	RNU47, U6	[122]
Ma/2016	Intrahepatic cholestasis of pregnancy, healthy pregnancy	Urine supernatant	Relative	Global expression level, cel-miR-1832, cel-miR-5594-5p	[123]
Duan/2016	Iga nephropathy patients, healthy controls, disease controls	Urine sediment	Relative	U6	[124]
Pavkovic/2016	Acetaminophen overdosed patients, healthy controls	Urine supernatant	Relative	Urinary creatinine concentration	[125]
Sasaki/2016	Bladder cancer patients, urinary tract infections, patients with benign diagnosis	Voided urine	Relative	miR-21-5p	[126]
Stuopelytė/ 2016	Prostate cancer patients, BHP patients	Urine sediment	Relative	Cel-miR-39	[127]
De Long/2015	Bladder cancer patients, healthy controls	Urine supernatant	Relative		[128]

(continued)

Table 1 (continued)

Author/year	Study subjects	Urine fraction	Quantification	Normalizer	Ref
				Without normalization/1/ Ct	
Sapre/2016	Bladder cancer patients, healthy controls	Voided urine	Relative	Urine osmolality	[129]
Weldon/2016	Parent/child, farmworker/non-farmworker pairs	Whole urine	Relative	MammU6	[130]
Delić/2016	Type II diabetes patients, type II diabetic nephropathy patients, healthy controls	Urine exosomes	Relative	miR-197-5p	[131]
Jia/2016	Type II diabetes patients, healthy controls	Urine exosomes	Relative	Cel-miR-39	[132]
Silakit/2017	Periductal fibrosis patients, cholangiocarcinoma patients	Urine supernatant	Relative	miR-16	[133]
Zhang/2016	Non-muscle invasive bladder cancer patients, cystitis patients, healthy donors	Urine supernatant	Relative	RNU6B	[134]
Perez-Hernandez/ 2015	Systemic lupus erythematosus patients, healthy controls	Cell-free urine, exosome-depleted supernatant	Relative	Median normalization, cel-miR-39	[135]
Sedlarikova/ 2015	Multiple myeloma patients, healthy donors	Urine supernatant	Relative	Cel-miR-39	[136]
Samsonov/ 2016	Prostate cancer patients, healthy controls	Urine exosomes	Relative	U6, hsa-miR-191-5p	[137]
Zhou/2015	Ovarian serous adenocarcinoma patients, patients with benign gynecological disease, healthy controls	Urine supernatant	Relative	RNU6	[138]
Wang/2015	Bladder cancer patients, healthy controls	Urine supernatant	Relative	Geometric mean of U6, RNU48	[139]
Solé/2015	Lupus nephritis patients, non-lupus chronic kidney disease patients, healthy controls	Urine exosomes	Relative	RNU6	[140]

Wang/2015	Iga nephropathy patients, patients with membranous nephropathy, patients with minimal changes disease, healthy subjects	Urine cell pellet	Relative	U6	[141]
Yun/2015	Prostate cancer patients, BHP patients	Urine supernatant	Absolute	Calibration curve	[142]
Erbes/2015	Breast cancer patients, healthy controls	Urine supernatant	Relative	miR-16, miR-26b	[143]
Stephan/2015	Prostate cancer patients, healthy controls	Urine cell pellet	Relative	miR-130b	[144]
Ramezani/2015	Focal segmental glomerulosclerosis patients, healthy controls	Urine exosomes	Relative	RNU6B	[145]
Eissa/2015	Bladder cancer patients, benign bladder lesions, healthy individuals	Urine sediment	Relative	U6	[146]
Korzeniewski/2015	Prostate cancer patients, healthy controls	Urine supernatant	Relative	Cel-miR-39	[147]
Giovannelli/2015	Jcpyv-DNA-positive, jcpyv-DNA-negative patients	Urine exosomes	Absolute	Calibration curve	[148]

miRNA that has been chosen as control is independent for studied biological variation. A good example is miR-16, which is frequently used as a reference gene [78] because it is highly expressed and relatively invariant across various samples. On the other hand, this miRNA occurs during hemolysis and is significantly downregulated in malignant prostate tissue. Normalization to miR-16 may thus lead to biased results as reported by Schaefer et al. [79]. Similarly, miR-93 and miR-191 were described as good normalizers for cancer tissue [80], while in the urine of bladder cancer patients, their level is significantly elevated [47], suggesting them as diagnostic biomarkers. Mentioned algorithms calculate stability score and intergroup and intragroup variances for specific conditions and therefore should mirror not only biological features but also technical variables.

Thus, the additional goal is the elimination of technological variability occurring during analysis. For this purpose, nonhuman or synthetic RNA such as *C. elegans* miRNA cel-miR-39 is spiked-in to the sample before the RNA isolation, which allows to monitor RNA purification and reverse transcription efficiency. It doesn't, however, correct for deviations in sampling and quality of samples.

Although a global agreement about miRNA reference gene in urine samples is still lacking, all mentioned aspects help to increase the accuracy of miRNA assessment. Identification of suitable reference controls for miRNA analysis in urine remains an important challenge.

4.2 New Quantification Methods

In order to achieve the detection of miRNA at a lower concentrations or to simplify the operation procedure, additional alternative approaches have been proposed. These methods differ in physical principles, signal detection, sample preparation, and others. Because of the broad range of technologies, the following list focuses only on methods validated within urinary miRNAs.

Numerous studies focus on direct detection of RNA without the need of reverse transcription and further amplification. In 2008, Geiss et al. reported a technology called the NanoString nCounter gene expression system, which captures and quantifies individual RNA transcripts including mRNAs or miRNAs [81]. Using fluorescently labeled molecular barcodes (reporter probe) and capture probe, the system can count miRNAs in various biological materials including urine [82]. Simple and sensitive direct detection method presented Min et al. who proposed a one-pot microRNA detection assay based on TPE-DNA bioprobes with only fluorogens and without quencher groups. Such arrangement should avoid difficulties with probe design such as the relative distances between the fluorogens and quencher groups [83]. Alternatively, molecular beacons were used for exosome miRNAs direct quantification in urine. The molecular beacon (MB) is a nano-sized bi-labelled (fluorescent and quencher) hairpin molecule. The loop within MB structure is complementary to the sequence of target miRNA; therefore, during hybridization hairpin-loop structure opens which results in fluorescence. As reported by Lee et al., this assay is able to successfully detect miRNAs in exosomes produced by prostate cancer cells as well as in urine of healthy controls [84].

Other large group of techniques represents biosensors utilizing optical or electrochemical detection. In urine, a technology combining surface plasmon resonance (SPR) sensor and a microfluidic chip was introduced in 2015. Due to the detection of collective oscillations of free electrons (surface plasmons) after target molecule hybridization, the method enabled quantify miRNA-16-5p in urine samples from patients with acute kidney injury [85]. Same authors extended this work in order to present a multiplex miRNA automated detection platform. Automatization is ensured by the use of gold nanoparticles instead of magnetic particles and transmission SPR sensing sensor. Interestingly authors showed that measurable signal was obtained after only 10 min of sample injection at a femtomolar concentration of the target miRNAs [86]. A different approach is electrochemical detection using a glassy carbon electrode with immobilized hybridization probes. In the presence of target miRNA, resistance is changing at the probe surface and miRNA concentration is obtained based on electrochemical measurements [87].

Less common are methods using colorimetric detection of urinary miRNAs. The approach by Nossier et al. is based on salt-induced aggregation citrate-capped gold nanoparticles (AuNPs), which may be prevented by miRNA or microRNA-probe hybrid due to AuNP stabilization. In the presence of the target miRNA, the nanoparticles are protected against salt-induced aggregation and keep the color unchanged [88]. The presence of miRNA can be then observed with the bare eye.

5 Conclusions

Urinary cell-free miRNAs are emerging as potential noninvasive biomarkers of urologic diseases and injury. Their implication in clinical practice suffers from lack of guidelines and recommendations for urine sampling and processing as well for miRNA isolation and quantification. Most of the detection methods are based on PCR technology for which proper data normalization is crucial. Since no universal normalizer is known, alternative approaches for urinary miRNAs quantification have been proposed. Improved detection together with standardized urine handling is needed to gain the full potential of urinary miRNAs.

Acknowledgments

This work was supported by grant NV18-03-00360 of the Czech Ministry of Health.

References

1. Lawrie CH, Gal S, Dunlop HM, Pushkaran B, Liggins AP, Pulford K, Banham AH, Pezzella F, Boultwood J, Wainscoat JS, Hatton CS, Harris AL (2008) Detection of elevated levels of tumour-associated microRNAs in serum of patients with diffuse large B-cell lymphoma. Br J Haematol 141:672–675
2. Hanke M, Hoefig K, Merz H, Feller AC, Kausch I, Jocham D, Warnecke JM, Sczakiel G (2010) A robust methodology to study urine microRNA as tumor marker: microRNA-126 and microRNA-182 are related to urinary bladder cancer. Urol Oncol 28:655–661
3. Mitchell PS, Parkin RK, Kroh EM, Fritz BR, Wyman SK, Pogosova-Agadjanyan EL, Peterson A, Noteboom J, O'Briant KC, Allen A, Lin DW, Urban N, Drescher CW, Knudsen BS, Stirewalt DL, Gentleman R, Vessella RL, Nelson PS, Martin DB, Tewari M (2008) Circulating microRNAs as stable blood-based markers for cancer detection. Proc Natl Acad Sci U S A 105:10513–10518
4. Valadi H, Ekstrom K, Bossios A, Sjostrand M, Lee JJ, Lotvall JO (2007) Exosome-mediated transfer of mRNAs and microRNAs is a novel mechanism of genetic exchange between cells. Nat Cell Biol 9:654–659
5. Camussi G, Deregibus MC, Bruno S, Grange C, Fonsato V, Tetta C (2011) Exosome/microvesicle-mediated epigenetic reprogramming of cells. Am J Cancer Res 1:98–110
6. Hessvik NP, Llorente A (2018) Current knowledge on exosome biogenesis and release. Cell Mol Life Sci 75:193–208
7. Zhou H, Yuen PS, Pisitkun T, Gonzales PA, Yasuda H, Dear JW, Gross P, Knepper MA, Star RA (2006) Collection, storage, preservation, and normalization of human urinary exosomes for biomarker discovery. Kidney Int 69:1471–1476
8. Keller S, Rupp C, Stoeck A, Runz S, Fogel M, Lugert S, Hager HD, Abdel-Bakky MS, Gutwein P, Altevogt P (2007) CD24 is a marker of exosomes secreted into urine and amniotic fluid. Kidney Int 72:1095–1102
9. Aubertin K, Silva AK, Luciani N, Espinosa A, Djemat A, Charue D, Gallet F, Blanc-Brude-O, Wilhelm C (2016) Massive release of extracellular vesicles from cancer cells after photodynamic treatment or chemotherapy. Sci Rep 6:35376
10. Zernecke A, Bidzhekov K, Noels H, Shagdarsuren E, Gan L, Denecke B, Hristov M, Koppel T, Jahantigh MN, Lutgens E, Wang S, Olson EN, Schober A, Weber C (2009) Delivery of microRNA-126 by apoptotic bodies induces CXCL12-dependent vascular protection. Sci Signal 2: ra81
11. Noferesti SS, Sohel MM, Hoelker M, Salilew-Wondim D, Tholen E, Looft C, Rings F, Neuhoff C, Schellander K, Tesfaye D (2015) Controlled ovarian hyperstimulation induced changes in the expression of circulatory miRNA in bovine follicular fluid and blood plasma. J Ovarian Res 8:81
12. Sohel MM, Hoelker M, Noferesti SS, Salilew-Wondim D, Tholen E, Looft C, Rings F, Uddin MJ, Spencer TE, Schellander K, Tesfaye D (2013) Exosomal and non-exosomal transport of extra-cellular micrornas in follicular fluid: implications for bovine oocyte developmental competence. PLoS One 8: e78505
13. Vickers KC, Palmisano BT, Shoucri BM, Shamburek RD, Remaley AT (2011) MicroRNAs are transported in plasma and delivered to recipient cells by high-density lipoproteins. Nat Cell Biol 13:423–433
14. Arroyo JD, Chevillet JR, Kroh EM, Ruf IK, Pritchard CC, Gibson DF, Mitchell PS, Bennett CF, Pogosova-Agadjanyan EL, Stirewalt DL, Tait JF, Tewari M (2011) Argonaute2 complexes carry a population of circulating microRNAs independent of vesicles in human plasma. Proc Natl Acad Sci U S A 108:5003–5008
15. Turchinovich A, Weiz L, Langheinz A, Burwinkel B (2011) Characterization of extracellular circulating microRNA. Nucleic Acids Res 39:7223–7233
16. Wang K, Zhang S, Weber J, Baxter D, Galas DJ (2010) Export of microRNAs and microRNA-protective protein by mammalian cells. Nucleic Acids Res 38:7248–7259
17. Kosaka N, Iguchi H, Yoshioka Y, Takeshita F, Matsuki Y, Ochiya T (2010) Secretory mechanisms and intercellular transfer of

microRNAs in living cells. J Biol Chem 285:17442–17452

18. Pigati L, Yaddanapudi SC, Iyengar R, Kim DJ, Hearn SA, Danforth D, Hastings ML, Duelli DM (2010) Selective release of microRNA species from normal and malignant mammary epithelial cells. PLoS One 5:e13515
19. Losche W, Scholz T, Temmler U, Oberle V, Claus RA (2004) Platelet-derived microvesicles transfer tissue factor to monocytes but not to neutrophils. Platelets 15:109–115
20. Ostrowski M, Carmo NB, Krumeich S, Fanget I, Raposo G, Savina A, Moita CF, Schauer K, Hume AN, Freitas RP, Goud B, Benaroch P, Hacohen N, Fukuda M, Desnos C, Seabra MC, Darchen F, Amigorena S, Moita LF, Thery C (2010) Rab27a and Rab27b control different steps of the exosome secretion pathway. Nat Cell Biol 12:19–30; sup pp 11–13
21. Villarroya-Beltri C, Gutierrez-Vazquez C, Sanchez-Cabo F, Perez-Hernandez D, Vazquez J, Martin-Cofreces N, Martinez-Herrera DJ, Pascual-Montano A, Mittelbrunn M, Sanchez-Madrid F (2013) Sumoylated hnRNPA2B1 controls the sorting of miRNAs into exosomes through binding to specific motifs. Nat Commun 4:2980
22. Collino F, Deregibus MC, Bruno S, Sterpone L, Aghemo G, Viltono L, Tetta C, Camussi G (2010) Microvesicles derived from adult human bone marrow and tissue specific mesenchymal stem cells shuttle selected pattern of miRNAs. PLoS One 5:e11803
23. Alexander M, Hu R, Runtsch MC, Kagele DA, Mosbruger TL, Tolmachova T, Seabra MC, Round JL, Ward DM, O'Connell RM (2015) Exosome-delivered microRNAs modulate the inflammatory response to endotoxin. Nat Commun 6:7321
24. Njock MS, Cheng HS, Dang LT, Nazari-Jahantigh M, Lau AC, Boudreau E, Roufaiel M, Cybulsky MI, Schober A, Fish JE (2015) Endothelial cells suppress monocyte activation through secretion of extracellular vesicles containing antiinflammatory microRNAs. Blood 125:3202–3212
25. Gracia T, Wang X, Su Y, Norgett EE, Williams TL, Moreno P, Micklem G, Karet Frankl FE (2017) Urinary exosomes contain MicroRNAs capable of paracrine modulation of tubular transporters in kidney. Sci Rep 7:40601
26. Chen T, Wang C, Yu H, Ding M, Zhang C, Lu X, Zhang CY, Zhang C (2019) Increased urinary exosomal microRNAs in children with idiopathic nephrotic syndrome. EBioMedicine 39:552–561
27. Whiteside TL (2017) The effect of tumor-derived exosomes on immune regulation and cancer immunotherapy. Future Oncol 13:2583–2592
28. Vella LJ (2014) The emerging role of exosomes in epithelial-mesenchymal-transition in cancer. Front Oncol 4:361
29. Wieckowski EU, Visus C, Szajnik M, Szczepanski MJ, Storkus WJ, Whiteside TL (2009) Tumor-derived microvesicles promote regulatory T cell expansion and induce apoptosis in tumor-reactive activated CD8+ T lymphocytes. J Immunol 1950(183):3720–3730
30. Al-Nedawi K, Meehan B, Micallef J, Lhotak V, May L, Guha A, Rak J (2008) Intercellular transfer of the oncogenic receptor EGFRvIII by microvesicles derived from tumour cells. Nat Cell Biol 10:619–624
31. Guescini M, Genedani S, Stocchi V, Agnati LF (2010) Astrocytes and Glioblastoma cells release exosomes carrying mtDNA. J Neural Transm (Vienna) 117:1–4
32. Hardin H, Helein H, Meyer K, Robertson S, Zhang R, Zhong W, Lloyd RV (2018) Thyroid cancer stem-like cell exosomes: regulation of EMT via transfer of lncRNAs. Lab Invest 98:1133–1142
33. Sempere LF, Christensen M, Silahtaroglu A, Bak M, Heath CV, Schwartz G, Wells W, Kauppinen S, Cole CN (2007) Altered MicroRNA expression confined to specific epithelial cell subpopulations in breast cancer. Cancer Res 67:11612–11620
34. Butz H, Nofech-Mozes R, Ding Q, Khella HWZ, Szabo PM, Jewett M, Finelli A, Lee J, Ordon M, Stewart R, Krylov S, Yousef GM (2016) Exosomal MicroRNAs are diagnostic biomarkers and can mediate cell-cell communication in renal cell carcinoma. Eur Urol Focus 2:210–218
35. Kirschner MB, van Zandwijk N, Reid G (2013) Cell-free microRNAs: potential biomarkers in need of standardized reporting. Front Genet 4:56
36. Armstrong DA, Dessaint JA, Ringelberg CS, Hazlett HF, Howard L, Abdalla MAK, Barnaby RL, Stanton BA, Cervinski MA, Ashare A (2018) Pre-analytical handling conditions and small RNA recovery from urine for miRNA profiling. J Mol Diagn 20:565–571
37. Mall C, Rocke DM, Durbin-Johnson B, Weiss RH (2013) Stability of miRNA in human urine supports its biomarker potential. Biomark Med 7:623–631
38. Beltrami C, Clayton A, Phillips AO, Fraser DJ, Bowen T (2012) Analysis of urinary

microRNAs in chronic kidney disease. Biochem Soc Trans 40:875–879

39. Liu Y, Gao G, Yang C, Zhou K, Shen B, Liang H, Jiang X (2014) Stability of miR-126 in urine and its potential as a biomarker for renal endothelial injury with diabetic nephropathy. Int J Endocrinol 2014:393109
40. Yun SJ, Jeong P, Kim WT, Kim TH, Lee YS, Song PH, Choi YH, Kim IY, Moon SK, Kim WJ (2012) Cell-free microRNAs in urine as diagnostic and prognostic biomarkers of bladder cancer. Int J Oncol 41:1871–1878
41. Lv LL, Cao Y, Liu D, Xu M, Liu H, Tang RN, Ma KL, Liu BC (2013) Isolation and quantification of microRNAs from urinary exosomes/microvesicles for biomarker discovery. Int J Biol Sci 9:1021–1031
42. Gheinani AH, Vögeli M, Baumgartner U, Vassella E, Draeger A, Burkhard FC, Monastyrskaya K (2018) Improved isolation strategies to increase the yield and purity of human urinary exosomes for biomarker discovery. Sci Rep 8:3945
43. Bail S, Swerdel M, Liu H, Jiao X, Goff LA, Hart RP, Kiledjian M (2010) Differential regulation of microRNA stability. RNA 16:1032–1039
44. Sethi P, Lukiw WJ (2009) Micro-RNA abundance and stability in human brain: specific alterations in Alzheimer's disease temporal lobe neocortex. Neurosci Lett 459:100–104
45. Turchinovich A, Weiz L, Burwinkel B (2012) Extracellular miRNAs: the mystery of their origin and function. Trends Biochem Sci 37:460–465
46. Salido-Guadarrama AI, Morales-Montor JG, Rangel-Escareno C, Langley E, Peralta-Zaragoza O, Cruz Colin JL, Rodriguez-Dorantes M (2016) Urinary microRNA-based signature improves accuracy of detection of clinically relevant prostate cancer within the prostate-specific antigen grey zone. Mol Med Rep 13:4549–4560
47. Juracek J, Peltanova B, Dolezel J, Fedorko M, Pacik D, Radova L, Vesela P, Svoboda M, Slaby O, Stanik M (2018) Genome-wide identification of urinary cell-free microRNAs for non-invasive detection of bladder cancer. J Cell Mol Med 22:2033–2038
48. Hanke M, Kausch I, Dahmen G, Jocham D, Warnecke JM (2007) Detailed technical analysis of urine RNA-based tumor diagnostics reveals ETS2/urokinase plasminogen activator to be a novel marker for bladder cancer. Clin Chem 53:2070–2077
49. Fredsoe J, Rasmussen AKI, Laursen EB, Cai Y, Howard KA, Pedersen BG, Borre M, Mouritzen P, Orntoft T, Sorensen KD (2019) Independent validation of a diagnostic noninvasive 3-microRNA ratio model (uCaP) for prostate cancer in cell-free urine. Clin Chem 65:540–548
50. Zhang F, Cheng X, Yuan Y, Wu J, Gao Y (2015) Urinary microRNA can be concentrated, dried on membranes and stored at room temperature in vacuum bags. PeerJ 3: e1082
51. Wachalska M, Koppers-Lalic D, van Eijndhoven M, Pegtel M, Geldof AA, Lipinska AD, van Moorselaar RJ, Bijnsdorp IV (2016) Protein complexes in urine interfere with extracellular vesicle biomarker studies. J Circulat Biomarkers 5:4
52. Puhka M, Nordberg ME, Valkonen S, Rannikko A, Kallioniemi O, Siljander P, Af Hallstrom TM (2017) KeepEX, a simple dilution protocol for improving extracellular vesicle yields from urine. Eur J Pharma Sci 98:30–39
53. Guduric-Fuchs J, O'Connor A, Camp B, O'Neill CL, Medina RJ, Simpson DA (2012) Selective extracellular vesicle-mediated export of an overlapping set of microRNAs from multiple cell types. BMC Genomics 13:357
54. Janas T, Janas MM, Sapon K, Janas T (2015) Mechanisms of RNA loading into exosomes. FEBS Lett 589:1391–1398
55. Weber JA, Baxter DH, Zhang S, Huang DY, Huang KH, Lee MJ, Galas DJ, Wang K (2010) The microRNA spectrum in 12 body fluids. Clin Chem 56:1733–1741
56. Martinez-Fernandez M, Paramio JM, Duenas M (2016) RNA detection in urine: from RNA extraction to good normalizer molecules. J Mol Diagn 18:15–22
57. Xu S, Nasr SH, Chen D, Zhang X, Sun L, Huang X, Qian C (2018) MiRNA extraction from cell-free biofluid using protein corona formed around carboxyl magnetic nanoparticles. ACS Biomater Sci Eng 4:654–662
58. Yasui T, Yanagida T, Ito S, Konakade Y, Takeshita D, Naganawa T, Nagashima K, Shimada T, Kaji N, Nakamura Y, Thiodorus IA, He Y, Rahong S, Kanai M, Yukawa H, Ochiya T, Kawai T, Baba Y (2017) Unveiling massive numbers of cancer-related urinary-microRNA candidates via nanowires. Sci Adv 3:e1701133
59. Soares Martins T, Catita J, Martins Rosa I, A B da Cruz E Silva O, Henriques AG (2018) Exosome isolation from distinct biofluids using precipitation and column-based approaches. PLoS One 13, e0198820
60. Zhang P, He M, Zeng Y (2016) Ultrasensitive microfluidic analysis of circulating exosomes

using a nanostructured graphene oxide/polydopamine coating. Lab Chip 16:3033–3042

61. Royo F, Diwan I, Tackett MR, Zuñiga P, Sanchez-Mosquera P, Loizaga-Iriarte A, Ugalde-Olano A, Lacasa I, Perez A, Unda M, Carracedo A, Falcon-Perez JM (2016) Comparative miRNA analysis of urine extracellular vesicles isolated through five different methods. Cancers (Basel) 8
62. Royo F, Zuniga-Garcia P, Sanchez-Mosquera P, Egia A, Perez A, Loizaga A, Arceo R, Lacasa I, Rabade A, Arrieta E, Bilbao R, Unda M, Carracedo A, Falcon-Perez JM (2016) Different EV enrichment methods suitable for clinical settings yield different subpopulations of urinary extracellular vesicles from human samples. J Extracell Vesicles 5:29497
63. Ku A, Ravi N, Yang M, Evander M, Laurell T, Lilja H, Ceder Y (2019) A urinary extracellular vesicle microRNA biomarker discovery pipeline; from automated extracellular vesicle enrichment by acoustic trapping to microRNA sequencing. PLoS One 14:e0217507
64. Ku A, Lim HC, Evander M, Lilja H, Laurell T, Scheding S, Ceder Y (2018) Acoustic enrichment of extracellular vesicles from biological fluids. Anal Chem 90:8011–8019
65. Mlcochova H, Hezova R, Stanik M, Slaby O (2014) Urine microRNAs as potential noninvasive biomarkers in urologic cancers. Urol Oncol 32(41):e41–e49
66. Kirschner MB, Kao SC, Edelman JJ, Armstrong NJ, Vallely MP, van Zandwijk N, Reid G (2011) Haemolysis during sample preparation alters microRNA content of plasma. PLoS One 6:e24145
67. Androvic P, Romanyuk N, Urdzikova-Machova L, Rohlova E, Kubista M, Valihrach L (2019) Two-tailed RT-qPCR panel for quality control of circulating microRNA studies. Sci Rep 9:4255
68. Balcells I, Cirera S, Busk PK (2011) Specific and sensitive quantitative RT-PCR of miRNAs with DNA primers. BMC Biotechnol 11:70
69. Benes V, Collier P, Kordes C, Stolte J, Rausch T, Muckentaler MU, Häussinger D, Castoldi M (2015) Identification of cytokine-induced modulation of microRNA expression and secretion as measured by a novel microRNA specific qPCR assay. Sci Rep 5:11590
70. Chen C, Ridzon DA, Broomer AJ, Zhou Z, Lee DH, Nguyen JT, Barbisin M, Xu NL, Mahuvakar VR, Andersen MR, Lao KQ, Livak KJ, Guegler KJ (2005) Real-time quantification of microRNAs by stem-loop RT-PCR. Nucleic Acids Res 33:e179
71. Schamberger A, Orban TI (2014) 3' IsomiR species and DNA contamination influence reliable quantification of microRNAs by stem-loop quantitative PCR. PLoS One 9: e106315
72. Androvic P, Valihrach L, Elling J, Sjoback R, Kubista M (2017) Two-tailed RT-qPCR: a novel method for highly accurate miRNA quantification. Nucleic Acids Res 45:e144
73. Xu M, Ye J, Yang D, Abdullah Al-maskri AA, Hu H, Jung C, Cai S, Zeng S (2019) Ultrasensitive detection of miRNA via one-step rolling circle-quantitative PCR (RC-qPCR). Anal Chim Acta 1077:208–215
74. Giraldez MD, Chevillet JR, Tewari M (2018) Droplet digital PCR for absolute quantification of extracellular microRNAs in plasma and serum: quantification of the cancer biomarker hsa-miR-141. Methods Mol Biol 1768:459–474
75. Wang C, Ding Q, Plant P, Basheer M, Yang C, Tawedrous E, Krizova A, Boulos C, Farag M, Cheng Y, Yousef GM (2019) Droplet digital PCR improves urinary exosomal miRNA detection compared to real-time PCR. Clin Biochem 67:54–59
76. Kroh EM, Parkin RK, Mitchell PS, Tewari M (2010) Analysis of circulating microRNA biomarkers in plasma and serum using quantitative reverse transcription-PCR (qRT-PCR). Methods 50:298–301
77. Gee HE, Buffa FM, Camps C, Ramachandran A, Leek R, Taylor M, Patil M, Sheldon H, Betts G, Homer J, West C, Ragoussis J, Harris AL (2011) The small-nucleolar RNAs commonly used for microRNA normalisation correlate with tumour pathology and prognosis. Br J Cancer 104:1168–1177
78. Muller V, Gade S, Steinbach B, Loibl S, von Minckwitz G, Untch M, Schwedler K, Lubbe K, Schem C, Fasching PA, Mau C, Pantel K, Schwarzenbach H (2014) Changes in serum levels of miR-21, miR-210, and miR-373 in HER2-positive breast cancer patients undergoing neoadjuvant therapy: a translational research project within the Geparquinto trial. Breast Cancer Res Treat 147:61–68
79. Schaefer A, Jung M, Miller K, Lein M, Kristiansen G, Erbersdobler A, Jung K (2010) Suitable reference genes for relative quantification of miRNA expression in prostate cancer. Exp Mol Med 42:749–758

80. Peltier HJ, Latham GJ (2008) Normalization of microRNA expression levels in quantitative RT-PCR assays: identification of suitable reference RNA targets in normal and cancerous human solid tissues. RNA 14:844–852
81. Geiss GK, Bumgarner RE, Birditt B, Dahl T, Dowidar N, Dunaway DL, Fell HP, Ferree S, George RD, Grogan T, James JJ, Maysuria M, Mitton JD, Oliveri P, Osborn JL, Peng T, Ratcliffe AL, Webster PJ, Davidson EH, Hood L, Dimitrov K (2008) Direct multiplexed measurement of gene expression with color-coded probe pairs. Nat Biotechnol 26:317–325
82. Armstrong DA, Green BB, Seigne JD, Schned AR, Marsit CJ (2015) MicroRNA molecular profiling from matched tumor and bio-fluids in bladder cancer. Mol Cancer 14:194
83. Min X, Zhuang Y, Zhang Z, Jia Y, Hakeem A, Zheng F, Cheng Y, Tang BZ, Lou X, Xia F (2015) Lab in a tube: sensitive detection of MicroRNAs in urine samples from bladder cancer patients using a single-label DNA probe with AIEgens. ACS Appl Mater Interfaces 7:16813–16818
84. Lee J, Kwon MH, Kim JA, Rhee WJ (2018) Detection of exosome miRNAs using molecular beacons for diagnosing prostate cancer. Artif Cells Nanomed Biotechnol 46:S52–s63
85. Mousavi MZ, Chen HY, Lee KL, Lin H, Chen HH, Lin YF, Wong CS, Li HF, Wei PK, Cheng JY (2015) Urinary micro-RNA biomarker detection using capped gold nanoslit SPR in a microfluidic chip. Analyst 140:4097–4104
86. Yeung WK, Chen HY, Sun JJ, Hsieh TH, Mousavi MZ, Chen HH, Lee KL, Lin H, Wei PK, Cheng JY (2018) Multiplex detection of urinary miRNA biomarkers by transmission surface plasmon resonance. Analyst 143:4715–4722
87. Smith DA, Newbury LJ, Drago G, Bowen T, Redman JE (2017) Electrochemical detection of urinary microRNAs via sulfonamide-bound antisense hybridisation. Sensors Actuators B Chem 253:335–341
88. Nossier AI, Abdelzaher H, Matboli M, Eissa S (2018) Dual approach for the colorimetric determination of unamplified microRNAs by using citrate capped gold nanoparticles. Mikrochim Acta 185:236
89. Khalid U, Newbury LJ, Simpson K, Jenkins RH, Bowen T, Bates L, Sheerin NS, Chavez R, Fraser DJ (2019) A urinary microRNA panel that is an early predictive biomarker of delayed graft function following kidney transplantation. Sci Rep 9:3584
90. Eissa S, Safwat M, Matboli M, Zaghloul A, El-Sawalhi M, Shaheen A (2019) Measurement of urinary level of a specific competing endogenous RNA network (FOS and RCAN mRNA/miR-324-5p, miR-4738-3p, / lncRNA miR-497-HG) enables diagnosis of bladder cancer. Urol Oncol 37:292.e219–292.e227
91. Hofbauer SL, de Martino M, Lucca I, Haitel A, Susani M, Shariat SF, Klatte T (2018) A urinary microRNA (miR) signature for diagnosis of bladder cancer. Urol Oncol 36:531.e531–531.e538
92. Ghorbanmehr N, Gharbi S, Korsching E, Tavallaei M, Einollahi B, Mowla SJ (2019) miR-21-5p, miR-141-3p, and miR-205-5p levels in urine-promising biomarkers for the identification of prostate and bladder cancer. Prostate 79:88–95
93. Prabu P, Rome S, Sathishkumar C, Gastebois C, Meugnier E, Mohan V, Balasubramanyam M (2019) MicroRNAs from urinary extracellular vesicles are non-invasive early biomarkers of diabetic nephropathy in type 2 diabetes patients with the 'Asian Indian phenotype'. Diabetes Metab 45:276–285
94. Perez-Hernandez J, Olivares D, Forner MJ, Ortega A, Solaz E, Martinez F, Chaves FJ, Redon J, Cortes R (2018) Urinary exosome miR-146a is a potential marker of albuminuria in essential hypertension. J Transl Med 16:228
95. Beltrami C, Simpson K, Jesky M, Wonnacott A, Carrington C, Holmans P, Newbury L, Jenkins R, Ashdown T, Dayan C, Satchell S, Corish P, Cockwell P, Fraser D, Bowen T (2018) Association of elevated urinary miR-126, miR-155, and miR-29b with diabetic kidney disease. Am J Pathol 188:1982–1992
96. Wang J, Zhu X, Xiong X, Ge P, Liu H, Ren N, Khan FA, Zhou X, Zhang L, Yuan X, Chen X, Chen Y, Hu C, Robertson ID, Chen H, Guo A (2018) Identification of potential urine proteins and microRNA biomarkers for the diagnosis of pulmonary tuberculosis patients. Emerg Microbes Infect 7:63
97. Catapano F, Domingos J, Perry M, Ricotti V, Phillips L, Servais L, Seferian A, Groot I, Krom YD, Niks EH, Verschuuren JJ, Straub V, Voit T, Morgan J, Muntoni F (2018) Downregulation of miRNA-29, −23

and −21 in urine of Duchenne muscular dystrophy patients. Epigenomics 10:875–889

98. Mytsyk Y, Dosenko V, Borys Y, Kucher A, Gazdikova K, Busselberg D, Caprnda M, Kruzliak P, Farooqi AA, Lubov M (2018) MicroRNA-15a expression measured in urine samples as a potential biomarker of renal cell carcinoma. Int Urol Nephrol 50:851–859
99. Petrozza V, Pastore AL, Palleschi G, Tito C, Porta N, Ricci S, Marigliano C, Costantini M, Simone G, Di Carlo A, Gallucci M, Carbone A, Fazi F (2017) Secreted miR-210-3p as non-invasive biomarker in clear cell renal cell carcinoma. Oncotarget 8:69551–69558
100. Xie Y, Jia Y, Cuihua X, Hu F, Xue M, Xue Y (2017) Urinary exosomal microRNA profiling in incipient type 2 diabetic kidney disease. J Diabetes Res 2017:6978984
101. Kao HW, Pan CY, Lai CH, Wu CW, Fang WL, Huang KH, Lin WC (2017) Urine miR-21-5p as a potential non-invasive biomarker for gastric cancer. Oncotarget 8:56389–56397
102. Millan O, Budde K, Sommerer C, Aliart I, Rissling O, Bardaji B, Matz M, Zeier M, Silva I, Guirado L, Brunet M (2017) Urinary miR-155-5p and CXCL10 as prognostic and predictive biomarkers of rejection, graft outcome and treatment response in kidney transplantation. Br J Clin Pharmacol 83:2636–2650
103. Fedorko M, Juracek J, Stanik M, Svoboda M, Poprach A, Buchler T, Pacik D, Dolezel J, Slaby O (2017) Detection of let-7 miRNAs in urine supernatant as potential diagnostic approach in non-metastatic clear-cell renal cell carcinoma. Biochem Med 27:411–417
104. Xu Y, Qin S, An T, Tang Y, Huang Y, Zheng L (2017) MiR-145 detection in urinary extracellular vesicles increase diagnostic efficiency of prostate cancer based on hydrostatic filtration dialysis method. Prostate 77:1167–1175
105. Mearini E, Poli G, Cochetti G, Boni A, Egidi MG, Brancorsini S (2017) Expression of urinary miRNAs targeting NLRs inflammasomes in bladder cancer. Onco Targets Ther 10:2665–2673
106. Gan L, Liu Z, Wei M, Chen Y, Yang X, Chen L, Xiao X (2017) MiR-210 and miR-155 as potential diagnostic markers for preeclampsia pregnancies. Medicine 96: e7515
107. Domenico TD, Joelsons G, Montenegro RM, Manfro RC (2017) Upregulation of microRNA 142-3p in the peripheral blood and urinary cells of kidney transplant recipients with post-transplant graft dysfunction. Braz J Med Biol Res 50:e5533
108. Min QH, Chen XM, Zou YQ, Zhang J, Li J, Wang Y, Li SQ, Gao QF, Sun F, Liu J, Xu YM, Lin J, Huang LF, Huang B, Wang XZ (2018) Differential expression of urinary exosomal microRNAs in IgA nephropathy. J Clin Lab Anal 32
109. Du L, Jiang X, Duan W, Wang R, Wang L, Zheng G, Yan K, Wang L, Li J, Zhang X, Pan H, Yang Y, Wang C (2017) Cell-free microRNA expression signatures in urine serve as novel noninvasive biomarkers for diagnosis and recurrence prediction of bladder cancer. Oncotarget 8:40832–40842
110. Li G, Zhao A, Peoch M, Cottier M, Mottet N (2017) Detection of urinary cell-free miR-210 as a potential tool of liquid biopsy for clear cell renal cell carcinoma. Urol Oncol 35:294–299
111. Liang S, Cai GY, Duan ZY, Liu SW, Wu J, Lv Y, Hou K, Li ZX, Zhang XG, Chen XM (2017) Urinary sediment miRNAs reflect tubulointerstitial damage and therapeutic response in IgA nephropathy. BMC Nephrol 18:63
112. Ingelmo-Torres M, Lozano JJ, Izquierdo L, Carrion A, Costa M, Gomez L, Ribal MJ, Alcaraz A, Mengual L (2017) Urinary cell microRNA-based prognostic classifier for non-muscle invasive bladder cancer. Oncotarget 8:18238–18247
113. Matsuzaki K, Fujita K, Jingushi K, Kawashima A, Ujike T, Nagahara A, Ueda Y, Tanigawa G, Yoshioka I, Ueda K, Hanayama R, Uemura M, Miyagawa Y, Tsujikawa K, Nonomura N (2017) MiR-21-5p in urinary extracellular vesicles is a novel biomarker of urothelial carcinoma. Oncotarget 8:24668–24678
114. Huang Z, Zhang Y, Zhou J, Zhang Y (2017) Urinary exosomal miR-193a can be a potential biomarker for the diagnosis of primary focal segmental glomerulosclerosis in children. Biomed Res Int 2017:7298160
115. Urquidi V, Netherton M, Gomes-Giacoia E, Serie D, Eckel-Passow J, Rosser CJ, Goodison S (2016) Urinary mRNA biomarker panel for the detection of urothelial carcinoma. Oncotarget 7:38731–38740
116. Foj L, Ferrer F, Serra M, Arevalo A, Gavagnach M, Gimenez N, Filella X (2017) Exosomal and non-exosomal urinary miRNAs in prostate cancer detection and prognosis. Prostate 77:573–583

117. Zou YF, Wen D, Zhao Q, Shen PY, Shi H, Zhao Q, Chen YX, Zhang W (2017) Urinary MicroRNA-30c-5p and MicroRNA-192-5p as potential biomarkers of ischemia-reperfusion-induced kidney injury. Exp Biol Med (Maywood) 242:657–667
118. Andreu Z, Otta Oshiro R, Redruello A, Lopez-Martin S, Gutierrez-Vazquez C, Morato E, Marina AI, Olivier Gomez C, Yanez-Mo M (2017) Extracellular vesicles as a source for non-invasive biomarkers in bladder cancer progression. Eur J Pharm Sci 98:70–79
119. Eissa S, Matboli M, Bekhet MM (2016) Clinical verification of a novel urinary microRNA panel: 133b, −342 and −30 as biomarkers for diabetic nephropathy identified by bioinformatics analysis. Biomed Pharmacother 83:92–99
120. Pospisilova S, Pazourkova E, Horinek A, Brisuda A, Svobodova I, Soukup V, Hrbacek J, Capoun O, Hanus T, Mares J, Korabecna M, Babjuk M (2016) MicroRNAs in urine supernatant as potential non-invasive markers for bladder cancer detection. Neoplasma 63:799–808
121. Khafaei M, Samie S, Mowla SJ, Alvanegh AG, Mirzaei B, Chavoshei S, Dorraj GS, Esmailnejad M, Tavallaie M, Nourani M (2015) Evaluation of miR-9 and miR-143 expression in urine specimens of sulfur mustard exposed patients. Interdiscip Toxicol 8:169–174
122. Zununi Vahed S, Omidi Y, Ardalan M, Samadi N (2017) Dysregulation of urinary miR-21 and miR-200b associated with interstitial fibrosis and tubular atrophy (IFTA) in renal transplant recipients. Clin Biochem 50:32–39
123. Ma L, Zhang XQ, Zhou DX, Cui Y, Deng LL, Yang T, Shao Y, Ding M (2016) Feasibility of urinary microRNA profiling detection in intrahepatic cholestasis of pregnancy and its potential as a non-invasive biomarker. Sci Rep 6:31535
124. Duan ZY, Cai GY, Bu R, Lu Y, Hou K, Chen XM (2016) Selection of urinary sediment miRNAs as specific biomarkers of IgA nephropathy. Sci Rep 6:23498
125. Pavkovic M, Robinson-Cohen C, Chua AS, Nicoara O, Cardenas-Gonzalez M, Bijol V, Ramachandran K, Hampson L, Pirmohamed M, Antoine DJ, Frendl G, Himmelfarb J, Waikar SS, Vaidya VS (2016) Detection of drug-induced acute kidney injury in humans using urinary KIM-1, miR-21, −200c, and −423. Toxicol Sci 152:205–213
126. Sasaki H, Yoshiike M, Nozawa S, Usuba W, Katsuoka Y, Aida K, Kitajima K, Kudo H, Hoshikawa M, Yoshioka Y, Kosaka N, Ochiya T, Chikaraishi T (2016) Expression level of urinary microRNA-146a-5p is increased in patients with bladder cancer and decreased in those after transurethral resection. Clin Genitourin Cancer 14:e493–e499
127. Stuopelyte K, Daniunaite K, Jankevicius F, Jarmalaite S (2016) Detection of miRNAs in urine of prostate cancer patients. Medicina (Kaunas) 52:116–124
128. Long JD, Sullivan TB, Humphrey J, Logvinenko T, Summerhayes KA, Kozinn S, Harty N, Summerhayes IC, Libertino JA, Holway AH, Rieger-Christ KM (2015) A non-invasive miRNA based assay to detect bladder cancer in cell-free urine. Am J Transl Res 7:2500–2509
129. Sapre N, Macintyre G, Clarkson M, Naeem H, Cmero M, Kowalczyk A, Anderson PD, Costello AJ, Corcoran NM, Hovens CM (2016) A urinary microRNA signature can predict the presence of bladder urothelial carcinoma in patients undergoing surveillance. Br J Cancer 114:454–462
130. Weldon BA, Shubin SP, Smith MN, Workman T, Artemenko A, Griffith WC, Thompson B, Faustman EM (2016) Urinary microRNAs as potential biomarkers of pesticide exposure. Toxicol Appl Pharmacol 312:19–25
131. Delic D, Eisele C, Schmid R, Baum P, Wiech F, Gerl M, Zimdahl H, Pullen SS, Urquhart R (2016) Urinary exosomal miRNA signature in Type II diabetic nephropathy patients. PLoS One 11: e0150154
132. Jia Y, Guan M, Zheng Z, Zhang Q, Tang C, Xu W, Xiao Z, Wang L, Xue Y (2016) miRNAs in urine extracellular vesicles as predictors of early-stage diabetic nephropathy. J Diabetes Res 2016:7932765
133. Silakit R, Loilome W, Yongvanit P, Thongchot S, Sithithaworn P, Boonmars T, Koonmee S, Titapun A, Khuntikeo N, Chamadol N, Techasen A, Namwat N (2017) Urinary microRNA-192 and microRNA-21 as potential indicators for liver fluke-associated cholangiocarcinoma risk group. Parasitol Int 66:479–485
134. Zhang X, Zhang Y, Liu X, Fang A, Wang J, Yang Y, Wang L, Du L, Wang C (2016) Direct quantitative detection for cell-free miR-155 in urine: a potential role in diagnosis and prognosis for non-muscle invasive bladder cancer. Oncotarget 7:3255–3266

135. Perez-Hernandez J, Forner MJ, Pinto C, Chaves FJ, Cortes R, Redon J (2015) Increased urinary exosomal microRNAs in patients with systemic lupus erythematosus. PLoS One 10:e0138618
136. Sedlarikova L, Besse L, Novosadova S, Kubaczkova V, Radova L, Stanik M, Krejci M, Hajek R, Sevcikova S (2015) MicroRNAs in urine are not biomarkers of multiple myeloma. J Negat Results Biomed 14:16
137. Samsonov R, Shtam T, Burdakov V, Glotov A, Tsyrlina E, Berstein L, Nosov A, Evtushenko V, Filatov M, Malek A (2016) Lectin-induced agglutination method of urinary exosomes isolation followed by mi-RNA analysis: application for prostate cancer diagnostic. Prostate 76:68–79
138. Zhou J, Gong G, Tan H, Dai F, Zhu X, Chen Y, Wang J, Liu Y, Chen P, Wu X, Wen J (2015) Urinary microRNA-30a-5p is a potential biomarker for ovarian serous adeno carcinoma. Oncol Rep 33:2915–2923
139. Wang J, Zhang X, Wang L, Dong Z, Du L, Yang Y, Guo Y, Wang C (2015) Downregulation of urinary cell-free microRNA-214 as a diagnostic and prognostic biomarker in bladder cancer. J Surg Oncol 111:992–999
140. Sole C, Cortes-Hernandez J, Felip ML, Vidal M, Ordi-Ros J (2015) miR-29c in urinary exosomes as predictor of early renal fibrosis in lupus nephritis. Nephrol Dial Transplant 30:1488–1496
141. Wang N, Bu R, Duan Z, Zhang X, Chen P, Li Z, Wu J, Cai G, Chen X (2015) Profiling and initial validation of urinary microRNAs as biomarkers in IgA nephropathy. PeerJ 3:e990
142. Yun SJ, Jeong P, Kang HW, Kim YH, Kim EA, Yan C, Choi YK, Kim D, Kim JM, Kim SK, Kim SY, Kim ST, Kim WT, Lee OJ, Koh GY, Moon SK, Kim IY, Kim J, Choi YH, Kim WJ (2015) Urinary MicroRNAs of prostate cancer: virus-encoded hsvl-miRH18 and hsv2-miR-H9-5p could be valuable diagnostic markers. Int Neurourol J 19:74–84
143. Erbes T, Hirschfeld M, Rucker G, Jaeger M, Boas J, Iborra S, Mayer S, Gitsch G, Stickeler E (2015) Feasibility of urinary microRNA detection in breast cancer patients and its potential as an innovative non-invasive biomarker. BMC Cancer 15:193
144. Stephan C, Jung M, Rabenhorst S, Kilic E, Jung K (2015) Urinary miR-183 and miR-205 do not surpass PCA3 in urine as predictive markers for prostate biopsy outcome despite their highly dysregulated expression in prostate cancer tissue. Clin Chem Lab Med 53:1109–1118
145. Ramezani A, Devaney JM, Cohen S, Wing MR, Scott R, Knoblach S, Singhal R, Howard L, Kopp JB, Raj DS (2015) Circulating and urinary microRNA profile in focal segmental glomerulosclerosis: a pilot study. Eur J Clin Investig 45:394–404
146. Eissa S, Habib H, Ali E, Kotb Y (2015) Evaluation of urinary miRNA-96 as a potential biomarker for bladder cancer diagnosis. Med Oncol 32:413
147. Korzeniewski N, Tosev G, Pahernik S, Hadaschik B, Hohenfellner M, Duensing S (2015) Identification of cell-free microRNAs in the urine of patients with prostate cancer. Urol Oncol 33:16.e17–16.e22
148. Giovannelli I, Martelli F, Repice A, Massacesi L, Azzi A, Giannecchini S (2015) Detection of JCPyV microRNA in blood and urine samples of multiple sclerosis patients under natalizumab therapy. J Neurovirol 21:666–670

Chapter 14

Improving Dendritic Cell Cancer Vaccine Potency Using RNA Interference

Stein Sæbøe-Larssen and Mouldy Sioud

Abstract

Dendritic cell cancer vaccines have already become a treatment modality for patients with various cancer types. However, the curative potential of this immunotherapy is limited by the existence of negative feedback mechanisms that control dendritic cells (DCs) and T-cell function. By inhibiting the expression of inhibitory factors using RNA interference technology, a new generation of DC vaccines was developed. Vaccine-stimulated T cells showed antitumor effects both in vitro and in cancer patients. Here, we describe the development and validation of a fully GMP-compliant production process of ex vivo DC cancer vaccines combined with the blockade of immunosuppressive pathways using small interfering RNAs. The protocol can be used for DC-based therapy for all cancer types.

Key words Small interfering RNAs, Immunosuppression, Indoleamine 2,3-dioxygenase, Checkpoint receptors, Immunotherapy, Dendritic cells, Cancer vaccines

1 Introduction

Therapeutic vaccines represent a viable option for active immunotherapy of cancers that aim to treat late-stage disease by using a patient's own immune system [1]. Despite considerable efforts to develop cancer vaccines, the clinical translation of this therapy into efficacious immunotherapy has been challenging for decades. In addition to the immunosuppressive tumor microenvironments, the immune counteracting mechanisms engaged during cancer vaccine responses may also compromise antitumor responses and contribute to tumor escape. Therefore, innovative strategies are mandatory for overcoming these tumor-dependent and independent immunoregulatory or immunosuppressive mechanisms or pathways to achieve beneficial clinical outcomes in cancer vaccine therapy (*see* Chapter 15). To induce effective immune responses against tumors, there is a need of inhibiting the expression of factors that dampen immune responses in patients. This could potentially lead

Mouldy Sioud (ed.), *RNA Interference and CRISPR Technologies: Technical Advances and New Therapeutic Opportunities*, Methods in Molecular Biology, vol. 2115, https://doi.org/10.1007/978-1-0716-0290-4_14,

to longer activation of DCs and T and NK cells and might be beneficial for cancer immunotherapy [2]. Among the immune inhibitory molecules expressed by dendritic cells (DC) themselves and tumor cells is indoleamine 2,3-dioxygenase (IDO), an enzyme that conveys immunosuppressive effects by degrading tryptophan, an essential amino acid required for T-cell proliferation and survival [3–6]. Depletion of tryptophan by IDO in the local microenvironment induces T-cell apoptosis and impairs normal T-cell function [6]. IDO-expressing DCs can also induce differentiation of naïve CD4+ T cells into regulatory T cells which suppress antitumor immunity [7]. Moreover, various tryptophan metabolites such as N-formylkynurenine and quinolinate suppress T-cell activation [3, 4]. These observations indicate that the regulation of tryptophan metabolism by IDO in DCs is a highly adaptable modulator of immunity. Indeed, injection of IDO-positive DC into mice suppressed the activation of antigen-specific T cells in the lymph nodes draining the injection site [3]. Effector T cells starved of tryptophan were unable to proliferate and enter into G1 cell cycle arrest. A promising strategy for enhancing antitumor immunity in cancer patients would be the blockade of IDO expression in DC cancer vaccines using RNA interference [8, 9]. To this end, we have developed several effective small interfering RNAs (siRNAs) targeting IDO that exhibited silencing potency at nanomolar concentrations [10, 11]. This chapter describes a facile protocol for in vitro synthesis of GMP siRNA using T7 RNA polymerase and the generation of siRNA-modified DC cancer vaccines under GMP-grade conditions.

2 Materials

2.1 Equipment

1. Fresco centrifuge.
2. Eppendorf centrifuge.
3. cleanLAF VFR 1206 BS.
4. Sterile bench.
5. Heating block.
6. Nanodrop apparatus.
7. Bioanalyzer.

2.2 Reagents

2.2.1 In Vitro Transcription

1. Ribomax large-scale RNA production system-T7.
2. DNA oligonucleotides.
3. TRIZOL reagent.
4. Chloroform.
5. Isoamyl alcohol.
6. Isopropanol.

7. 75% ethanol, absolute standard diluted in DEPC-treated sterile water.
8. Diethyl pyrocarbonate (DEPC), approx. 97% NMR.
9. Deionized and DEPC-treated sterile water.
10. 1.5 ml sterile centrifugation tube.
11. 15 ml centrifuge tubes.
12. 1.0 ml cryotubes.

2.2.2 Dendritic Cell Generation

1. VueLife FEP Teflon cell culture bags.
2. CellGro DC medium.
3. VLE-RPMI medium.
4. Granulocyte/macrophage colony-stimulating factor (GM-CSF).
5. Interleukin-4 (IL-4).
6. Interleukin-1β (IL-1β).
7. Interleukin-6 (IL-6).
8. Tumor necrosis factor-α (TNF-α).
9. Prostaglandin E2 (PgE2).
10. ECM 830 square-wave electroporator (BTX).
11. 4 mm (0.8 ml) electroporation cuvettes (BTX).

3 Methods

3.1 In Vitro Generation of siRNAs

Although chemically synthesized siRNAs can be obtained from commercial sources (*see* Chapter 2), below a technically simple, quick, and cost-effective method for the production of GMP-siRNAs that makes use of in vitro transcription is described. The siRNA sequences are transcribed from chemically made DNA oligonucleotides containing the T7 promoter (*see* **Note 1**). Alternatively, other DNA templates can be used (*see* **Note 2**). It should be noted that in vitro transcribed siRNA molecules contain 5′ triphosphate group that efficiently activates the RIG-1 pathway and can be used as gene silencing and immunostimulatory agent (adjuvant) for cancer immunotherapy ([8], *see* Chapter 16). As a working example, a siRNA targeting IDO was selected and in vitro transcribed using the following set of DNA oligonucleotides.

Sense strand:

Oligo 1: 5′`CGTTTAATACGACTCACTATAGGTCCGTGAGTTTGTCCTTTCAA`3′.

Oligo 2: 5′`TTGAAAGGACAAACTCACGGACCTATAGTGAGTCGTATTAAACG`3′.

Antisense strand:

Oligo 1: 5′CGTTTAATACGACTCACTATAGAAAGGACAAACTCACGGACTG3′.

Oligo 2: 5′CAGTCCGTGAGTTTGTCCTTTCTATAGTGAGTCGTATTAAACG3′.

3.2 Oligonucleotide Annealing

Prior to transcription, the DNA oligonucleotides must be annealed to form double-stranded DNA, and siRNA sense and antisense strands are then transcribed from the double-stranded DNA template. When designing primers containing the T7 promoter sequence, it is recommended to add one or two GG after the T7 promoter sequence. Transcription efficiency has been shown to be reduced if the start site was not optimized (*see* **Notes 1** and **2**). Usually annealed oligonucleotides are used for short transcripts such as siRNAs.

Double-stranded DNA template for the sense strand:

5′-CGTTTAATACGACTCACTATA**G**$^{+1}$GTCCGTGAGTTTGTCCTTTCAA-3′

3′-GCAAATTATGCTGAGTGATATCC AGGCACTCAAACAGGAAAGTT-5′

Double-stranded DNA template for the antisense strand:

5′-CGTTTAATACGACTCACTATA**G**$^{+1}$AAAGGACAAACTCACGGACTG-3′

3′-GCAAATTATGCTGAGTGATATC TTTCCTGTTTGAGTGCCTGAC-5′

Underlined sequence corresponds to the consensus sequence of T7 RNA polymerase with indication of the +1 nucleotide (bold; corresponds to the first nucleotide in the transcript).

3.3 In Vitro Transcription of siRNA

For short template like siRNA, we recommend to increase the polymerase concentration and reaction time, because transcription of small RNAs is largely dependent on the number of initiation events. For 1 ml RNA synthesis reaction, combine the following components:

1. 1000 μl mRNA synthesis.

60 μl	rATP of 100 mM to 6 mM
60 μl	rCTP of 100 mM to 6 mM
60 μl	rUTP of 100 mM to 6 mM
60 μl	rGTP of 100 mM to 6 mM
200 μl	5× T7-transcription buffer
50 μg	Annealed DNA oligonucleotides with T7 promoter
100 μl	T7 RNA polymerase-mixture (MIX)
x	DEPC-water
1000 μl	Total

2. Mix properly with a pipette and incubate 37 °C for 4–6 h. In this protocol, both sense and antisense siRNA strands are transcribed in the same transcription reaction. After transcription, they will anneal to form siRNA duplexes, which are more stable than single-stranded siRNAs (*see* **Note 3**).
3. To eliminate DNA template, add RQ1 RNase-free DNase to a concentration of 0.5 U/μg of template DNA. Incubate for 15 min at 37 °C.
4. With respect to long transcripts such as mRNA-encoding tumor antigens used for vaccination, similar conditions are used for in vitro transcription (*see* **Note 4**).

3.4 Purification of Transcribed siRNAs

1. Transfer the transcription reaction (1 ml) into 15 ml centrifuge tube and add 9 ml TRIZOL. Shake the tube well for 15 s and incubate 5 min at room temperature.
2. Add 1.8 ml chloroform per ml TRIZOL. Shake the tube well for 15 s and incubate 5 min at RT.
3. Centrifuge at 12,000 × *g* for 15 min at 4 °C.
4. Transfer aqueous phase (top layer) containing the RNA to a new centrifuge tube. Be careful not to disrupt the interphase pellet.
5. Add 1 volume isopropanol and mix well by inverting the tube several times and incubate for 20 min at RT.
6. Centrifuge at 12,000 × *g* for 30 min at 4 °C with the RNA pellet facing outward. Mark the tube orientation in order to easily localize the RNA pellet.
7. Remove the supernatant without disturbing the RNA pellet and add 1 ml ice-cold RNase-free 75% ethanol.
8. Centrifuge at 12,000 × *g* for 5 min at 4 °C. Gently remove the supernatant with a pipette.
9. Centrifuge at 12,000 × *g* for 1 min at 4 °C and remove the remaining supernatant with a pipette.
10. Dry the RNA pellet under a lamp, and then dissolve it in 100 μl DEPC-MQ water.
11. Transfer 2 μl of the RNA mixture to a new tube for quality control and add 18 μl DEPC-MQ water.
12. Store siRNA preparation at −70 °C until use.
13. Measure siRNA concentration by nanodrop (*see* **Note 5**).
14. Dilute siRNA in DEPC-MQ water to the desired concentration and make aliquots. Remove a sample (5 μl) from one aliquot for contamination test.
15. Store siRNA at −70 °C.

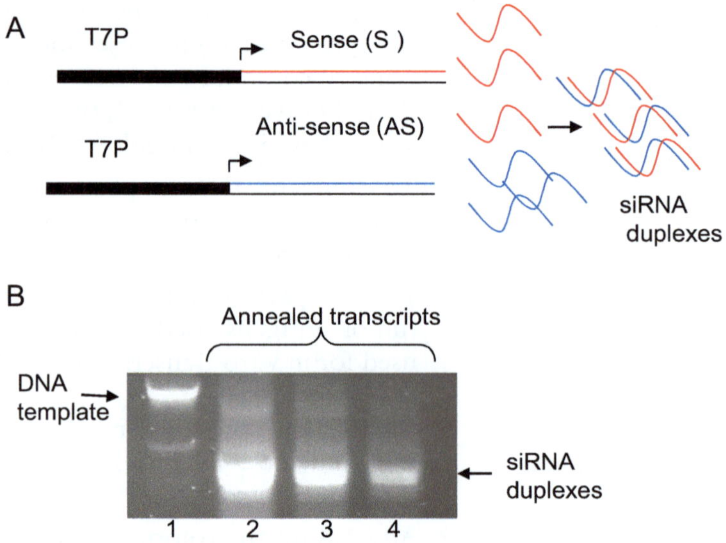

Fig. 1 Characterization of IDO siRNA. (**a**) Schematic representation of in vitro transcription of siRNAs from DNA templates. (**b**) Analysis of siRNA duplexes. Subsequent to in vitro transcription and siRNA precipitation, IDO siRNA was analyzed by a 15% PAGE gel. Lanes 2, 3, and 4 contain various siRNA concentrations. Lane 1 contains the DNA template

3.5 Quality Control of siRNA

1. Analyze siRNA preparation by gel electrophoresis to verify that DNA was removed and that the siRNA synthesis is of satisfactory quality (*see* **Note 6**, Fig. 1a, b).
2. Furthermore, you can use the bioanalyzer from Agilent Technologies to monitor siRNA quality.

3.6 DC Generation and Transfection with mRNA-Encoding Tumor Antigens and siRNAs

1. Obtain a monocyte-enriched cell fraction from peripheral blood. In a clinical setting, this is usually done by harvesting peripheral blood mononuclear cells (PBMCs) by leukapheresis, followed by enrichment of monocytes by elutriation. In a research lab, it is more common to isolate PBMCs using Lymphoprep and remove T and B cells by immunomagnetic depletion. A freeze-thaw cycle may be used to remove granulocytes.
2. Culture cells for 4 days at 10^6 cells/ml in CellGro DC medium supplemented with 2500 U/ml GM-CSF and 1000 U/ml IL-4 for differentiation to immature DC. VueLife Teflon culture bags or similar should be used to avoid cell adherence.
3. Prepare culture medium for maturation of transfected cells (1 ml/10^6 cells); CellGro DC medium supplemented with 2500 U/ml GM-CSF, 1000 U/ml IL-4, 10 ng/ml IL-1β, 1000 U/ml IL-6, 10 ng/ml TNFα, and 1 μg/ml PgE2. Store it in a 37 °C incubator.

4. Prepare a 10× concentrate of the mRNA, or a mixture of several RNAs, in nuclease-free water (10× concentration of each component, compared to final concentration in electroporation). The concentration of mRNA in electroporation is usually in the range 10–50 μg/ml, depending on the desired level of protein expression and/or the sensitivity of the method used to verify protein expression. At this stage, add the in vitro transcribed siRNA (around 10 μg) to the mRNA sample.
5. Collect DC cells by centrifugation (600 × *g*; 5 min) and wash once in RPMI medium. Then resuspend the cells in RPMI at a concentration 4×10^7 cells/ml, or lower.
6. When using standard 4 mm (0.8 ml) electroporation cuvettes, dispense 0.72 ml cell suspension to new tubes. The final volume in a 4 mm cuvette should be 0.8 ml to produce an optimal voltage drop.
7. Pulse parameters when using a square-wave electroporator should be 500 V for 2 ms. Lower voltage/longer duration may be used, but expect more aggregation of cells on the electrodes (*see* **Note 7**).
8. Add 80 μl mRNA and siRNA mixture to a tube with cells. Mix well with a pipette set at 0.8 ml, transfer to the cuvette (avoid foam and air bubbles), and apply the pulse.
9. Loosen cells that may have attached to the electrodes with a pipette and transfer cells to maturation medium. Incubate for 1 day at 37 °C.
10. Cryopreserve cells in separate vaccine doses with the desired number of cells until use (*see* **Note 8**). At the stage you may verify the gene silencing in DC preparations. Figure 2 illustrates the gene silencing potency of IDO siRNA in 6 DC cancer vaccine preparations.
11. Therapeutic DC cancer vaccines aim to eliminate cancer cells via activation and enhancement of the patient immune system against cancer cells (*see* **Note 9**).

4 Notes

1. The T7 promoter has the requirement of a guanosine (G) in the position +1 position. Therefore, in vitro transcribed siRNAs must begin with at least one G. Based upon these roles, the siRNA site requirement is G-N17C (site length is 19 nt). Furthermore, the T7 RNA polymerase functions best if the first three nucleotides of the RNA are G-purine-purine.

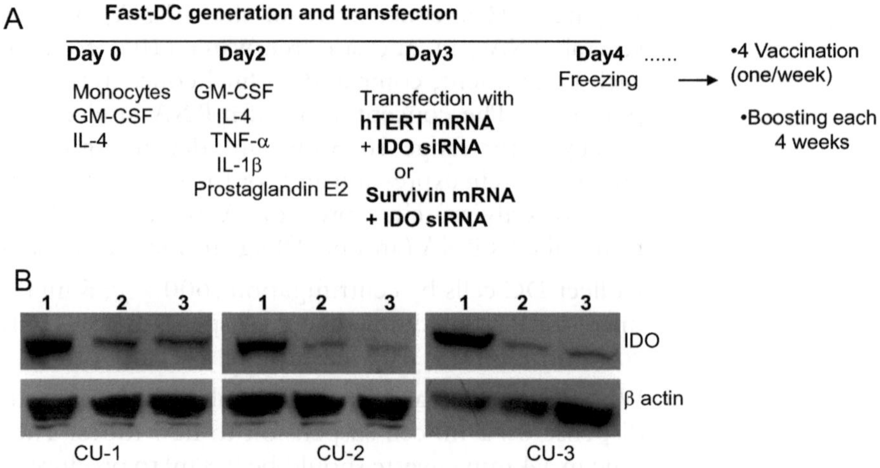

Fig. 2 (**a**) Schematic illustration of the vaccine design. (**b**) Gene silencing in fast DC cancer vaccines. Fast DC vaccine preparations from three patients (CU-1, CU-2, CU-3) were transfected with IDO siRNA in combination with mRNA-encoding human telomerase reverse transcriptase (hTERT) or survivin tumor antigen as illustrated in (**a**). After 24 h post-transfection time, IDO expression was investigated by Western blots, and the remaining cells were cryopreserved into vaccine doses. Lanes 1, 2 and 3 correspond to mock DCs, DCs transfected with IDO siRNA, and mRNA encoding for hTERT, DCs transfected with IDO siRNA, and mRNA encoding for survivin, respectively

2. In addition to DNA oligonucleotides, siRNAs can be transcribed from PCR products containing T7 promoter sequence. In vitro transcription will produce single-stranded or double-stranded RNA directly if both PCR primers contain the T7 promoter sequence.
3. To confer stoichiometric annealing of the sense and antisense strands, test various DNA template concentrations. Notably, some DNA sequences are transcribed better than others. Alternatively, antisense and sense strands can be transcribed separately. After transcription and RNA precipitation, measure RNA concentrations and then anneal equal amount of each strand in 0.5× transcription buffer. Heat the sample at 70 °C for 5 min and then incubate at 37 °C for 1 h.
4. For in vitro transcription of mRNA encoding a tumor antigen, template DNA (cDNA) or a cloned DNA fragment inserted into a vector can be used. During synthesis, a 7-methylguanylate structure, also known as a "cap" such as ARCA (anti-reverse cap analog), is added to the 5′ end of mRNA, and it is essential for efficient translation. ARCA contains an additional 3′-O-methyl group on the 7-methylguanosine to ensure incorporation in the correct orientation (m7Gppp5'N). Therefore, add an mRNA cap analog to the transcription reaction. The cap analog is incorporated as the first nucleotide of the transcript.

5. Around 2 mg of double-stranded siRNA can be generated from 10 reactions (10 ml) using the described in vitro transcription protocol.
6. Gel electrophoresis of the transcribed siRNA or mRNA should confer one single, sharp band. If not, RNA might be degraded. RNA quantity is measured at 260 nm. The ratio of the readings at 260 and 280 nm provides an estimate of the purity of the mRNA and should be between 1.9 and 2.1. If not, RNA might be contaminated. After aliquoting and freezing, avoid multiple freeze-thaw cycles of the siRNAs.
7. RNA electroporation is a rapid and highly efficient technique for transfection of DCs without altering their immunophenotype, maturation potential, migration capacity, and T-cell stimulatory potential. Furthermore, electroporated mature DCs can be cryopreserved without loss of these characteristics. The technique has been validated by different groups for loading DCs with antigen-encoding mRNAs, and several clinical studies using mRNA electroporated DCs have been reported.
8. DCs for vaccination must be produced in a clean room in accordance with the Good Manufacturing Practice (GMP). The antigens used for our vaccine were hTERT and survivin, two universal tumor antigens. DCs were generated from autologous blood monocytes and then transfected with mRNA-encoding hTERT or survivin proteins in combination with siRNA targeting IDO as described previously [11].
9. The immunization with the engineered IDO-silenced DC vaccines was tolerated in all patients. Additionally, T-cell responses against hTERT and survivin tumor antigens were seen in all patients, leading to improved patient outcomes [9]. Hence, the immunization with IDO-silenced DC vaccines is safe and attractive for enhancing the potency of cancer vaccines.

Acknowledgments

This work was supported by the Norwegian Cancer Society.

References

1. Steinman RM, Banchereau J (2007) Taking dendritic cells into medicine. Nature 449:245–252
2. Hoos A, Britten CM, Huber C, O'Donnell-Tormey J (2011) A methodological framework to enhance the clinical success of cancer immunotherapy. Nat Biotechnol 29:867–870
3. Munn DH, Sharma MD, Lee JR, Jhaver KG, Johnson TS et al (2002) Potential regulatory function of human dendritic cells expressing indoleamine 2,3-dioxygenase. Science 297:1867–1870
4. Hwu P, Du MX, Lapointe R, Do M, Taylor M-W, Young HA (2000) Indoleamine 2,3-dioxygenase production by human dendritic cells results in the inhibition of T cell proliferation. J Immunol 164:3596–3599

5. Braun D, Longman RS, Albert ML (2005) A two-step induction of indoleamine 2,3 dioxygenase (IDO) activity during dendritic cell maturation. Blood 106:2375–2381
6. Munn DH, Sharma MD, Hou D (2004) Expression of indoleamine 2,3-dioxygenase by plasmacytoid dendritic cells in draining-draining lymph nodes. J Clin Invest 114:280–290
7. Jonuleit H, Schmitt E, Stenbrink K, Enk AH (2001) Dendritic cells as a tool to induce anergic and regulatory T cells. Trends Immunol 22:394–400
8. Sioud M (2019) Releasing the immune system brakes using siRNAs enhances cancer immunotherapy. Cancers 11, cancers11020176
9. Sioud M (2011) Promises and challenges in developing RNAi as a research tool and therapy. Methods Mol Biol 703:173–187
10. Flatekval GF, Sioud M (2010) Modulation of dendritic cell function and maturation with mono- and bifunctional small interfering RNAs targeting indoleamine 2,3-dioxygenase. Immunology 128:e837–e848
11. Sioud M, Saebøe-Larssen S, Hetland TE, Kaern J, Mobergslien A, Kvalheim G (2013) Silencing of indoleamine 2,3-dioxygenase enhances dendritic cell immunogenicity and antitumour immunity in cancer patients. Int J Oncol 43:280–288

Chapter 15

Unleashing the Therapeutic Potential of Dendritic and T Cell Therapies Using RNA Interference

Mouldy Sioud

Abstract

Therapeutic dendritic cell (DC) cancer vaccines work to boost the body's immune system to fight a cancer. Although this type of immunotherapy often leads to the activation of tumor-specfic T cells, clinical responses are fairly low, arguing for the need to improve the design of DC-based vaccines. Recent studies revealed a promising strategy of combining DC vaccines with small interfering RNAs (siRNAs) targeting immunosuppressive signals such as checkpoint receptors. Similarly, incorporating checkpoint siRNA blockers in adoptive T-cell therapy to amplify cytotoxic T lymphocyte responses is now being tested in the clinic. The development of the next generation of cancer immunotherapies using siRNA technology will hopefuly benefit patients with various cancer types including those who did not respond to current therapies. This review highlights the latest advances in RNA interference technology to improve the therapeutic efficacy of DC cancer vaccines and T cell therapy.

Key words Dendritic cells, Cancer immunotherapy, Gene silencing, Adoptive cell therapy, RNA interference, Interferon response, Checkpoint inhibitors

1 Introduction

Since their discovery by Ralph Steinman over 45 years ago [1], there has been a major focus in using dendritic cells (DCs) in cancer vaccines. DCs function as sentinels at peripheral tissues where they uptake, process, and present pathogen- or host-derived antigenic peptides to naïve T lymphocytes at the lymphoid organs in the context of major histocompatibility molecules. The response of DCs to many different stimuli is reflected by the expression of a large panel of pathogen recognition receptors (PPRs), including Toll-like receptors (TLRs), nucleotide-binding oligomerization domain (NOD-like) receptors, and RIG-1-like helicases [1, 2]. Additionally, they express various cell surface C-type lectins, chemokine receptors, cytokines, and other key regulatory factors such as interleukin-(IL)-12, which drives the generation of CD8+ cytotoxic T lymphocytes (CTLs) and enhances the activation of

Mouldy Sioud (ed.), *RNA Interference and CRISPR Technologies: Technical Advances and New Therapeutic Opportunities*, Methods in Molecular Biology, vol. 2115, https://doi.org/10.1007/978-1-0716-0290-4_15,

cytotoxic natural killer (NK) cells [2]. Although multiple antigen-presenting cells are able to cross-present antigens to T cells, DCs are the most efficient cells in vivo. In this process, exogenous antigens, which are usually presented by MHC class II, are presented in conjunction with MHC class I molecules, resulting in the induction of CTL responses against tumor cells and also against pathogens that do not infect DCs [1, 2]. Most human DC subsets are capable to cross-present antigens, although other factors like the antigen formulation and the mode of delivery can also influence cross-presentation [2].

In general, the DC's life cycle can be divided into two stages: the immature and the mature stage. Immature DCs reside in the periphery and are specialized in antigen recognition and uptake. The activation of immature DCs by microbial- or certain host-derived molecules, known as danger signals, induces DC maturation, a process that lead to the expression of co-stimulatory molecules CD80, CD86, and CD40 that are necessary for immune activation; otherwise, DCs remain in an immature or semi-mature state that can induce immune tolerance to the captured antigen [1]. Depending on the nature and time course of activation, DCs can express co-inhibitory molecules such as PD-L1 (B7-H1, CD274) and PD-L2 (B7-DC, CD273) [3–5]. The expression of these cell surface molecules, known as checkpoint inhibitors, by DCs is part of the normal function of the immune system in order to protect host tissues from collateral damage during responses to pathogens (Fig. 1). In addition to cell surface receptors, DCs also control the strength and duration of immune responses through the expression of cytokines such as IL-10, transforming growth factor-β (TGF-β), and metabolic enzymes such as indoleamine 2,3-dioxygenase (IDO) [6–9]. DCs also respond to negative signals that are delivered by other immune cells, particularly Treg cells. Among the signals, IL-10 and TGF-β produced by Tregs dampen the antigen-presenting capabilities of DCs by inhibiting the expression of MHC and co-stimulatory molecules ([6, 10], Fig. 2). In contrast to T cells, Treg cells constitutively express high levels of CTLA-4. Binding of CTLA-4 to CD80 and CD86 expressed by DCs downregulates their expression and induces the expression of the immunosuppressive molecules IDO and B7-H4 in DCs. Additionally, binding of LAG3 on Treg cells to MHC II molecules on immature DCs activates an inhibitor signaling pathway that suppresses DC maturation.

Although the potent capacity of these negative mechanisms to protect the host from autoimmunity and tissue damage has been well documented, they might suppress antitumor immunity where sustained T-cell activation and proliferation are important [2, 5, 11]. Hence, several co-inhibitory signals, like those transmitted by cytotoxic T lymphocyte-associated antigen 4 (CTLA-4) interaction with B7 molecules (CD80/CD86) or those involving the

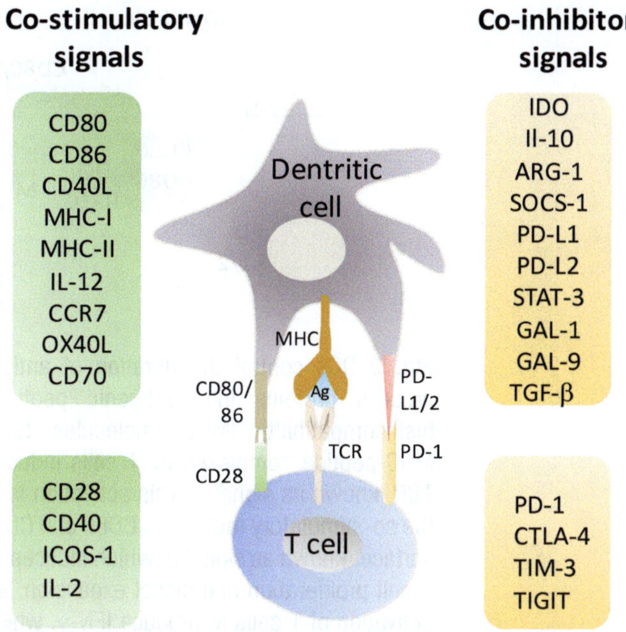

Fig. 1 Important role of dendritic cells in deciding immunity versus tolerance. Dendritic cells express several co-stimulatory factors (e.g., CD80, CD86, CD40, IL-12, CCR7) that facilitate T-cell activation. To regulate to the magnitude of immune responses, DC also express co-inhibitory molecules such as PD-1 ligands, IDO, and IL-10. Antigen presentation by DC without an additional co-stimulatory signal (e.g., CD80/CD86) or in combination with a co-inhibitory signal (e.g., PD-L1/2) can lead to tolerance induction. Studies in the field of T-cell regulation identified the CTLA-4 and PD-1 as inhibitory pathways that restrain T cells from full and persistent activation and proliferation under normal physiologic conditions

interaction of programmed cell death protein 1 (PD-1) with its ligands PD-L1 and PD-L2, should be taken into consideration during DC vaccine and adoptive cell therapy (ACT) design. CTLA-4 is a CD28-related protein expressed by activated T cells that interacts with CD80/CD86 but plays an opposing role to that of CD28 causing the suppression of previously activated T cells [4]. Similarly, the interaction of PD-1 expressed by activated T cells with its ligands PD-L1 and PD-L2 on surface DCs leads to inhibition of T-cell activation. Both PD-1 ligands are upregulated in response to inflammatory cytokines such as interferon (INF)-γ and IL-10. PD-L1 appears to be overexpressed in various cell types, including tumor cells, whereas PD-L2 is more usually overexpressed in DCs [12]. Given the role played by DCs and T cells in tumor immunity, the current engineering strategies for DC cancer vaccines and ACT should include inhibitors against immune suppressive cytokines, checkpoint ligands, and other suppressive factors. Harnessing the unique capacity of RNAi to turn off-gene

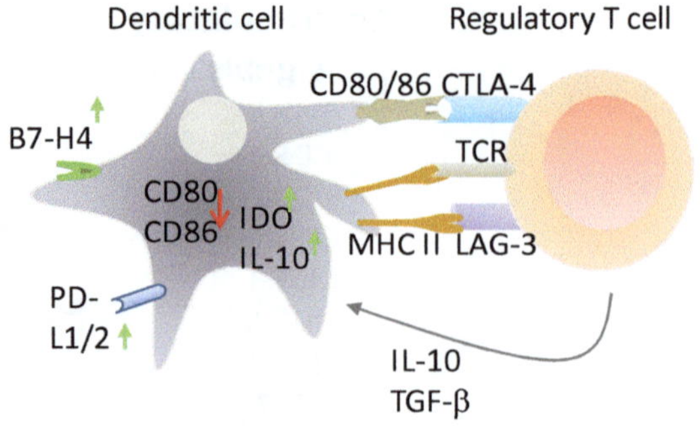

Fig. 2 DCs control the duration of antigen-specific immune response. DCs process proteins into antigenic peptides that are presented on major histocompatibility (MHC) molecules to T cells (**a**). The recognition of MHC-peptide complexes by T cells induces antigen-specific signaling via the TCR, known as signal 1. This activation is amplified by the interaction between the co-stimulatory molecules CD80 and CD86 on DC surface with CD28 on T-cell surface, known as signal 2, which induces the expression of IL-2. IL-2 facilitates T-cell proliferation and clonal expansion. IL-12 produced by DCs promotes the activation of T cells to produce IFN-γ, which favors the generation of CTLs and enhances the activation of cytotoxic NK cells. IFN-γ also induces the expression of IDO by DCs, resulting in their conversion into tolerogenic DCs (**b**). In general, IDO+ DCs polarize naïve T cells into Treg cells, a population of CD4+ T cells that inhibits, rather than promotes, immune responses. Moreover, IDO-expressing DCs convert tryptophan into toxic metabolites with immunosuppressive activity on lymphocytes

expression in immune cells is emerging as a key strategy for the development of new generation of cancer immunotherapies (*see* Chapters 14 and 16). Achieving an effective and durable CTL response remains the ultimate goal of cancer immunotherapies.

2 Monocyte-Derived DCs

As indicated above, DCs have unique characteristics that have made them an ideal choice for active immunization against cancer. However, their low number in the blood (around 0.1–1.0% of blood mononuclear cells) limits their direct use in vaccination strategies. Owing to their abundance in peripheral blood, monocytes are an attractive and easily available cell source for DC generation [2, 13, 14]. Most investigators used leukapheresis to obtain peripheral blood mononuclear cells (PBMCs) from patients. After isolation from PBMCs, monocytes can be differentiated into immature DCs by culturing them for 2–6 days in the presence of GM-CSF and IL-4. Subsequently, the immature DCs are exposed to specific maturation cocktails (consisting of various immunostimulatory cytokines and Toll-like receptor agonists) for the development of a mature phenotype before injection to patients or cryopreservation

[2]. Although various types of DC-based vaccines have been generated, the DC-based vaccine most commonly used so far involves the loading of DCs with a source of tumor-associated antigens (TAAs) followed by their stimulation with defined maturation cocktails [2, 13]. While these ex vivo-generated DCs express high levels of co-stimulatory molecules that are required for T-cell activation, DC maturation also results in the upregulation of co-inhibitory molecules that may dampen their immunogenic function [15]. Among the inhibitory factors expressed by monocyte-derived DCs are TGF-β, IL-10, PD-L1, PD-L2, suppressor of cytokine signaling (SOCS) 1, and IDO [2, 7, 15, 16]. To escape the immune system control, tumor cells also express co-inhibitory molecules and cytokines, including PD-L1, IDO, and HLA-G that have immuno-tolerogenic properties and inhibit CTL and NK cell lytic functions [17]. Hence, new strategies for overcoming the immune suppression induced by tumor microenvironment and/or initiated by DCs may enhance the immunogenicity and potentiate the efficacy of DC cancer vaccines.

To enhance and sustain DC immunogenic function, one may combine mRNA encoding for tumor antigens with siRNA targeting co-inhibitory molecules using electroporation methods (*see* Chapter 14). A potential additional advantage of this delivery strategy is that multiple siRNAs and mRNA encoding tumor antigens can be loaded onto DCs at the same time. Moreover, using transient RNA-based transfection of DC or T cells, sustained severe side effects could be avoided. Once inside the cell, double-stranded siRNAs are extremely stable. Indeed, several studies indicate that dilution due to cell division, and not intracellular siRNA half-life, governs the duration of gene silencing. DCs are relatively stable and do not proliferate. Hence, we expect sustained long-term gene silencing in DC cancer vaccines.

3 Targeting Indoleamine 2,3-Dioxygenase

Among the immunosuppressive factors that have received attention is IDO, an enzyme involved in the catabolism of tryptophan, an essential amino acid that is required for T-cell survival and proliferation, into the stable metabolite kynurenines [18–20]. Kynurenines can directly generate an immunosuppressive environment by promoting the conversion of CD4+ T cells into Treg cells [18]. Previous work showed that transgenic DCs with high levels of IDO expression and tryptophan metabolites were able to permanently suppress allogeneic T-cell proliferation in vitro [21]. Moreover, IDO was found to be essential for successful allogeneic pregnancy, suggesting that it is important in tolerance induction under normal physiological conditions [18]. Similarly, upregulation of IDO expression in DCs resulted in the induction of a long-lasting allograft tolerance in combination with a locally restricted immunosuppressive environment [22]. With respect to cancer, the catabolism of tryptophan in tumor cells mediated by IDO has been increasingly identified as a critical microenvironment factor involved in supporting immune escape through suppression of antitumor immunity [23]. Collectively, these results indicate that the depletion of tryptophan by IDO together with an increase in the production of toxic tryptophan metabolites inhibits effector T-cell function and induces immune suppressive Treg cells, a major challenge for cancer immunotherapy.

In addition to T-cell priming, DCs also control the quality and the duration of T-cell activation via the production of cytokines. Once naïve T cells are effectively primed in the lymph nodes, pro-inflammatory cytokines such as interferon (INF)-γ secreted by activated T cells induce IDO expression in DCs (Fig. 3a) [18, 21], resulting in their conversion into tolerogenic DCs (Fig. 3b). Moreover, reverse signaling via B7 molecules (CD80/86) after interaction with CD28 on T cells can induce IDO expression in DCs [18]. Ligation of CD80/86 on DC with CTLA-4 on T cells also induces IDO expression [24]. In the case of DC cancer vaccines, IDO expression can occur during in vitro maturation of monocyte-derived DCs as well as in vivo after T-cell activation [15, 25]. Under our ex vivo culture conditions, all DC maturation cocktails induced IDO [26, 27], thus highlighting the urgent need for the identification of new therapeutic agents that can enhance the immunogenic function of DC cancer vaccines.

A strategy for enhancing the potency of DC cancer vaccines would be the blockade of IDO expression in DCs prior to reinfusion into patients. We have previously demonstrated that DCs modified with siRNAs against IDO strongly activated T cells as compared to their unmodified counterparts [15]. Moreover, IDO-silenced DCs expressed high levels of co-stimulatory molecules similar to control DCs, indicating maintenance of their mature and stimulatory phenotype. In preclinical studies, Zheng and colleagues showed that an efficient antitumor response can be induced with an IDO-silenced DC vaccine in a syngeneic mouse model for breast cancer [28]. The vaccination resulted in the induction of tumor antigen-specific CTL response and decrease in the number of circulating Treg cells.

Having demonstrated that IDO-silenced DCs are functional in vitro, we next investigated the feasibility, safety, and immunogenicity of the vaccine in four patients with metastatic ovarian cancer using our GMP-compatible DC production pipeline [29, 30]. Fast DC preparations from patients were transfected ex vivo with IDO siRNA along with mRNA encoding the human reverse transcriptase subunit of telomerase (hTERT) or survivin using the electroporation method [29]. After, the cells were grown for 24 h with cytokines facilitating maturation (TNF-α, IL-6, and prostaglandin E2)

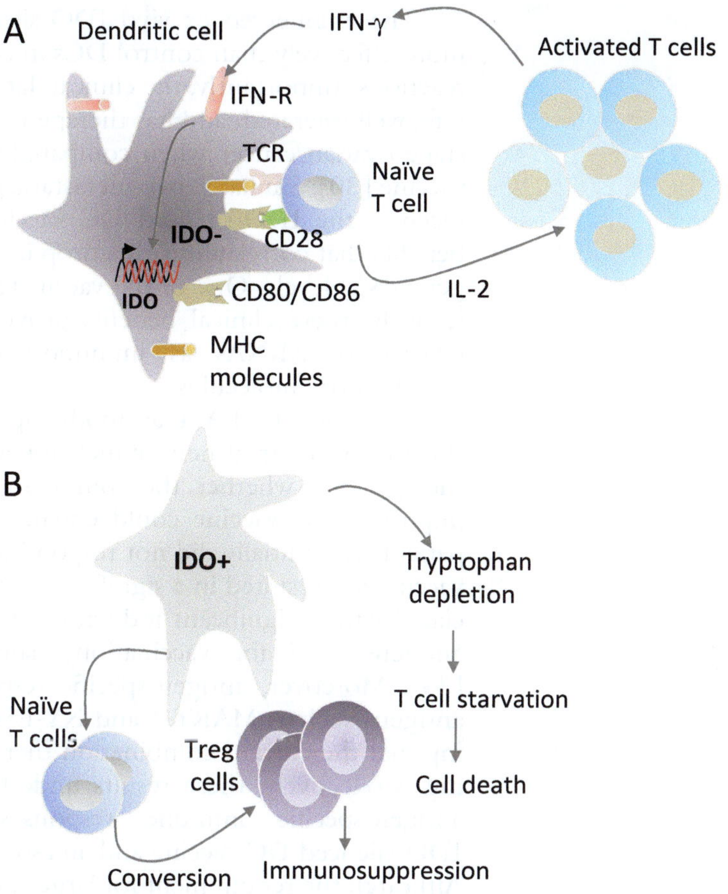

Fig. 3 Combined gene silencing and TLR activation by a single siRNA sequence. Most siRNA molecules, delivered by cationic lipids, nanoparticles, or cell-type-specific delivery agents, begin their intracellular trafficking in the endosomal vesicles where they can activate Toll-like receptors, resulting in cytokine production and eventually DC maturation. After endosomal escape, siRNA molecules enter the RNAi pathway to exert their silencing function as illustrated in Fig. 2

and then cryopreserved into separate vaccine doses. It should be noted that hTERT is an attractive, almost universal antigen target in cancer being overexpressed in the majority of human tumors and absent in most normal adult tissues [31]. Moreover, hTERT is expressed in cancer cells with stem cell-like properties, and targeting this antigen could therefore be an important tool to eliminate these cells, which are not easily killed by conventional therapy [32]. DCs modified to express whole tumor antigens will present multiple, even unidentified epitopes in many HLA alleles, which will recruit a wide spectrum of T cells. With respect to antigen loading, mRNA is safer and easier to handle and is chosen by many investigators for inclusion in DC vaccines.

The engineered ex vivo IDO-silenced DCs activated T cells more effectively than control DCs in allogeneic mixed lymphocyte reactions. Importantly, the clinical data revealed that the vaccine is safe, well tolerated, and has therapeutic potential even in advanced stage ovarian cancer when compared to matured "standard" DC vaccine [30]. Similarly, two metastatic prostate cancer patients who received the IDO-silenced DC vaccine in 2014 showed clinical benefits that correlated with a drop in serum prostate-specific antigen (PSA) levels. One of the vaccinated patients is still alive. Collectively, these clinical benefits provide a clear evidence for the efficacy of siRNAs as immunotherapy that target immune co-inhibitory molecules.

Since anti-CTLA-4 antibody (ipilimumab) was approved as therapy for the treatment of melanoma patients [33], we addressed the question whether the combination of ipilimumab with our improved DC vaccine could enhance the antitumor effects in a patient who initially did not respond to the antibody therapy. The vaccination resulted in a significant antitumor effect that was associated with a significant induction of T-cell responses against the antigens used for vaccination, namely, hTERT and survivin [34]. Moreover, antigen-specific responses to melanoma tumor antigens such as MART-1 and NY-ESO-1 were induced, suggesting that the current combination therapy induced epitope spreading. Collectively, these results underline the success in inducing antigen-specific immune responses after vaccination with IDO-silenced DC vaccine and an association with clinical benefits. Although the requirement for large and well-designed randomized trials to assess the efficacy of IDO-silenced DC vaccine is clear, case studies can make important contributions as instructive scientific observations.

4 Targeting CTLA-4, PD-1, and PD-1 Ligands

Under normal conditions, the stimulation of T cells via the T-cell receptor (TCR) leads to intracellular signaling that promotes T-cell activation and proliferation [3]. Following activation, T cells express on their surface CTLA-4, a membrane receptor that antagonize T-cell function through inhibition of CD28 signaling by competing for their shared ligands B7 molecules [3, 4]. CTLA-4 has a higher affinity for B7 molecules than CD28 and serves as a natural inhibitor to terminate T-cell priming by DCs in the lymph nodes. Effector T cells also upregulate the expression of PD-1, a negative regulator of T-cell function [5]. Unlike CTLA-4, PD-1 signaling pathway is a checkpoint involved in controlling peripheral tissue damage after inflammatory immune responses. Excessive induction of PD-1 on T cells in the setting of chronic antigen exposure inhibits T-cell activation, as a result of T-cell exhaustion.

Hence, during immunotherapy the interactions between PD-1 and its ligands, PD-L1 and PD-L2, expressed by APCs such as DCs or tumor cells are expected to block the immune responses against tumors. Similar to CTLA-4, the PD-1 signaling pathway is also involved in extrinsic suppression of T-cell activation by Treg cells [35].

Agents such as monoclonal antibodies (mAbs) and siRNAs targeting these inhibitory receptors could substantially boost the efficacy of DC vaccines or restore the killing potency of tumor-infiltrating lymphocytes (TIL). Indeed, antibodies that bind to either CTLA-4 or PD-1, and thereby alleviate the immune inhibition indicated above, have generated promising clinical data in patients with melanoma or lung cancer [36–40]. However, only a fraction of patients responded durably to the therapy. Furthermore, severe side effects were detected in a high proportion of patients, especially in those treated with anti-CTLA-4 therapy as compared to anti-PD1 therapy [40]. These effects are more likely due to the polyclonal activation of autoreactive T cells due to systemic administration of the antibodies. Hence, additional targeted strategies designed to inhibit the expression of these co-inhibitory molecules are warranted. In this respect, Hobo et al. used RNAi to down-regulate the expression of PD-1 ligands in DCs [41]. PD-L1 and PD-L2-silenced DCs enhanced T-cell proliferation in vitro. Similarly, silencing PD-L1 in DCs enhanced the antitumor effects of bladder cancer antigen-loaded DC vaccines [42]. SiRNA silencing of PD-1 ligands on DC vaccines also enhanced the expansion of minor histocompatibility antigen (MiHA)-specific CD8+ T cells in mice [43]. In addition to electroporation, recently the authors used a new formulation of cationic lipids (named SAINT-18) that is compatible with GMP manufacturing to deliver PD-L1 and PD-L2 siRNAs to DCs [44, 45]. Interestingly, PD-L-silenced DCs loaded with mRNA encoding for MiHA showed enhanced MiHA-specific T-cell activating potential than control DCs. Overall, the data indicate that silencing of PD-1 ligands in DCs can enhance their immunogenic function, leading to the induction of stronger antigen-specific CTL responses in in vitro models and antitumor immunity in various mouse cancer models. Given the encouraging in vitro data, a phase I/II study in currently underway to investigate the clinical effect of a MiHA-loaded PD-L silenced DC vaccine (NCT02528682).

5 Targeting SOCS-1 and STAT-3 Transcription Factors

SOCS-1, a member of the SOCS and cytokine-inducible SH2 family of intracellular proteins, has also emerged as a critical inhibitory molecule for controlling the response to cytokines and antigen presentation by DCs [46]. SOCS-1 is induced by cellular activation and serves as a negative feedback mechanism for a range of cytokines. A number of studies in mice have documented the important role of SOCS-1 in modulating the extent of immune stimulation by DCs suggesting that its inhibition may enhance vaccine-induced immune responses. Indeed, silencing SOCS-1 in DCs enhanced antigen presentation, T-cell priming, and antitumor immunity [47, 48]. More recently, Wand and colleagues tested the therapeutic efficacy of a SOCS-1-silenced DC vaccine loaded with two tumor-associated antigens, survivin and MUC-1, in patients with relapsed acute leukemia (AL) after allogenic hematopoietic stem cell transplantation (allo-HSCT) [49]. Relapse after allo-HSCT is the leading cause of treatment failure for AL patients [50]. The designed DC vaccine was tolerated and induced antigen-specific CTL responses capable of selectively killing leukemic cells. Interestingly, clinical data showed that 10 out of 12 early relapsed patients who received the vaccine exhibited complete remission. Moreover, patients treated with the vaccine had no grade 3 or 4 graft-versus-host disease, again indicating that the treatment with the vaccine is safe. Hence, the inhibition of SOCS-1 expression in DCs could be exploited as a novel adjuvant strategy to boost the potency of DC cancer vaccines (NCT01956630). Although this clinical success is very encouraging for siRNA inclusion in DC vaccines, defining how the vaccine impacts the durability and memory properties of CD8+ T cells will be critical for understanding the potential of this vaccine to provide a durable mechanism of immunity in setting of cancer relapse.

In addition to SOCS-1, the induction of signal transducer and activator of transcription 3 (STAT-3) in DCs by tumor-derived factors renders DCs tolerogenic and suppresses their antitumor potency. PD-L1 expression in DC is also regulated by STAT-3 signaling pathway [51]. Hence, silencing STAT3 in DCs should be beneficial for cancer immunotherapy. A number of therapeutic strategies have explored selective inhibition of STAT-3 signaling using various agents such as small-molecule inhibitors, dominant-negative STAT-3 mutants, and siRNAs. Systemic administration of STAT-3 inhibitors not only inhibited tumor growth but also resulted in the downregulation of immunosuppressive cytokine expression and induction of functional DCs and CTL responses. Similarly, STAT-3 gene silencing restored DC maturation and enhanced CTL responses to tumors [52]. The siRNA strategy was also applied to CD204, an immunosuppressive scavenger receptor in DCs [53]. CD204-silenced DCs were injected into the tumors followed by local radiotherapy, resulting in enhanced antitumor effects. Gene silencing of galectin-1 and galectin-3 in DCs also enhanced their capacity to stimulate T-cell activation and IFN-γ production [54]. In addition to the regulation of inflammatory responses, galectin-1 and galectin-3 may function as a negative regulator of T-cell activation by increasing the TCR activation threshold as both proteins have been shown to interact with CD45, CD7, and CD3 molecules within the immunological synapses [55]. Normal T cells undergo a process of TCR desensitization before entering the secondary lymphoid tissues. This TCR tuning modulates the intensity of TCR signaling and is thought to be especially important for cells with relatively high self-reactivity [56]. In principle, lowering the activation threshold of TCR increases the expansion of T cells, which is beneficial in immunotherapy.

6 Targeting IL-10

Although IL-10 is involved in various aspects of the anti-inflammatory process, its main function is related to downregulation of T-cell function [57]. Moreover, IL-10 increases PD-L1 expression in macrophages and dendritic cells. Various stimuli, in particular Toll-like receptor (TLRs) agonists, have been explored as adjuvants in different vaccination strategies or in ex vivo DC maturation cocktails. Although they efficiently induce DC maturation and IL-12 expression, they also induce IL-10 expression [58]. Whether produced by DC themselves or present in the tumor microenvironment, IL-10 conditions human DC to acquire a tolerogenic phenotype favoring the induction of T-cell anergy and suppressive function. Moreover, early in vitro studies have identified IL-10 production by DC to be an important mechanism in driving T-cell anergy and suppression [59]. Hence, the development of agents that stimulate DC maturation and simultaneously inhibit the expression of IL-10 may facilitate the design of better DC vaccines.

We and others have previously demonstrated that formulation of siRNAs in lipid-based delivery vehicles, which deliver siRNAs to endosomal compartments in which TLR7 and TLR8 reside, activates innate immunity, leading to type I interferon and pro-inflammatory cytokine production [60]. These unwanted effects present a key challenge for siRNA therapeutics. However, there are clinical situations, such as viral infection and cancer, in which this adjuvant effect of siRNAs could be beneficial. In fact, TLR activation was more likely responsible for the promising early clinical trial data for ocular antiangiogenic siRNAs and topically applied siRNAs respiratory syncytial virus [61]. Fortunately, site-specific chemical modifications that also reduce sequence-specific off-target effects largely inhibited the activation of TLRs by siRNAs [62–64]. In contract to lipid delivery, we demonstrated that siRNAs delivered via electroporation to the cytosol did not activate innate immunity, again supporting the involvement of endosomal TLRs in siRNA sensing by innate immunity [60]. Because certain siRNA sequences activated TLR-7 and TLR-8 [60, 65, 66], we

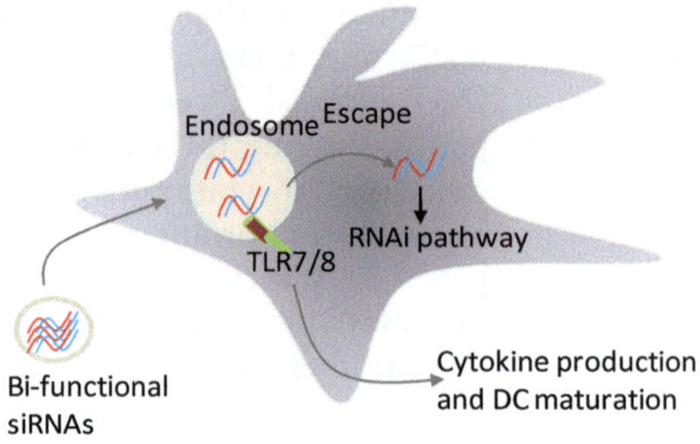

Fig. 4 Induction of tolerogenic DCs by Treg cells. A number of different factors/signals delivered by Treg cells might act in concert to convert immunogenic DCs into tolerogenic DCs. In addition to direct interaction, Treg can produce IL-10 and TGF-β, which inhibit the function of DC and therefore the generation of effector T cells (*see* text)

hypothesized that a combinatorial strategy that blocks the expression of immunosuppressive factors and simultaneously activates innate immunity via TLRs could increase the efficacy of DC vaccines (Fig. 4). In this respect, we have shown that targeting IL-10 expression by a bifunctional siRNA can block IL-10 expression and activates TLR-7/8 signaling in human monocytes and monocyte-derived immature DCs [67]. The inhibition of IL-10 and deliberate activation of immature DCs with a single siRNA sequence opens the possibility of generating mature DCs without the addition of external cytokines. This strategy of DC maturation may be extended into a therapeutic vaccination setting. In this respect, we have developed a new vaccination protocol where immature DCs were loaded with tumor antigens and transfected with an IL-10 siRNA capable of co-coordinately activating DCs via TLR7/8 and silencing IL-10 expression ([68], *see* Chapter 16). Using a rat model of acute myeloid leukemia, we found that the DC vaccine was more effective in stimulating CTL responses against leukemia cells when compared to control vaccine. Thus, the in vivo application of a DC vaccine, loaded with bifunctional IL-10 siRNA, not only promoted DC maturation and enhanced CTL responses but also suppressed leukemia cell growth in vivo.

Previous studies have shown that the adjuvant effect of CpG oligonucleotides is TLR-9 dependent [58, 69]. In addition to IL-12, CpG DNA motifs also induced the expression of IL-10 and TGF-β in vitro and in vivo. While IL-12 is critical for the orchestration of cellular immunity, both IL-10 and TGF-β promote immune tolerance. Hence, blocking IL-10 at the time of treatment with CpG DNA may enhance antitumor effects. In this respect, simultaneous immunotherapy with CpG DNA and IL-10 siRNA enhanced immune protection of an idiotype DNA vaccine in a prophylactic murine model of B-cell lymphoma as compared to control treatments [70]. Similarly, IL-10 gene silencing in DCs induced stronger CTL responses against the human melanoma antigen MART-1 as compared to unmodified DCs [71]. More recently, Ahn et al. showed that co-targeting IL-10 receptor and TGF-β in DCs by siRNAs enhanced the immunogenic function of DCs, resulting in a significant antitumor effect against highly immune resistant tumor cells that secrete more IL-10 and TGF-β than the parental tumor cells [72]. Other studies also reported that co-administration of IL-10-specific siRNA-loaded nanocarriers and a DC vaccine enhanced antitumor immune responses and inhibited tumor growth and metastasis [73]. Collectively, these findings indicate that targeting IL-10 gene expression with siRNA in DCs before or at the time of immunization can increase vaccine-induced CTL responses and block tumor growth in animal models. Hence, combining IL-10 siRNA with DC vaccines should be a strategy of choice to enhance vaccine efficacy. This conclusion is further supported by the fact that clinical and strong immunological responses in prostate cancer patients correlated with low expression of IL-10 in DC preparations [74].

7 Targeting INF-γ-Induced Proteasome Proteases

Notably, the development of an effective cancer vaccine requires effective possessing and presentation of tumor antigen to T cells. The standard proteasome is considered to be central in peptide generation for MHC class I molecules that present peptides primarily derived from proteins located in the cytosolic compartment of the cell [75]. The catalytic activity in charge for the cleavage of the peptides is exerted by three proteases of the β-ring subunits, namely, β1, β2, and β5 [75]. In the presence of IFN-γ, these three catalytic subunits are replaced by their homologous subunits β1i (LMP2), β2i (MECL1), and β5i (LMP7), resulting on the formation of so-called immunoproteasome that is more efficient at processing antigenic epitopes [75, 76]. Immature DCs usually express both proteasome types, whereas mature dendritic cells express only the immunoproteasome and therefore only present peptides generated by the immunoproteasome [77]. Most, if not all, non-self-proteins including microbial proteins are processed by the immunoproteasome. However, several self-proteins such as melanoma antigens used in DC vaccination are not processed by the immunoproteasome [76]. The induction of the immunoproteasome in mature DCs may be one mechanism to control the development of autoimmunity by preventing the processing of self-proteins. Given that most TAA are self-proteins, this safeguard mechanism may limit the effectiveness of DC cancer vaccines. Hence, the knockdown of LMP2, LMP7, and MECL1 could be used to enhance cancer vaccine potency. By targeting the expression of these

IFN-γ-inducible proteases by siRNAs, Dannull and colleagues showed that TAAs can be processed into peptides that are more relevant for the induction of effective anti-melanoma immunity [77]. Although the levels of gene silencing are not remarkable, the proteasome-modified DCs stimulated superior anti-melanoma immunity in vitro. Based on these results, the authors have initiated a phase I clinical trial in which patients with metastatic melanoma were vaccinated with mature autologous DCs transfected with RNAs encoding the malanoma antigens MART-1, MAGE3, gp100, and tyrosinase in combination with siRNAs against the three inducible immunoproteasome subunits indicated above [78]. Vaccination stimulated antigen-specific T-cell responses in all patients, which peaked after three to four vaccinations and sustained high only in patients who received siRNA-modified DCs. Two patients with active metastatic disease at the time of vaccination with siRNA-modified DCs showed clinical benefit. One patient showed a complete clinical response and remained free of active disease 20 months after the end of vaccination (NCT00672542). Although more clinical data are needed, the data support the use of the proteasome-modified DCs in cancer immunotherapy.

8 Enhancing the Functionality of T Cells

The generation of T cells with superior cytotoxic activity is essential for developing effective immunotherapeutic strategies against infectious agents and cancers. Similar to DCs, the activation of T cells is negatively regulated by various intrinsic co-inhibitory factors such as CTLA-4, PD-1, and TCR-adaptor proteins (Fig. 1). Hence, ACT must address these various inhibitory barriers in order to enhance therapy response. ACT is a potent method of harnessing autologous immune cells, allowing for ex vivo manipulation of T cells or natural killer cells prior to their reinfusion into the patients [79, 80]. However, ex vivo cultured T cells appear to upregulate the expression of certain inhibitory factors such as CTLA-4 and PD-1, which dampen their in vivo function. Combining ACT with antibodies targeting checkpoint inhibitors is an attractive possibility already pursued in clinical trials [81]. However, such combination may result in systemic serious adverse events caused by the antibodies acting on patient autoreactive T cells. Moreover, cytosolic inhibitory factors are difficult to target with antibodies. An alternative option would be the ex vivo inhibition of these inhibitory factors by siRNAs. As for DCs, a versatile method that could potentially be used for siRNA delivery to T cells is the ex vivo route, whereby blood T cells or TIL could be isolated from a patient, transfected with siRNAs, and then infused into the same patient. In our experience, the use of standard wave electroporation method can deliver siRNAs to human blood T cells with close to 95% transfection efficiency and with a little effect on cell viability and function. Nucleofection of siRNA into T cells also resulted in significant gene silencing [82, 83]. Furthermore, the introduction of chemical modifications into siRNAs extended gene silencing in T cells [82]. Indeed, the authors showed that transfected T cells remain functional and maintain siRNA-induced CD4 knockdown for up to 2 weeks after transfer into recipient mice. With respect to immunotherapy, T cells transfected with siRNAs against PD-1 showed an increased cytotoxic effect against PD-L1-expressing melanoma cells [84]. Similarly, gene silencing of CTLA-4 in T lymphocytes from hepatitis B-positive patients resulted in a stronger CTL response against virus-infected cells [85]. These in vitro study successes are encouraging and should set the stage for in vivo investigations.

Previous studies have shown that the attachment of a 3′ cholesterol group to the sense strand of siRNA facilitated its uptake by human cells [86]. Chemically modified, self-delivered siRNAs provide an alternative RNAi strategy because they enable siRNAs to enter cells without the need of a transfection reagent (*see* Chapter 3). This type of siRNAs has been successfully used in RNAi experiments in a variety of primary cells, including macrophages, T cells, and in vivo models, thus holds great potential for clinical applications [87]. More recently, Ligtenberg et al. used this strategy to deliver PD-1-specific siRNA to TIL [88]. Interestingly, T cells expanded in the presence of siRNA exhibited high cytotoxic activities against autologous tumors as compared to control TIL. Notably, clinical trials have shown promising results with TIL therapy in patients with melanoma [79]. Empowering T-cell function by blocking PD-1 expression could easily be adopted into ACT protocols. Moreover, potential resistance to PD-1 therapy may be prevented by silencing additional T-cell inhibitory molecules such as TIM-3, LAG-3, and TIGIT. In line with this notion, dual inhibition of IDO and PD-L1 or PD-1 in T cells resulted in more enhanced T-cell responses in vitro when compared to monogene silencing [89].

In addition to TIL, T cells engineered to express defined TCRs specific for tumor antigens or chimeric antigen receptors (CAR) hold great promise for the treatment of metastatic cancers [80]. CD19-targeting CAR T cells are the first gene therapy drug approved by the US Food and Drug Administration for the treatment of B-cell lymphomas [90]. Although CAR-based ACTs were clinically successful in the treatment of B-cell malignancies as documented in multiple independent clinical trials, so far the responses in solid tumors are still limited. The immunosuppressive environment in solid tumors, in conjunction with inhibitory receptors on the engineered T cells, is thought to negatively influence the efficacy of CAR T cells. Additionally, in vitro-expanded T cells upregulate the expression of PD-1 that could be suppressed by siRNA [84]. A recent study demonstrated that CAR T cells transfected with siRNAs targeting PD-1 alone or in combination with siRNA targeting CTLA-4 produced more effector cytokines and showed a significant increase in cytotoxicity toward melanoma cells expressing PD-L1 as compared to control cells [91].

As mentioned above, certain proteins involved in the regulation TCR signaling are certainly good targets for inhibition by RNAi technology. For example, the E3 ubiquitin ligase Cbl-b and the src homology 2 domain containing protein tyrosine phosphatase (SHP-1) negatively regulate TCR activation [92]. Both proteins are cytosolic and therefore not amenable to antibody-mediated therapies. Like CTLA-4 knockout mice, mice deficient in *cblb* gene are very susceptible to spontaneous and antigen-induced experimental autoimmune diseases [93]. Therefore, inhibition of Cbl-b expression in T cells may enhance their antitumor potency. In this respect, silencing of Cbl-b expression in primary murine CD8+ T cells with siRNAs via nucleofection, followed by adoptive transfer of the cells into recipient mice, enhanced the effects of an anticancer vaccine [94]. SHP-1 is a negative regulator of antigen-dependent activation and proliferation of T cells in part by reducing T-cell/DC interactions [95]. RNAi inhibition of SHP-1 in T cells enhanced the antitumor activity of adoptively transferred T cells [96]. These two examples show that targeting cytosolic inhibitors involved in TCR tuning can enhance or restored T-cell function. Although we do not yet know which combination therapies are optimal for clinical translation, the performed preclinical studies with siRNAs are very encouraging and should overcome some of the obstacles associated with adoptive immunotherapy for cancer patients.

In addition to ex vivo delivery, several studies have shown that siRNAs can be delivered to immune cells in vivo. For example, antibody fragment fused to protamine or to other nucleic acid-binding peptides has been used to deliver siRNA and knock down gene expression in tumors, disseminated blood cells, and organs such as the lung [97]. Moreover, siRNAs were targeted to 4-1BB-expression CD8+ T cells by conjugation to a 4-1BB-binding oligonucleotide aptamer [98]. 4-1BB is a major immune stimulatory receptor transiently expressed on activated CD8+ T cells [99]. The data indicate that 4-1BB aptamer delivery of CD25 siRNA to activated circulating CD8+ T cells in vivo is efficient and specific, leading to enhanced memory differentiation and improved control of tumor growth. Overall, these studies show that receptor-directed antibodies or aptamers can deliver siRNAs to circulating CD4 and CD8 T cells, further supporting the therapeutic use of RNAi [100].

9 Concluding Remarks

The major goal of vaccination against cancer is the generation of high avidity antigen-specific CTLs and long-lived T-memory cells. While the current DC cancer vaccines induced antigen-specific immune responses, the duration and magnitude of these responses are weak and clinical benefits have been limited. The knockdown of inhibitory signals by siRNAs is a viable method to make DCs- or T-cell-mediated immunotherapies more effective. Most of the targeted inhibitory molecules have been successful in that their inhibition by siRNAs resulted in strong CTL responses. What was remarkable was the objective clinical responses obtained with either IDO- or SOCS-1-silenced DC vaccine. Because of the manufacturing simplicity, the introduction of chemically made siRNAs or shRNAs into DCs along with mRNA encoding TAAs or neoantigens through electroporation is a straightforward strategy that is not associated with additional manufacturing costs. Moreover, a variety of lipid-based siRNA delivery methods and chemical modifications offer an alternative strategy to electroporation, in particular self-delivered siRNAs. These chemically modified molecules can enter immune cells without needing a delivery reagent and can be produced under GMP conditions. The effective targeting of siRNAs to circulating blood T cells suggests that siRNAs could be tailored to provide clinical benefit as cancer immunotherapies. Additionally, nanocarriers are valuable for potentiating the simultaneous delivery of antigens and immunostimulatory siRNAs to DCs ex vivo or in situ. An immunostimulatory adjuvant is essential for generation of an effective immune response. Overall, the data discussed in this review demonstrate that it is feasible to use siRNAs to boost DC immunogenic function and T-cell cytotoxic function, and therefore they should lay the groundwork for designing novel immunotherapeutic strategies for the treatment of advanced cancers. Additionally, improved analysis of T cell and autoantibody specifities to identify antigens for vaccinations should facilitate the applications of such therapies in oncology and autoimmune diseases [101–103].

Acknowledgments

This work was supported by the Norwegian Cancer Society.

References

1. Liu K, Nussenzweig MC (2010) Origin and development of dendritic cells. Immunol Rev 234:45–54
2. Palucka K, Banchereau J, Mellman I (2010) Designing vaccines based on biology of human dendritic cell subsets. Immunity 33:464–478

3. Chen L (2004) Co-inhibitory molecules of the B7-CD28 family in the control of T-cell immunity. Nat Rev Immunol 4:336–347
4. Schwartz RH (1992) Costimulation of T lymphocytes: the role of CD28, CTLA-4, and B7/BB1 in interleukin-2 production and immunotherapy. Cell 71:1065–1068
5. Fife BT, Pauken KE, Eagar TN, Obu T, Wu J, Tang Q, Azuma M, Krummel MF, Bluestone JA (2009) Interaction between PD-1 and PD-L-1 promote tolerance by blocking the TcR-induced stop signal. Nat Immunol 10:1185–1192
6. Chung CY, Ysebaert D, Berneman ZN, Cools N (2013) Dendritic cells: cellular mediators for immunological tolerance. Clin Dev Immunol 2013:972865
7. Munn DH, Mellor AL (2007) Indolamine 2,3-dioxygenase and tumor-induced tolerance. J Clin Invest 17:1147–1154
8. Corinti S, Albanesis C, la Salsa A, Pastore S, Girolomoni G (2001) Regulatory activity of autocrine IL-10 on dendritic cell functions. J Immunol 166:4312–4320
9. Yu CI, Becker C, Wang Y, Marches F, Helft J, Leboeuf M, Anguiano E, Pourpe S, Goller K, Pascual V, Banchereau J, Merad M, Palucka K (2013) Human CD1c+ dendritic cells drive the differentiation of CD103+ CD8+ mucosal effector T cells via the cytokine TGF-β. Immunity 38:818–830
10. Mahnke K, Johnson TS, Ring A, Enk AH (2007) Tolerogenic dendritic cells and regulatory T cells: a two-way relationship. J Dermatol 46:159–167
11. Zhao H, Liao X, Kang Y (2017) Tregs: where we are and what comes next? Front Immunol 8:1578. eCollection 2017
12. Zheng P, Zhou Z (2015) Human cancer immunotherapy with PD-1/PD-L1 blockade. Biomark Cancer 7(Suppl 2):15–18. eCollection
13. Anguille S, Smits EL, Lion E, van Tendeloo VF, Berneman ZN (2014) Clinical use of dendritic cells for cancer therapy. Lancet Ocol 15: e257–e267
14. Palucka K, Banchereau J (2013) Human dendritic cell subsets in vaccination. Curr Opin Immunol 25:396–402
15. Flatekval GF, Sioud M (2009) Modulation of dendritic cell maturation and function with mono- and bifunctional small interfering RNAs targeting indoleamine 2,3-dioxygenase. Immunology 128: e837–e848
16. Steinman RM, Hawiger D, Nussenzweig MC (2003) Tolerogenic dendritic cells. Annu Rev Immunol 21:685–711
17. Beatty GL, Gladney WL (2015) Immune escape mechanisms as a guide for cancer immunotherapy. Clin Cancer Res 15:687–692
18. Munn DH, Mellor AL (2004) IDO expression by dendritic cells: tolerance and tryptophan catabolism. Nat Rev Immunol 4:762–774
19. Munn DH, Sharma MD, Hou D, Baban B, Lee JR, Antonia SJ et al (2004) Expression of indoleamine 2,3-dioxygenase by plasmacytoid dendritic cells in draining-draining lymph nodes. J Clin Invest 114:280–290
20. Fallarino F, Grohmann U, Vacca C et al (2002) T cell apoptosis by tryptophan catabolism. Cell Death Differ 9:1069–1077
21. Terness P, Bauer TM, Röse L, Dufter C, Watzlik A, Simon H, Opelz G (2002) Inhibition of allogeneic T cell proliferation by indoleamine 2,3-dioxygenase-expressing dendritic cells: mediation of suppression by tryptophan metabolites. J Exp Med 96:447–457
22. Liu H, Liu L, Liu K, Bizargity P, Hancock WW, Visner GA (2009) Reduced cytotoxic function of effector CD8+ T cells is responsible for indoleamine 2,3-dioxygenase-dependent immune suppression. J Immunol 183:1022–1031
23. Platten M, Wick W, Van den Eynde BJ (2012) Tryptophan catabolism in cancer: beyond IDO and tryptophan depletion. Cancer Res 72:5435–5440
24. Grohmann U, Orabona C, Fallarino F, Vacca C, Calcinaro F, Falorni A, Candeloro P, Belladonna ML, Bianchi R, Fioretti MC, Puccetti P (2002) CTLA-4-Ig regulates tryptophan catabolism in vivo. Nat Immunol 3:1097–1101
25. Wobster M, Voigt H, Houben R, Eggert AO, Freiwald M, Kaemmerer U et al (2007) Dendritic cell based antitumor vaccination: impact of functional indoleamine 2,3-dioxygenase expression. Cancer Immunol Immunother 56:1017–1024
26. Furset G, Fløisand Y, Sioud M (2007) Impaired expression of indoleamine 2, 3-dioxygenase in monocyte-derived dendritic cells in response to Toll-like receptor-7/8 ligands. Immunology 123:263–271
27. Flatekval GF (2019) Insights into siRNA-sensing by the immune system and immunotherapeutic strategies using siRNA. PhD thesis, University of Oslo, December 2019.

28. Zheng X, Koropatnick J, Chen D et al (2013) Silencing IDO in dendritic cells: a novel approach to enhance cancer immunotherapy in a murine breast cancer model. Int J Cancer 132:967–977
29. Sæbø-Larsen S, Fossberg E, Gaudernack G (2002) mRNA-based electrotransfection of human dendritic cells and induction of cytotoxic T lymphocyte responses against the telomerase catalytic subunit (hTERT). J Immunol Methods 259:191–203
30. Sioud M, Saebøe-Larssen S, Hetland TE, Kaern J, Mobergslien A, Kvalheim G (2013) Silencing of indoleamine 2,3-dioxygenase in dendritic cell cancer vaccines: evaluation *in vitro* and in cancer patients. Int J Oncol 43:280–288
31. Kim NW, Piatyszek MA, Prowse KR, Harley CB, West MD, Ho PL, Coviello GM, Wright WE, Weinrich SL, Shay JW (1994) Specific association of human telomerase activity with immortal cells and cancer. Science 266:2011–2015
32. Marian CO, Cho SK, McEllin BM, Maher EA, Hatanpaa KJ, Madden CJ, Mickey BE, Wright WE, Shay JW, Bachoo RM (2010) The telomerase antagonist, imetelstat, efficiently targets glioblastoma tumor-initiating cells leading to decreased proliferation and tumor growth. Clin Cancer Res 16:154–163
33. Sharma P, Allison JP (2015) The future of immune checkpoint therapy. Science 348:56–61
34. Sioud M, Nyakas M, Sæbøe-Larssen S, Mobergslien A, Aamdal S, Kvalheim G (2016) Diversification of antitumour immunity in a patient with metastatic melanoma treated with ipilimumab and an IDO-silenced dendritic cell vaccine. Case Rep Med 2016:9639585
35. Francisco LM, Sage PT, Sharpe AH (2010) The PD-1 pathway in tolerance and autoimmunity. Immunol Rev 236:219–242
36. Schadendorf D, Hodi FS et al (2015) Pooled analysis of long-term survival data from phase II and phase III trials of ipilimumab in unresectable or metastatic melanoma. J Clin Oncol 33:1889–1894
37. Ohaegbulam KC, Assal A, Lazar-Molnar E, Yao Y, Zang X (2015) Human cancer immunotherapy with antibodies to the PD-1 and PD-L1 pathway. Trends Mol Med 21:24–33
38. Snyder A, Makarov V, Merghoub T, Yuan J, Zaretsky JM, Desrichard A et al (2014) Genetic basis for clinical response to CTLA-4 blockade in melanoma. N Engl J Med 371:2189–2199
39. Rizvi NA, Hellmann MD, Snyder A, Kvistborg P, Makarov V, Havel JJ et al (2015) Cancer immunology. Mutational landscape determines sensitivity to PD-1 blockage in non-small cell lung cancer. Science 348:124–128
40. Boussiotis VA (2014) Somatic mutations and immunotherapy outcome with CTLA-4 blockade in melanoma. N Engl J Med 371:2230–2232
41. Hobo W, Novobrantseva TI, Fredrix H, Wong J, Milstein S, Epstein-Barash H et al (2013) Improving dendritic cell vaccine immunogenicity by silencing PD-1 ligands using siRNA-lipid nanoparticles combined with antigen mRNA electroporation. Cancer Immunol Immunother 62:285–297
42. Wang S, Wang Y, Liu J, Shao S, Li X, Gao J et al (2014) Silencing B7-H1 enhances the anti-tumor effect of bladder cancer antigen-loaded dendritic cell vaccine in vitro. Onco Targets Ther 7:1389–1396
43. Hobo W, Mams F, Adistry N, de Witte T, Schaap N, van der Voort R et al (2010) siRNA silencing of PD-L1 and PD-L2 on dendritic cells augments expansion and function of minor histocompatibility antigen-specific CD8+ T cells. Blood 116:4501–4511
44. Roeven MW, Hobo W, van der Voort R, Fredrix H, Norde WJ, Teijgeler K, Ruiters MH, Schaap N, Dolstra H (2015) Efficient nontoxic delivery of PD-L1 and PD-L2 siRNA into dendritic cell vaccines using the cationic lipid SAINT-18. J Immunother 38:145–154
45. Van den Bergh JMJ, Smits ELJM, Berneman ZN, Hutten TJA, De Reu H, Van Tendeloo VFI, Dolstra H, Lion E, Hobo W (2017) Monocyte-derived dendritic cells with silenced PD-1 ligands and transpresenting interleukin-15 stimulate strong tumor-reactive T-cell expansion. Cancer Immunol Res 5:710–715
46. Kubo M, Handa T, Yoshimura A (2003) Suppressors of cytokine signaling and immunity. Nat Immunol 4:1169–1176
47. Shen L, Evel-Kabler K, Strube R, Chen SY (2004) Silencing of SOCS1 enhances antigen presentation by dendritic cells and antigen-specific anti-tumor immunity. Nat Biotechnol 22:1546–1552
48. Subramanya S, Armant M, Salkowitz JR (2010) Enhanced induction of HIV-specific cytotoxic T lymphocytes by dendritic cell-tar geted delivery of SOCS-1 siRNA. Mol Ther 10:2028–2037

49. Wang D, Huang XF, Hong B, Song XT, Hu L, Jiang M, Zhang B et al (2018) Efficacy of intracellular immune checkpoint-silenced DC vaccine. JCI Insight 3. https://doi.org/10.1172/jci.insight.98368.
50. Short NJ, Rytting ME, Cortes JE (2018) Acute myeloid leukaemia. Lancet 392:593–606
51. Wölfe SJ, Strebovsky J, Bartz H et al (2011) PD-L1 expression on tolerogenic APCs is controlled by STAT-3. Eur J Immunol 41:413–424
52. Alshansan A, Haddadi A, Hamdy S et al (2010) STAT3 silencing in dendritic cells by siRNA polyplexes encapsulated in PLGA nanoparticles for the modulation of anticancer immune response. Mol Pharm 7:1643–1654
53. Guo C, Yi H, Yu X, Zuo D, Qian J, Yang G, Foster BA, Subjeck JR, Sun X, Mikkelsen RB, Fisher PB, Wang XY (2012) In situ Vaccination with CD204 gene-silenced dendritic cell, not unmodified dendritic cell, enhances radiation therapy. Mol Cancer Ther 11:2331–2339
54. Mobergslien A, Sioud M (2012) Galectin-1 and -3 gene silencing in immature and mature dendritic cells enhances T cell activation and interferon-γ production. J Leucocyte Biol 91:461–467
55. Chen HY, Fermin A, Vardhana S, Weng IC, Lo KF, Chang EY, Maverakis E, Yang RY, Hsu DK, Dustin ML, Liu FT (2009) Galectin-3 negatively regulates TCR-mediated CD4+ T-cell activation at the immunological synapse. Proc Natl Acad Sci USA 106:14496–14501
56. Grossman Z, Singer A (1996) Tuning of activation threshold explains flexibility in the selection and development of T cells in the thymus. Proc Natl Acad Sci USA 93:14747–14752
57. Mittal SK, Roche PA (2015) Suppression of antigen presentation by IL-10. Curr Opin Immunol 34:22–27
58. Reis e Sousa C (2004) Toll-like receptors and dendritic cells: for whom the bug tolls. Semin Immunol 16:27–34
59. Jonuleit H, Schmitt E, Schuler G, Knop J, Enk AH (2000) Induction of interleukin 10-producing, nonproliferating CD4(+) T cells with regulatory properties by repetitive stimulation with allogeneic immature human dendritic cells. J Exp Med 192:1213–1222
60. Sioud M (2005) Induction of inflammatory cytokines and interferon responses by double-stranded and single-stranded siRNAs is sequence-dependent and requires endosomal localization. J Mol Biol 348:1079–1010
61. Kleinman ME, Yamada K, Takeda A, Chandrasekaran V, Nozaki M, Baffi JZ et al (2008) Sequence- and target-independent angiogenesis suppression by siRNA via TLR3. Nature 452:591–597
62. Sioud M (2006). Single-stranded small interfering RNA are more immunostimulatory than their double-stranded counterparts: a central role for 2′-hydroxyl uridines in immune responses. Eur J Immunol 36:1222–1230
63. Cekaite L, Furset G, Hovig E, Sioud M (2007) Gene expression analysis in blood cells in response to unmodified and 2-′-modified siRNAs reveals TLR-dependent and independent effects. J Mol Biol 365:90–108
64. Sioud M, Furset G, Cekaite L (2007) Suppression of immunostimulatory siRNA-driven innate immune activation by 2′-modified RNAs. Biochem Biophys Res Commun 36:122–126
65. Hornung V, Guenthner-Biller M, Bourquin C et al (2005) Sequence-specific potent induction of IFN-alpha by short interfering RNA in plasmacytoid dendritic cells through TLR7. Nat Med 11:263–270
66. Judge AD, Sood V, Shaw JR, Fang D, McClintock K, MacLachlan I (2005) Sequence-dependent stimulation of the mammalian innate immune response by synthetic siRNA. Nat Biotechnol 23:457–462
67. Furset G, Sioud M (2007) Design of bifunctional siRNAs: combining immunostimulation and gene-silencing in one single siRNA molecule. Biochem Biophys Res Comm 352:642–649
68. Iversen PO, Semaeva E, Sørensen DR, Wiig H, Sioud M (2009) Dendritic cells loaded with tumor antigens and a dual immunostimulatory and anti-interleukin 10-specific small interference RNA prime T lymphocytes against leukemic cells. Transl Oncol 2:242–246
69. Campbell JD (2017) Development of the CpG Adjuvant 1018: a case study. Methods Mol Biol 1494:15–27
70. Pradhan P, Qin H, Leleux JA, Gwak D, Sakamaki I, Kwak LW, Roy K (2014) The effect of combined IL10 siRNA and CpG ODN as pathogen-mimicking microparticles on Th1Th2 cytokine balance in dendritic cells and protective immunity against B cell lymphoma. Biomaterials 35:5491–5504

71. Chhabra A, Chakraborty NG, Mukherji B (2008) Silencing of endogenous IL-10 in human dendritic cells leads to the generation of an improved CTL response against human melanoma associated antigenic epitope, MART-$1_{27\text{-}35}$. Clin Immunol 126:251–259

72. Ahn YH, Hong SO, Kim JH, Noh KH, Song KH, Lee YH, Jeon JH, Kim DW, Seo JH, Kim TW (2015) The siRNA cocktail targeting interleukin 10 receptor and transforming growth factor-β receptor on dendritic cells potentiates tumour antigen-specific CD8+ T cell immunity. Clin Exp Immunol 181:164–178

73. Kim JH, Kang TH, Noh KH et al (2011) Blocking the immunosuppressive axis with small interfering RNA targeting interleukin 10 receptor enhances dendritic cell-based vaccine potency. Clin Exp Immunol 165:180–189

74. Castello L, Sabation M, Ren J, Terabe M, Khuu H, Wood LV et al (2017) Expression of CD14, IL10, and tolerogenic signature in dendritic cells inversely correlate with clinical and immunologic response to TARP vaccination in prostate cancer patients. Clin Cancer Res 23:3352–3364

75. Van den Eynde BJ, Morel S (2001) Differential processing of class-I-restricted epitopes by the standard proteasome and the immunoproteasome. Curr Opin Immunol 13:147–153

76. Heink S, Fricke B, Ludwig D, Kloetzel PM, Kruger E (2006) Tumor cell lines expressing the proteasome subunit isoform LMP7E1 exhibited immunoproteasome deficiency. Cancer Res 66:649–652

77. Dannull J, Lesher D-T, Holzknecht R, Qi W, Hanna G, Seigler H, Tyler DS, Pruitt SK (2007) Immunoproteasome down-regulation enhances the ability of dendritic cells to stimulate antitumor immunity. Blood 110:4341–4350

78. Dannull J, Haley NR, Archer G, Nair S, Boczkowski D, Harper M, De Rosa N et al (2013) Melanoma immunotherapy using mature DCs expressing the constitutive proteasome. J Clin Invest 123:3135–3145

79. Rosenberg SA, Restifo NP, Yang JC, Morgan RA, Dudley ME (2008) Adoptive cell transfer: a clinical path to effective cancer immunotherapy. Nat Rev Cancer 8:299–330

80. June CH (2007) Adoptive T cell therapy for cancer in the clinic. J Clin Invest 117:1466–1476

81. Hout R, Schultz LM, Marabelle A, Kohrt H (2015) T-cell-based immunotherapy: adoptive cell transfer and checkpoint inhibition. Cancer Immunol Res 3:1115–1122

82. Mantei A, Rutz S, Janke M et al (2008) siRNA stabilization prolongs gene knockdown in primary T lymphocytes. Eur J Immunol 38:2616–2625

83. Freeley M, Long A (2013) The two hit hypothesis: an improved method for siRNA-mediated gene silencing in stimulated primary human T cells. J Immunol Methods 396:116–127

84. Borkner L, Kaiser A, van de Kasteele W, Andreesen R, Mackensen A, Haanen JB, Schumacher TN, Blank C (2010) RNA interference targeting programmed death receptor-1 improves immune functions of tumor-specific T cells. Cancer Immunol Immunother 59:1173–1183

85. Yu Y, Wu H, Tang Z, Zang G (2009) CTLA4 silencing with siRNA promotes deviation of Th1/Th2 in chronic hepatitis B patients. Cell Mol Immunol 6:123–127

86. Soutschek J, Akinc A, Bramlage B, Charisse K, Constien R et al (2004) Therapeutic silencing of an endogenous gene by systemic administration of modified siRNAs. Nature 432:173–178

87. Shmushkovich T, Monopoli KR, Homsy D, Leyfer D et al (2018) Functional features defining the efficacy of cholesterol-conjugated, self-deliverable, chemically modified siRNAs. Nucleic Acids Res 46:10905–10916

88. Ligtenberg MA, Pico de Coaña Y, Shmushkovich T, Yoshimoto Y, Truxova I, Yang Y, Betancur-Boissel M, Eliseev AV, Wolfson AD, Kiessling R (2018) Self-delivering RNAi targeting PD-1 improves tumor-specific T cell functionality for adoptive cell therapy of malignant melanoma. Mol Ther 26:1482–1493

89. Sioud M (2014) Engineering better immunotherapies via RNA interference. Hum Vaccin Immunother 10:3165–3174

90. Chow VA, Shadman M, Gopal AK (2018) Translating anti-CD19 CAR T-cell therapy into clinical practice for relapsed/refractory diffuse large B-cell lymphoma. Blood 132:777–781

91. Simon B, Harrer D, Schuler-Thurner B, Schaft N, Schulaer G, Dörrie UU (2018) The siRNA-mediated downregulation of PD-1 alone or simultaneously with CTL-4 shows enhanced in vitro CAR-T-cell functionality for further clinical development towards the potential use in immunotherapy of melanoma. Exp Dermatol 27:769–778

92. Pike KA, Tremblay ML (2013) Regulating naive and memory CD8 T cell homeostasis -a role for protein tyrosine phosphatases. FEBS J 280:432–444
93. Jeon MS, Atfield A, Venuprasad K, Krawczyk C, Sarao R, Elly C, Yang C, Arya S, Bachmaier K et al (2004) Essential role of the E3 ubiquitin ligase Cbl-b in T cell anergy induction. Immunity 21:167–177
94. Hinterleitner R, Gruber T, Pfeifhofer-Obermair C, Lutz-Nicoladoni C et al (2012) Adoptive transfer of siRNA Cblb-silenced CD8+ T lymphocytes augments tumour vaccine efficacy in a B16 melanoma model. PLoS One 7:e44295
95. Sathish JG, Dolton G, Leroy FG, Matthews RJ (2007) Loss of Src homology region 2 domain-containing protein tyrosine phosphatase-1 increases CD8+ T cell-APC conjugate formation and is associated with enhanced in vivo CTL function. J Immunol 178:330–337
96. Stromnes IM, Fowler C, Casamina CC, Georgopolos CM, McAfee MS, Schmitt TM, Tan X, Kim TD, Choi I, Blattman JN et al (2012) Abrogation of Src homology region 2 domain-containing phosphatase 1 in tumor-specific T cells improves efficacy of adoptive immunotherapy by enhancing the effector function and accumulation of short-lived effector T cells in vivo. J Immunol 189:1812–1825
97. Kim HJ, Kim A, Miyata K, Kataoka K (2016) Recent progress in development of siRNA delivery vehicles for cancer therapy. Adv Drug Deliv Rev 104:61–77
98. Rajagopalan A, Berezhny A, Schrand B, Puplampu-Dove Y, Gilboa E (2017) Aptamer-targeted attenuation of Il-2 signaling in CD8+ T cells enhances antitumor immunity. Mol Ther 25:54–61
99. Wang C, Lin GH, McPherson AJ, Watts TH (2009) Immune regulation by 4-1BB and 4-1BBL: complexities and challenges. Immunol Rev 229:192–215
100. Sioud M (2011) Promises and challenges in developing RNAi as a research toll and therapy. Methods Mol Biol 703:173–187
101. Dybwad A, Førre O, Natvig JB, Sioud M (1995) Structural characterization of peptides that bind synovial-fluid antibodies from RA patients—a novel strategy for identification of disease-related epitopes using a random peptide library. Clin Immunol Immunopathol 75 (1):45–50
102. Sioud M, Kjeldsen-Kragh J, Suleyman S, Vinje O, Natvig JB, Førre O (1992) Limited heterogeneity of T cell receptor variable region gene usage in juvenile rheumatoid arthritis synovial T cells. Eur J Immunol 22:2413–2418
103. Dybwad A, Bogen B, Natvig JB, Førre O, Sioud M (1995) Peptide phage libraries can be an efficient tool for identifying antibody ligands for polyclonal antisera. Clin Exp Immunol 102:438–442

Chapter 16

Harnessing the Antiviral-Type Responses Induced by Immunostimulatory siRNAs for Cancer Immunotherapy

Per Ole Iversen and Mouldy Sioud

Abstract

When introduced into endosomes via cationic lipids, certain small interfering RNA (siRNA) sequences activate the interferon signaling pathways in immune cells such as dendritic cells (DCs), known as the most efficient antigen-presenting cells of the immune system. Human immature DCs produced high levels of the immune-response protein interferon-α and tumor necrosis factor- α upon incubation with siRNA/lipid formulations, resulting in their maturation and expression of co-stimulatory molecules like CD80, CD86, and CD40 on the cell surface. These molecules are used by mature DCs to co-stimulate T cells during antigen presentation in lymphoid organs. Ex vivo loading of immature DCs with DOTAP-formulated immunostimulatory siRNAs and tumor antigens has proven effective as a cancer vaccine in a rat model of acute myeloid leukemia. Here, we describe this new vaccination strategy that targets tumor cells by activating DCs and blocking the expression of immunosuppressive factors.

Key words Dendritic cells, Cancer vaccines, Immunotherapy, Immunosuppression, RNA interference, Small interfering RNAs, Lipid formulations

1 Introduction

Dendritic cells (DCs) are the most potent professional antigen-presenting cells (APCs). In order to promote immunity rather than tolerance, DCs require a robust maturation signal that leads to the expression of co-stimulatory molecules CD80, CD86, and CD40 ligand on the surface of DCs [1]. Modulation of other co-stimulatory molecules, such as CD70, or pro-inflammatory factors, such as IL12 and CCR7, also enhances DC immunogenic function [1]. While TLR ligands are essential for the generation of effective antitumor responses, they also induce the expression of certain immuno suppressive molecules such as IL-10, which suppresses T-cell function [2]. This inhibitory feedback mechanism may hamper effective immunity against tumor cells. Hence, the development of agents that stimulate DC maturation through

Mouldy Sioud (ed.), *RNA Interference and CRISPR Technologies: Technical Advances and New Therapeutic Opportunities*, Methods in Molecular Biology, vol. 2115, https://doi.org/10.1007/978-1-0716-0290-4_16,

TLRs and simultaneously inhibit the expression of immunosuppressive factors may enhance the potency of cancer vaccines.

Among the gene-silencing approaches, RNA interference (RNAi) offers a novel opportunity for blocking the expression of immunosuppressive factors in vitro and in vivo ([3–5], *see* Chapter 1). Initially, the RNAi pathway was thought to be nonfunctional in mammalian cells, where double-stranded RNAs longer than 30 nucleotides induce a nonspecific antiviral response known as the interferon (IFN) pathway [3]. In 2001, it has been reported that small interference (si) RNA, when introduced into mammalian cells, did not trigger the IFN pathway but effectively silenced gene expression in a sequence-dependent manner in HEK cells [3]. However, we have shown that certain siRNA sequences can bind to TLRs expressed by immune cells and activate the interferon pathway [6–10]. Hence, the development of siRNAs that stimulate DC maturation via TLRs and simultaneously inhibit the expression of immunosuppressive factors such an IL10 may facilitate the design of better DC vaccines. This new generation of bifunctional siRNAs can be used to enhance antitumor immunity during vaccination. Here, we describe the development of a cationic liposome vaccine composed of the broadly used lipid DOTAP, tumor antigens, and an immunostimulatory IL-10 siRNA for use in cancer immunotherapy [11, 12].

2 Materials and Facilities

1. Brown Norway rats.
2. Leukemic cells originating from Brown Norway rats.
3. Small interfering RNAs. There are a variety of online tools that can used to design siRNAs that can be ordered from various commercial sources.
4. Cationic liposomes *N*-[1-(2,3-diolecoyloxy)]-*N,N,N*-trimethyl-ammonium propane methylsulfate (DOTAP). Other commercially equivalent liposomes such Lipofectamine may be used.
5. Transfection buffer: 20 mM HEPES, 150 mM NaCl, pH 7.4.
6. Rat granulocyte-macrophage colony-stimulating factor.
7. Rat interleukin 4.
8. RPMI medium supplemented with antibiotics and 10% FCS.
9. X-vivo 15 medium.
10. Cell culture flasks or dishes.
11. PBS buffer with no additives.
12. Nuclease-free water.

13. Commercially available cytokine ELISA kits.
14. Ficoll-Hypaque gradients.
15. Sterile cell culture facilities.
16. Facilities for working with animals will also be required.
17. In-house access to basic chemicals.

3 Methods

3.1 Immuno-stimulatory siRNAs

To activate TLRs, the anti-IL-10 siRNA was designed to contain a significant number of uridines, found to activate human TLR7/8 [8]. A 2′-*O*-methyl uridine-modified siRNA targeting mouse basigin was used as an inactive control because this modification abolishes siRNA recognition by TLRs [9]. The target site sequences of the siRNA are as follows:

- Active siRNA, 5′-ACGCUGUCAUCGAUUUCUC-3′.
- Inactive (control) siRNA, 5′-GAGGCAAUCACCAAUAGCA-3′.

3.2 Preparation of Leukemic Cell Protein Lysates

1. Leukemic cells from the Brown Norwegian rat model of human AML were suspended in PBS buffer at a density of 1×10^7 cells/ml, frozen at −80 °C, and disrupted by four freeze-thaw cycles. After, the cell suspensions were sonicated and then centrifuged for 10 min at 12,000 × *g*.
2. Subsequent to centrifugation, supernatants (protein lysates) were collected, passed through a 0.2-μm filter, and then incubated at 95 °C for 10 min to inactivate nucleases and any remaining cells.
3. Protein concentrations were determined by the standard Bradford method, and samples were stored at −20 °C until use.

3.3 Generation of Immature (i) DCs

Blood cells from healthy BN rats were separated on Ficoll-Hypaque gradients, and then monocytes were isolated by adherence to plastic. To generate iDCs, monocytes were cultured in complete RPMI medium supplemented with granulocyte-macrophage colony-stimulating factor (50 ng/ml) and IL-4 (100 ng/ml) to generate immature (i) DCs. On day 5, iDCs were transfected with either active or inactive siRNA as indicated below.

3.4 Transfection of iDCs

1. Mix 50-μg DOTAP and 100 μl of transfection buffer in a polystyrene tube.
2. Mix 20-μg siRNA in 50-μl transfection buffer in 1.5-ml microcentrifuge tube.

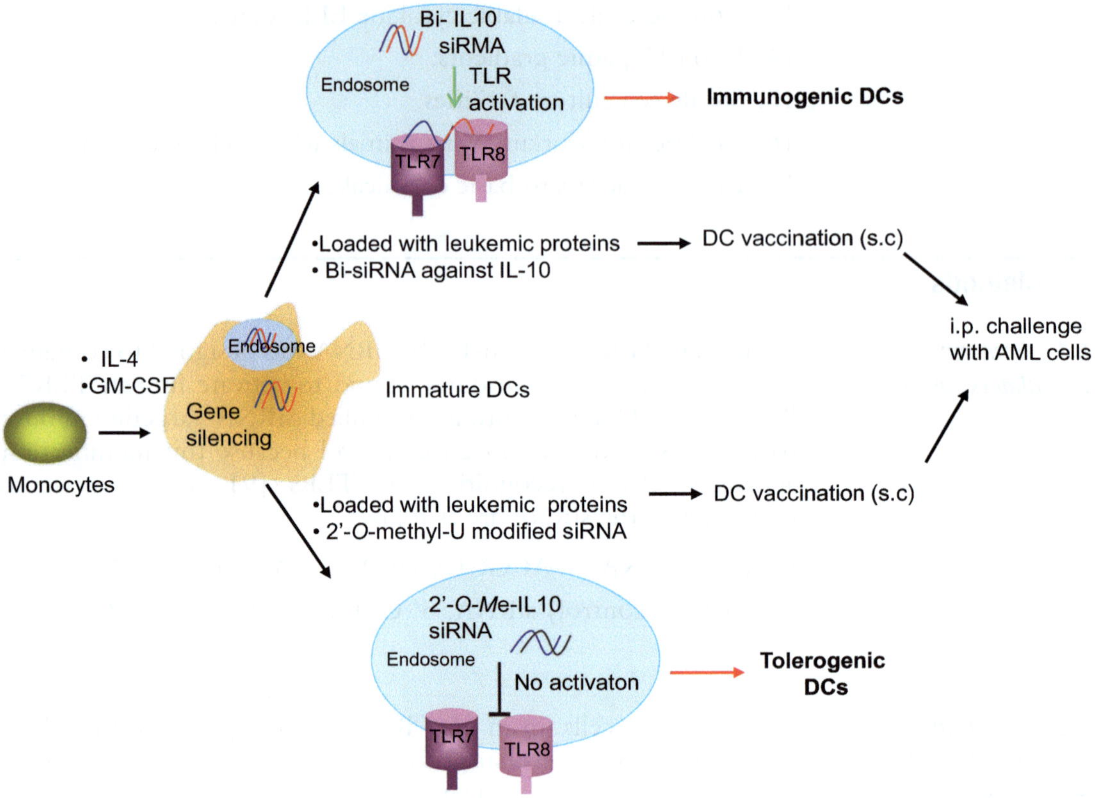

Fig. 1 Schematic illustration of the vaccine design

3. Transfer the siRNA solution to the polystyrene tube containing DOTAP, mix well, and incubate 15 min at RT.
4. Add to the mixture 50–100 μg of protein lysates, mix, and continue incubation for 30 min at RT. Total volume around 200 μl.
5. Add the transfection-ready siRNA mixture dropwise to iDCs (2×10^6 cells) growing in 4- to 5-ml complete RPMI medium.
6. Rock the culture dish or flask to eventually distribute the complexes, and then incubate at 37 °C, 5% CO_2 for 18 h.
7. Similarly, transfect iDCs with an inactive siRNA (control).
8. The following day, harvest the cells, resuspend them in the appropriate X-vivo 15 medium, and then proceed to immunization as illustrated in Fig. 1. Usually we use between 1 and two million cells per vaccination in 100- to 200-μl X-vivo medium. Therefore, the DC cultures should be adjusted based on the number of animal to be immunized. For 10 rats, we transfect around 15–20 million iDCs in 10 ml medium.

3.5 AML Model and Vaccination

We used the BN model of AML because it reliably mimics several important features of human AML including development and progression of the disease as well as response to chemotherapy [13].

1. Syngeneic BN rats ($n = 10$ per group) were randomly assigned to be injected subcutaneously with one to two million DCs transfected with either active or inactive siRNA and loaded with the leukemic protein lysates as described in Subheading 2, **item 3**.
2. Vaccination was repeated three times (once a week), before rats were transplanted with leukemic cells. As a control, unvaccinated leukemic rats were included.
3. The rats were killed with an intraperitoneal overdose of barbiturate when they developed obvious signs of disease.

3.6 Evaluation of the Vaccine

To examine if DCs transfected with the active siRNA could induce a T-cell-specific response, rats were immunized with DCs loaded with leukemic protein lysates and siRNA as outlined in Fig. 1. T-cell cytotoxicity against leukemic cells was determined 4 weeks later. T cells from rats given DCs with active siRNA had a profound increase in cytolytic capacity against leukemia cells in vitro when compared with T cells from rats given DCs with inactive siRNA, demonstrating induction of T-cell-specific antileukemic activity (Fig. 2a). The cytotoxicity of T cells sampled from healthy rats was similar to those from rats given DCs transfected with inactive siRNA.

Overall, these results indicate that cellular immunity to certain leukemic proteins can be induced by activating TLRs and blocking IL-10 expression in iDC vaccine. We next examined whether IL-10-silenced DCs would induce antileukemic immunity in vivo. Kaplan-Meier analysis revealed a survival advantage among rats given DCs transfected with active siRNA compared with those given DCs transfected with inactive siRNA (Fig. 2b; $P = 0.035$, log-rank test) [11].

In conclusion, the siRNA efficiently reduced IL-10 expression by silencing its gene in DCs. In addition, transfection of DCs with the siRNA enhanced production of TNFα, indicating stimulation of TLRs 7/8. Importantly, vaccination with DCs harboring this siRNA and leukemic antigens activated T-cell-mediated killing of leukemic cells as well as enhanced survival and reduced the leukemic cell burden and metastasis in a rat model of AML [11]. Together, these results provide preclinical evidence that lipid delivery of immunostimulatory siRNAs and tumor antigens can break self-tolerance to tumors and thus open the door for better vaccine design.

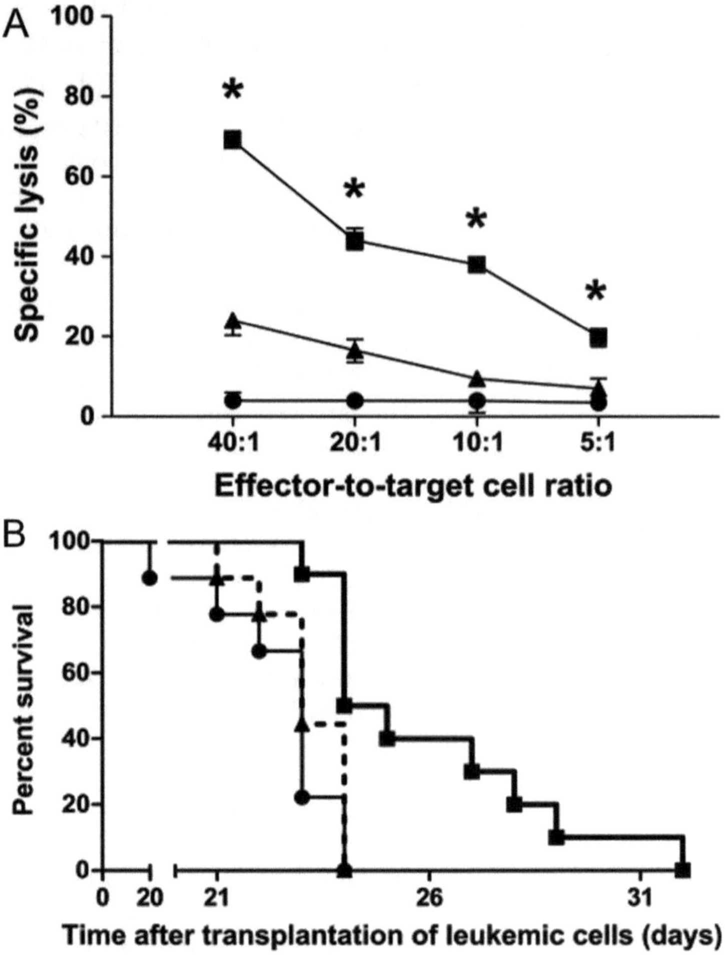

Fig. 2 (**a**) Cytotoxicity. Lysis of autologous leukemic cells was increased at all effector-to-target cell rations tested when splenic T cells from leukemic rats vaccinated with iDC transfected with active siRNA (squares) were compared with those given inactive siRNA (triangles). The specific lysis of autologous leukemic cells given anti-MHC class I antibodies (circles) was insignificant. Values are mean and SEM, $n = 10$. $*P < 0.05$. (**b**) Increased survival on vaccination with iDC loaded with anti-IL-10 siRNA and tumor antigens. Leukemic rats were vaccinated with iDCs transfected with active siRNA (thick filled line) lived longer compared with those given either inactive siRNA (thin solid line) or with untreated rats (dashed line)

Acknowledgments

This work was supported by the Norwegian Cancer Society.

References

1. Banchereau J, Paczesny S, Blanco P, Bennett L, Pascual V, Fay J et al (2003) Dendritic cells: controllers of the immune system and a new promise for immunotherapy. Ann N Y Acad Sci 987:180–187
2. Blander JM, Medzhitov R (2006) On regulation of phagosome maturation and antigen presentation. Nat Immunol 7:1029–1035
3. Fire A, Xu S, Montgomery MK, Kostas SA, Driver SE, Mello CC (1998) Potent and specific genetic interference by double-stranded RNA in Caenorhabditis elegans. Nature 391:806–811
4. Elbashir SM, Harborth J, Lendeckel W, Yalcin A, Weber K, Tuschl T (2001) Duplexes of 21-nucleotide RNAs mediate RNA interference in cultured mammalian cells. Nature 411:494–498
5. Sioud M (2004) Therapeutic siRNAs. Trends Pharmacol Sc 25:22–28
6. Sioud M (2005) Induction of inflammatory cytokines and interferon responses by double-stranded and single-stranded siRNAs is sequence-dependent and requires endosomal localization. J Mol Biol 348:1079 1090
7. Furset G, Sioud M (2007) Design of bifunctional siRNAs: combining immunostimulation and gene-silencing in one single siRNA molecule. Biochem Biophys Res Commun 352:642–649
8. Sioud M (2006) Single-stranded small interfering RNA are more immunostimulatory than their double-stranded counterparts: a central role for 2′-hydroxyl uridines in immune responses. Eur J Immunol 36:1222–1230
9. Sioud M, Furset G, Cekaite L (2007) Suppression of immunostimulatory siRNA-driven innate immune activation by 2′-modified RNAs. Biochem Biophys Res Commun 361:122–126
10. Sioud M (2011) Promises and challenges in developing RNAi as a research tool and therapy. Methods Mol Biol 703:173–187
11. Iversen PO, Semaeva E, Sorensen DR, Wiig H, Sioud M (2009) Dendritic cells loaded with tumor antigens and a dual immunostimulatory and anti-interleukin 10-specific small interference RNA prime T lymphocytes against leukemic cells. Transl Oncol 2:242–246
12. Iversn PO, Sørensen DR, Tronstad KJ, Gudbrandsen OA, Rustan AC, Berge RK, Drevon CA (2006) A bioactivity modified fatty acid improves survival and impairs metastasis in preclinical models of acute leukemia. Clin Cancer Res 12:3525–3531
13. Martens AC, van Bekkum DW, Hagenbeek A (1990) The BN acute myelocytic leukemia (BNML) (a rat model for studying human acute myeloid leukemia (AML)). Leukemia 4:241–257

Chapter 17

Cancer Immunotherapy: Targeting Tumor-Associated Macrophages by Gene Silencing

Karin Zins and Dietmar Abraham

Abstract

Tumor-associated macrophages (TAMs) are representing a major leukocyte population in solid tumors. Macrophages are very heterogeneous and plastic cells and can acquire distinct functional phenotypes ranging from antitumorigenic to immunosuppressive tumor-promoting M2-like TAMs, depending on the local tissue microenvironment (TME). TAMs express cytokines, chemokines, growth factors, and extracellular matrix (ECM) modifying factors, and the cross talk with the TME regulates pathways involved in the recruitment, polarization, and metabolism of TAMs during tumor progression. Due to their crucial role in tumor growth and metastasis, selective targeting of TAM for the treatment of cancer with therapeutic agents that promote phagocytosis or suppress survival, proliferation, trafficking, or polarization of TAMs may prove to be beneficial in cancer therapy. In this chapter, we will discuss TAM biology and current strategies for the targeting of TAMs using small interfering RNA (siRNA)-based drugs. In the past few years, advances in the field of nanomedicine pave the way for the development of siRNA-based drugs as an additional class of personalized cancer immuno-nanomedicines. Fundamental challenges associated with this group of therapeutics include the development process, delivery system, and clinical translation for siRNA-based drugs.

Key words Tumor microenvironment, Matrix metalloproteinases, Extracellular matrix, Growth factors, Tumor-associated macrophages, Metastasis, Nanoparticle, RNA interference, Small interfering RNA

1 The Tumor Microenvironment: The Split Nature of Stromal Inflammation

The tumor microenvironment (TME) plays a crucial role during tumor progression. The TME is composed of the extracellular matrix (ECM), tissue-resident cells, and recruited immune cells that can constitute a large portion of the tumor mass in certain solid tumors [1, 2]. Basically, infiltrating immune cells are part of the cancer immune surveillance recognizing and targeting tumor cells in the body before the tumor becomes clinically apparent. In established tumors, lymphocyte recruitment and effector function are also needed for an effective immune response [3]. Moreover,

Mouldy Sioud (ed.), *RNA Interference and CRISPR Technologies: Technical Advances and New Therapeutic Opportunities*, Methods in Molecular Biology, vol. 2115, https://doi.org/10.1007/978-1-0716-0290-4_17,

non-resolving inflammation caused by cancer cells within the TME is considered a hallmark of cancer [4], and various types of immune cells can be found in all stages of tumor development ranging from subtle infiltration to strong inflammation. However, preclinical and clinical data of the past years that have shaped our current understanding of the cross talk between malignant cells and the tumor stroma and tumor-infiltrating leukocytes showing that inflammation caused by the tumor cells and the surrounding stroma even promotes tumor growth and progression [1, 2, 5]. Many studies have demonstrated the important tumor-promoting effects of the innate immune system, within the stroma. Protumorigenic processes such as angiogenesis, ECM remodeling, and immune evasion are mediated by tumor-associated macrophages (TAMs), tumor-associated neutrophils (TANs), and immature dendritic cells (DCs). In contrast, the recruitment of cytotoxic macrophages and neutrophils, NK cells, and mature DCs results in elimination of tumor cells in primary sites and of disseminated tumor cells [6].

Though immune cells are still supposed to react to the presence of tumor cells at the earliest stages of tumor onset by activating T cells and macrophages, thereby reducing the incidence of cancer [7], the immune TME is modified to a protumor environment promoting its progression while suppressing any immune cell-mediated cytotoxicity after progressing past the initial state [8]. Thus, the tumor-associated inflammatory response reveals a paradoxical leukocyte behavior of enhancing tumor initiation, progression, and ultimately metastasis [9]. On the one hand, the local immune suppression commonly found within the TME is a major obstacle to anticancer therapy [3]. On the other hand, since the tumor stroma cells are more genetically stable than malignant cells, which are characterized by genetic mutations, (infiltrating) stromal cells represent promising targets for therapeutic intervention.

From a more general perspective, inflammation is also linked to induction of cancer development involving the interplay of multiple cell-intrinsic and cell-extrinsic processes, which include genomic instability, proliferative responses, and reprogramming of the stromal environment, which support malignant transformation [10]. Examples for such inflammatory processes increasing the risk of developing malignancies are inflammatory bowel disease, pelvic inflammatory disease, and reflux esophagitis [10]. In accordance with this findings, inhibition of inflammation has been suggested to reduce the risk of developing cancer [11, 12].

Besides tumor cells themselves, different inflammatory immune cells, including macrophages, dendritic cells, neutrophils, fibroblasts, and T cells, modulate the inflammatory environment of tumors by the secretion of soluble growth factors, cytokines, and chemoattractants, which render inflammatory cells suppressive against anticancer T-cell responses [6, 10, 13]. In this context, substantial clinical and experimental evidence indicates that macrophages have a major regulatory role in promoting tumor growth and progression [14].

2 Macrophages in the TME and Their Roles in Tumor Progression

2.1 Macrophages

Macrophages belong to the mononuclear phagocytic system and are a group of terminally differentiated cells that play crucial roles in tissue homeostasis, inflammation, as well as protection against pathogenic infections [15]. The current understanding of the origins of macrophages is that most tissue-resident macrophages are not derived from bone marrow progenitors that extravasate into tissues, as previously thought. Instead a new model of macrophage development is emerging, suggesting that at least three different embryonic sources (erythro-myeloid progenitors, in the yolk sac and fetal liver, and macrophage/dendritic cell progenitor cells in the bone marrow) can differentiate into tissue-specific macrophages [16, 17]. The presence of "resident" embryonic macrophage populations that develop independently of bone marrow hematopoietic stem cells and persist throughout life in most tissues suggests that these tissues harbor pre-macrophages that can proliferate and generate mature macrophages [18]. It is also increasingly clear that distinct transcriptional regulators control the development of tissue-specific macrophage phenotypes [19]. For example, the transcription factor cMYB is required for bone marrow-derived cells, while it is not required for yolk sac-derived macrophages [20]. Nonetheless, the majority of macrophages is regulated by the transmembrane receptor colony-stimulating factor 1 receptor (CSF1R; CD115) via its ligands colony-stimulating factor-1 (CSF1) [21] and interleukin-34 (IL34) [22]. The CSF1R controls the differentiation and survival of almost all macrophages regardless of their origins, and both, CSF1 and IL34, can promote polarization of macrophages into immunosuppressive M2 macrophages [23]. Despite the requirement for CSF-CSF1R, there are tissue-specific growth factors and chemokines as well as different tissue microenvironments that specify tissue-specific macrophage diversity [24].

As macrophages are among the most heterogeneous and plastic cell types, they can acquire distinct functional phenotypes depending on the tissue microenvironment. Various environmental signals activate a broad array of intracellular transcriptional networks in mature macrophages resulting in the functional polarization ranging from proinflammatory responses to immune regulation, parasite control, and tissue remodeling. The M1/M2 nomenclature for macrophages classified macrophages into two polarized phenotypes, the classically activated M1 macrophages, which exhibit inflammatory functions, versus the alternatively activated M2 macrophages, which exhibit anti-inflammatory functions [25].

However, the M1/M2 classification of macrophages is now considered an oversimplified approach that does not adequately describe the full spectrum of macrophage phenotypes. Rather, macrophages can exhibit phenotypes in a spectrum of possible forms with two extremes defined by M1 and M2 [26]. M1-polarized macrophages are activated by IFN-γ and Toll-like receptor (TLR) ligand lipopolysaccharide (LPS) leading to the secretion of high amounts of the proinflammatory cytokines IL1, IL6, IL12, and tumor necrosis factor (TNF) and the production of reactive nitrogen and reactive oxygen species. In addition, they have efficiently present antigen and consequently promote Th1 response. Consequently, M1 macrophages are considered strongly microbicidal and antitumoral [25]. In contrast, the M2 phenotype is activated by IL4, IL13, IL10, and CSF1 and characterized by low production of proinflammatory cytokines, high production of anti-inflammatory cytokines IL10 and COX-2 derived prostaglandin E2 (PGE2), high expression of mannose (CD206) and scavenger (CD163) receptors and increased release of ornithine and polyamines through the arginase pathway. Thus, M1/M2 polarization mirrors the Th1-Th2 polarization of T cells [25, 27]. M1 and M2 macrophages are also distinct in their chemokine profiles. M1 macrophages characteristically express high levels of C-X-C motif chemokine ligand 9 (CXCL9), CXCL10, and C-C motif chemokine ligand 5 (CCL5) that amplify cytotoxic T-cell activation and resistance to intracellular pathogens and tumors. M2 chemokines (e.g., CCL2, CCL17, and CCL22) promote recruitment of regulatory T cells (Treg), tissue repair, remodeling, and tumor progression [27].

Notably, TAMs do not fit entirely into the criteria for M1 or M2 macrophages [28]. Macrophages acquire the TAM phenotype due to the local cytokine environment and promote tumor survival, growth, and metastasis [29]. TAMs largely originate from bone marrow monocytes that are recruited through inflammatory signals released by cancer cells [30, 31], but in gliomas and pancreatic cancer, TAMs can also derive from erythro-myeloid progenitors developed in the yolk sac at the embryonic stage [32–34]. Importantly, the TME promotes differentiation of these macrophages to varying protumoral TAM phenotypes depending on their location in the tumor [5] (Fig. 1). From a therapeutic perspective, different origins of TAMs might be of clinical relevance, as they could allow independent targeting of macrophage subpopulations in the tumor tissue [5].

2.2 Macrophages in the TME

Macrophages are a major component of the leukocyte infiltrate in tumors, and they have a dominant role as regulators and drivers of cancer-related inflammation [35]. Meanwhile, it is increasingly clear that TAMs have important tumor-promoting effects within the stroma, which replaced the old dominant view that

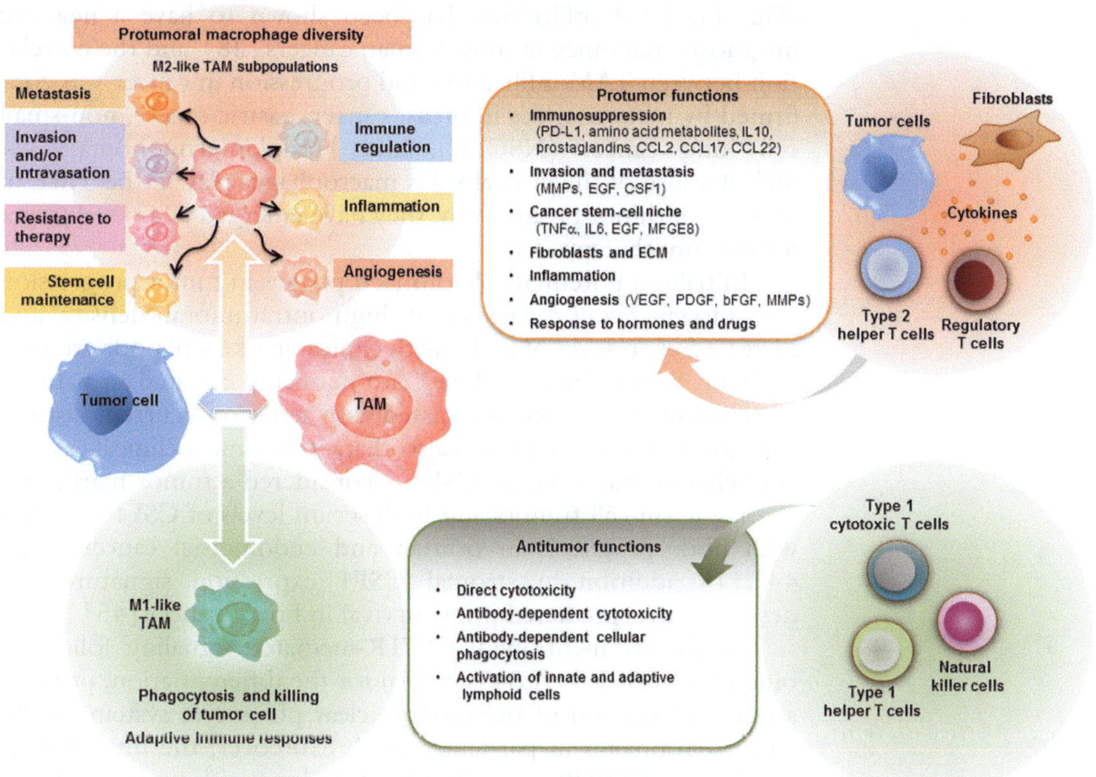

Fig. 1 Tumor-associated macrophages in cancer progression. Tumor-associated macrophages (TAMs) can exert dual function in patients with cancer, depending on the activation state and therapy. During tumor progression, protumor functions prevail, driving tumor progression and resistance to therapy. TAMs can be polarized to adopt different protumoral functions depending on the environment, as indicated. These activities promote tumor initiation through inflammation, tumor progression to malignancy via increasing angiogenesis, immunosuppression, invasion, intravasation, and at distant sites tumor cell extravasation and persistent growth. Each of these functions is performed by a subpopulation of macrophages with a different transcriptome and cell surface markers, representing M2-like TAM subpopulations. Negative signals delivered by tumor cells to macrophages through CD47 can prevent their participation in tumor killing. In contrast, blocking tumor cell-macrophage interactions can unleash macrophage-mediated tumor cell phagocytosis and killing by inducing an antitumor M1-like TAM phenotype. Adapted from [5, 215]. *bFGF* basic fibroblast growth factor, *CCL* C-C motif ligand, *CSF1* colony-stimulating factor 1, *ECM* extracellular matrix, *EGF* epidermal growth factor, *IL* interleukin, *MFGE8* milk fat globule epidermal growth factor, *MMPs* matrix metalloproteinases, *PDGF* platelet-derived growth factor, *PD-L1* programmed death ligand 1, *TNFα* tumor necrosis factor alpha, *VEGF* vascular endothelial growth factor

macrophages were tumoricidal [5]. At the primary tumor site, macrophages enhance tumor cell growth directly by promoting angiogenesis and tumor cell chemotherapy resistance and indirectly by inducing immune dysfunctions through interactions with other immune cells within the TME, resulting in immune evasion of cancer cells [28, 36]. The basis for promotion of tumor progression to malignancy by TAMs is their interaction with cancer cells [5]

(Fig. 1). TAM infiltration has been shown to have a negative prognostic relevance in most human cancers [28], and the correlation between TAM infiltration and progression in cancer was confirmed by meta-analyses in breast cancer, gastric cancer, non-small cell lung cancer (NSCLC), and Hodgkin lymphoma [37–40]. Because of the critical roles macrophages play in the TME in support of tumor progression, they have emerged as a novel target for immunotherapy.

Initially, a potential role of macrophages in tumor progression was shown by correlations of high intratumoral density and increased expression of CSF1 with clinical markers of poor prognosis [41]. Serum levels of CSF1 are markedly elevated in patients with endometrial cancer associated with active or recurrent disease. Increased serum CSF1 levels also characterize most clinical cases of epithelial ovarian cancers. CSF1 is considered a tumor marker for ovarian germ cell tumors and high serum levels of CSF1 correlate with poor prognosis in ovarian and endometrial cancers [42–44]. In addition, a stromal CSF1 expression signature was described that predicted poor survival in breast cancer [45].

As already mentioned, CSF1R-mediated signaling following binding of CSF1 or IL34 is crucial for the differentiation, proliferation, and survival of the mononuclear phagocyte system (MPS) and macrophages in particular [46]. Moreover, the intratumoral presence of CSF1R+ macrophages correlates with poor survival in breast, gastric, oral, ovarian, bladder, and thyroid cancers but not in colorectal cancer [47], and expression of CSF1R combined with CD68-positive macrophage infiltration is considered an independent predictor of short progression-free survival in Hodgkin lymphoma [48]. Consequently, targeting CSF1R signaling in tumor-promoting TAM represents has been shown to be an attractive strategy to eliminate these cells [49, 50].

The role of IL34 in cancer progression is still largely unknown. IL34 shows overlapping functions with, but no sequence similarity to, CSF1 and binds to a different binding site on CSF1R with a significantly higher affinity than CSF1 [22]. Recent studies show protumorigenic functions of IL34 in the tumor microenvironment via multiple effects on both cancer cells and myeloid cells [23]. IL34 produced by chemoresistant cancer cells was able to sustain the immunosuppressive functions of TAMs and promote the survival of cancer cells [51]. However, the effects of IL34 appear to be dependent on the type of tumor as recent study in breast cancer suggested that correlation of IL34 gene expression with survival was dependent on the molecular breast cancer subtype and was not associated with myeloid cell infiltration while directly regulating breast cancer cell migration and signaling [52].

2.3 TAMs in Solid Tumors

2.3.1 Cancer Initiation

Chronic inflammation or oncogene activation results in the expression of key proinflammatory transcription factors within tumor cells such as nuclear factor-κB (NF-κB), signal transducer, and activator of transcription 3 (STAT3) and hypoxia-inducible factor 1α (HIF1α). Activation of these factors mediate the expression of cytokines and chemokines (including TNFα and IL6) leading to the recruitment of macrophages and other immune cells further supporting the inflammatory responses within the tumor microenvironment [53]. As a result of the cross talk between cells within the TME, macrophages produce proinflammatory factors such as IL-6, TNF, and interferon-γ (IFN-γ), growth factors including epidermal growth factor (EGF) and WNTs, proteases, and reactive oxygen species (ROS) and nitrogen species to sustain inflammation and initiate tumorigenesis by triggering DNA mutations [5, 54, 55]. It is assumed that macrophages are mainly tumoricidal in the first stages of tumor formation, but T-helper 2 (TH2) cells in the microenvironment of the growing tumor "educate" macrophages to become protumoral [5].

2.3.2 Neo-angiogenesis

Tumor progression requires the creation of a newly formed vascular network. TAMs also promote tumor angiogenesis and lymphangiogenesis by producing growth factors, in particular members of the vascular endothelium growth factor (VEGF) family, angiogenic CXC chemokines (CXCL8 and CXCL12), secretion of growth factors, and production of proteases thereby exacerbating the transition to malignancy [5, 49, 56]. Remodeling of the ECM is crucial to both angiogenesis and tumorigenesis and primarily involves the matrix metalloprotease (MMP) family of proteolytic enzymes. MMPs degrade the ECM, including the basement membrane and in conjunction with soluble growth factors, and foster the migration and proliferation of endothelial cells. This process promotes angiogenesis and also allows tumors to spread locally and distantly [57]. Modification of the ECM by M2-polarized macrophages occurs through production of highly angiogenic matrix metalloproteinase-9 zymogen (proMMP-9) associated with shutdown of their tissue inhibitor of metalloproteinases-1 (TIMP1) gene expression [58].

Additional factors such as WNT7B, transforming growth factor-β (TGFβ), TNF, and thymidine phosphorylase contribute to the angiogenic process by recruitment and activation of endothelial cells or other cells such as fibroblasts or pericytes that further support the generation of vascular networks in the microenvironment [5]. The fact that TAM density and blood vessel density correlate and the identification of a angiopoietin 1 receptor (TIE2) expressing subpopulation of TAMs that lie in close association with vessels confirmed the central role of TAMs in tumor angiogenesis [59], as their depletion inhibited tumor progression [60], highlighting the connection between these two components of the tumor microenvironment.

In this context, hypoxia is an important environmental cue that shapes the function of TAMs. Due to rapid growth, tumors are prone to develop hypoxic areas leading to hypoxia-inducible factor (HIF) activation in cells, which increases expression of VEGF, CXCL12, CXCR4, basic fibroblast growth factor (bFGF), and CXCL8. Thus, in an HIF-dependent manner, these factors attract macrophages to hypoxic areas and activate a proangiogenic program [61, 62].

Furthermore, hypoxia-induced semaphorin 3A (Sema3A) recruits TAMs into avascular areas by triggering vascular endothelial growth factor receptor 1 (VEGFR1) phosphorylation through the associated holoreceptor, composed of neuropilin-1 (Nrp1) and PlexinA1/PlexinA4. Importantly, deletion of the Nrp1 gene in macrophages resulted in the retention of TAMs in normoxic areas, ablating their proangiogenic and immunosuppressive functions suggesting that the localization of macrophages in the tumor microenvironment, regulated by Sema3A/Nrp-1 signaling, plays an important role in their functional states [63].

Moreover, the relationship between hypoxic TAMs and endothelial cells can also be a determinant of metastasis [64]. Hypoxia-induced expression of regulated in development and DNA damage response 1 (REDD1), an inhibitor of mechanistic target of rapamycin (mTOR), strongly inhibits glucose uptake and glycolysis in TAMs via mTOR inhibition. As a result of this REDD1-mediated mTOR inhibition, TAMs use less glucose, which will feed instead the neighboring endothelial cells, leading to the formation of abnormal, leaky vascular network. REDD1-deficient TAMs normalized glycolytic metabolism by relieving mTOR inhibition and reduced glucose availability to endothelial cells leading to the formation of mature, pericyte-covered functional vessels. The normalized vascular structure had no impact on primary tumor growth but prevented leaky vessels, hypoxia, and metastasis, thus establishing a causative link between TAM metabolism in hypoxia and tumor angiogenesis [64, 65].

2.3.3 Tumor Invasion and Metastasis

TAMs are involved in creation of an invasive TME, which enables highly invasive tumor cells to move from the primary tumor and intravasate into the blood or lymphatic vessels, where they need to avoid eradication by the immune system, survive in the circulation, and migrate to distant sites to establish micrometastases in hostile environments [66]. Tumor cell migration can be induced by TAMs via a paracrine loop between EGF secreting macrophages and CSF1 secreting tumor cells, leading to a synergistic relationship between migration of both TAMs and tumor cells, since CSF1 induces migration of macrophages and EGF migration of tumor cells [67, 68]. An important biological barrier in invasion and metastasis is the basement membrane. During cancer progression, this membrane is degraded by proteolytic enzymes. TAMs also produce cathepsins and matrix-remodeling enzymes that stimulate this process and increase intravasation by the upregulation of MMP9 [67, 69, 70] and a variety of other MMPs, like MMP1, MMP2, MMP7, and MMP12 in a TNFα-dependent manner [71]. Additionally, macrophages can induce an epithelial-to-mesenchymal transition (EMT) in malignant cells [72, 73]. Other studies have also shown that macrophages play key roles in the maintenance of lung metastasis [31, 74].

Macrophages have been identified as powerful drivers of intravasation from invasive breast cancer tumors via the establishment of tumor microenvironments of metastasis (TMEM) consisting of perivascular TAMs, cancer cells that have heightened motility, and endothelial cells [75]. Similar structures identified in clinical samples led to the identification of a prognostic clinical score that predicts the chance of metastasis in breast cancer regardless of tumor stage or clinical subtype [76]. Transcriptional profiling of invasive TAMs also revealed upregulation of key components of the WNT and hedgehog gene families [77, 78], and WNT7B was identified as one of the key factors secreted by TAMs to promote tumor progression and metastasis, in part through effects on angiogenesis but also through promotion of tumor cell invasion [79].

Recently, the role of macrophages in the process of dissemination during evolutionary early stages of breast cancer progression has been investigated. In premalignant lesions, CCL2 produced by cancer cells and myeloid cells attracts CD206+/Tie2+ macrophages and stimulate these macrophages to produce Wnt1 that in turn leads to disruption of E-cadherin junctions between early cancer cells and early dissemination. Before tumors form, these events result in early dissemination microenvironments that propel active intravasation and affect metastasis formation [80]. In addition, a distinct population of macrophages, metastasis-associated macrophages (MAMs), and increase extravasation of tumor cells, they can send survival and growth signals to tumor cells and inhibit anti-cytotoxic T cells, thereby promoting late steps of the metastasis [31, 74, 81, 82]. Importantly, ablation of these MAMs by inhibiting FMS-like tyrosine kinase 1 (Flt1, also known as VEGFR1) or blockade of their recruitment results in inhibition of metastasis in mice [31, 81, 83], suggesting that they represent therapeutic targets in metastatic disease.

Mechanistically, metastatic cells attract bone marrow-derived classical monocytes through a CCL2-CCR2 mechanism [31]. Once extravasated, these monocytes differentiate into MAMs [81] through engagement of a chemokine cascade that involves CCR1-CCL3 autocrine signaling, and the cross talk between MAMs and metastatic cancer cells improves MAM retention. In addition, CSF1 expression in MAMs promotes metastatic growth [74]. MAMs found in the lungs of mice with metastatic disease are transcriptionally different from resident macrophages [84] and are characterized by the expression of the markers CD11b, VEGFR1, CXCR3, and CCR2 [74, 81]. Macrophages in the lung interact physically with vascular cell adhesion protein 1 (VCAM1)-expressing metastatic breast cancer cells and support their survival through an AKT-dependent mechanism as well as by inducing epithelial to mesenchymal transition by producing versican, an extracellular matrix proteoglycan [85, 86]. MAMs have characteristics of monocytic myeloid-derived suppressor cell phenotype and inhibit cytotoxic T cells through reactive oxygen species (ROS)-mediated but checkpoint ligand-independent mechanism, suggesting that they are also involved in the maintenance of an immunosuppressive metastatic tumor microenvironment [84].

2.3.4 Immunosuppression by TAMs

Macrophages can potentially mount a robust antitumoral response. In general, there are two types of immunologic responses, T-helper 1 (TH1)-based and T-helper 2 (TH2)-based responses. In the TH1 response, macrophages undergo M1-type activation, synthesizing proteases, NO, H_2O_2, and other chemicals used to eliminate viruses and bacteria. Cancer cells have also been shown to be destroyed by the TH1 response. In the TH1-mediated anticancer response, IFN-γ, IL12, IL2, and TNFα play major roles [87, 88]. TAMs also can support the adaptive immune response through presentation of tumor antigens and the production of chemokines and cytokines to recruit and activate cytotoxic CD8+ T cells and natural killer (NK) cells. Importantly, antitumoral macrophage activities are restricted by the TH2 cell dominance in the TME. Macrophages undergo M2-like activation, which profoundly affects macrophage functions such that their phenotypes resemble those involved in tissue development and repair, with a consequent suppression of antitumoral response [14, 89]. Tumor progression is associated with suppression of immunity, and factors in the tumor microenvironment derived from cancer cells, stromal cells, or infiltrating leukocytes modulate TAMs to a tumor-promoting mode (Fig. 1). TAM-mediated suppression of the adaptive immune system occurs through various mechanisms.

TAMs lack the proper ability for antigen presentation, making them unsuitable for eliciting an immune response by other immune cells. They can express inhibitory receptors, including nonclassical major histocompatibility complex class I (MHC-I) molecules such as HLA-E and HLA-G, which are able to negatively modulate the activation of NK cells and T cells by interaction with CD94 and leukocyte immunoglobulin-like receptor subfamily B member 1 (LIR1, also known as ILT2), respectively [90].

Importantly, TAMs express T-cell immune checkpoint ligands, such as Programmed cell death ligand 1 (PD-L1), PD-L2, B7–1 (CD80), and B7–2 (CD86), which directly inhibit T-cell functions [91, 92]. Furthermore, they secrete IL10 and TGFβ, while being unable to secrete the proinflammatory cytokine IL12, a lack of which leads to induction of T-regulatory (Treg) cells. Together, this contributes to the maintenance of a strong immunosuppressive microenvironment by inhibiting CD4+ (Th1 cells and Th2 cells) and CD8+ T cells, by inducing regulatory T (Treg)-cell expansion, and by suppressing the activity of monocytes [5, 93, 94]. TAMs also secrete anti-inflammatory cytokines such as CCL2, CCL3, CCL4, CCL5, and CCL20 further contributing to the recruitment of Treg cells in the TME [14]. TAMs also directly inhibit T-cell cytotoxicity through depletion of the amino acid L-arginine by the release of arginase 1 (Arg1). Arg1 cleaves arginine to form ornithine and urea. Arginine is necessary for the expression of the invariant ζ-chain of the T-cell receptor (TCR) complex, and in environments of depleted arginine via Arg1 for example, T-cell function is inhibited [14, 95]. Similarly, depletion of tryptophan or production of tryptophan metabolites by indoleamine 2,3-dioxygenase (IDO) expressed by macrophages can inhibit cytotoxic T cells [96, 97].

3 Methods of TAM Targeting

The identification of subpopulations of macrophages with different markers and phenotypes that are involved in several of the hallmarks of cancer, including angiogenesis, cell invasion and migration, and suppression of antitumor immune responses described above predestines macrophages to attractive multifactorial treatment targets. Thus, clinical trials with therapeutic agents (small molecules, antibodies) that promote protumor responses, phagocytosis or suppress survival, proliferation, recruitment, or polarization of TAMs are currently underway. These early results offer the promise of improved cancer outcomes [5, 98] (Fig. 2).

TAMs are also attractive therapeutic targets for combination therapies including standard of care (radiotherapy, chemotherapy) and immunotherapy [5]. However, the effect of radiotherapy, designed to target cancer cells by inducing DNA damage in malignant cells, on TAMs is not fully understood [99]. In contrast, it is well known that macrophages can limit the efficacy of chemotherapy either directly, by mechanisms that involve direct contact between macrophages and tumor cells or through the secretion of soluble products, or indirectly by modulating the immune system [100]. Initially, antibody-mediated inhibition of CSF1R reversed resistance to chemotherapy in a mouse xenograft model of human

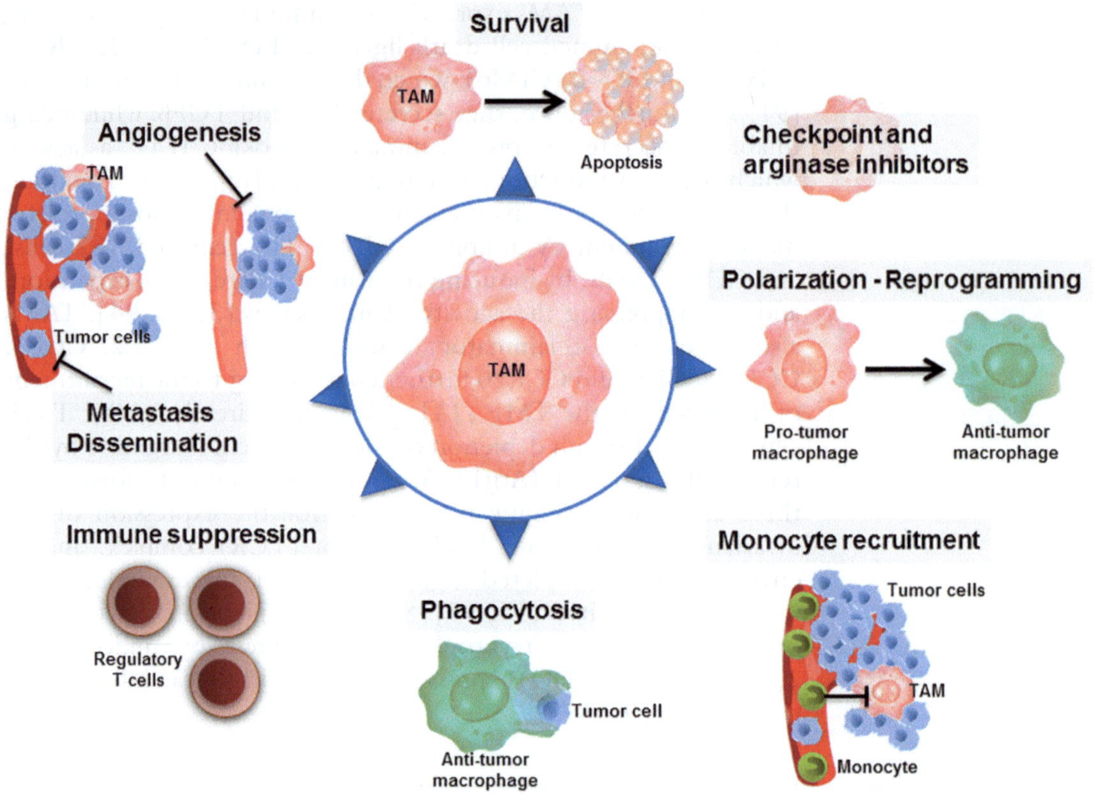

Fig. 2 Tumor-associated macrophage-targeting strategies. Preclinical studies in mice or therapeutic trials in humans have identified and targeted following critical activities of protumoral tumor-associated macrophage (TAM): (1) Survival: for example, inhibitors of CSF1R can suppress the survival of TAMs, thereby reducing tumor growth. (2) Angiogenesis: inhibition of macrophage-derived VEGF can suppress tumor angiogenesis and tumor progression. (3) Metastasis, dissemination: blocking protumoral functions involved in promotion of intravasation, metastatic extravasation, and/or dissemination of malignant cells. (4) Immune suppression: inhibition of the tumor suppressive microenvironment. (5) Immune checkpoint control and metabolism: targeting checkpoints and the metabolism of macrophages using checkpoint inhibitors, anti-immunosuppressive cytokine and/or protein inhibitors to allow cytotoxic T-cell activity, or arginase inhibitors via targeting the PI3K/Akt/mTOR myeloid signaling pathway. (6) Polarization: reprogramming TAMs toward a proinflammatory phenotype promoting tumor suppression. (7) Monocyte recruitment: inhibiting recruitment of the monocytic progenitors of TAMs and thereby suppressing macrophage accumulation in tumors. (8) Phagocytosis: promoting phagocytosis of tumor cells by macrophages, for example, by inhibition of CD47. Targeting key receptors, cytokines/growth factors, or signaling proteins can inhibit these TAM properties and suppress tumor growth and tumor progression

breast cancer suggesting a role of TAMs in mediating resistance to chemotherapy [101]. A study showing that tumors from cancer patients who received neoadjuvant therapy had a much larger infiltrate of CD45+CD11b+CD14+ macrophages than those from patients who received only surgery confirmed these findings [102]. The increase of TAMs following chemotherapy is mostly the result of monocyte recruitment from peripheral circulation and, to a lesser extent, proliferation of tissue-resident macrophages. Moreover, chemotherapy treatment can also lead to the rapid accumulation of proangiogenic TAMs in the tumor microenvironment, thereby supporting cancer metastasis, and at the same time limiting the efficacy of antiangiogenic therapies [100]. TAMs are also responsible for the increased production of cathepsins that leads to increased lymphangiogenesis and metastasis after paclitaxel treatment and treatment of colorectal cancer with 5-fluorouracil results in the release of the diamine putrescine by TAMs, which confers resistance to apoptosis on cancer cells [5].

3.1 Immunotherapy

Programmed cell death protein 1 (PD-1) is an immune checkpoint receptor that is upregulated on activated T cells for the induction of immune tolerance. The ligand for PD-1, PD-L1, in turn is frequently expressed by tumor cells to limit antitumor immune responses. Anti-PD-1 or anti-PD-L1 therapies have shown notable clinical efficacy in patients with a variety of cancers by promoting the activation of tumor-specific cytotoxic T cells (CTLs) [103, 104].

TAMs express high levels of PD-1, and PD-1 increases with stage of disease. Moreover, PD-1 expression on TAMs correlates with decreased phagocytosis of cancer cells, and therefore, TAM-specific PD1 inhibition reduces tumor growth [105]. PD-1 expression inhibits numerous immune cell subsets in the tumor microenvironment, including T cells, B cells, NK cells, and dendritic cells, and immunosuppressive TAMs thus constitute new cellular targets for anti-PD-1 or anti-PD-L1 therapies to improve the efficacy of checkpoint blockade [105]. Moreover, cells of the monocyte-macrophage lineage or lineages, including TAMs, express the ligands for the inhibitory receptor programmed cell death protein-1 (PD-L1 and PD-L2) and CTLA-4 [106].

However, many cancer patients have exhibited resistance to immunotherapies [107]. Consequently, novel therapeutics are in clinical development targeting tumor-associated myeloid cell recruitment, survival, or polarization due to their potential in reversing immunosuppression and synergizing with checkpoint inhibitors to activate CTLs aiming to suppress tumor growth [98]. In addition, clinical trials combining checkpoint inhibitors and anti-TAM agents (such as anti-CSF1R antibodies) are currently ongoing in different solid tumors [5]. Recently, Clever-1 (also known as Stabilin-1), a molecule expressed in M2-polarized macrophages and in TAMs, was shown to serve as a checkpoint in TAMs [108].

Targeting Clever-1 increased antitumor activity in transplanted mouse tumor models by unleashing a protective CD8 T-cell response. Of interest, Clever-1 targeting resulted in decreased PD-L1 expression both in TAMs and cancer cells, indicating that it can block this mechanism of immunotherapeutic resistance. Accordingly, inhibition or inactivation of Clever-1 synergized with anti-PD-1 checkpoint blockade therapy only in high PD-L1 expressing immunogenic breast and colon tumor models, but not in the poorly immunogenic lung tumor model [108].

3.2 TAM Depletion: Controlling Macrophage Proliferation and Survival

The CSF1R plays a major role in regulating macrophage proliferation and survival. Consequently, the dependence on CSF1R signaling makes this an attractive target to selectively deplete macrophages. In fact, CSF1/CSF1R were one of the first macrophage molecules to be targeted for cancer therapy in preclinical cancer studies [101, 109–111]. Mechanistically, targeting the CSF1R may result in suppression of macrophage accumulation in cancer by depleting TAMs, decreased macrophage recruitment as well as vessel density, and increased CTL activation [98]. The use of a CSF1R kinase inhibitor to target TAMs in a glioblastoma mouse model reduced immunosuppressive macrophage/microglia polarization, decreased tumor cell proliferation, and increased apoptosis, resulting in reduced tumor progression and prolonged survival. Notably, half of all mice developed resistance to therapy suggesting that not all macrophages depend on CSF1R signaling for their survival [112]. Different antibodies and compounds targeting the CSF1/CSF1R axis suggest a favorable safety profile, making them attractive combination partners for both standard treatment modalities and immunotherapies [5, 50]. In contrast, no agents targeting IL34, the second ligand for CSF1R, are in clinical development so far.

3.3 Inhibition of TAM Recruitment

Several tumor-derived chemoattractants, in particular chemokines CCL2 (MCP-1), CCL3 [macrophage inflammatory protein (MIP) 1α], CCL4 (MIP1β), and CXCL12 (stromal cell-derived factor 1α; SDF-1α), as well as IL6 and IL1β, and growth factors CSF1 and VEGF-A, and their receptors promote myeloid cell recruitment to tumors by activating integrins particularly α4β1, which then allow extravasation of myeloid cells from the circulation into tumors [98, 113].

As numerous cytokines and their receptors can promote leukocyte invasion into tumors, inhibition of tumor inflammation by blocking only selected chemoattractants or their receptors may be difficult. For example, macrophage accumulation in the tumors is often mediated by monocytic recruitment through CCL2 and its chemokine receptor CCR2 [5]. However, clinical trials for CCL2-neutralizing antibodies and a CCR2-targeting small-molecule inhibitor were largely disappointing. Alternatively, inhibition of molecules required for retention of monocytes, such as CCL3 [74], or identification of signaling pathways that are common to many chemoattractant receptors might be more efficient inhibitors of myeloid cell trafficking. Recently, CCL8 produced by TAMs was shown to increase monocyte infiltration and the number of protumoral TAMs as well as generating an immunosuppressive microenvironment and increasing the malignancy of tumor cells, thus constituting another potential target [114].

Notably, diverse chemoattractants, such as SDF-1α, VEGFα, and IL1β, can lead to activation of the phosphoinositide 3-kinase-γ (PI3Kγ) signaling pathway via receptor tyrosine kinases (RTKs), G protein-coupled receptors (GPCRs), Toll-like receptor (TLR)/IL1R, and TLR ligands/IL1β, respectively. PI3Kγ is the most highly expressed PI3K isoform in myeloid cells and functions downstream of these chemoattractant-receptor pairs to promote myeloid cell recruitment to tumors [115].

Following receptor-mediated activation of Ras, PI3Kγ then activates Bruton's tyrosine kinase (BTK), phospholipase C (PLC)γ, and RasGrp/CalDAG-GEFs, which activate Ras-related protein (Rap)1a and, subsequently, its effector, Rap1-GTP-interacting adapter molecule (RIAM) at the cell membrane of cells [115–117]. RIAM then can facilitate the interactions of paxillin and talin with the integrin α4β1 and the actin cytoskeleton to induce activation of α4β1 by inducing conformational changes that unmask ligand binding sites [115]. Finally, activated α4β1 can bind to the endothelial receptor VCAM-1 to promote leukocyte trafficking across blood vessels [115]. In accordance with these mechanistic findings, inhibition of IL1R, CXCR4, PI3Kγ, BTK, PLCγ, Rap1a, RIAM, or integrin α4β1 has been shown to reduce the intratumoral infiltration of myeloid cells, tumor growth, and metastasis in several mouse tumor models [98].

3.4 Tissue-Resident Macrophages and Bone-Marrow-Derived Macrophages in Tumors

Macrophages in tumors can accumulate in tumors by at least two distinct mechanisms. Tissue-resident macrophages (TRMs) develop from macrophage progenitors that are seeded into tissues during embryonic development and thus accumulate in tumors by proliferation from tissue-resident precursors. In contrast, bone-marrow-derived macrophages (BMDMs) develop from hematopoietic stem cells that generate myeloid progenitors and then mature into monocytes, which need to be recruited to the tumor [98]. Accordingly, TRMs express a unique set of biomarkers compared with BMDMs, suggesting that improving cancer immunotherapy requires selective therapeutic approaches targeting of both, BMDMs and TRMs, to limit macrophage content in tumors in order to reduce immunosuppressive cytokine production [5, 98].

3.5 Reprogramming of TAMs

This approach offers the advantage of reprogramming macrophages from protumoral one to antitumoral one and at the same time avoiding the drawbacks and long-term toxicity of depleting all macrophages [5]. Toll-like receptors (TLRs) stimulate innate immune responses by polarizing macrophages toward a proinflammatory phenotype following activation by bacterial LPS and viral nucleic acids (RNA or DNA). For this reason, different TLR synthetic ligands have been tested in cancer models as strategies to elicit antitumor immune responses by switching TAMs to tumoricidal macrophages in the TME [118]. Intratumoral delivery of TLR7 and TLR9 agonists caused increased monocyte infiltration in the tumor and macrophage repolarization in a mammary tumor mouse model [119]; similar results were obtained with a TLR7/TLR8 agonist (3M-052) that induced macrophage repolarization and tumor regression in a mouse model of subcutaneous melanoma [120]. Two TLR7 ligands (imiquimod and 852A) and one TLR9 ligand (IMO-2055) are being tested for their antitumoral properties on several cancers in clinical trials [5]. However, the effects of TLR agonists can be limited due to induction of PD-L1 expression in macrophages, resulting in an inflammation-dampening effect. Thus, strategies to prolong the effects of TLR 4, 7/8, and 9 agonists, such as in combination with PD-1 blockade, are under investigation in solid tumor clinical trials [98].

CD40 is a receptor that belongs to the TNF receptor superfamily, is expressed on antigen-presenting cells, and promotes proinflammatory macrophage polarization. Specifically, CD40 ligation in human or murine monocytes or macrophages stimulates secretion of proinflammatory cytokines such as IL1, IL12, TNFα, MIP1α, and NO [121, 122]. Agonistic anti-CD40 antibodies exerted tumor inhibitory effects in several tumor mouse models [123]. TAM treatment with CD40 agonists in combination with anti-CSF1R antibodies resulted in a profound TAM reprogramming before their depletion and elicits effective T-cell responses even in tumors that were nonresponsive to immune checkpoints inhibitors as well as tumor regression in mouse tumor models [121, 124, 125]. Anti-CD40 antibodies and recombinant CD40 ligands such as CP-870,893, RO7009789, APX005M, ADC-1013, SGN-40, and SEA-CD40 are currently under investigation in several clinical trials for solid tumors as single agents, or in combination with chemotherapy or immunotherapy [98].

Other strategies to repolarize macrophages include targeting other signaling molecules, transcription factors, and miRNAs. For example, BTK acts downstream of PI3Kγ in macrophages and is activated in TAMs from pancreatic ductal adenocarcinoma [117]. Inhibition of BTK with the inhibitor ibrutinib (PCI-32765) stimulated proinflammatory macrophage polarization, fosters CD8(+) T-cell cytotoxicity, and suppressed tumor growth in murine pancreatic ductal adenocarcinoma-bearing mice [117]. Inhibitors of other macrophage signaling proteins are also undergoing clinical testing. For instance, histone deacetylases (HDACs) may also be useful targets for cancer immunotherapy, despite the fact that they are not selective inhibitors. HDACs are responsible for removing the acetyl groups on histones, a crucial process in epigenetic regulation of gene expression. TMP195, a class IIA HDAC inhibitor [126], has been reported to increase infiltration of CD11b + myeloid cells from blood into the tumor, where they differentiated into antitumoral macrophages expressing proinflammatory cytokines [127]. Moreover, repolarized TAMs synergized with checkpoint inhibition of PD-1 to reduce tumor burden and metastasis in a mouse model of breast cancer suggesting that class IIA HDAC inhibitors can reprogram macrophages in the tumor [127, 128].

Receptor-interacting serine/threonine protein kinase 1 (RIPK1) is a key regulator of inflammatory signaling and cell death pathways [129]. Targeting TAMs with the small-molecule inhibitor GSK547 repolarized TAMs toward a proinflammatory state with increased MHCII, TNFα, and IFN-γ expression, increased infiltration and activation of CTLs, and reduced tumor growth in pancreatic ductal adenocarcinoma mouse models [130]. Moreover, RIPK1 inhibition synergized with anti-PD1 antibody treatment to suppress tumor growth in mouse models [130] and a combination of the RIPK1 inhibitor GSK3145095 and pembrolizumab (anti-PD1 antibody) are currently underway [98].

Transcription factors are another target for macrophage repolarization, such as Janus kinase 2/signal transducer and activator of transcription 3 (JAK2/STAT3). STAT3 is activated in myeloid cells where it can promote immunosuppression. Depletion of STAT3 in myeloid cells resulted in repolarization of TAMs, increased infiltration of CTLs, and decreased tumor burden, in a colorectal cancer mouse model [131]. The STAT3 inhibitor TTI-101 is already undergoing clinical testing in patients with advanced cancers. Alternatively, targeting JAK2, the major activator of STAT3 in myeloid cells, can be a key strategy to inhibit STAT3 [98].

ROS have also been shown to play a critical role in macrophage differentiation. Pretreatment of monocytes with the antioxidant butylated hydroxyanisole inhibited polarization of monocytes toward the M2 type, but not the M1 type, suggesting that blocking ROS can be a potential cancer therapy [132].

Inhibition of microRNA (miRNA) activity or targeting specific miRNAs may constitutes another approach of repolarizing TAMs [5]. miRNAs are small noncoding RNAs that regulate transcription and translation in a sequence-specific manner, and they are processed by the RNase-III enzyme DICER [133]. DICER inhibition reprogrammed TAMs to become antitumoral by expressing an IFN-γ-STAT1 signature, fostered the recruitment of activated cytotoxic T lymphocytes (CTLs) to the tumors, and was also associated with a better response to immune checkpoint blockade [134].

3.6 Promoting Phagocytosis

CD47 is a membrane protein that binds to the myeloid inhibitory immunoreceptor SIRPα (also termed CD172a or SHPS-1) on macrophages [135]. Engagement of SIRPα by CD47 provides a downregulatory signal that protects host cell from phagocytosis by macrophages, and CD47 therefore functions as a "don't-eat-me" signal [136]. Antibody-mediated inhibition of CD47 promotes macrophage-mediated phagocytosis of tumor cells and can initiate an antitumor T-cell response [137]. CD47 is the target of several new therapeutic agents to stimulate phagocytosis of tumor cells, including the human recombinant protein TTI-621, that blocks binding of SIRPα to CD47 and anti-CD47 antibodies [98].

3.7 Targeting the Metabolism

Metabolic shifts play critical roles in the control of macrophage activation and polarization and altering the metabolic programs of TAMs by controlling macrophage metabolic regulators such as mTOR, and Akt could be a useful strategy for repolarizing macrophages to promote antitumor effector functions [138, 139]. In this context, targeting Clever-1 was associated with a change of TAM metabolic state with increased glycolysis and mTOR activity associated with a switch in the TAM phenotype toward an M1-like profile [108]. Clever-1 belongs to the family of scavenger receptors and mediates the uptake and targeting for degradation of endogenous proteins like oxidized low-density lipoproteins and the extracellular matrix glycoprotein, secreted protein acidic, and rich in cysteine (SPARC) [140].

The immunosuppressive macrophage phenotype is defined by the production of ornithine and polyamines through the arginase pathway as well as by expression of Th2 cytokines that include IL10 [141, 142]. IL4 stimulation of macrophages induces specific metabolic alterations that are critical for supporting the antitumor activity, namely, increased fatty acid oxidation and oxidative phosphorylation [14].

Furthermore, PI3Kγ inhibition may constitute a treatment strategy to modulate metabolic changes to inhibit cancer progression. The PI3K/Akt/mTOR myeloid signaling pathway promotes L-arginine metabolism that in turn promotes TAM-mediated immunosuppression. TAMs upregulate expression of arginase 1 (Arg1), which catalyzes the conversion of L-arginine to ornithine and urea. Inhibition of PI3Kγ can suppress myeloid cell Arg1 [143], and deletion of the murine gene for PI3Kγ inhibits Arg1 expression and promotes the expression of the enzyme NOS instead, which catalyzes the formation of the tumoricidal free radical, NO, from L-arginine [141, 144].

In addition, the acidic, nutrient poor microenvironment of tumors directly drives macrophages to adopt immunosuppressive phenotypes, induced by lactate, a by-product of glycolytic tumor cells [145]. Intratumoral acidosis-induced G protein-coupled receptor (GPCR)-dependent expression of the transcriptional repressor ICER in TAMs that led to their functional polarization toward a noninflammatory, protumorigenic phenotype [146]. Accordingly, targeting the proton transporter V-ATPase, which is responsible for the acidification of cells compartments, with the proton pump inhibitor omeprazole reduced macrophage immunosuppressive polarization and delayed hepatocellular carcinoma tumor cell growth [147].

Besides that, it is clear that hypoxia regulates TAMs in cancer progression and therapy as well. Normoxic areas are associated with proinflammatory macrophages, while hypoxic regions are associated with anti-inflammatory macrophages [148]. On the one hand, the phenotype in macrophages populating hypoxic areas is influenced by fine-tuning of hypoxia-responsive genes that are involved in metabolism, angiogenesis, and metastasis, thereby fostering tumor progression [56].

On the other hand, LPS- and IFN-γ-stimulated macrophages display a classically activated phenotype that is characterized by enhanced glycolysis, flux through the pentose phosphate pathway, fatty acid synthesis, and a truncated tricarboxylic acid cycle [149, 150].

4 Nanomedical Approaches for Tackling TAMs by Gene Silencing

RNA interference (RNAi) is the biologic process by which RNA molecules induce sequence-specific inhibition of target gene expression or translation [151]. In 2001, it was shown that 21 and 22 nucleotide (nt) double-stranded RNAs (small interfering RNAs, siRNAs) could induce RNAi silencing in mammalian cells without eliciting nonspecific interferon responses [152]. These siRNAs soon became a powerful method for silencing specific genes in biological research [153, 154].

The ability of siRNA to silence target genes with high efficiency and specificity, and the versatility of siRNAs, has stimulated efforts to develop these molecules as therapeutic agents [155]. Thus, the major advantage of siRNA is the high specificity in a sequence-specific manner resulting in a restricted choice of targets. By contrast, small-molecule inhibitors can act over multiple targets [156]. Moreover, the prospect of suppressing genes encoding proteins that are "undruggable" [155, 157, 158] by classical small molecules has attracted particular interest within oncology, where many important targets have proven undruggable [155]. The RNAi-based strategy also benefits from the establishment of comprehensive nucleotide sequence databases, forming a solid basis for the investigation of RNAi-based drugs [159–161]. Moreover, chemically synthesized siRNAs can efficiently silence gene expression without altering the host genetic material, and the delivery of synthetic siRNAs can be adapted to specific clinical needs [162].

Although gene silencing approaches such as RNAi are widely used in preclinical models, the first clinical trials using unmodified siRNAs resulted in immune-related toxicities and questionable RNAi effects [163, 164]. A second wave of clinical trials using systemic administration of siRNA to patients with solid cancers using a targeted, nanoparticle (NP) delivery system achieved considerable progress, by confirming RNAi effects in humans from systemically delivered nanoparticles [165], but also showed significant, dose-limiting toxicities and insufficient therapeutic efficacy [166]. Nevertheless, several early-phase clinical trials have reported clinical responses in cancer patients following RNAi therapies providing the proof of principle that delivering RNAi nucleotides into tumors is feasible [155].

4.1 Development of RNAi Drugs

Despite the encouraging results of these clinical trials, RNAi-based therapeutics face numerous challenges involving efficiency of tumoral delivery of siRNAs, choice of RNAi targets and off-target RNAi activity, and safety, which currently limit the RNAi technology to reach its maximal potential in clinics [155]. Successful systemic nonviral siRNA delivery is a key issue in clinical applications of RNAi, and an effective nanocarrier addresses the challenges of delivering naked siRNA/miRNA, pharmacodynamics-related challenges in targeting specificity, pharmacokinetics-related challenges in systemic circulation, cellular uptake and endosomal escape, and innate immune stimulation but also offers "smart" targeted delivery [161, 166].

Avoiding nonspecific toxicity of RNAi therapeutics is an additional critical issue. Clinical RNAi drug development faces different sources of toxicity [166]: Immunogenic reactions to RNAi triggers due to sensing of dsRNAs by PKR, TLR-3 and TLR7, a problem, which can be largely avoided by siRNAs with extensive 2′-O-methyl base modifications [163, 167, 168]. In addition, immunogenic and nonimmunogenic toxic effects of excipient chemicals of nanoparticle drug formulations can be a substantial problem causing dose-limiting toxicities of nanoparticle-delivered drug candidates [169]. Clinical trials show that problems can arise directly from excipient components, from metabolic breakdown of the excipients, or from degradation already during storage [169], or from intravenous immunological infusion reactions [170]. An unintended RNAi activity due to off-target RNA silencing can occur from unintended seed region matches between RNAi guide strands and nontargeted mRNAs [171], a problem that can be only overcome by thorough screening of target sites together with extensive testing [166]. Finally, the systemic delivery of siRNAs to the body can lead to accumulation of RNAi drugs in non-target tissues resulting in a toxic on-target RNAi activity in these tissues [166].

In short, addressing structural motifs, sequence selection, chemical formulation of RNAi triggers, and selection and engineering of delivery routes and excipients are the key points for the development of RNAi drug formulations [166]. The lack of biomarkers associated with biological response following treatment with RNAi therapeutics further complicates optimal clinical trial design [155]. Despite these challenges, significant progress has been made in the fields of cancer, and a variety of carriers including liposomes, lipid nanoparticles, aptamers, and antibodies have been used to deliver siRNA molecules [162, 172, 173].

4.2 Targeting Tumor-Associated Macrophages: Nanoparticle RNAi Delivery Systems

Safe and efficient delivery systems are needed to transport RNAi-based therapeutics. The carriers can generally be divided into viral and nonviral vectors. The advances of viral vectors are efficient delivery, sustained gene silencing, and ease of expression of multiple copies of RNAi molecules from one transcript [174]. The main drawbacks of using virus vectors are its immunogenicity and cytotoxicity that pose safety concerns. An additional cause of concern over using viral gene transfer vehicle is mutagenesis caused by viral genome insertion, which can disrupt the expression of a tumor suppression gene or activate an oncogene leading to the malignant transformation of cells [175]. By contrast, the major advantage of using nonviral vectors is its biosafety, low cost, and ease of production [176]. Thus, great efforts have been undertaken to introduce synthetic vehicles for the in vivo delivery of noncoding RNAs (ncRNAs). Nanoparticles are the most common choice for the delivery of RNAi via nonviral carriers [177]. A wide range of materials has been explored for systemic in vivo delivery of siRNA-loaded nanoparticles aiming at providing serum stability, functional tenability, mitigation of interactions with non-target cells, enhancing cell entry and endosome escape, resisting renal clearance, and generating low toxicity and immunogenicity [161]. The systems for targeting TAMs can be grouped by composition into liposomes, polymers, and inorganic nanoparticles [161, 178]. The principal mechanisms of macrophage-based anticancer therapy rely on (1) inhibition of TAM recruitment to the tumor, (2) direct killing of TAMs, and (3) reprogramming of TAMs from their M2-like protumoral phenotype to an M1-like antitumoral phenotype with a proinflammatory response [179] (Table 1).

4.3 Lipid-Based Nanoparticles

Liposomes are one of the most studied and frequently used carriers in nanomedicine. Due to their composition, these particles enable the encapsulation of hydrophobic, as well as hydrophilic compounds, and they have been shown to be excellent carriers for a multitude of different active ingredients, including siRNA [180, 181].

4.3.1 Targeting TAMs: CD163 Receptor

M2 macrophages overexpress different types of receptors, including the scavenger receptor CD163, which makes them excellent targets for drug delivery. Zhang and Palmer described hemoglobin (Hb)-modified liposome uptake into macrophages via CD163 receptor-mediated endocytosis. With Hb on the surface of liposomes, the complex demonstrated specificity toward macrophages,

Table 1
Gene silencing strategies for inhibition of TAM recruitment, survival, and reprogramming and macrophage-targeted delivery

Strategies	siRNA target	Nanomaterial	Ref.
Inhibition of recruitment	CCR2 siRNA	*Liposome* Cationic liposome	[183]
	CCR2 siRNA	*Polymer* Cationic nanoparticle	[194]
	CX3CL1 siRNA	*Polymer* 7C1 nanoparticle	[193]
	CCL5 siRNA	*Polymer/nanoparticle* Silicon nanoparticles loaded with cationic polymer	[195]
	VEGF siRNA	*Inorganic nanoparticles* M2pep-modified AuNPs	[205]
	MIF siRNA	*Polymer* Glucan-based nanoparticle	[189]
Interference with survival	CSF1R siRNA	*Polymer* α-Peptide-linked with M2pep	[197]
Reprogramming	STAT3 siRNA	*Liposome* Cationic liposomes	[184]
	Anti-IL10 and anti-IL10RA ODN	*Polymer* PEG-modified alginate	[200]
	IκBα siRNA	*Polymer* Mannosylated polymer nanoparticles	[199]
Macrophage-targeted (siRNA) delivery	siRNA cargo	*Polymeric micelles* Mannosylated polymer nanoparticles (CD206, mannose receptor-targeted siRNA delivery)	[190]
	Potential siRNA cargo	*Liposome* Hemoglobin (Hb)-modified liposomes targeting macrophages	[182]

CCL5 C-C motif chemokine ligand 5, *CCR2* C-C chemokine receptor type 2 (CCR2 or CD192), *CSF1R* colony-stimulating factor-1 receptor, *CX3CL1* C-X3-C motif chemokine 1 (fractalkine), *IκBα* NF-kappa-B inhibitor alpha, *IL10* interleukin 10, *IL10RA* interleukin 10 receptor subunit alpha, *MIF* macrophage migration inhibitory factor, *STAT3* signal transducer and activator of transcription 3, *VEGF* vascular endothelial growth factor

while any agents can be loaded into the liposome to influence the function of the engulfing macrophages [182].

4.3.2 Inhibition of TAM Recruitment to the Tumor

Targeting monocytes, either in the blood, in the bone marrow, or in the lymphoid organs, can reduce the generation of TAMs [179]. Since unprotected siRNA is rapidly degraded, various delivery systems based on liposomes have been developed. The first nanotechnology-based design to inhibit the recruitment of monocytes was based on anti-CCR2 siRNA-loaded lipid nanoparticles in order to reduce the expression of the chemokine receptor CCR2, which is required for the recruitment of monocytes to the tumor

[183]. The final lipid-siRNA weight ratio of the nanoparticle was 7:1 with a siRNA entrapment efficiency of 95%. The in vivo distribution of these nanoparticles after intravenous injection was analyzed by siRNA labeled with a near-infrared fluorochrome. The blood half-life of the siRNA was 8.1 min and accumulated highest in the spleen, but there was also considerable accumulation in the liver and bone marrow, where they delivered siRNA into the Ly6C-high monocytes, which are precursors of TAMs. The siRNA-loaded liposome decreased the number of TAMs and reduced tumor growth in xenograft tumor models [183].

4.3.3 Reprogramming of Macrophages

Protumoral TAMs are generated involving STAT3 signaling cascade, suggesting that inhibition of STAT3 can convert them to antitumor M1 type macrophages. The effect of STAT3 inhibition was recently tested in a mouse melanoma model by iontophoretic co-delivery of curcumin and anti-STAT3 siRNA using cationic liposomes. Curcumin was encapsulated in 1,2-dioleoyl-3-trimethylammonium propane (DOTAP)-based cationic liposomes and then complexed with STAT3 siRNA. The curcumin-loaded liposome-STAT3 siRNA complex was prepared at a DOTAP/siRNA ratio of 10:1, with a final diameter of 192.6 ± 9.0 nm. Coadministration of curcumin and STAT3 siRNA effectively controlled tumor progression [184].

4.4 Polymer-Based Nanoparticles

Polymer-based nanoparticles have a high drug-loading efficiency, and their surface can be easily modified making them an attractive tool for the development of cancer therapies.

4.4.1 Targeting TAMs

Polymers such as N-(2-hydroxypropyl) methacrylamide (HPMA) can be used as drug carriers to improve biocompatibility, prolong drug retention time, and reduce systemic toxicity associated with conventional chemotherapy. Zimel et al. used a HPMA copolymer conjugated with Alexa Fluor 647 and folic acid (P-Alexa647-FA). The strategy of incorporating folic acid into the construct is based upon the finding that activated macrophages in general, including TAMs, upregulate the folate receptor beta (FOLR2) on the cell surface [185].

The found that HPMA copolymers can target TAMs in tumors and that HPMA was specifically taken up $CD11b^{+}CD68^{+}$ TAMs in both primary and metastatic breast tumors associated with reversed TAM polarization [186].

In another approach, a glucan-based carrier was developed for the delivery of siRNA to macrophages [187]. Glucans display pathogen-associated molecular patterns (PAMPs), which are recognized and internalized by macrophages [188]. The authors developed a water-soluble, nontoxic β-$(1 \rightarrow 3)$-$(1 \rightarrow 4)$-glucan (BG34). This glucan was found to mediate receptor-induced internalization by primary macrophages via interaction with CD11b [187]. In a later study, siRNA against migration inhibitory factor (MIF) was incorporated into this glucan, resulting in 80–120 nm nanoparticles. Systemic injection of these nanoparticles led to reduced MIF expression in TAMs and specific tumor accumulation in 4T1 mammary tumor-bearing mice [189].

By using polymers, tri-block copolymer nanoparticles were developed for the delivery of siRNA to macrophages. Macrophage-specific RNAi delivery was created by pH-responsive polymeric micelles that were mannosylated to achieve CD206 (mannose receptor)-targeted siRNA delivery. In these particles, the core consists of a hydrophobic pH-responsive block, which is capable of disrupting endosomes at low pH (endosomal escape) and enables cytoplasmatic delivery of siRNA. The second block consists of cationic 2-(dimethylamino) ethyl methacrylate (DMAEMA) polymers for condensation of nucleic acids within the particle, thus carrying and protecting the cargo. The third block presents azide groups at the corona of the formed micelles, enabling the immobilization of alkyne-functionalized ligands onto the micelles, such a mannose as targeting molecule. The mannose-targeted siRNA (MnNP) carriers generated a robust knockdown of a model gene in primary macrophages and increased the delivery of siRNA 13-fold into these cells as compared to cancer cells [190]. These mannosylated polymer nanoparticles then were effectively delivered to TAMs in spontaneously formed, primary mammary tumors, and they enhanced delivery of labeled nucleotides to ovarian tumor TAMs compared to nontargeted particles. The nanoparticles produced no significant toxicity when used in vivo [191].

In general, lack of selectivity is still the major challenge to be addressed when looking for targets on the surface of TAMs, as these molecules or receptors are also present on other cells. For example, CD206 is expressed by dendritic cells, specific lymphatic or endothelial cells, or in macrophages in other locations [179].

4.4.2 Inhibition of TAM Recruitment to the Tumor

Jung et al. generated 7C1 nanoparticles [192] containing CX3CL1 siRNA to block monocyte recruitment and improve the efficiency of anti-VEGF therapy in colorectal cancer. They showed that an anti-VEGFR2 therapy upregulated the expression of CX3CL1, leading to the recruitment of CX3CR1 expressing Ly6C^{lo} monocytes, which then attracts neutrophils via CXCL5, resulting in the formation of an immunosuppressive microenvironment with a reduction of cytotoxic T cells [193]. Treatment with 7C1-CX3CL1 siRNA reduced the CX3CL1 level in vivo and significantly inhibited monocyte recruitment to anti-VEGF-treated tumors.

Another way to inhibit monocyte-recruitment is to disturb CCL2-CCR2 pathways. Monocyte-targeted, siCCR2-encapsulated cationic nanoparticles were shown to accumulate in monocytes and modified the tumor immunosuppressive microenvironment in an orthotopic murine breast cancer model [194].

A similar approach has been recently presented by Ban et al. through targeting of the CCL5-CCR5 axis, a major regulator of immunosuppressive myeloid cells. They developed bone marrow-targeted biodegradable mesoporous silicon nanoparticles (MSVs) loaded with liposomal CCL5-siRNA and decorated on their surface with an E-selectin thioaptamer (ESTA), expressed on the bone marrow endothelium. CCL5-siRNA-loaded MSV nanoparticles were delivered intravenously to 4T1 tumor-bearing mice and caused reprograming of immunosuppressive myeloid cells, reduced the number of TAMs, and significantly reduced of tumor growth associated with increased CD8+ T-cell infiltration in the tumor. These antitumor immune responses were significantly enhanced by the combination of the CCL5-siRNA-loaded-MSVs, targeting CCL5 in the bone marrow, with the CCR5 inhibitor Maraviroc®, blocking CCR5 in the periphery [195].

4.4.3 Elimination of TAMs

Qian et al. developed TAM-targeting nanoparticles (M2NPs) loaded with anti-CSF1R siRNA, with α-peptide (a scavenger receptor B type 1-binding peptide) linked with M2pep (an M2 macrophage-binding peptide) on the particle surface. The M2 peptide is a useful tool for modulating murine M2 macrophages [196]. This approach aimed at specifically blocking the survival signal of TAMs and to deplete them from melanoma tumors. Indeed, the M2NP-loaded CSF1R siRNA delivery resulted in a significant elimination of TAMs, decreased tumor size, and prolonged survival. Furthermore, this system inhibited the production of IL10 and TGFβ while increasing expression of proinflammatory cytokines IL12 and IFN-γ and enhancing CD8+ T-cell infiltration in the tumor microenvironment. Moreover, this delivery system downregulated the expression of PD-1 and TIM3 on infiltrating CD8+ T cells and enhanced their IFN-γ secretion indicating the restoration of T-cell immune function [197].

4.4.4 Reprogramming of TAMs in the Tumor

Only few nanotechnological approaches have been explored to achieve an effective and long-lasting functional reeducation of TAMs [198].

Targeting TAMs by using nanoparticles with a mannosylated surface and loaded with IκBα siRNA was used to restore the classical NF-κB activity in TAMs. Following delivery of IκBα siRNA to macrophages by mannosylated endosomal escape nanoparticles (MnNPs) resulted in activation of the NF-κB pathway and inducing a proinflammatory, immunogenic phenotype in the transfected macrophages. Thus, MnNP with IκBα siRNA could represent a strategy to reprogram TAMs to generate antitumor functions in macrophages for cancer therapy [199].

Huang et al. developed a nucleic acid drug delivery system to target and to reprogram TAMs by combining CpG oligonucleotides (ODN), anti-IL10 ODN, and anti-IL10 receptor ODN [200]. Targeting of TAMs was achieved by using galactosylated cationic dextran (gal-C-dextran), which can associate with ODN to form a stable nanocomplex to target TAMs. This strategy is based on the finding that TAMs express high levels of macrophage galactose-type lectin (Mgl) [201]. Dextran is widely used as a component of drug-delivering nanoparticles, and it is an alternative for PEGylation to avoid nanoparticle-opsonin interactions [202]. In addition, PEG-histidine-modified alginate (PHA), a pH-sensitive material, allowed the nanocarriers to preferentially deliver and release the anti-IL10 and anti-IL10R ODN in TAMs in the acidic tumor microenvironment, i.e., to hide the nanocomplexes from macrophage uptake in other tissues. The lower intratumoral pH is caused by H^+ ions generated by tumor cells in a hypoxic environment during the glycolysis process, especially in low-flow tumors due to insufficient drainage [203, 204]. The intravenous administration of these nanocomplexes in an allograft hepatoma murine model suppressed the protumor functions of TAMs with concomitant stimulation of their antitumor activities [200].

Another TAM-modulating therapy developed used a combination of the TAM-targeting M2 peptide and PEGylated gold nanoparticles (15 nm), loaded with siRNA directed against VEGF in a lung cancer model. Treatment with this formulation resulted in reduced recruitment of TAMs and a reeducation of the TME. The stimulation of a host immune response was critical for reduced tumor size and increased animal survival [205].

Overall, major challenges for the targeting and reeducation of TAM are the specificity of ligands to exclusively reach macrophages in tumors and not other populations of macrophages as well as the resilient nature of the immunosuppressive microenvironment, as it often takes time to mount a sufficient, long-lasting immune response to combat cancer. This underlines the necessity to develop strategies, which effectively reprogram TAMs in a long-lasting manner [179].

4.5 Inorganic Nanoparticles

In addition to liposomes and polymers, there are some reports describing inorganic nanoparticles with small sizes, large surface areas, and various surface modifications for siRNA delivery to TAMs.

By studying the polarization of macrophages upon exposure to differently sized titanium dioxide (TiO_2) particles, it was concluded the size of the nanoparticle is an important factor that affects macrophage polarization, which may potentially guide the design of new nanoparticles to treat diseases such as cancer [206]. In a lung cancer model, M2 peptide (M2pep)-functionalized Au nanoparticles loaded with VEGF siRNA were designed for cancer immunotherapy by targeting TAMs [205]. The release of siRNA from both RNAi- (passive targeting) and RNAi-M2pep-AuNPs (active targeting) was evaluated. AuNPs were decorated with M2pep and thiol-siRNA via a PEG linker with a mean diameter of 23.2 (±2.8) nm. Both RNAi-AuNPs and M2pep-RNAi-AuNPs were found to be stable in serum for over 48 h with only a 25% RNA release. Additionally, both nanoparticles were taken up by lung cancer cells, producing a robust reduction of VEGF expression. Silencing VEGF in both TAMs and cancer cells using M2-targeted siRNA-NPs (RNAi-M2pep-AuNPs) turned out to be more efficient in tumor size reduction than just silencing VEGF in cancer cells using RNAi-AuNPs via passive targeting. Accordingly, M2pep-RNAi-AuNPs showed an increased reduction of the number of TAMs compared to RNAi-AuNPs [205].

4.6 Additional Challenges for the Efficacy of Cancer Nanomedicine

TAMs are considered as potential therapeutic targets for cancer therapy on the one hand, and on the other, the high phagocytic activity of the mononuclear phagocytic system, mainly liver and lung macrophages, have been identified as major responsible for the undesirable uptake and clearance of intravenously injected nanomedicines. Thus, further studies are needed to optimize the biodistribution profile of nanomedicines in preclinical animal models and patients [179].

A different problem for the efficient transport of nanomedicines to TAMs can be found in the characteristics of the TME, specifically an abnormal vasculature, hypoxia, and the composition of the ECM in the tumor tissue, which are highly influenced by the tumor immunosuppressive microenvironment [179]. The ECM of the tumor prevents the penetration of nanoparticle-based therapeutics, antibodies, or other drugs; is mainly present in highly stromal tumors, such as pancreatic cancer; and it consists mainly of fibroblasts, endothelial cells and immune cells, embedded into a net of ECM formed by hyaluronan, collagen and fibronectin [207–209]. Of interest, TAMs promote tumor development by the remodeling of its ECM composition and structure by instructing the deposition, cross-linking, and linearization of collagen fibers during tumor progression [210].

Furthermore, the abnormal tumor vasculature fails to deliver enough oxygen and nutrients to the growing tumor mass resulting in avascular and poorly oxygenated, hypoxic tumor regions. These deep tumor regions with no access to oxygen not only hamper the delivery of therapeutic agents but also block the activity of anticancer therapeutics, such as radiotherapy, which require oxygen molecules for generation of ROS and induction of apoptosis [211]. A recent report showed that the reprogramming of immunosuppressive cells through immune checkpoint blockade treatments results in the tumor vessel normalization and reduction of hypoxia [212]. Likewise, TAM targeting may result in tumor vessel normalization, reduction of hypoxia, and depletion of the excessive extracellular matrix present in highly stromal tumors. Consequently, elimination of immunosuppressive TAM and/or their reprogramming into antitumoral macrophages has been proposed as a strategy to improve the delivery of nanomedicines into the tumor [179, 213].

5 Conclusions and Perspectives

Since the discovery of RNA interference and its utilization for gene knockdown in mammalian cells, significant progress has been achieved in the development of siRNA therapeutics. siRNA is a promising therapeutic tool against various diseases, and with the approval of patisiran, systemic delivery of RNAi therapeutics to the liver is now a clinical reality [166]. In the field of cancer, a number of siRNA-based therapeutics have been tested in clinical trials involving patients with solid tumors providing proof-of-principle for RNAi-based therapeutics in humans. These trials are based on nanoparticle delivery systems to transport therapeutic siRNA to tumors following systemic administration and showed that functional siRNA can be successfully delivered to human tumors. Importantly, safe target gene inhibition can be achieved in patients, when systemically treated twice per week with either lipid- or polymer-based nanoparticle formulations containing siRNAs at doses in the 0.5–1.0 mg siRNA per kg range [169]. Importantly, the applied siRNA therapeutics (both naked and chemically modified siRNA) are well tolerated [169].

Despite its significant potential benefits, safe, efficient, and target oriented delivery of siRNA is one of the major challenges in siRNA therapeutics. To address these issues, the properties of nanoparticle formulations for delivery of RNAi therapeutics such as size, shape, charge, material, surface functionalization, and choice of adjuvants must be tackled to achieve precise delivery, improved stability, targeted biodistribution, favorable pharmacokinetics, and low systemic toxicity [178].

Key challenges for the systematic delivery of RNAi therapeutics in general include avoiding both renal and reticuloendothelial clearance, enhancing extravasation and tissue perfusion, and improving endosomal escape [214]. As already mentioned, macrophages of the MPS are largely responsible for the undesirable uptake of nanomedicines in the bloodstream. However, strategies aiming to avoid the uptake of therapeutics by MPS-macrophages would also lead to reduced uptake of nanoparticles by the intended target cells, in this case TAMs. A potential solution to solve

MPS-clearance considers the therapeutic manipulation of MPS-macrophages prior to the systemic injection of nanomedicines [179]. However, the design of "pre-programmed" myeloid cells could also offer new therapeutic opportunities with intrinsic antitumor properties which could be activated when they infiltrate the tumor [179].

General problems that can arise specifically in the TME include the binding of therapeutic nanoparticles by endogenous proteins that affect distribution, metabolism, and excretion of NPs and an impenetrable ECM barrier. Furthermore, intracellular trafficking and chemical transformations of internalized nanoparticles play an important role in intracellular toxicity of NPs and therapeutic efficacy [178]. TAM-targeted (RNAi-based) nanomedicines could help to solve the problem of penetration of nanomedicines into stromal tumors by depleting the excessive extracellular matrix and the reeducation of the immunosuppressive cells in the TME to induce potent antitumor immune responses [179].

Thus, a better understanding is required to understand the full details of nanomedicine-macrophage interactions in the TME and for the clinical translation of TAM-targeting immuno-nanomedicines. Despite these challenging problems, innovative technologies under development made by nanotechnologists and immunologists in the field of cancer have the potential to lead to the development of new RNAi-based therapeutics to provide more personalized and targeted cancer therapies.

References

1. Liotta LA, Kohn EC (2001) The microenvironment of the tumour-host interface. Nature 411:375–379
2. Brigati C, Noonan DM, Albini A, Benelli R (2002) Tumors and inflammatory infiltrates: friends or foes? Clin Exp Metastasis 19:247–258
3. Lanca T, Silva-Santos B (2012) The split nature of tumor-infiltrating leukocytes: implications for cancer surveillance and immunotherapy. Oncoimmunology 1:717–725
4. Hanahan D, Weinberg RA (2011) Hallmarks of cancer: the next generation. Cell 144:646–674
5. Cassetta L, Pollard JW (2018) Targeting macrophages: therapeutic approaches in cancer. Nat Rev Drug Discov 17(12):887–904
6. Gonzalez H, Hagerling C, Werb Z (2018) Roles of the immune system in cancer: from tumor initiation to metastatic progression. Genes Dev 32:1267–1284
7. Dunn GP, Old LJ, Schreiber RD (2004) The three Es of cancer immunoediting. Annu Rev Immunol 22:329–360
8. Gajewski TF, Schreiber H, Fu YX (2013) Innate and adaptive immune cells in the tumor microenvironment. Nat Immunol 14:1014–1022
9. Quail DF, Joyce JA (2013) Microenvironmental regulation of tumor progression and metastasis. Nat Med 19:1423–1437
10. Elinav E, Nowarski R, Thaiss CA, Hu B, Jin CC, Flavell RA (2013) Inflammation-induced cancer: crosstalk between tumours, immune cells and microorganisms. Nat Rev Cancer 13:759–771
11. Flossmann E, Rothwell PM (2007) Effect of aspirin on long-term risk of colorectal cancer: consistent evidence from randomised and observational studies. Lancet 369:1603–1613
12. Tougeron D, Sha D, Manthravadi S, Sinicrope FA (2014) Aspirin and colorectal cancer: back to the future. Clin Cancer Res 20:1087–1094
13. Kalluri R, Zeisberg M (2006) Fibroblasts in cancer. Nat Rev Cancer 6:392–401

14. Noy R, Pollard JW (2014) Tumor-associated macrophages: from mechanisms to therapy. Immunity 41:49–61
15. Petty AJ, Yang Y (2017) Tumor-associated macrophages: implications in cancer immunotherapy. Immunotherapy 9:289–302
16. Perdiguero EG, Geissmann F (2016) The development and maintenance of resident macrophages. Nat Immunol 17:2–8
17. Lavin Y, Mortha A, Rahman A, Merad M (2015) Regulation of macrophage development and function in peripheral tissues. Nat Rev Immunol 15:731–744
18. Mass E, Ballesteros I, Farlik M et al (2016) Specification of tissue-resident macrophages during organogenesis. Science 353:pii: aaf4238
19. Gautier EL, Shay T, Miller J et al (2012) Gene-expression profiles and transcriptional regulatory pathways that underlie the identity and diversity of mouse tissue macrophages. Nat Immunol 13:1118–1128
20. Schulz C, Gomez Perdiguero E, Chorro L et al (2012) A lineage of myeloid cells independent of Myb and hematopoietic stem cells. Science 336:86–90
21. Chitu V, Stanley ER (2006) Colony-stimulating factor-1 in immunity and inflammation. Curr Opin Immunol 18:39–48
22. Lin H, Lee E, Hestir K et al (2008) Discovery of a cytokine and its receptor by functional screening of the extracellular proteome. Science 320:807–811
23. Baghdadi M, Umeyama Y, Hama N et al (2018) Interleukin-34, a comprehensive review. J Leukoc Biol 104:931–951
24. Okabe Y, Medzhitov R (2014) Tissue-specific signals control reversible program of localization and functional polarization of macrophages. Cell 157:832–844
25. Sica A, Mantovani A (2012) Macrophage plasticity and polarization: in vivo veritas. J Clin Invest 122:787–795
26. Xue J, Schmidt SV, Sander J et al (2014) Transcriptome-based network analysis reveals a spectrum model of human macrophage activation. Immunity 40:274–288
27. Mantovani A, Biswas SK, Galdiero MR, Sica A, Locati M (2013) Macrophage plasticity and polarization in tissue repair and remodelling. J Pathol 229:176–185
28. Qian BZ, Pollard JW (2010) Macrophage diversity enhances tumor progression and metastasis. Cell 141:39–51
29. Pollard JW (2004) Tumour-educated macrophages promote tumour progression and metastasis. Nat Rev Cancer 4:71–78
30. Arwert EN, Harney AS, Entenberg D et al (2018) A unidirectional transition from migratory to perivascular macrophage is required for tumor cell intravasation. Cell Rep 23:1239–1248
31. Qian BZ, Li JF, Zhang H et al (2011) CCL2 recruits inflammatory monocytes to facilitate breast-tumour metastasis. Nature 475:222–U129
32. Zhu Y, Herndon JM, Sojka DK et al (2017) Tissue-resident macrophages in pancreatic ductal adenocarcinoma originate from embryonic hematopoiesis and promote tumor progression. Immunity 47:597–597
33. Chen Z, Feng X, Herting CJ et al (2017) Cellular and molecular identity of tumor-associated macrophages in glioblastoma. Cancer Res 77:2266–2278
34. Bowman RL, Klemm F, Akkari L et al (2016) Macrophage ontogeny underlies differences in tumor-specific education in brain malignancies. Cell Rep 17:2445–2459
35. Mantovani A, Marchesi F, Malesci A, Laghi L, Allavena P (2017) Tumour-associated macrophages as treatment targets in oncology. Nat Rev Clin Oncol 14:399–416
36. Biswas SK, Mantovani A (2010) Macrophage plasticity and interaction with lymphocyte subsets: cancer as a paradigm. Nat Immunol 11:889–896
37. Yin S, Huang J, Li Z et al (2017) The prognostic and clinicopathological significance of tumor-associated macrophages in patients with gastric cancer: a meta-analysis. PLoS One 12:e0170042
38. Zhao XX, Qu JK, Sun YC et al (2017) Prognostic significance of tumor-associated macrophages in breast cancer: a meta-analysis of the literature. Oncotarget 8:30576–30586
39. Mei J, Xiao Z, Guo C et al (2016) Prognostic impact of tumor-associated macrophage infiltration in non-small cell lung cancer: a systemic review and meta-analysis. Oncotarget 7:34217–34228
40. Guo B, Cen H, Tan X, Ke Q (2016) Meta-analysis of the prognostic and clinical value of tumor-associated macrophages in adult classical Hodgkin lymphoma. BMC Med 14:159
41. Lewis CE, Pollard JW (2006) Distinct role of macrophages in different tumor microenvironments. Cancer Res 66:605–612
42. Scholl SM, Bascou CH, Mosseri V et al (1994) Circulating levels of colony-stimulating factor 1 as a prognostic indicator in 82 patients with epithelial ovarian cancer. Br J Cancer 69:342–346
43. Price FV, Chambers SK, Chambers JT et al (1993) Colony-stimulating factor-I in

primary ascites of ovarian-cancer is a significant predictor of survival. Am J Obstet Gynecol 168:520–527
44. Smith HO, Anderson PS, Kuo DY et al (1995) The role of colony-stimulating factor 1 and its receptor in the etiopathogenesis of endometrial adenocarcinoma. Clin Cancer Res 1:313–325
45. West RB, Nuyten DS, Subramanian S et al (2005) Determination of stromal signatures in breast carcinoma. PLoS Biol 3:e187
46. Stanley ER, Chitu V (2014) CSF-1 receptor signaling in myeloid cells. Cold Spring Harb Perspect Biol 6:pii: a021857
47. Zhang QW, Liu L, Gong CY et al (2012) Prognostic significance of tumor-associated macrophages in solid tumor: a meta-analysis of the literature. PLoS One 7:e50946
48. Steidl C, Lee T, Shah SP et al (2010) Tumor-associated macrophages and survival in classic Hodgkin's lymphoma. N Engl J Med 362:875–885
49. Aharinejad S, Sioud M, Lucas T, Abraham D (2007) Target validation using RNA interference in solid tumors. Methods Mol Biol 361:227–238
50. Cannarile MA, Weisser M, Jacob W, Jegg AM, Ries CH, Ruttinger D (2017) Colony-stimulating factor 1 receptor (CSF1R) inhibitors in cancer therapy. J Immunother Cancer 5:53
51. Baghdadi M, Endo H, Takano A et al (2018) High co-expression of IL-34 and M-CSF correlates with tumor progression and poor survival in lung cancers. Sci Rep 8:418
52. Zins K, Heller G, Mayerhofer M, Schreiber M, Abraham D (2018) Differential prognostic impact of interleukin-34 mRNA expression and infiltrating immune cell composition in intrinsic breast cancer subtypes. Oncotarget 9:23126–23148
53. Crusz SM, Balkwill FR (2015) Inflammation and cancer: advances and new agents. Nat Rev Clin Oncol 12:584–596
54. Moore RJ, Owens DM, Stamp G et al (1999) Mice deficient in tumor necrosis factor-alpha are resistant to skin carcinogenesis. Nat Med 5:828–831
55. Canli O, Nicolas AM, Gupta J et al (2017) Myeloid cell-derived reactive oxygen species induce epithelial mutagenesis. Cancer Cell 32:869–883, e865
56. Henze AT, Mazzone M (2016) The impact of hypoxia on tumor-associated macrophages. J Clin Invest 126:3672–3679
57. Stetler-Stevenson WG, Yu AE (2001) Proteases in invasion: matrix metalloproteinases. Semin Cancer Biol 11:143–152
58. Zajac E, Schweighofer B, Kupriyanova TA et al (2013) Angiogenic capacity of M1-and M2-polarized macrophages is determined by the levels of TIMP-1 complexed with their secreted proMMP-9. Blood 122:4054–4067
59. De Palma M, Venneri MA, Galli R et al (2005) Tie2 identifies a hematopoietic lineage of proangiogenic monocytes required for tumor vessel formation and a mesenchymal population of pericyte progenitors. Cancer Cell 8:211–226
60. Mazzieri R, Pucci F, Moi D et al (2011) Targeting the ANG2/TIE2 axis inhibits tumor growth and metastasis by impairing angiogenesis and disabling rebounds of proangiogenic myeloid cells. Cancer Cell 19:512–526
61. Sica A, Schioppa T, Mantovani A, Allavena P (2006) Tumour-associated macrophages are a distinct M2 polarised population promoting tumour progression: potential targets of anti-cancer therapy. Eur J Cancer 42:717–727
62. Sica A, Allavena P, Mantovani A (2008) Cancer related inflammation: the macrophage connection. Cancer Lett 267:204–215
63. Casazza A, Laoui D, Wenes M et al (2013) Impeding macrophage entry into hypoxic tumor areas by Sema3A/Nrp1 signaling blockade inhibits angiogenesis and restores antitumor immunity. Cancer Cell 24:695–709
64. Wenes M, Shang M, Di Matteo M et al (2016) Macrophage metabolism controls tumor blood vessel morphogenesis and metastasis. Cell Metab 24:701–715
65. Mantovani A, Locati M (2016) Macrophage metabolism shapes angiogenesis in tumors. Cell Metab 24:887–888
66. Kitamura T, Qian BZ, Pollard JW (2015) Immune cell promotion of metastasis. Nat Rev Immunol 15:73–86
67. Wyckoff J, Wang WG, Lin EY et al (2004) A paracrine loop between tumor cells and macrophages is required for tumor cell migration in mammary tumors. Cancer Res 64:7022–7029
68. Wyckoff JB, Wang Y, Lin EY et al (2007) Direct visualization of macrophage-assisted tumor cell intravasation in mammary tumors. Cancer Res 67:2649–2656
69. Goswami S, Sahai E, Wyckoff JB et al (2005) Macrophages promote the invasion of breast carcinoma cells via a colony-stimulating factor-1/epidermal growth factor paracrine loop. Cancer Res 65:5278–5283
70. Hernandez L, Smirnova T, Kedrin D et al (2009) The EGF/CSF-1 paracrine invasion loop can be triggered by heregulin beta 1 and CXCL12. Cancer Res 69:3221–3227

71. Hagemann T, Robinson SC, Schulz M, Trumper L, Balkwill FR, Binder C (2004) Enhanced invasiveness of breast cancer cell lines upon co-cultivation with macrophages is due to TNF-alpha dependent up-regulation of matrix metalloproteases. Carcinogenesis 25:1543–1549
72. Liu CY, Xu JY, Shi XY et al (2013) M2-polarized tumor-associated macrophages promoted epithelial-mesenchymal transition in pancreatic cancer cells, partially through TLR4/IL-10 signaling pathway. Lab Investig 93:844–854
73. Bonde AK, Tischler V, Kumar S, Soltermann A, Schwendener RA (2012) Intratumoral macrophages contribute to epithelial-mesenchymal transition in solid tumors. BMC Cancer 12:35
74. Kitamura T, Qian BZ, Soong D et al (2015) CCL2-induced chemokine cascade promotes breast cancer metastasis by enhancing retention of metastasis-associated macrophages. J Exp Med 212:1043–1059
75. Harney AS, Arwert EN, Entenberg D et al (2015) Real-time imaging reveals local, transient vascular permeability, and tumor cell intravasation stimulated by TIE2hi macrophage-derived VEGFA. Cancer Discov 5:932–943
76. Rohan TE, Xue X, Lin HM et al (2014) Tumor microenvironment of metastasis and risk of distant metastasis of breast cancer. J Natl Cancer Inst 106:pii: dju136
77. Ojalvo LS, King W, Cox D, Pollard JW (2009) High-density gene expression analysis of tumor-associated macrophages from mouse mammary tumors. Am J Pathol 174:1048–1064
78. Ojalvo LS, Whittaker CA, Condeelis JS, Pollard JW (2010) Gene expression analysis of macrophages that facilitate tumor invasion supports a role for Wnt-signaling in mediating their activity in primary mammary tumors. J Immunol 184:702–712
79. Yeo EJ, Cassetta L, Qian BZ et al (2014) Myeloid WNT7b mediates the angiogenic switch and metastasis in breast cancer. Cancer Res 74:2962–2973
80. Linde N, Casanova-Acebes M, Sosa MS et al (2018) Macrophages orchestrate breast cancer early dissemination and metastasis. Nat Commun 9:21
81. Qian B, Deng Y, Im JH et al (2009) A distinct macrophage population mediates metastatic breast cancer cell extravasation, establishment and growth. PLoS One 4:e6562
82. Zhao L, Lim SY, Gordon-Weeks AN et al (2013) Recruitment of a myeloid cell subset (CD11b/Gr1 mid) via CCL2/CCR2 promotes the development of colorectal cancer liver metastasis. Hepatology 57:829–839
83. Qian BZ, Zhang H, Li J et al (2015) FLT1 signaling in metastasis-associated macrophages activates an inflammatory signature that promotes breast cancer metastasis. J Exp Med 212:1433–1448
84. Kitamura T, Doughty-Shenton D, Cassetta L et al (2017) Monocytes differentiate to immune suppressive precursors of metastasis-associated macrophages in mouse models of metastatic breast cancer. Front Immunol 8:2004
85. Chen Q, Zhang XH, Massague J (2011) Macrophage binding to receptor VCAM-1 transmits survival signals in breast cancer cells that invade the lungs. Cancer Cell 20:538–549
86. Gao D, Joshi N, Choi H et al (2012) Myeloid progenitor cells in the premetastatic lung promote metastases by inducing mesenchymal to epithelial transition. Cancer Res 72:1384–1394
87. Onoe K, Yanagawa Y, Minami K, Iijima N, Iwabuchi K (2007) Th1 or Th2 balance regulated by interaction between dendritic cells and NKT cells. Immunol Res 38:319–332
88. Sica A, Larghi P, Mancino A et al (2008) Macrophage polarization in tumour progression. Semin Cancer Biol 18:349–355
89. Pollard JW (2009) Trophic macrophages in development and disease. Nat Rev Immunol 9:259–270
90. Morandi F, Pistoia V (2014) Interactions between HLA-G and HLA-E in physiological and pathological conditions. Front Immunol 5:394
91. Santarpia M, Karachaliou N (2015) Tumor immune microenvironment characterization and response to anti-PD-1 therapy. Cancer Biol Med 12:74–78
92. Buchbinder EI, Desai A (2016) CTLA-4 and PD-1 pathways: similarities, differences, and implications of their inhibition. Am J Clin Oncol 39:98–106
93. Sica A, Saccani A, Bottazzi B et al (2000) Autocrine production of IL-10 mediates defective IL-12 production and NF-kappa B activation in tumor-associated macrophages. J Immunol 164:762–767

94. Allavena P, Mantovani A (2012) Immunology in the clinic review series; focus on cancer: tumour-associated macrophages: undisputed stars of the inflammatory tumour microenvironment. Clin Exp Immunol 167:195–205
95. Baniyash M (2004) TCR zeta-chain downregulation: curtailing an excessive inflammatory immune response. Nat Rev Immunol 4:675–687
96. Mellor AL, Keskin DB, Johnson T, Chandler P, Munn DH (2002) Cells expressing indoleamine 2,3-dioxygenase inhibit T cell responses. J Immunol 168:3771–3776
97. Mbongue JC, Nicholas DA, Torrez TW, Kim NS, Firek AF, Langridge WH (2015) The role of indoleamine 2, 3-dioxygenase in immune suppression and autoimmunity. Vaccines (Basel) 3:703–729
98. Pathria P, Louis TL, Varner JA (2019) Targeting tumor-associated macrophages in cancer. Trends Immunol 40:310–327
99. Madani I, De Neve W, Mareel M (2008) Does ionizing radiation stimulate cancer invasion and metastasis? Bull Cancer 95:292–300
100. Sanchez LR, Borriello L, Entenberg D, Condeelis JS, Oktay MH, Karagiannis GS (2019) The emerging roles of macrophages in cancer metastasis and response to chemotherapy. J Leukoc Biol 106:259–274
101. Paulus P, Stanley ER, Schafer R, Abraham D, Aharinejad S (2006) Colony-stimulating factor-1 antibody reverses chemoresistance in human MCF-7 breast cancer xenografts. Cancer Res 66:4349–4356
102. DeNardo DG, Brennan DJ, Rexhepaj E et al (2011) Leukocyte complexity predicts breast cancer survival and functionally regulates response to chemotherapy. Cancer Discov 1:54–67
103. Ribas A, Wolchok JD (2018) Cancer immunotherapy using checkpoint blockade. Science 359:1350
104. Wei SC, Duffy CR, Allison JP (2018) Fundamental mechanisms of immune checkpoint blockade therapy. Cancer Discov 8:1069–1086
105. Gordon SR, Maute RL, Dulken BW et al (2017) PD-1 expression by tumour-associated macrophages inhibits phagocytosis and tumour immunity. Nature 545:495–499
106. Mantovani A, Allavena P (2015) The interaction of anticancer therapies with tumor-associated macrophages. J Exp Med 212:435–445
107. Vajaitu C, Draghici CC, Solomon I et al (2018) The central role of inflammation associated with checkpoint inhibitor treatments. J Immunol Res 2018:4625472
108. Viitala M, Virtakoivu R, Tadayon S, Rannikko J, Jalkanen S, Hollmen M (2019) Immunotherapeutic blockade of macrophage clever-1 reactivates the CD8(+) T-cell response against immunosuppressive tumors. Clin Cancer Res. https://doi.org/10.1158/1078-0432.CCR-18-3016
109. Aharinejad S, Abraham D, Paulus P et al (2002) Colony-stimulating factor-1 antisense treatment suppresses growth of human tumor xenografts in mice. Cancer Res 62:5317–5324
110. Aharinejad S, Paulus P, Sioud M et al (2004) Colony-stimulating factor-1 blockade by antisense oligonucleotides and small interfering RNAs suppresses growth of human mammary tumor xenografts in mice. Cancer Res 64:5378–5384
111. Quail DF, Joyce JA (2017) Molecular pathways: deciphering mechanisms of resistance to macrophage-targeted therapies. Clin Cancer Res 23:876–884
112. Pyonteck SM, Akkari L, Schuhmacher AJ et al (2013) CSF-1R inhibition alters macrophage polarization and blocks glioma progression. Nat Med 19:1264
113. Schmid MC, Avraamides CJ, Foubert P et al (2011) Combined blockade of integrin-alpha 4 beta 1 plus cytokines SDF-1 alpha or IL-1 beta potently inhibits tumor inflammation and growth. Cancer Res 71:6965–6975
114. Cassetta L, Fragkogianni S, Sims AH et al (2019) Human tumor-associated macrophage and monocyte transcriptional landscapes reveal cancer-specific reprogramming, biomarkers, and therapeutic targets. Cancer Cell 35:588–602, e510
115. Schmid MC, Avraamides CJ, Dippold HC et al (2011) Receptor tyrosine kinases and TLR/IL1 Rs unexpectedly activate myeloid cell PI3K gamma, a single convergent point promoting tumor inflammation and progression. Cancer Cell 19:715–727
116. Schmid MC, Franco I, Kang SW, Hirsch E, Quilliam LA, Varner JA (2013) PI3-kinase gamma promotes Rap1a-mediated activation of myeloid cell integrin alpha4beta1, leading to tumor inflammation and growth. PLoS One 8:e60226
117. Gunderson AJ, Kaneda MM, Tsujikawa T et al (2016) Bruton tyrosine kinase-dependent immune cell cross-talk drives pancreas cancer. Cancer Discov 6:270–285

118. Kaczanowska S, Joseph AM, Davila E (2013) TLR agonists: our best frenemy in cancer immunotherapy. J Leukoc Biol 93:847–863
119. Le Mercier I, Poujol D, Sanlaville A et al (2013) Tumor promotion by intratumoral plasmacytoid dendritic cells is reversed by TLR7 ligand treatment. Cancer Res 73:4629–4640
120. Singh M, Khong H, Dai ZM et al (2014) Effective innate and adaptive antimelanoma immunity through localized TLR7/8 activation. J Immunol 193:4722–4731
121. Hoves S, Ooi CH, Wolter C et al (2018) Rapid activation of tumor-associated macrophages boosts preexisting tumor immunity. J Exp Med 215:859–876
122. Lum HD, Buhtoiarov IN, Schmidt BE et al (2006) Tumoristatic effects of anti-CD40 mAb-activated macrophages involve nitric oxide and tumour necrosis factor-alpha. Immunology 118:261–270
123. Lum HD, Buhtoiarov IN, Schmidt BE et al (2006) In vivo CD40 ligation can induce T cell-independent antitumor effects that involve macrophages. J Leukoc Biol 79:1181–1192
124. Wiehagen KR, Girgis NM, Yamada DH et al (2017) Combination of CD40 agonism and CSF-1R blockade reconditions tumor-associated macrophages and drives potent antitumor immunity. Cancer Immunol Res 5:1109–1121
125. Perry CJ, Munoz-Rojas AR, Meeth KM et al (2018) Myeloid-targeted immunotherapies act in synergy to induce inflammation and antitumor immunity. J Exp Med 215:877–893
126. Lobera M, Madauss KP, Pohlhaus DT et al (2013) Selective class IIa histone deacetylase inhibition via a nonchelating zinc-binding group. Nat Chem Biol 9:319–325
127. Guerriero JL, Sotayo A, Ponichtera HE et al (2017) Class IIa HDAC inhibition reduces breast tumours and metastases through anti-tumour macrophages. Nature 543:428
128. Guerriero JL (2018) Macrophages: the road less traveled, changing anticancer therapy. Trends Mol Med 24:472–489
129. Ofengeim D, Yuan JY (2013) Regulation of RIP1 kinase signalling at the crossroads of inflammation and cell death. Nat Rev Mol Cell Biol 14:727–736
130. Wang W, Marinis JM, Beal AM et al (2018) RIP1 kinase drives macrophage-mediated adaptive immune tolerance in pancreatic cancer. Cancer Cell 34:757–774, e757
131. Pathria P, Gotthardt D, Prchal-Murphy M et al (2015) Myeloid STAT3 promotes formation of colitis-associated colorectal cancer in mice. Oncoimmunology 4:e998529
132. Zhang Y, Choksi S, Chen K, Pobezinskaya Y, Linnoila I, Liu ZG (2013) ROS play a critical role in the differentiation of alternatively activated macrophages and the occurrence of tumor-associated macrophages. Cell Res 23:898–914
133. Ha M, Kim VN (2014) Regulation of microRNA biogenesis. Nat Rev Mol Cell Biol 15:509–524
134. Baer C, Squadrito ML, Laoui D et al (2016) Suppression of microRNA activity amplifies IFN-gamma-induced macrophage activation and promotes anti-tumour immunity. Nat Cell Biol 18:790–802
135. Okazawa H, Motegi SI, Ohyama N et al (2005) Negative regulation of phagocytosis in macrophages by the CD47-SHPS-1 system. J Immunol 174:2004–2011
136. Barclay AN, van den Berg TK (2014) The interaction between signal regulatory protein alpha (SIRP alpha) and CD47: structure, function, and therapeutic target. Annu Rev Immunol 32(32):25–50
137. Tseng D, Volkmer JP, Willingham SB et al (2013) Anti-CD47 antibody-mediated phagocytosis of cancer by macrophages primes an effective antitumor T-cell response. Proc Natl Acad Sci U S A 110:11103–11108
138. Covarrubias AJ, Aksoylar HI, Horng T (2015) Control of macrophage metabolism and activation by mTOR and Akt signaling. Semin Immunol 27:286–296
139. Porta C, Sica A, Riboldi E (2018) Tumor-associated myeloid cells: new understandings on their metabolic regulation and their influence in cancer immunotherapy. FEBS J 285:717–733
140. Mantovani A, Bonecchi R (2019) One clever macrophage checkpoint. Clin Cancer Res 25 (11):3202–3204
141. Kaneda MM, Messer KS, Ralainirina N et al (2017) PI3K gamma is a molecular switch that controls immune suppression. Nature 542:124–124
142. Foubert P, Kaneda MM, Varner JA (2017) PI3K gamma activates integrin alpha(4) and promotes immune suppressive myeloid cell polarization during tumor progression. Cancer Immunol Res 5:957–968
143. Kaneda MM, Cappello P, Nguyen AV et al (2016) Macrophage PI3K gamma drives pancreatic ductal adenocarcinoma progression. Cancer Discov 6:870–885
144. Thomas AC, Mattila JT (2014) "Of mice and men": arginine metabolism in macrophages. Front Immunol 5:479

145. Colegio OR, Chu NQ, Szabo AL et al (2014) Functional polarization of tumour-associated macrophages by tumour-derived lactic acid. Nature 513:559
146. Bohn T, Rapp S, Luther N et al (2018) Tumor immunoevasion via acidosis-dependent induction of regulatory tumor-associated macrophages. Nat Immunol 19:1319–1329
147. Kuchuk O, Tuccitto A, Citterio D et al (2018) pH regulators to target the tumor immune microenvironment in human hepatocellular carcinoma. Oncoimmunology 7(7): e1445452
148. Carmona-Fontaine C, Deforet M, Akkari L, Thompson CB, Joyce JA, Xavier JB (2017) Metabolic origins of spatial organization in the tumor microenvironment. Proc Natl Acad Sci U S A 114:2934–2939
149. Buck MD, Sowell RT, Kaech SM, Pearce EL (2017) Metabolic instruction of immunity. Cell 169:570–586
150. Geeraerts X, Bolli E, Fendt SM, Van Ginderachter JA (2017) Macrophage metabolism as therapeutic target for cancer, atherosclerosis, and obesity. Front Immunol 8:289
151. Fire A, Xu SQ, Montgomery MK, Kostas SA, Driver SE, Mello CC (1998) Potent and specific genetic interference by double-stranded RNA in *Caenorhabditis elegans*. Nature 391:806–811
152. Elbashir SM, Harborth J, Lendeckel W, Yalcin A, Weber K, Tuschl T (2001) Duplexes of 21-nucleotide RNAs mediate RNA interference in cultured mammalian cells. Nature 411:494–498
153. Rettig GR, Behlke MA (2012) Progress toward in vivo use of siRNAs-II. Mol Ther 20:483–512
154. Bobbin ML, Rossi JJ (2016) RNA interference (RNAi)-based therapeutics: delivering on the promise? Annu Rev Pharmacol Toxicol 56:103–122
155. Wu SY, Lopez-Berestein G, Calin GA, Sood AK (2014) RNAi therapies: drugging the undruggable. Sci Transl Med 6:240ps7
156. Arkin MR, Tang YY, Wells JA (2014) Small-molecule inhibitors of protein-protein interactions: progressing toward the reality. Chem Biol 21:1102–1114
157. Hopkins AL, Groom CR (2002) The druggable genome. Nat Rev Drug Discov 1:727–730
158. Finan C, Gaulton A, Kruger FA et al (2017) The druggable genome and support for target identification and validation in drug development. Sci Transl Med 9:pii: eaag1166
159. Kozak K (2013) Annotation and specificity of existing genome-wide small interfering RNA libraries. Nucleic Acid Ther 23:71–80
160. Mohr SE, Perrimon N (2012) RNAi screening: new approaches, understandings, and organisms. WIREs RNA 3:145–158
161. Chen X, Mangala LS, Rodriguez-Aguayo C, Kong X, Lopez-Berestein G, Sood AK (2018) RNA interference-based therapy and its delivery systems. Cancer Metastasis Rev 37:107–124
162. Sioud M (2019) Releasing the immune system brakes using sirnas enhances cancer immunotherapy. Cancers (Basel) 11:pii: E176
163. Kleinman ME, Yamada K, Takeda A et al (2008) Sequence- and target-independent angiogenesis suppression by siRNA via TLR3. Nature 452:591–U591
164. DeVincenzo J, Lambkin-Williams R, Wilkinson T et al (2010) A randomized, double-blind, placebo-controlled study of an RNAi-based therapy directed against respiratory syncytial virus. Proc Natl Acad Sci U S A 107:8800–8805
165. Davis ME, Zuckerman JE, Choi CHJ et al (2010) Evidence of RNAi in humans from systemically administered siRNA via targeted nanoparticles. Nature 464:1067–U1140
166. Setten RL, Rossi JJ, Han SP (2019) The current state and future directions of RNAi-based therapeutics. Nat Rev Drug Discov 18 (6):421–446
167. Garcia MA, Gil J, Ventoso I et al (2006) Impact of protein kinase PKR in cell biology: from antiviral to antiproliferative action. Microbiol Mol Biol Rev 70:1032
168. Robbins M, Judge A, Liang L, McClintock K, Yaworski E, MacLachlan I (2007) 2′-O-methyl-modified RNAs act as TLR7 antagonists. Mol Ther 15:1663–1669
169. Zuckerman JE, Davis ME (2015) Clinical experiences with systemically administered siRNA-based therapeutics in cancer. Nat Rev Drug Discov 14:843–856
170. Szebeni J, Simberg D, Gonzalez-Fernandez-A, Barenholz Y, Dobrovolskaia MA (2018) Roadmap and strategy for overcoming infusion reactions to nanomedicines. Nat Nanotechnol 13:1100–1108
171. Janas MM, Schlegel MK, Harbison CE et al (2018) Selection of GalNAc-conjugated siRNAs with limited off-target-driven rat hepatotoxicity. Nat Commun 9:723
172. Kim HJ, Kim A, Miyata K, Kataoka K (2016) Recent progress in development of siRNA delivery vehicles for cancer therapy. Adv Drug Deliv Rev 104:61–77

173. Wang T, Shigdar S, Al Shamaileh H et al (2017) Challenges and opportunities for siRNA-based cancer treatment. Cancer Lett 387:77–83
174. Grimm D (2009) Small silencing RNAs: state-of-the-art. Adv Drug Deliv Rev 61:672–703
175. Giacca M, Zacchigna S (2012) Virus-mediated gene delivery for human gene therapy. J Control Release 161:377–388
176. Glover DJ, Lipps HJ, Jans DA (2005) Towards safe, non-viral therapeutic gene expression in humans. Nat Rev Genet 6:299–U229
177. Keller M (2009) Nanomedicinal delivery approaches for therapeutic siRNA. Int J Pharm 379:210–211
178. Ovais M, Guo M, Chen C (2019) Tailoring Nanomaterials for targeting tumor-associated macrophages. Adv Mater 31(19):e1808303
179. Andon FT, Digifico E, Maeda A et al (2017) Targeting tumor associated macrophages: the new challenge for nanomedicine. Semin Immunol 34:103–113
180. Allen TM, Cullis PR (2013) Liposomal drug delivery systems: from concept to clinical applications. Adv Drug Deliv Rev 65:36–48
181. Binnemars-Postma K, Storm G, Prakash J (2017) Nanomedicine strategies to target tumor-associated macrophages. Int J Mol Sci 18:pii: E979
182. Zhang N, Palmer AF (2012) Liposomes surface conjugated with human hemoglobin target delivery to macrophages. Biotechnol Bioeng 109:823–829
183. Leuschner F, Dutta P, Gorbatov R et al (2011) Therapeutic siRNA silencing in inflammatory monocytes in mice. Nat Biotechnol 29:1005–1010
184. Jose A, Labala S, Ninave KM, Gade SK, Venuganti VVK (2018) Effective skin cancer treatment by topical co-delivery of curcumin and STAT3 siRNA using cationic liposomes. AAPS PharmSciTech 19:166–175
185. Nogueira E, Gomes AC, Preto A, Cavaco-Paulo A (2016) Folate-targeted nanoparticles for rheumatoid arthritis therapy. Nanomed Nanotechnol 12:1113–1126
186. Zimel MN, Horowitz CB, Rajasekhar VK et al (2017) HPMA-copolymer nanocarrier targets tumor-associated macrophages in primary and metastatic breast cancer. Mol Cancer Ther 16:2701–2710
187. Zhang M, Kim JA (2012) Effect of molecular size and modification pattern on the internalization of water soluble beta-(1→3)-(1→4)-glucan by primary murine macrophages. Int J Biochem Cell B 44:914–927
188. Hirayama D, Iida T, Nakase H (2018) The phagocytic function of macrophage-enforcing innate immunity and tissue homeostasis. Int J Mol Sci 19:pii: E92
189. Zhang M, Gao YX, Caja K, Zhao BC, Kim JA (2015) Non-viral nanoparticle delivers small interfering RNA to macrophages in vitro and in vivo. PLoS One 10:e0118472
190. Yu SS, Lau CM, Barham WJ et al (2013) Macrophage-specific RNA interference targeting via "click", mannosylated polymeric micelles. Mol Pharm 10:975–987
191. Ortega RA, Barham WJ, Kumar B et al (2015) Biocompatible mannosylated endosomal-escape nanoparticles enhance selective delivery of short nucleotide sequences to tumor associated macrophages. Nanoscale 7:500–510
192. Dahlman JE, Barnes C, Khan OF et al (2014) In vivo endothelial siRNA delivery using polymeric nanoparticles with low molecular weight. Nat Nanotechnol 9:648–655
193. Jung K, Heishi T, Khan OF et al (2017) Ly6Clo monocytes drive immunosuppression and confer resistance to anti-VEGFR2 cancer therapy. J Clin Invest 127:3039–3051
194. Shen S, Zhang Y, Chen KG, Luo YL, Wang J (2018) Cationic polymeric nanoparticle delivering CCR2 siRNA to inflammatory monocytes for tumor microenvironment modification and cancer therapy. Mol Pharm 15:3642–3653
195. Ban Y, Mai JH, Li X et al (2017) Targeting autocrine CCL5-CCR5 axis reprograms immunosuppressive myeloid cells and reinvigorates antitumor immunity. Cancer Res 77:2857–2868
196. Cieslewicz M, Tang JJ, Yu JL et al (2013) Targeted delivery of proapoptotic peptides to tumor-associated macrophages improves survival. Proc Natl Acad Sci U S A 110:15919–15924
197. Qian Y, Qiao S, Dai YF et al (2017) Molecular-targeted immunotherapeutic strategy for melanoma via dual-targeting nanoparticles delivering small interfering rna to tumor-associated macrophages. ACS Nano 11:9536–9549
198. Andon FT, Alonso MJ (2015) Nanomedicine and cancer immunotherapy – targeting immunosuppressive cells. J Drug Target 23:656–671
199. Ortega RA, Barham W, Sharman K, Tikhomirov O, Giorgio TD, Yull FE (2016) Manipulating the NF-kappaB pathway in macrophages using mannosylated, siRNA-delivering nanoparticles can induce immunostimulatory and tumor cytotoxic functions. Int J Nanomedicine 11:2163–2177

200. Huang Z, Zhang ZP, Jiang YC et al (2012) Targeted delivery of oligonucleotides into tumor-associated macrophages for cancer immunotherapy. J Control Release 158:286–292
201. Biswas SK, Gangi L, Paul S et al (2006) A distinct and unique transcriptional program expressed by tumor-associated macrophages (defective NF-kappa B and enhanced IRF-3/STAT1 activation). Blood 107:2112–2122
202. Wasiak I, Kulikowska A, Janczewska M et al (2016) Dextran nanoparticle synthesis and properties. PLoS One 11:e0146237
203. Vaupel P, Kallinowski F, Okunieff P (1989) Blood-flow, oxygen and nutrient supply, and metabolic microenvironment of human-tumors – a review. Cancer Res 49:6449–6465
204. Gerweck LE, Seetharaman K (1996) Cellular pH gradient in tumor versus normal tissue: potential exploitation for the treatment of cancer. Cancer Res 56:1194–1198
205. Conde J, Bao CC, Tan YQ et al (2015) Dual targeted immunotherapy via in vivo delivery of biohybrid RNAi-peptide nanoparticles to tumor-associated macrophages and cancer cells. Adv Funct Mater 25:4183–4194
206. Schoenenberger AD, Schipanski A, Malheiro V et al (2016) Macrophage polarization by titanium dioxide (TiO_2) particles: size matters. ACS Biomater Sci Eng 2:908–919
207. Cox TR, Erler JT (2014) Molecular pathways: connecting fibrosis and solid tumor metastasis. Clin Cancer Res 20:3637–3643
208. Miao L, Lin CM, Huang L (2015) Stromal barriers and strategies for the delivery of nanomedicine to desmoplastic tumors. J Control Release 219:192–204
209. Adiseshaiah PP, Crist RM, Hook SS, McNeil SE (2016) Nanomedicine strategies to overcome the pathophysiological barriers of pancreatic cancer. Nat Rev Clin Oncol 13:750–765
210. Afik R, Zigmond E, Vugman M et al (2016) Tumor macrophages are pivotal constructors of tumor collagenous matrix. J Exp Med 213:2315–2331
211. Jahanban-Esfahlan R, de la Guardia M, Ahmadi D, Yousefi B (2018) Modulating tumor hypoxia by nanomedicine for effective cancer therapy. J Cell Physiol 233: 2019–2031
212. Tian L, Goldstein A, Wang H et al (2017) Mutual regulation of tumour vessel normalization and immunostimulatory reprogramming. Nature 544:250
213. Silva VL, Al-Jamal WT (2017) Exploiting the cancer niche: tumor-associated macrophages and hypoxia as promising synergistic targets for nano-based therapy. J Control Release 253:82–96
214. Dowdy SF (2017) Overcoming cellular barriers for RNA therapeutics. Nat Biotechnol 35:222–229
215. Mantovani A, Longo DL (2018) Macrophage checkpoint blockade in cancer – back to the future. N Engl J Med 379:1777–1779

Chapter 18

Use of RNA Interference with TCR Transfer to Enhance Safety and Efficiency

Nicholas Paul Casey, Jon Amund Kyte, and Hiroshi Fujiwara

Abstract

The gene transfer of T-cell receptors (TCRs) is an attractive strategy for adoptive cell therapy, allowing the transfer of reactivity against antigens that may not otherwise engender an immune response. The TCRs recognize intracellular or extracellular antigens presented in the context of MHC class I or II, respectively. This broadens the range of targets considerably, compared to antibodies and chimeric antigen receptors, that are generally confined to surface antigens. However, TCR transfer must overcome some technical hurdles, relating to interference with endogenous α- and β-TCR chains and competition with other existing TCR infrastructure of T cells. In this review, we will outline the challenges facing TCR gene transfer and compare several approaches to address them. We will then focus upon one of the most promising amongst these—RNA interference—and detail the methods involved in designing and using this technology.

Key words T cell receptor, TCR gene transfer, RNA interference, Small interfering RNAs, immunotherapy, Adoptive cell therapy

1 Introduction

T-cell receptor (TCR) gene transfer combines the techniques of gene therapy with adoptive cell therapy and allows the generation of useful numbers of target-reactive cells in patients with cancers and refractory viral diseases. The ability of TCRs to recognize intracellular neoantigens offers great promise for personalized medicine. At present, there are a number of solutions to the technical challenges facing this field.

1.1 TCR Gene Transfer in Cancer Immunotherapy

T-cell receptors (TCRs) are heterodimers composed of an α- and a β-chain expressed on the cell surface of T lymphocytes. TCRs recognize and bind to short, processed peptides, of intracellular origin presented by a major histocompatibility complex (MHC) molecule, on antigen-presenting cells (APC) and target disease cells [1, 2]. Such interactions may be acutely sensitive, with TCRs

Mouldy Sioud (ed.), *RNA Interference and CRISPR Technologies: Technical Advances and New Therapeutic Opportunities*, Methods in Molecular Biology, vol. 2115, https://doi.org/10.1007/978-1-0716-0290-4_18,

capable of mediating activation from very limited numbers of peptide-MHC (pMHC) complexes (6, 7). The transmembrane TCR associates with the various subunits of the CD3 complex within the effector cells. The CD3 subunits each possess intracellular immunoreceptor tyrosine-rich activation motifs (ITAMs) [3], which are phosphorylated by CD4 or CD8 co-receptor molecules [4]. Co-stimulation via CD28 is also required for activation [5].

In general, TCRs are sensitive to mutated peptides derived from cancer cells, so-called neoantigens [6–8], and thus contribute to immune surveillance of cancer [9]. It is apparent that this ongoing surveillance contributes to the elimination of numerous mutated, potential "precancerous" cells, before they undergo transformation and form tumors [10, 11]. On the other hand, tumors that do form usually escape from immune surveillance by a process of immunoediting. This evasion may be reduced via immunogenicity, or resistance to, or suppression of, reactive immune cells [12, 13].

Lymphocytes that traffic to and infiltrate tumors (known as tumor-infiltrating lymphocytes, or "TILs") may be isolated from the suppressive microenvironment, expanded ex vivo, and reinfused into the patient [14–16]. Tumor-reactive cytotoxic T lymphocytes (CTLs) may also be isolated and expanded from the peripheral blood of cancer-bearing patients [17]. These approaches have resulted in durable and positive effects in cancers with a high mutational burden, in particular in melanoma [6]. However, the timely provision of sufficient numbers of TILs for therapeutic use is typically challenging, as their preparation and expansion is time-consuming and labor-intensive [18]. Furthermore, such cells are often not available for all patients [19].

1.2 TSAs and TAAs

Another hurdle to the use of endogenous T cells is that tumors that have undone immunoediting will frequently lack tumor-specific antigens (TSAs). Often, however, they will express normal antigens at elevated levels. Such antigens are referred to as tumor-associated antigens (TAAs). Certain TAA proteins are not expressed in normal adult tissues, being expressed only by the immuno-privileged testis, or by cancer tissues. These are referred to as cancer-testis antigens (CTAs) and have garnered much attention as targets for cancer immunotherapy [20]. Such self-antigens are tolerated as normal by the immune system, so the isolation and use of endogenous CTLs reactive against these targets is particularly challenging [21]. Nevertheless, the isolation of TCRs against these targets is possible via certain experimental methods, and so autologous CTLs with the desired reactivity may subsequently be generated by TCR gene transfer [22–24]. Thus TCR gene transfer is a powerful tool for the rapid generation of T cells with known reactivity, facilitating rapid production of therapeutic quantities of antigen-specific T cells [25]. Consequently, the transfer of TCR-encoding sequences to patient-derived T cells forms the second branch of adoptive cell transfer (after administration of TILs). A third branch involves targeting of surface antigens by transfer of sequences encoding for chimeric antigen receptors (*see*, e.g., [26, 27]), but this branch will not be covered in the current review.

1.3 Mispairing and How to Avoid It

TCR gene transfer—the expression of transgenic α- and β-TCR chains—poses an inherent problem. CD8+ cytotoxic T cells, and CD4+ helper T cells, are often the first choice for effectors of immune-gene therapy. However, these cell types also express endogenous TCRs. Mispairing between the endogenous and transgenic α- and β-chains has been demonstrated to result in unexpected neoreactivity [28], as well as lethal autoimmunity in mouse models [29]. These drawbacks have resulted in a variety of designs that are intended to promote the correct pairing of transgenic TCR chains.

1.3.1 Preferential Association

TCRs derived from mice have previously been used to transfer specificity to human immune cells and have been tolerated [30]. Retention of only the constant domains from the murine TCR chains resulted in preferential association and pairing of the hybrid TCR chains and a more stable association with the CD3 complex [31, 32]. These advantages were retained by minimally murinized sequences, in which only four amino acids in the α-chain constant region, and five in the β-chain constant region, were of murine origin [33]. The inclusion of additional cysteine residues to the constant domains of the transgenic TCR chains and the resulting disulphide bonds are further able to strengthen the association between the transgenic chains [34, 35]. Targeted substitution of amino acids, to generate positive and negative electrostatic charges on the TCR chains, has also been used to favor correct association [36]. Equimolar expression of the α- and β-transgenic chains has been achieved with the use of self-cleaving peptides [37, 38]. This and codon optimization for improved expression [39] have resulted in considerable improvements in transgenic TCR formation and expression. The incorporation of leucine-rich zipper domains, derived from c-Jun and v-Fos proteins, results in strong association between TCR chains. This was originally developed in a soluble TCR format [40], but is highly effective in the context of surface-bound molecules also. In fact zipper domains are one of the stronger options in this category, with respect to the correct pairing of TCR chains and the lysis and IFNγ production by gene-modified T cells in response to targets [41].

Finally, the strategy of "domain swapping"—exchanging the transmembrane domains in the transgenic chains—would not directly prevent mispairing, but would render mispaired TCRs unable to assemble CD3 complexes and thus nonfunctional [42]. It must be noted, of course, that various of these modifications may be used in combination [43, 44]. However, while these designs may encourage expression, and/or correct pairing, of the transgenic TCR chains, they may not completely eliminate the risk of mispairing [45, 46].

1.3.2 Linking Domains

Rather than expressing the α- and β-chains as two separate molecules, expression of the Vα, Vβ, and Cβ TCR domains as a single chain was effective in forcing the association of the Vα and Vβ domains, precluding interaction with the endogenous domains [47–50]. However, linkage in this manner necessitates modifications in one or the other of the V regions, which may affect targeting or immunogenicity of the transgene. Indeed, despite higher surface expression of the single-chain form, Willemsen and coauthors found reduced binding to the target peptide-MHC complex, compared with a two-chain equivalent [47].

The normal repertoire of CD3 subunits (CD3ζ(x2), CD3ε (x2), CD3δ, CD3γ) in T cells is limited. The introduction of additional TCR chains forces competition for the CD3 complex between the endogenous and transgenic TCR chains. Another strategy to address this competition is by the expression of additional copies of CD3 [51]. Alternately, the use of murine TCR constant domains may be advantageous, as these evidently have a higher affinity for human CD3 than do the human TCR constant domains [31]. However, the linked-domain designs—whether as a single Vα-Vβ-Cβ-CD3ζ design, or as a two-chain Vα-Cα-CD3ζ + Vβ-Cβ-CD3ζ design—by incorporating a CD3ζ domain, eliminate certain competition for CD3 elements. The fusion of CD3ζ into linked TCR chains has been taken a step further, with the additional incorporation of CD28 domains, resulting in "CAR-like TCRs" [52, 53]. These latter designs are two-chain models, avoiding the modifications needed for linked chains. Instead, they utilize disulphide bonds, as well as ribosomal-skipping/self-cleaving 2A peptides, to achieve equimolar co-expression. This greatly reduces heterodimerization, but again, it is not eliminated completely [54].

1.3.3 Soluble TCRs/Dual-Engaging Forms

Another approach that sidesteps the risk of mispairing of cell-surface TCR chains is the development of soluble TCRs [54]. Akin to monoclonal antibodies, soluble TCRs have been used for labelling, and have also been conjugated to therapeutic molecules, for targeted delivery [54]. They have also been conjugated to anti-CD3 scFv (single-chain fragment variable) molecules [55], in the manner of bispecific T-cell engagers [56]. Soluble TCRs have certain technical (and commercial) advantages; however, TCRs generally have a lower affinity than antibodies, and so soluble forms usually require affinity enhancement [57]. Consequently, the pool of dual-engaging, soluble TCRs, such as ImmTACs (immune mobilizing monoclonal T-cell receptors against cancer, developed by Immunocore), is currently limited. Of course, the challenges of mispairing and CD3 competition would be avoided if the endogenous TCR chains were simply eliminated from the equation. This is the objective of the following approaches.

1.4 Gene Editing

Stable elimination of endogenous TCRs—which contribute to graft-versus-host disease—is also desirable for the purposes of "off-the-shelf" therapies, in which a single cell line may be administered to various patients, with reduced need for detailed matching [58, 59]. Screening and characterization of a single cell lineage is compatible with the inherently mutagenic mechanisms of gene editing, as edited cells with undesirable characteristics may be eliminated in the course of development. "Off-the-shelf" strategies are also compatible with the goals of TCR transfer [60]. Gene editing also allows substitution of a transgenic TCR into the locus of the endogenous genes (e.g., TRAC), which has proven beneficial with respect to transgene expression and the functionality of the modified immune cells ([61]; *see* Chapters 22 and 23).

A series of gene editing technologies are under development for application to cancer immunotherapy, namely, meganucleases [62], ZFNs (zinc finger nucleases) [63], TALENs (transcription activator-like effector-based nucleases) [58, 64], and CRISPR/Cas (clustered regularly interspersed short palindromic repeats) [61, 65–67]. These technologies generally function by cleaving the genomic DNA at characteristic, even sequence-specific locations. These double-strand breaks may then be repaired by nonhomologous end joining (NHEJ), or by homology-directed repair (HDR) using the allele on the other chromosome as a template. NHEJ is more potentially genotoxic than HDR, and so the use of "nicking" nuclease variants—which cut only a single strand—is often favored, as such breaks are predominantly repaired by HDR, rather than NHEJ [68].

Provision of novel, transgenic templates to the HDR pathway allows incorporation of transgenes to specific genomic locations. In this manner, substitution of the transgenic TCR may be inserted into an endogenous TCR locus (e.g., TRAC), resulting in benefits with respect to transgene expression and the functionality of the modified immune cells [61]. CRISPR/Cas-9 system has become a clear front-runner amongst gene editing technologies, due to its power, flexibility, and simplicity of use (*see* Chapters 1 and 19). However, despite continual improvements to fidelity and safety of CRISPR technology [69, 70], extensive screening of modified cells is required, to address unintended and off-target effects, which remain an ongoing challenge [71, 72]. As a form of gene therapy, gene editing is markedly less efficient compared with other gene therapy methods, such as retroviral transduction. While efficiency rates for CRIPSR gene editing are generally improving, certain loci will remain hard to target [66]. As such, it must be noted that gene editing is currently at a numerical disadvantage, with respect to the provision of cell-based therapies.

1.5 RNA Interference

Until the "holy grail" of universally applicable, off-the-shelf cell therapies becomes more widely applicable and available to patients, we must examine more mature technologies. RNA interference (RNAi) is a candidate for the downregulation of endogenous TCRs, which is available via multiple delivery options (*see* Chapter 1). In contrast to gene editing approaches, RNAi does not directly interact with genomic DNA.

RNAi is thought to have evolved as a cellular defense system against double-stranded RNA (dsRNA) sequences, which are key to the replication of certain virus classes [73]. RNAi acts posttranscriptionally, by degrading mRNA in a sequence-specific manner, and it is possible to trigger this silencing via synthetic short-interfering RNAs (siRNAs) [74]. Short-hairpin RNAs (shRNAs), which can be expressed as a single molecule, after processing by the intracellular "Dicer" machinery, will also trigger RNAi [75]. shRNAs may be expressed from plasmids or vectors, and there is evidence that this form may result in less off-target activity than siRNAs (thought to result in part from siRNAs becoming involved in the microRNA (miRNA) pathway) [76].

1.5.1 siTCR Designs

To downregulate expression of the endogenous TCR proteins, RNAi molecules (whether siRNA or shRNA) are designed to target the wild-type mRNA of the constant domain of the endogenous TCR sequences. The transgenic TCR sequences are codon-optimized to include silent mutations, such that the amino acid sequence of the constant domains remains unchanged, but the nucleotide sequence (mRNA) is no longer recognized by the RNAi molecules.

A number of "siTCR" designs exist, in both retroviral [77, 78] and transient [79] formats. In each of these protocols, one RNAi sequence was directed against the α constant domain, and one against the β constant domain. These designs resulted in >60% knockdown of the relevant TCR transcripts in human cells [77] and >80% in mouse cells [78]. In both studies, the inclusion of a second RNAi sequence against each of α- and β-chains did not give any additive effect. Electroporation of siRNAs in human cells gave >70% knockdown [79] and was able to eliminate mispairing. It must be noted that in none of these studies was there 100% knockdown of the endogenous TCRs. Nevertheless, these designs were considered to be sufficiently safe and beneficial to enter clinical trials and have proceeded to clinical trials [80, 81].

1.5.2 Delivery

Transient delivery, of the siRNA molecules and a transgenic TCR sequence, does not directly affect the genome and so avoids the risks or insertional mutagenesis [82]. Transient transfection also offers greater simplicity, as TCRs may be expressed from electroporated mRNA. Thus, simpler laboratory facilities are required, at least for experimental testing. The shorter turnaround times with transient delivery may also be more attractive for personalized therapy with TCRs. The downside is that the transient downregulation of endogenous TCRs, and expression of transgenic TCRs, starts to diminish after 3 days [79]. While retroviral transduction does carry the risk of insertional mutagenesis, it is a powerful and widely used technology, resulting in stable transgene expression in rapidly proliferating cell populations. While methods with other integrating characteristics, such as retrotransposons (e.g., [83]), are being investigated for TCR transfer [84], there is evidence that mature T cells are evidently singularly resistant to oncogenic transformation [85]. Thus, as the technology has matured and developed, and the risks more completely understood, retroviral gene therapy has gained acceptance [86]. For example, the FDA-approved CAR T-cell therapies utilize a retroviral platform [87]. It is worth noting here that if retroviral vectors are chosen for delivery of the TCRs, RNAi sequences may be incorporated into the same vector (as shRNAs), with no additional mutagenic risk.

Several clinical trials with novel TCRs and CARs have produced unpredicted and serious adverse events [88, 89]. This has promoted the interest in mRNA-based transient retargeting of T cells, by use of TCRs/CARs encoded by mRNA [90, 91]. This may be useful for safe testing of novel receptors, as well as settings where only a transient effect is desired, e.g., for elimination of residual disease up-front of allogeneic transplant for hematological cancers. In the context of mRNA-based TCRs, the use of siRNA to counter mispairing may offer a completely transient and nonintegrating alternative, which may carry considerable advantages for safe and cost-effective clinical development. It may also be possible to use siRNA to render the retargeted T cells resistant to mechanisms of immune suppression.

1.6 RNA Interference vs. Gene Editing

A variety of strategies are available to enhance TCR gene transfer therapies. The best approaches for reducing mispairing, and competition for the CD3 machinery, are methods that downregulate or eliminate the endogenous TCR molecules. The best candidate methods to meet these criteria are gene editing and RNA interference. The former necessarily involves risk to the integrity of the genome, whereas the latter has the option of being delivered transiently, without cutting of the genomic DNA. At the present time, and for the foreseeable future, RNAi also offers a strong numerical advantage with respect to the rapid provision of gene-modified effector cells. Furthermore, the siTCR technology is relatively mature, with multiple clinical trials completed or ongoing. Thus, the methods outlined here are highly recommended as an adjunct to TCR transfer protocols. The following section describes a protocol using retroviral vectors and shRNAs.

2 Materials

We will here outline the materials used to relate to the scope of the subsequent methods. That is, we do not describe the materials required for in vivo/mouse experiments. Instead, we will describe materials and methods for transfection and transduction of T cells (cell lines and PBMCs) with siRNAs. Gene-modified T cells may then be used in vivo as per the appropriate protocols.

2.1 Facilities

These methods will require access to common or standard biomedical research facilities:

1. Sterile cell culture facilities suitable for retroviral work. Retroviral production and use requires a higher biosafety level than nonviral work. Access to such a facility is therefore necessary. Such facilities should (and generally do) generally include many items of equipment required for this work, such as a laminar flow cabinet, incubators for cell culture, and centrifuges.
2. Thermal cyclers (PCR machines) are necessary. In-house access to a qPCR is also beneficial. These steps may be outsourced if necessary.
3. Facilities for transformation and culture of bacteria will also be required, for cloning plasmids and vectors.

2.2 Cell Culture

1. Complete DMEM: Dulbecco's Modified Eagle's Medium with L-glutamine, 10% FCS, and antibiotics.
2. DMEM with 1% HyClone: Dulbecco's Modified Eagle's Medium with L-glutamine, 1% HyClone serum, and antibiotics.
3. Cytotoxic T-lymphocyte (CTL) culture media: culture media optimized for T-lymphocyte culture (e.g., GT-T551, Takara Bio) supplemented with 1% autologous plasma, 0.2% human serum albumin, antibiotic (e.g., Fungizone, or penicillin/streptomycin), and 600 IU/mL interleukin-2.
4. PBS(−): phosphate-buffered saline, with no additives.
5. T cells were activated with OKT3, an anti-CD3 monoclonal antibody.
6. Retronectin: Use at a final concentration of 20 μg/mL in PBS (−).

7. Anti-CD3 antibody OKT3: Use at a final concentration of 1 μg/mL in PBS(−).
8. Trypsin-EDTA: Standard concentrations, from commercial suppliers.

2.3 Cells

1. Peripheral blood mononuclear cells (PBMCs) were isolated from healthy donors who gave their informed consent and were used fresh or frozen. PBMCs were cultured in CTL culture media.
2. Various cell lines (e.g., HEK293, HEK293T, PG13) cells are available from ATCC (*see* **Note 1**).

2.4 Synthetic Peptides

Testing of new TCR designs should include appropriate controls. These typically include HLA-matched and mismatched cell types and target peptide-positive and peptide-negative cells. It is also optimal to test reactivity against a cell type with endogenous expression of the target peptide. Peptides of a specified sequence may be obtained from various commercial sources.

2.5 Transfection

For transfection with DNA plasmids and vectors, commercial regents such as Lipofectamine, FuGene, or Xtremegene are convenient and efficient. Transfection reagents are usually delivered in a serum-free medium such as OptiMEM.

2.6 Plasticware and Equipment for Molecular Cloning and PCRs

1. Nuclease-free pipette tips (10 μL, 200 μL, 1000 μL). Use of barrier tips can help prevent cross-contamination.
2. Nuclease-free PCR tubes (200 μL).
3. Nuclease-free centrifuge tubes (1.5–1.8 mL).
4. Thermal cycler. The PCR steps used in this protocol are relatively simple, although a thermal cycler capable of performing qPCR will be required.

2.7 Custom Oligos and Longer DNA Sequences

At various points in the protocol, it is necessary to obtain custom DNA sequences (for PCR primers and longer sequences for cloning). These can usually be obtained from commercial sources (e.g., Eurofins Genomics, Macrogen, Integrated DNA Technologies). In-house sequencing services, if available, are also suitable.

We note that for the protocol described here, the shRNA/miRNA designs, which will be expressed from plasmids or vectors, should be ordered as dsDNA sequences.

2.8 Plasmids and Vectors

1. The original protocol used "pSplice" vectors. It is advantageous to express the siTCR sequence from an intronic location, so a vector with splice donor and splice acceptor sites is desirable. Similar vectors are available from AddGene, or researchers may choose to use a vector they are already familiar with.

2. For the expression of shRNA alone (e.g., for initial testing of designs), the pSilencer plasmid series contain polymerase III (PolIII) promoters, suitable for expression of shRNAs. They also include suitable restriction sites to facilitate cloning and marker genes to identify cells carrying the plasmid.
3. In order to produce infectious virus particles for transducing cells, accessory plasmids must be co-transfected into packaging cells. These plasmids may be obtained from AddGene and prepared in-house. Commercial kits are also available.

2.9 Flow Cytometry

Expression of each TCR chain (e.g., "wild-type" α- and β-chains, or the codon-optimized α- and β-chains) may be done by standard flow cytometric techniques. Conjugated antibodies are available for α and β subtypes, and by their use, it is possible to quantitate expression of each TCR type.

2.10 Kits for Isolation and Quantitation of RNA and DNA

Kits for the isolation of mRNA, DNA, and plasmid DNA are readily available from various sources.

For quantitation of mRNA levels by quantitative reverse transcription polymerase chain reaction (qRT-PCR), a SYBR-green based kit is the most flexible. SYBR-green kits are used with custom primers, and so multiple genes may be assessed with a single kit. Suitable kits will also include the necessary reagents for reverse transcription of mRNA.

3 Methods

3.1 Design shRNAs

1. On the chosen provider's website, design algorithms will be available. The default settings should be appropriate.
2. Input the constant regions (TRAC and TRBC1/TRBC2) as the target sequence. The algorithm will output a number of siRNA sequences for testing (*see* **Note 2**).
3. Order sequences compatible with chosen cloning methods (e.g., restriction digestion, or InFusion cloning). If there is an option to design shRNAs, this is preferred for expression from a plasmid or vector.

3.2 Testing shRNA Designs

1. Synthesized shRNA sequences are typically provided in plasmid form, either as DNA, or in *E. coli*. Whichever form they are provided in, it is strongly advisable to expand the plasmids via bacterial culture and plasmid preparation.
   (a) If provided as DNA, the plasmids must be redissolved (unless supplied as a solution) and transformed into competent *E. coli* strains (commercially available).
   (b) Plasmids containing *E. coli* must be expanded by standard methods, and then the plasmid prepared with a plasmid DNA kit.

2. If the shRNA sequence may be supplied in a general plasmid, it must be cloned into a suitable expression plasmid for testing (e.g., pSilencer). This may usually be done by standard restriction digestion cloning. Otherwise, shRNA sequences may be supplied in an expression plasmid for further use. Again, cloned plasmids should be expanded, and sequenced if necessary, before proceeding.
3. If a suitable cell line, expressing TCRs of a known wild-type sequence is available, these may be used here. Otherwise such a cell line may be generated by stable expression of a wild-type TCR. This entails construction of a TCR expression vector, as described in Subheadings 3.3–3.6. The key point is that TCR with wild-type TCR constant domains are required for testing shRNA sequences. For such purposes, an easily transfectable or transducible cell line, such as HEK293, may be used.
4. A modified, intronic miRNA cluster is optimal for the expression of multiple shRNAs from a single cassette. After the selection of 2–4 suitable shRNAs (1–2 each for TRAC and TRBC), a custom miRNA incorporating these sequences instead of the original miRNA sequences might be best ordered as a whole fragment, for cloning into the retroviral vector (below).

3.3 TCR Sequences and Codon-Optimization

The purpose of the siTCR design is to facilitate the expression of therapeutic, transgenic TCRs. We assume that TCR sequences for these will be available, at least in silico.

As a model for the endogenous TCRs, wild-type constant sequences are used. For the transgenic TCRs, it will be necessary to use codon-optimized constant domains that are unaffected by the siRNA sequences chosen in Subheading 3.2 (above). Codon-optimized constant domains, and the transgenic TCR sequences, might best be ordered from commercial providers (e.g., Eurofins Genomics). These may be designed as a single large sequence that might be cloned into a retroviral backbone in a single step, or as smaller stretches may be ordered separately, if backbones containing useful parts are available. Furthermore, these sequences may be designed to be cloned in by restriction digestion, or by PCR-based cloning. Restriction digestion is a widely known method, so in this protocol we will describe the methods for "InFusion," PCR-based cloning.

3.4 InFusion Cloning

We will here describe the general methods for InFusion cloning.

For PCR-based cloning, a high-fidelity Taq polymerase is highly desirable. Master mix kits are available from many sources, but the InFusion HD Cloning Plus kit includes a high-fidelity Taq, as well as a proprietary enzyme that mixes for fusion of the separate PCR products.

This method can in principle be used for all cloning options. That is, the production of shRNA-expressing plasmids, the generation of TCR-expressing plasmids/vectors, and the final production of miRNA-expressing, codon-optimized TCR-expressing vectors.

This method may be used for cloning shRNA sequences into pSilencer expression plasmids, as well as cloning of miRNA cluster and TCR sequences into retroviral vector backbones (*see* **Note 3**).

3.4.1 Preparation of DNA

Resuspend primers at 100 μM with sterile nuclease-free water (SNFW). Make working primer stocks of 2.5 μM (i.e., 1/40 dilution).

Use 0.1–5 ng of each template (per 25 μL reaction).

Example InFusion PCR reaction 25 μL

2× Premix	12.5 μL
DNA	~1 μL
FP	2.5 μL
RP	2.5 μL
dH2O	~6.5 μL

Thermal cycler program

1. 98 °C	3 mins
2. 98 °C	10s
3. 55 °C	5 s (or 15 s)
4. 72 °C	50s (5 s/kb)
Repeat 30× (**steps 2–4**)	
5. 72 °C	7 min

It is advisable to run PCR products on an agarose gel, to confirm the size, and also for isolation of the correct PCR product, prior to the InFusion reaction. However, in principle, and the Cloning Enhancer step is used, PCR clean-up is not necessary.

3.4.2 Cloning Enhancer

Add 2-μL cloning enhancer per 5 μL of PCR reaction product.

On a thermal cycler, incubate

1. 15 min (*20 if >100 ng template DNA in PCR reaction*) at 37 °C.
2. Then 15 min at 80 °C.

Use immediately or store at −20 °C.

Use the product of the cloning enhancer reaction at a maximum of 4 μL per 10 μL final InFusion reaction.

3.4.3 InFusion Reaction

1. 5× In-Fusion HD Enzyme Premix 2 μL.
2. Linearized vector *x* μL (<0.5 kb, 10–50 ng; 0.5–10 kb, 50–100 ng; >10 kb, 50–200 ng).
3. Purified PCR fragment *x* μL (<10 kb, 50–100 ng; >10 kb, 50–200 ng).
4. dH_2O (as needed) *x* μL.

Complete to 10-μL total volume.

Please note that for reactions with larger volumes of vector and PCR insert (>7 μL of vector + insert), double the amount of enzyme premix, and add dH_2O for a total volume of 20 μL. In other words, ratio of DNA to buffer is unimportant.

Incubate the reaction for 15 min at 50 °C, and then place on ice. Proceed directly to transformation of *E. coli*. Alternatively, cloning reactions can be stored at −20 °C until use.

After the culture of *E. coli* colonies by standard methods, individual colonies may be cultured, mini-prepped, and characterized by diagnostic digestion. Candidate colonies may then be sequenced, as per standard methods.

3.5 Preparation of HEK293 Packaging Cells

The following is relevant to transient transfection of adherent cells for the purpose of transient expression of genes. Thus it applies to transfection of TCRs or shRNAs for initial testing and to transfection of viral vector and accessory plasmids, for generation of supernatant containing infectious particles.

1. Thaw frozen HEK293 cells in a 37 °C water bath.
2. Upon thawing, transfer the cells in freezing medium to a sterile tube containing 5- to 10-mL prewarmed complete DMEM.
3. Spin down the cells at 300 × *g*, for 5 min, at 37 °C.
4. Note: freshly thawed cells are fragile and should be handled with care. Nevertheless, rates of cell death upon thawing may be high (e.g., up to 50%) in the days after thawing, depending upon the methods and solution used for freezing.
5. After trypsinization, cells will begin to readhere in the presence of serum, so cells should not be left in alternate containers for extended periods.
6. Remove supernatant and resuspend cells in an appropriate volume of complete DMEM.
7. Seed cells in a suitable container for recovery/initial expansion. T75 flasks allow for easy handling and maintenance of adherent cells and have a capacity that provides a good number of cells for most purposes. Of course, culture volumes may be scaled up or down, as appropriate.

8. Depending upon the initial cell numbers, and the scale of transfection planned, after 1 day or so, cells should be adherent and proliferating. Do not let HEK293 cells exceed ~70–80% confluency.
9. [Note: HEK293 cells (and variants) should be maintained below 70–80% confluency.] Also, certain cell types (e.g., HEK293, "Phoenix" cells) are much more sensitive to passage number than others (e.g., HEK293T). Sensitive cell types should not be used beyond passage 10, so careful planning of transfection experiments is necessary.
10. For trypsinization, wash cells 2× with prewarmed PBS(−) to remove serum-containing medium.
11. Add Trypsin-EDTA at 1–2 mL for a T75 flask and return to incubator. Check after ~5 min. If cells have not detached into single cells (or very small clusters), return to incubator for another 5 min.
12. Dilute trypsinized cells with PBS(−) if desired, and then count cells.
13. Seed cells in a suitable cell culture vessel. Appropriate seeding density will depend upon cell type and culture vessel, but 300,000–500,000 cells in a 30-mm dish is usually appropriate. This can be fine-tuned and may be scaled up as appropriate.

3.6 Retroviral Production

3.6.1 Seed Packaging Cells (Day 1)

Seed HEK293T cells into a 6-well plate at 600,000 cells/well, in 2-mL complete DMEM/well.

3.6.2 Transfect Packaging Cells (Day 2)

Prepare reagents and packaging plasmids in the following amounts:

X-tremeGene 9	4.5 μL
OptiMEM	150 μL
Accessory plasmid A	0.666 μg
Envelope plasmid B	0.666 μg
Retroviral plasmid	0.666 μg

Incubate this mixture for 15 min at room temperature.

During incubation of the mixture, change the medium on the HEK293 cells for fresh complete, prewarmed DMEM.

After incubation, add the mixture dropwise to each well, aiming for maximum and even distribution across the well. Gently move the culture vessel up and down and then left and right. Do not swirl.

3.6.3 Change Medium (Day 3)

Replace transfection mixture with DMEM with 1% HyClone.
4 mL for a 6-cm dish, 2 mL/well (of a 6-well plate).

3.6.4 First Harvest (Day 4)

Harvest supernatant and replace with fresh DMEM with 1% HyClone. Store supernatant overnight at 4 °C.

3.6.5 Second Harvest (Day 5)

Harvest supernatant and pool with the first harvest. Spin down supernatant for 5 min (900 × *g*) and then filter with a sterile 0.45-μm cartridge filter. Aliquot into cryotubes, and store at −80 °C.

3.7 Transduction of PBMCs with Retroviral Supernatant

This protocol may also be used to transduce other cell lines, although Retronectin may not be necessary for adherent cells. Activation will not be necessary for the expansion of immortal cell lines.

3.7.1 Activation and (Early) Expansion of PBMCs

T cells will be activated by OKT3 antibody, as this molecule binds to the CD3 co-receptor. After 2 days' culture CD3+ cells—CD4 and CD8 T cells—will preferentially expand. Alternatively, it may be desirable to enrich for, e.g., CD3- or CD8-positive cells using magnetic beads at the beginning of the culture period.

1. Prepare OCT3-coated plates by diluting OKT3 antibody in PBS (−) at a concentration of 1 mg/mL.
2. Dispense to a 24-well plate at a volume of 500 μL/well.
3. Incubate plates at 4 °C overnight, or at room temperature for 2 h.
4. Remove OKT3-PBS solution immediately before adding cells. Do not allow the plates to dry out.
5. Add PBMCs at 500,000–1,000,000 cells per well, in 500–1000 μL CTL culture medium. Culture at 7 °C with 5% CO_2.
6. After 2 days, wash and count cells, and proceed to transduction.

3.7.2 Retroviral Transduction of PBMCs or Cell Lines

1. Prepare Retronectin-coated plates by diluting Retronectin in PBS(−) at a concentration of 20 g/mL.
2. Dispense to a 24-well plate at a volume of 500 μL/well.
3. Incubate plates at 4 °C overnight, or at room temperature for 2 h.
4. Before adding cells, remove Retronectin-PBS solution and add 500 μL of PBS with 2% FCS. Incubate at room temperature for 30 min before removing the 2% FCS/PBS, and then add cells. PBMCs may be transduced at 300,000–500,000 cells per well, in 500 μL plus 500 μL vector supernatant.

5. "Spinoculate" cells at 900 × *g* for 60 min, at 37 °C, and then return to incubator.
6. After 4–6 h, remove 500 μL of medium, and replace with 500 μL of fresh media.
7. The following day, a second transduction may be performed, if desired, or necessary for a higher titer. Retronectin plates may be prepared in advance, as per **step 1** above. Thereafter perform the second transduction from **step 4** above.
8. At 2–3 days' post-transduction, check for expression of the transgenes, for example, by flow cytometry.

3.8 Coculture Assays

In order to test the reactivity of TCR-bearing T cells against target peptide-MHC complexes (pMHC), it is necessary to coculture effector cells against suitable target cells. Various assays for reactivity may be used (e.g., CD107a degranulation, INFγ and TNFα secretion, caspase-3 assay). These are standard methods, and we will not detail the various protocols here.

It is appropriate to test TCRs against cells that are naturally pMHC-positive. However, it is also appropriate to conduct dose/response curves. For this it will be necessary to use cells that are HLA-matched but are negative for the target peptide. These cells should then be preincubated with the target peptide for 2–12 h at 37 °C before coculture. Doses will range from 0.001 to 10 μM, plus a 0-μM control.

4 Notes

1. HEK293 cells (and their derivatives) are an adherent cell line that is easily transfected and transduced. Therefore, these cells were used to express "wild-type" and codon-optimized TCRs. Thus, HEK293 cells are an easy model for testing siRNA designs. Various variants of the original HEK293 line are available (such as 293T, which expresses the large T antigen and so express certain plasmids at higher rates). HEK-293T cells are a reliable and widely used cell line for vector supernatant production. It is common to transfect adherent cells in a circular culture dish, for example, a 6-well plate, or 60- or 100-mm culture dishes. Culture dishes should be sterile and pretreated for cell culture.
2. Several siRNA designs for downregulation of TCRs have been published [77–79]. These are as follows:
   Targeting human TCRα constant domain;

   5′-GTAAGGATTCTGATGTGTA-3′, and 5′-CAACAAATCT GACTTTGCA-3′.

   Targeting human TCRβ constant domain;

5′-CCACCATCCTCTATGAGAT-3′, and 5′-CCATGGTCAA GAGAAAGGA-3′ [77].

Targeting murine TCRα constant domain;

5′-TTTAGGTTCATATCTGTTTCA-3′.

Targeting murine TCRβ constant domain;

5′-CTGATGTTCTGTGTGACAGGT-3′ [78].

Novel designs may also be tested, and there are many algorithms available online for this purpose. They are typically supported by various commercial entities, who are also able to provide the siRNA molecules themselves, in various formats.

Here we have described methods for transfection and transduction of plasmid and retroviral vectors, utilizing shRNA designs. For testing siRNAs, and delivering plasmids, alternative methods such as TransIT-TKO and electroporation may be used. Details of the protocols for these methods are available from the relevant commercial suppliers.

miRNA clusters are an effective design for the expression of multiple siRNAs from a single cassette, driven by a Pol-II promoter. As noted above, the insertion of a custom miRNA cluster in an intronic location also results in superior transgene expression. Previous work has found that for this purpose, the miRNA cluster miR-17-20 is suitable. This sequence is customized by replacing the existing miRNA inverted repeats with the selected siRNA sequences. Unless such sequences are already available, it is likely convenient to design this custom miRNA cluster in silico, and have it synthesized as a single sequence, for cloning into the vector.

3. An online tool to assist with design of InFusion primers is available here: https://www.takarabio.com/learning-centers/cloning/in-fusion-cloning-tools.

This tool outputs primer sequences based upon the intended fusion point. It does not take into account other primer characteristics (e.g., GC content, Tm). In our experience, under the recommended conditions, more such primers are effective.

References

1. Kuhns MS, Davis MM, Garcia KC (2006) Deconstructing the form and function of the TCR/CD3 complex. Immunity 24 (2):133–139. https://doi.org/10.1016/j.immuni.2006.01.006
2. Wucherpfennig KW, Gagnon E, Call MJ, Huseby ES, Call ME (2010) Structural biology of the T-cell receptor: insights into receptor assembly, ligand recognition, and initiation of signaling. Cold Spring Harb Perspect Biol 2 (4):a005140. https://doi.org/10.1101/cshperspect.a005140
3. Kane LP, Lin J, Weiss A (2000) Signal transduction by the TCR for antigen. Curr Opin Immunol 12(3):242–249
4. Artyomov MN, Lis M, Devadas S, Davis MM, Chakraborty AK (2010) CD4 and CD8 binding to MHC molecules primarily acts to

enhance Lck delivery. Proc Natl Acad Sci U S A 107(39):16916–16921. https://doi.org/10.1073/pnas.1010568107

5. Esensten JH, Helou YA, Chopra G, Weiss A, Bluestone JA (2016) CD28 costimulation: from mechanism to therapy. Immunity 44 (5):973–988. https://doi.org/10.1016/j.immuni.2016.04.020
6. Robbins PF, Lu YC, El-Gamil M, Li YF, Gross C, Gartner J, Lin JC, Teer JK, Cliften P, Tycksen E, Samuels Y, Rosenberg SA (2013) Mining exomic sequencing data to identify mutated antigens recognized by adoptively transferred tumor-reactive T cells. Nat Med 19(6):747–752. https://doi.org/10.1038/nm.3161
7. Schumacher TN, Schreiber RD (2015) Neoantigens in cancer immunotherapy. Science 348 (6230):69–74. https://doi.org/10.1126/science.aaa4971
8. Parkhurst M, Gros A, Pasetto A, Prickett T, Crystal JS, Robbins P, Rosenberg SA (2017) Isolation of T-cell receptors specifically reactive with mutated tumor-associated antigens from tumor-infiltrating lymphocytes based on CD137 expression. Clin Cancer Res 23 (10):2491–2505. https://doi.org/10.1158/1078-0432.CCR-16-2680
9. Muenst S, Laubli H, Soysal SD, Zippelius A, Tzankov A, Hoeller S (2016) The immune system and cancer evasion strategies: therapeutic concepts. J Intern Med 279(6):541–562. https://doi.org/10.1111/joim.12470
10. Fridman WH (2018) From cancer immune surveillance to cancer immunoediting: birth of modern immuno-oncology. J Immunol 201(3):825–826. https://doi.org/10.4049/jimmunol.1800827
11. Ribatti D (2017) The concept of immune surveillance against tumors. The first theories. Oncotarget 8(4):7175–7180. https://doi.org/10.18632/oncotarget.12739
12. Mittal D, Gubin MM, Schreiber RD, Smyth MJ (2014) New insights into cancer immunoediting and its three component phases—elimination, equilibrium and escape. Curr Opin Immunol 27:16–25. https://doi.org/10.1016/j.coi.2014.01.004
13. Shankaran V, Ikeda H, Bruce AT, White JM, Swanson PE, Old LJ, Schreiber RD (2001) IFNgamma and lymphocytes prevent primary tumour development and shape tumour immunogenicity. Nature 410(6832):1107–1111. https://doi.org/10.1038/35074122
14. Rosenberg SA, Packard BS, Aebersold PM, Solomon D, Topalian SL, Toy ST, Simon P, Lotze MT, Yang JC, Seipp CA et al (1988) Use of tumor-infiltrating lymphocytes and interleukin-2 in the immunotherapy of patients with metastatic melanoma. A preliminary report. N Engl J Med 319(25):1676–1680. https://doi.org/10.1056/NEJM198812223192527
15. Restifo NP, Dudley ME, Rosenberg SA (2012) Adoptive immunotherapy for cancer: harnessing the T cell response. Nat Rev Immunol 12 (4):269–281. https://doi.org/10.1038/nri3191
16. Dudley ME, Wunderlich JR, Robbins PF, Yang JC, Hwu P, Schwartzentruber DJ, Topalian SL, Sherry R, Restifo NP, Hubicki AM, Robinson MR, Raffeld M, Duray P, Seipp CA, Rogers-Freezer L, Morton KE, Mavroukakis SA, White DE, Rosenberg SA (2002) Cancer regression and autoimmunity in patients after clonal repopulation with antitumor lymphocytes. Science 298(5594):850–854. https://doi.org/10.1126/science.1076514
17. Yee C, Savage PA, Lee PP, Davis MM, Greenberg PD (1999) Isolation of high avidity melanoma-reactive CTL from heterogeneous populations using peptide-MHC tetramers. J Immunol 162(4):2227–2234
18. Dudley ME, Wunderlich JR, Shelton TE, Even J, Rosenberg SA (2003) Generation of tumor-infiltrating lymphocyte cultures for use in adoptive transfer therapy for melanoma patients. J Immunother 26(4):332–342
19. Rosenberg SA, Restifo NP, Yang JC, Morgan RA, Dudley ME (2008) Adoptive cell transfer: a clinical path to effective cancer immunotherapy. Nat Rev Cancer 8(4):299–308. https://doi.org/10.1038/nrc2355
20. Scanlan MJ, Gure AO, Jungbluth AA, Old LJ, Chen YT (2002) Cancer/testis antigens: an expanding family of targets for cancer immunotherapy. Immunol Rev 188:22–32
21. Theobald M, Biggs J, Hernandez J, Lustgarten J, Labadie C, Sherman LA (1997) Tolerance to p53 by A2.1-restricted cytotoxic T lymphocytes. J Exp Med 185(5):833–841. https://doi.org/10.1084/jem.185.5.833
22. Kuball J, Schuler M, Antunes Ferreira E, Herr W, Neumann M, Obenauer-Kutner L, Westreich L, Huber C, Wolfel T, Theobald M (2002) Generating p53-specific cytotoxic T lymphocytes by recombinant adenoviral vector-based vaccination in mice, but not man. Gene Ther 9(13):833–843. https://doi.org/10.1038/sj.gt.3301709
23. Dembic Z, Haas W, Weiss S, McCubrey J, Kiefer H, von Boehmer H, Steinmetz M (1986) Transfer of specificity by murine alpha and beta T-cell receptor genes. Nature 320 (6059):232–238

24. Morgan RA, Dudley ME, Wunderlich JR, Hughes MS, Yang JC, Sherry RM, Royal RE, Topalian SL, Kammula US, Restifo NP, Zheng Z, Nahvi A, de Vries CR, Rogers-Freezer LJ, Mavroukakis SA, Rosenberg SA (2006) Cancer regression in patients after transfer of genetically engineered lymphocytes. Science 314(5796):126–129. https://doi.org/10.1126/science.1129003
25. Schumacher TN (2002) T-cell-receptor gene therapy. Nat Rev Immunol 2(7):512–519. https://doi.org/10.1038/nri841
26. Eshhar Z, Waks T, Bendavid A, Schindler DG (2001) Functional expression of chimeric receptor genes in human T cells. J Immunol Methods 248(1–2):67–76
27. Barrett DM, Grupp SA, June CH (2015) Chimeric antigen receptor- and TCR-modified T cells enter main street and wall street. J Immunol 195(3):755–761. https://doi.org/10.4049/jimmunol.1500751
28. van Loenen MM, de Boer R, Amir AL, Hagedoorn RS, Volbeda GL, Willemze R, van Rood JJ, Falkenburg JH, Heemskerk MH (2010) Mixed T cell receptor dimers harbor potentially harmful neoreactivity. Proc Natl Acad Sci U S A 107(24):10972–10977. https://doi.org/10.1073/pnas.1005802107
29. Bendle GM, Linnemann C, Hooijkaas AI, Bies L, de Witte MA, Jorritsma A, Kaiser AD, Pouw N, Debets R, Kieback E, Uckert W, Song JY, Haanen JB, Schumacher TN (2010) Lethal graft-versus-host disease in mouse models of T cell receptor gene therapy. Nat Med 16 (5):565–570. https://doi.org/10.1038/nm.2128. 561 p. Following 570
30. Cohen CJ, Zheng Z, Bray R, Zhao Y, Sherman LA, Rosenberg SA, Morgan RA (2005) Recognition of fresh human tumor by human peripheral blood lymphocytes transduced with a bicistronic retroviral vector encoding a murine anti-p53 TCR. J Immunol 175(9):5799–5808
31. Cohen CJ, Zhao Y, Zheng Z, Rosenberg SA, Morgan RA (2006) Enhanced antitumor activity of murine-human hybrid T-cell receptor (TCR) in human lymphocytes is associated with improved pairing and TCR/CD3 stability. Cancer Res 66(17):8878–8886. https://doi.org/10.1158/0008-5472.CAN-06-1450
32. Voss RH, Kuball J, Engel R, Guillaume P, Romero P, Huber C, Theobald M (2006) Redirection of T cells by delivering a transgenic mouse-derived MDM2 tumor antigen-specific TCR and its humanized derivative is governed by the CD8 coreceptor and affects natural human TCR expression. Immunol Res 34 (1):67–87. https://doi.org/10.1385/IR:34:1:67
33. Sommermeyer D, Uckert W (2010) Minimal amino acid exchange in human TCR constant regions fosters improved function of TCR gene-modified T cells. J Immunol 184 (11):6223–6231. https://doi.org/10.4049/jimmunol.0902055
34. Boulter JM, Glick M, Todorov PT, Baston E, Sami M, Rizkallah P, Jakobsen BK (2003) Stable, soluble T-cell receptor molecules for crystallization and therapeutics. Protein Eng 16 (9):707–711
35. Kuball J, Dossett ML, Wolfl M, Ho WY, Voss RH, Fowler C, Greenberg PD (2007) Facilitating matched pairing and expression of TCR chains introduced into human T cells. Blood 109(6):2331–2338. https://doi.org/10.1182/blood-2006-05-023069
36. Voss RH, Willemsen RA, Kuball J, Grabowski M, Engel R, Intan RS, Guillaume P, Romero P, Huber C, Theobald M (2008) Molecular design of the Calphabeta interface favors specific pairing of introduced TCRalphabeta in human T cells. J Immunol 180(1):391–401
37. Szymczak AL, Workman CJ, Wang Y, Vignali KM, Dilioglou S, Vanin EF, Vignali DA (2004) Correction of multi-gene deficiency in vivo using a single 'self-cleaving' 2A peptide-based retroviral vector. Nat Biotechnol 22 (5):589–594. https://doi.org/10.1038/nbt957
38. Yang S, Cohen CJ, Peng PD, Zhao Y, Cassard L, Yu Z, Zheng Z, Jones S, Restifo NP, Rosenberg SA, Morgan RA (2008) Development of optimal bicistronic lentiviral vectors facilitates high-level TCR gene expression and robust tumor cell recognition. Gene Ther 15 (21):1411–1423. https://doi.org/10.1038/gt.2008.90
39. Scholten KB, Kramer D, Kueter EW, Graf M, Schoedl T, Meijer CJ, Schreurs MW, Hooijberg E (2006) Codon modification of T cell receptors allows enhanced functional expression in transgenic human T cells. Clin Immunol 119(2):135–145. https://doi.org/10.1016/j.clim.2005.12.009
40. Willcox BE, Gao GF, Wyer JR, O'Callaghan CA, Boulter JM, Jones EY, van der Merwe PA, Bell JI, Jakobsen BK (1999) Production of soluble alphabeta T-cell receptor heterodimers suitable for biophysical analysis of ligand binding. Protein Sci 8(11):2418–2423. https://doi.org/10.1110/ps.8.11.2418
41. Foley KC, Spear TT, Murray DC, Nagato K, Garrett-Mayer E, Nishimura MI (2017) HCV T cell receptor chain modifications to enhance expression, pairing, and antigen recognition in T cells for adoptive transfer. Mol Ther

Oncolytics 5:105–115. https://doi.org/10.1016/j.omto.2017.05.004

42. Bethune MT, Gee MH, Bunse M, Lee MS, Gschweng EH, Pagadala MS, Zhou J, Cheng D, Heath JR, Kohn DB, Kuhns MS, Uckert W, Baltimore D (2016) Domain-swapped T cell receptors improve the safety of TCR gene therapy. elife 5. https://doi.org/10.7554/eLife.19095
43. Cohen CJ, Li YF, El-Gamil M, Robbins PF, Rosenberg SA, Morgan RA (2007) Enhanced antitumor activity of T cells engineered to express T-cell receptors with a second disulfide bond. Cancer Res 67(8):3898–3903. https://doi.org/10.1158/0008-5472.CAN-06-3986
44. Hart DP, Xue SA, Thomas S, Cesco-Gaspere-M, Tranter A, Willcox B, Lee SP, Steven N, Morris EC, Stauss HJ (2008) Retroviral transfer of a dominant TCR prevents surface expression of a large proportion of the endogenous TCR repertoire in human T cells. Gene Ther 15(8):625–631. https://doi.org/10.1038/sj.gt.3303078
45. Reus K, Mayer J, Sauter M, Zischler H, Muller-Lantzsch N, Meese E (2001) HERV-K(OLD): ancestor sequences of the human endogenous retrovirus family HERV-K(HML-2). J Virol 75 (19):8917–8926. https://doi.org/10.1128/JVI.75.19.8917-8926.2001
46. Reuss S, Sebestyen Z, Heinz N, Loew R, Baum C, Debets R, Uckert W (2014) TCR-engineered T cells: a model of inducible TCR expression to dissect the interrelationship between two TCRs. Eur J Immunol 44 (1):265–274. https://doi.org/10.1002/eji.201343591
47. Willemsen RA, Weijtens ME, Ronteltap C, Eshhar Z, Gratama JW, Chames P, Bolhuis RL (2000) Grafting primary human T lymphocytes with cancer-specific chimeric single chain and two chain TCR. Gene Ther 7 (16):1369–1377. https://doi.org/10.1038/sj.gt.3301253
48. Chung S, Wucherpfennig KW, Friedman SM, Hafler DA, Strominger JL (1994) Functional three-domain single-chain T-cell receptors. Proc Natl Acad Sci U S A 91 (26):12654–12658
49. Voss RH, Thomas S, Pfirschke C, Hauptrock B, Klobuch S, Kuball J, Grabowski M, Engel R, Guillaume P, Romero P, Huber C, Beckhove P, Theobald M (2010) Coexpression of the T-cell receptor constant alpha domain triggers tumor reactivity of single-chain TCR-transduced human T cells. Blood 115(25):5154–5163. https://doi.org/10.1182/blood-2009-11-254078
50. Knies D, Klobuch S, Xue SA, Birtel M, Echchannaoui H, Yildiz O, Omokoko T, Guillaume P, Romero P, Stauss H, Sahin U, Herr W, Theobald M, Thomas S, Voss RH (2016) An optimized single chain TCR scaffold relying on the assembly with the native CD3-complex prevents residual mispairing with endogenous TCRs in human T-cells. Oncotarget 7(16):21199–21221. https://doi.org/10.18632/oncotarget.8385
51. Ahmadi M, King JW, Xue SA, Voisine C, Holler A, Wright GP, Waxman J, Morris E, Stauss HJ (2011) CD3 limits the efficacy of TCR gene therapy in vivo. Blood 118 (13):3528–3537. https://doi.org/10.1182/blood-2011-04-346338
52. Govers C, Sebestyen Z, Roszik J, van Brakel M, Berrevoets C, Szoor A, Panoutsopoulou K, Broertjes M, Van T, Vereb G, Szollosi J, Debets R (2014) TCRs genetically linked to CD28 and CD3epsilon do not mispair with endogenous TCR chains and mediate enhanced T cell persistence and anti-melanoma activity. J Immunol 193(10):5315–5326. https://doi.org/10.4049/jimmunol.1302074
53. Walseng E, Koksal H, Sektioglu IM, Fane A, Skorstad G, Kvalheim G, Gaudernack G, Inderberg EM, Walchli S (2017) A TCR-based chimeric antigen receptor. Sci Rep 7(1):10713. https://doi.org/10.1038/s41598-017-11126-y
54. Walseng E, Walchli S, Fallang LE, Yang W, Vefferstad A, Areffard A, Olweus J (2015) Soluble T-cell receptors produced in human cells for targeted delivery. PLoS One 10(4): e0119559. https://doi.org/10.1371/journal.pone.0119559
55. Liddy N, Bossi G, Adams KJ, Lissina A, Mahon TM, Hassan NJ, Gavarret J, Bianchi FC, Pumphrey NJ, Ladell K, Gostick E, Sewell AK, Lissin NM, Harwood NE, Molloy PE, Li Y, Cameron BJ, Sami M, Baston EE, Todorov PT, Paston SJ, Dennis RE, Harper JV, Dunn SM, Ashfield R, Johnson A, McGrath Y, Plesa G, June CH, Kalos M, Price DA, Vuidepot A, Williams DD, Sutton DH, Jakobsen BK (2012) Monoclonal TCR-redirected tumor cell killing. Nat Med 18(6):980–987. https://doi.org/10.1038/nm.2764
56. Baeuerle PA, Kufer P, Bargou R (2009) BiTE: teaching antibodies to engage T-cells for cancer therapy. Curr Opin Mol Ther 11(1):22–30
57. Oates J, Jakobsen BK (2013) ImmTACs: novel bi-specific agents for targeted cancer therapy. Oncoimmunology 2(2):e22891. https://doi.org/10.4161/onci.22891
58. Poirot L, Philip B, Schiffer-Mannioui C, Le Clerre D, Chion-Sotinel I, Derniame S,

Potrel P, Bas C, Lemaire L, Galetto R, Lebuhotel C, Eyquem J, Cheung GW, Duclert A, Gouble A, Arnould S, Peggs K, Pule M, Scharenberg AM, Smith J (2015) Multiplex genome-edited T-cell manufacturing platform for "off-the-shelf" adoptive t-cell immunotherapies. Cancer Res 75 (18):3853–3864. https://doi.org/10.1158/0008-5472.CAN-14-3321

59. Osborn MJ, Webber BR, Knipping F, Lonetree CL, Tennis N, DeFeo AP, McElroy AN, Starker CG, Lee C, Merkel S, Lund TC, Kelly-Spratt KS, Jensen MC, Voytas DF, von Kalle C, Schmidt M, Gabriel R, Hippen KL, Miller JS, Scharenberg AM, Tolar J, Blazar BR (2016) Evaluation of TCR gene editing achieved by TALENs, CRISPR/Cas9, and megaTAL nucleases. Mol Ther 24(3):570–581. https://doi.org/10.1038/mt.2015.197

60. Mensali N, Dillard P, Hebeisen M, Lorenz S, Theodossiou T, Myhre MR, Fane A, Gaudernack G, Kvalheim G, Myklebust JH, Inderberg EM, Walchli S (2019) NK cells specifically TCR-dressed to kill cancer cells. EBioMedicine 40:106–117. https://doi.org/10.1016/j.ebiom.2019.01.031

61. Legut M, Dolton G, Mian AA, Ottmann OG, Sewell AK (2018) CRISPR-mediated TCR replacement generates superior anticancer transgenic T cells. Blood 131(3):311–322. https://doi.org/10.1182/blood-2017-05-787598

62. Certo MT, Gwiazda KS, Kuhar R, Sather B, Curinga G, Mandt T, Brault M, Lambert AR, Baxter SK, Jacoby K, Ryu BY, Kiem HP, Gouble A, Paques F, Rawlings DJ, Scharenberg AM (2012) Coupling endonucleases with DNA end-processing enzymes to drive gene disruption. Nat Methods 9(10):973–975. https://doi.org/10.1038/nmeth.2177

63. Provasi E, Genovese P, Lombardo A, Magnani Z, Liu PQ, Reik A, Chu V, Paschon DE, Zhang L, Kuball J, Camisa B, Bondanza A, Casorati G, Ponzoni M, Ciceri F, Bordignon C, Greenberg PD, Holmes MC, Gregory PD, Naldini L, Bonini C (2012) Editing T cell specificity towards leukemia by zinc finger nucleases and lentiviral gene transfer. Nat Med 18(5):807–815. https://doi.org/10.1038/nm.2700

64. Berdien B, Mock U, Atanackovic D, Fehse B (2014) TALEN-mediated editing of endogenous T-cell receptors facilitates efficient reprogramming of T lymphocytes by lentiviral gene transfer. Gene Ther 21(6):539–548. https://doi.org/10.1038/gt.2014.26

65. Georgiadis C, Preece R, Nickolay L, Etuk A, Petrova A, Ladon D, Danyi A, Humphryes-Kirilov N, Ajetunmobi A, Kim D, Kim JS, Qasim W (2018) Long terminal repeat CRISPR-CAR-coupled "universal" T cells mediate potent anti-leukemic effects. Mol Ther 26(5):1215–1227. https://doi.org/10.1016/j.ymthe.2018.02.025

66. Roth TL, Puig-Saus C, Yu R, Shifrut E, Carnevale J, Li PJ, Hiatt J, Saco J, Krystofinski P, Li H, Tobin V, Nguyen DN, Lee MR, Putnam AL, Ferris AL, Chen JW, Schickel JN, Pellerin L, Carmody D, Alkorta-Aranburu G, Del Gaudio D, Matsumoto H, Morell M, Mao Y, Cho M, Quadros RM, Gurumurthy CB, Smith B, Haugwitz M, Hughes SH, Weissman JS, Schumann K, Esensten JH, May AP, Ashworth A, Kupfer GM, Greeley SAW, Bacchetta R, Meffre E, Roncarolo MG, Romberg N, Herold KC, Ribas A, Leonetti MD, Marson A (2018) Reprogramming human T cell function and specificity with non-viral genome targeting. Nature 559 (7714):405–409. https://doi.org/10.1038/s41586-018-0326-5

67. Albers JJ, Ammon T, Gosmann D, Audehm S, Thoene S, Winter C, Secci R, Wolf A, Stelzl A, Steiger K, Ruland J, Bassermann F, Kupatt C, Anton M, Krackhardt AM (2019) Gene editing enables T-cell engineering to redirect antigen specificity for potent tumor rejection. Life Sci Alliance 2(2). https://doi.org/10.26508/lsa.201900367

68. Davis L, Maizels N (2014) Homology-directed repair of DNA nicks via pathways distinct from canonical double-strand break repair. Proc Natl Acad Sci U S A 111(10):E924–E932. https://doi.org/10.1073/pnas.1400236111

69. Kleinstiver BP, Pattanayak V, Prew MS, Tsai SQ, Nguyen NT, Zheng Z, Joung JK (2016) High-fidelity CRISPR-Cas9 nucleases with no detectable genome-wide off-target effects. Nature 529(7587):490–495. https://doi.org/10.1038/nature16526

70. Chen JS, Dagdas YS, Kleinstiver BP, Welch MM, Sousa AA, Harrington LB, Sternberg SH, Joung JK, Yildiz A, Doudna JA (2017) Enhanced proofreading governs CRISPR–Cas9 targeting accuracy. Nature. https://doi.org/10.1038/nature24268

71. Kosicki M, Tomberg K, Bradley A (2018) Repair of double-strand breaks induced by CRISPR-Cas9 leads to large deletions and complex rearrangements. Nat Biotechnol. https://doi.org/10.1038/nbt.4192

72. Rezza A, Jacquet C, Le Pillouer A, Lafarguette F, Ruptier C, Billandon M, Isnard Petit P, Trouttet S, Thiam K, Fraichard A, Cherifi Y (2019) Unexpected genomic rearrangements at targeted loci associated with

CRISPR/Cas9-mediated knock-in. Sci Rep 9 (1):3486. https://doi.org/10.1038/s41598-019-40181-w

73. Fire A, Xu S, Montgomery MK, Kostas SA, Driver SE, Mello CC (1998) Potent and specific genetic interference by double-stranded RNA in *Caenorhabditis elegans*. Nature 391 (6669):806–811. https://doi.org/10.1038/35888

74. Elbashir SM, Harborth J, Lendeckel W, Yalcin A, Weber K, Tuschl T (2001) Duplexes of 21-nucleotide RNAs mediate RNA interference in cultured mammalian cells. Nature 411 (6836):494–498. https://doi.org/10.1038/35078107

75. Paddison PJ, Caudy AA, Bernstein E, Hannon GJ, Conklin DS (2002) Short hairpin RNAs (shRNAs) induce sequence-specific silencing in mammalian cells. Genes Dev 16 (8):948–958. https://doi.org/10.1101/gad.981002

76. Rao DD, Senzer N, Cleary MA, Nemunaitis J (2009) Comparative assessment of siRNA and shRNA off target effects: what is slowing clinical development. Cancer Gene Ther 16 (11):807–809. https://doi.org/10.1038/cgt.2009.53

77. Okamoto S, Mineno J, Ikeda H, Fujiwara H, Yasukawa M, Shiku H, Kato I (2009) Improved expression and reactivity of transduced tumor-specific TCRs in human lymphocytes by specific silencing of endogenous TCR. Cancer Res 69(23):9003–9011

78. Bunse M, Bendle GM, Linnemann C, Bies L, Schulz S, Schumacher TN, Uckert W (2014) RNAi-mediated TCR knockdown prevents autoimmunity in mice caused by mixed TCR dimers following TCR gene transfer. Mol Ther 22(11):1983–1991. https://doi.org/10.1038/mt.2014.142

79. Campillo-Davo D, Fujiki F, Van den Bergh JMJ, De Reu H, Smits E, Goossens H, Sugiyama H, Lion E, Berneman ZN, Van Tendeloo V (2018) Efficient and non-genotoxic RNA-based engineering of human T cells using tumor-specific T cell receptors with minimal TCR mispairing. Front Immunol 9:2503. https://doi.org/10.3389/fimmu.2018.02503

80. Tawara I, Kageyama S, Miyahara Y, Fujiwara H, Nishida T, Akatsuka Y, Ikeda H, Tanimoto K, Terakura S, Murata M, Inaguma Y, Masuya M, Inoue N, Kidokoro T, Okamoto S, Tomura D, Chono H, Nukaya I, Mineno J, Naoe T, Emi N, Yasukawa M, Katayama N, Shiku H (2017) Safety and persistence of WT1-specific T-cell receptor gene-transduced lymphocytes in patients with AML and MDS. Blood 130 (18):1985–1994. https://doi.org/10.1182/blood-2017-06-791202

81. Kageyama S, Ikeda H, Miyahara Y, Imai N, Ishihara M, Saito K, Sugino S, Ueda S, Ishikawa T, Kokura S, Naota H, Ohishi K, Shiraishi T, Inoue N, Tanabe M, Kidokoro T, Yoshioka H, Tomura D, Nukaya I, Mineno J, Takesako K, Katayama N, Shiku H (2015) Adoptive transfer of MAGE-A4 T-cell receptor gene-transduced lymphocytes in patients with recurrent esophageal cancer. Clin Cancer Res 21(10):2268–2277. https://doi.org/10.1158/1078-0432.CCR-14-1559

82. Hacein-Bey-Abina S, Von Kalle C, Schmidt M, McCormack MP, Wulffraat N, Leboulch P, Lim A, Osborne CS, Pawliuk R, Morillon E, Sorensen R, Forster A, Fraser P, Cohen JI, de Saint BG, Alexander I, Wintergerst U, Frebourg T, Aurias A, Stoppa-Lyonnet D, Romana S, Radford-Weiss I, Gross F, Valensi F, Delabesse E, Macintyre E, Sigaux F, Soulier J, Leiva LE, Wissler M, Prinz C, Rabbitts TH, Le Deist F, Fischer A, Cavazzana-Calvo M (2003) LMO2-associated clonal T cell proliferation in two patients after gene therapy for SCID-X1. Science 302 (5644):415–419. https://doi.org/10.1126/science.1088547

83. Yoshida J, Akagi K, Misawa R, Kokubu C, Takeda J, Horie K (2017) Chromatin states shape insertion profiles of the piggyBac, Tol2 and Sleeping Beauty transposons and murine leukemia virus. Sci Rep 7:43613. https://doi.org/10.1038/srep43613

84. Clauss J, Obenaus M, Miskey C, Ivics Z, Izsvak Z, Uckert W, Bunse M (2018) Efficient non-viral T-cell engineering by sleeping beauty minicircles diminishing DNA toxicity and miRNAs silencing the endogenous T-cell receptors. Hum Gene Ther 29(5):569–584. https://doi.org/10.1089/hum.2017.136

85. Newrzela S, Cornils K, Li Z, Baum C, Brugman MH, Hartmann M, Meyer J, Hartmann S, Hansmann ML, Fehse B, von Laer D (2008) Resistance of mature T cells to oncogene transformation. Blood 112(6):2278–2286. https://doi.org/10.1182/blood-2007-12-128751

86. Marcucci KT, Jadlowsky JK, Hwang WT, Suhoski-Davis M, Gonzalez VE, Kulikovskaya I, Gupta M, Lacey SF, Plesa G, Chew A, Melenhorst JJ, Levine BL, June CH (2018) Retroviral and lentiviral safety analysis of gene-modified T cell products and infused HIV and oncology patients. Mol Ther 26 (1):269–279. https://doi.org/10.1016/j.ymthe.2017.10.012

87. Ni B (2017) First-ever CAR T-cell therapy approved in U.S. Cancer Discov 7(10):OF1.

https://doi.org/10.1158/2159-8290.CD-NB2017-126

88. Linette GP, Stadtmauer EA, Maus MV, Rapoport AP, Levine BL, Emery L, Litzky L, Bagg A, Carreno BM, Cimino PJ, Binder-Scholl GK, Smethurst DP, Gerry AB, Pumphrey NJ, Bennett AD, Brewer JE, Dukes J, Harper J, Tayton-Martin HK, Jakobsen BK, Hassan NJ, Kalos M, June CH (2013) Cardiovascular toxicity and titin cross-reactivity of affinity-enhanced T cells in myeloma and melanoma. Blood 122(6):863–871. https://doi.org/10.1182/blood-2013-03-490565

89. Morgan RA, Chinnasamy N, Abate-Daga D, Gros A, Robbins PF, Zheng Z, Dudley ME, Feldman SA, Yang JC, Sherry RM, Phan GQ, Hughes MS, Kammula US, Miller AD, Hessman CJ, Stewart AA, Restifo NP, Quezado MM, Alimchandani M, Rosenberg AZ, Nath A, Wang T, Bielekova B, Wuest SC, Akula N, McMahon FJ, Wilde S, Mosetter B, Schendel DJ, Laurencot CM, Rosenberg SA (2013) Cancer regression and neurological toxicity following anti-MAGE-A3 TCR gene therapy. J Immunother 36(2):133–151. https://doi.org/10.1097/CJI.0b013e3182829903

90. Almasbak H, Walseng E, Kristian A, Myhre MR, Suso EM, Munthe LA, Andersen JT, Wang MY, Kvalheim G, Gaudernack G, Kyte JA (2015) Inclusion of an IgG1-Fc spacer abrogates efficacy of CD19 CAR T cells in a xenograft mouse model. Gene Ther 22(5):391–403. https://doi.org/10.1038/gt.2015.4

91. Kyte JA, Fane A, Pule M, Gaudernack G (2019) Transient redirection of T cells for adoptive cell therapy with telomerase-specific T helper cell receptors isolated from long term survivors after cancer vaccination. Oncoimmunology 8(4):e1565236. https://doi.org/10.1080/2162402X.2019.1565236

Chapter 19

CRISPR/Cas9 Guide RNA Design Rules for Predicting Activity

Kasidet Hiranniramol, Yuhao Chen, and Xiaowei Wang

Abstract

A critical stage in performing gene editing experiments using the CRISPR/Cas9 system is the design of guide RNA (gRNA). In this chapter, we conduct a review of the current gRNA design rules for maximizing on-target Cas9 activity while minimizing off-target activity. In addition, we present some of the currently available computational tools for gRNA activity prediction and assay design.

Key words sgRNA, CRISPR, Cas9, Genome editing, Machine learning

1 Introduction

The CRISPR/Cas9 system is a powerful gene editing tool which has received widespread usage in the scientific community ([1], *see* Chapter 1). Originally discovered in bacteria as a viral defense pathway, this RNA-guided endonuclease system has since been engineered to suit a variety of scientific applications [2]. The endogenous system consists of a CRISPR RNA (crRNA) and a trans-activating crRNA (tracrRNA) which are processed by RNase III to form a dual-RNA heteroduplex [3]. The crRNA/tracrRNA complexes with Cas9, directing it to target and cleave foreign DNA. An adaptation of the natural system was developed by engineering a synthetic single-guide RNA (sgRNA) containing both crRNA and tracrRNA sequences [4]. This resulted in a simplified two-component CRISPR/Cas9 system that forgoes the heteroduplex assembly and processing steps. Precise targeting can thus be achieved simply by synthesizing an sgRNA with a guide domain (gRNA) complementary to the target sequence. In addition, the CRISPR system generates complete knockout at the DNA level, an effect which could not be achieved by the commonly used RNA interference protocols. The resulting advances and increased accessibility of gene editing have led to improved protocols for studies of gene function, large-scale genetic screens, and clinical therapies.

Mouldy Sioud (ed.), *RNA Interference and CRISPR Technologies: Technical Advances and New Therapeutic Opportunities*, Methods in Molecular Biology, vol. 2115, https://doi.org/10.1007/978-1-0716-0290-4_19,

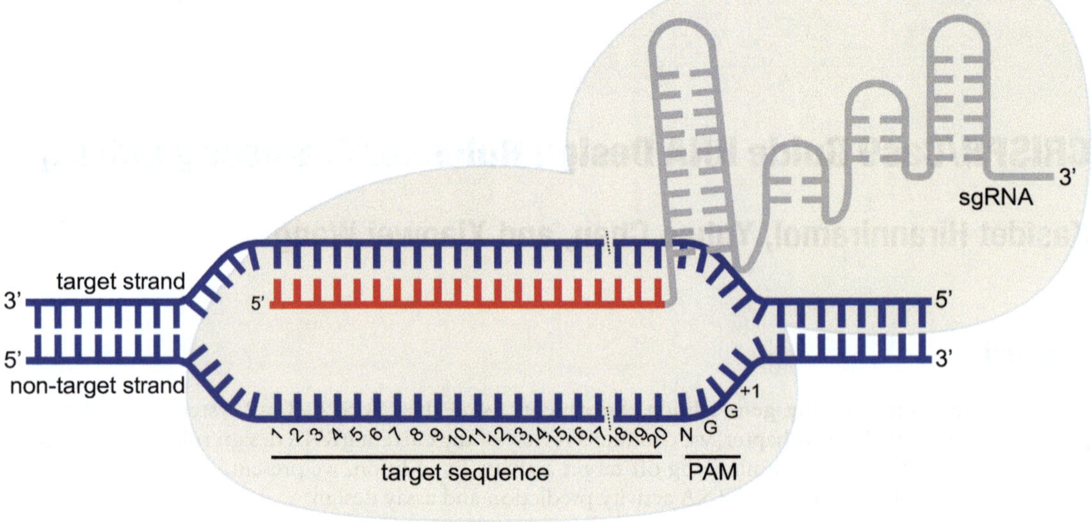

Fig. 1 The CRISPR/Cas9 system. Cas9 in complex with sgRNA, containing a gRNA domain (red) and scaffold (grey), induces a double-stranded break 3 nt upstream of the PAM

The Cas9 protein described in this chapter refers to the widely used *Streptococcus pyogenes* Cas9. Its gRNA is a 20-nucleotide (nt) sequence at the 5′ end of the sgRNA and is analogous to crRNA in the prokaryotic system. The gRNA hybridizes with its complementary DNA sequence, guiding Cas9 to produce a double-strand break between positions 17 and 18 of the gRNA [4] (Fig. 1). In addition to determining the precise cleavage location, variations in the gRNA sequence also impact the cleavage efficiency [5–13]. Poor gRNA designs can result in insufficient or no Cas9 activity, severely reducing the sensitivity of assays as important gene candidates are missed due to failed editing. In small-scale genome engineering, gRNAs can be tested for efficiency and redesigned, but a computational prediction tool is necessary to avoid excessive experimental screening. Another important issue arising from gRNA design is the prediction of off-target editing which has been observed in sequences similar to the intended target [11, 12, 14, 15]. Such specificity issues introduce confounding variables to experiments and can lead to false hits in genetic screens. Off-target editing must be thoroughly characterized to prevent any unintended and potentially harmful effects.

Cas9 activity needs to be quantified for a large number of gRNAs in order to assess specific contributions from hundreds of gRNA features. Studies often use proliferation assays to gather this information, in which successful knockout confers measurable survival pressure [5, 7, 11]. Other tactics include using sequencing-based quantification, fluorescent reporter readouts, and flow cytometry-based cell sorting [9, 10, 13]. Off-target activity can be assessed using similar methods, by introducing gRNAs that have various mismatches to the potential targets [11, 12, 14, 15]. After a gRNA dataset with known editing efficiency is established, statistical analysis can be performed to identify the most impactful features. Features that have been investigated include nucleotide composition, sequence features, structural features, and genomic context [5–13].

Machine learning approaches have been commonly employed to generate computational predictions of Cas9/sgRNA activity. These methods such as SVM, logistic regression, and deep learning automatically extract information from a combination of the feature data in order to produce a mathematical model which can then be used to predict Cas9 activity for any designed gRNA. The design process is thus a workflow consisting of identifying potential gRNAs in a desired gene and generating predictions for each gRNA, choosing only the gRNAs with high predicted efficacy and specificity.

In this chapter, we review the most significant gRNA features for predicting Cas9 efficacy as well as targeting specificity. We focus on features derived from datasets utilizing a U6 promotor, which is the most commonly used promotor for gRNA screening. We also present selected computational tools for gRNA design and describe the strategies used to produce predictions.

2 Cas9 Efficiency

Many sgRNA features modulating Cas9 activity have been discovered from large-scale gRNA efficiency screens. The following sections discuss the most significant and consistent features affecting cleavage efficiency. A summary of the discussed features can be found in Table 1.

2.1 Protospacer Adjacent Motif (NGG)

The protospacer adjacent motif (PAM) is a short sequence in the target DNA, immediately downstream of the gRNA matching sequence. PAMs have been studied extensively, as they are required in DNA interrogation by Cas9 as well as in triggering endonuclease activity [16, 17]. The canonical PAM is an NGG trinucleotide sequence, i.e., any nucleotide followed by two guanines (G) [4]. Specific interactions between Cas9 and the PAM are required for binding of target DNA and local strand separation prior to cleavage [17]. Even a single mismatch in the PAM results in at least 74% reduced Cas9 activity, so gRNAs should always be designed to target sequences containing an NGG PAM [11]. While the identity of the N is flexible, studies have reported that cytosines (C) are favored and thymines (T) are disfavored in this position [6, 10, 12]. An extended PAM may also be considered by including the +1 nucleotide (Fig. 1). In this additional position, T is favorable and G is unfavorable.

Table 1
sgRNA features that are related to on-target knockout activities

Enriched features	Depleted features
Nucleotide content	
A count	U, G count
A in the middle	GG count
AG, CA, AC, UA	GGG count
	UU, GC count
Position-specific nucleotides	
G in position 20	U in positions 17–20
A in position 20	C in position 20
Purines in position 19	G in position 16
C in position 18	T in PAM
C in position 16	G in +1 position
C in PAM T in +1 position	
Thermodynamic and structural features	
GC in first ten bases	GC >80% or GC <35%
GC in positions 4–8	Stable self-folding
Low Tm in middle of gRNA	Stable DNA/RNA duplex
GC in positions 15–20	
Accessibility in positions 18–20	
Genomic context	
Target near 5′ end of coding seq	Targets in 5′ or 3′ UTR
Target chromatin more open	Targets in introns
	Target methylated
Motifs	
NGG PAM	poly-N, TT-motif, GCC-motif

Tm melting temperature

2.2 gRNA Sequence Motifs

Sequence motifs in the gRNA which prevent efficient Cas9 editing have also been identified. Certain motifs may interfere at the stage of gRNA synthesis or transcription as opposed to direct interference with Cas9 cleavage. Therefore, researchers should carefully consider whether the following properties apply to the experimental methods used, especially for RNA polymerase-specific effects. Decreased efficiency in oligo synthesis commonly occurs when there are repetitive nucleotides [6]. Specifically, cleavage efficiency is significantly decreased when the gRNA contains a poly-N motif, defined as five contiguous A, five contiguous C, four contiguous G, or four contiguous uracils (U). In addition to oligo synthesis effects, a successfully synthesized GGGG sequence may form a guanine tetraplex structure which interferes with target recognition [18]. On the other hand, a UUUU motif can also interfere with transcription by RNA polymerase III which terminates upon recognition of a repetitive U sequence [19]. Edit efficiency is also decreased when the 3′ end of the target sequence has a "TT motif," defined as either a three pyrimidine sequence containing a TT dinucleotide or a four pyrimidine sequence containing at least two Ts [20]. This could potentially be related to the tetra-T sequence at the 5′ end of the scaffold sequence. The TT motif could extend the T-rich sequence, resulting in a transcription termination signal and decreased gene editing. A T5A mutated scaffold can be used to overcome suppressed transcription and rescue Cas9 activity in this case. At the same position, a "GCC motif," which is defined as DGCC or GCCT, decreases efficiency at the Cas9 targeting stage irrespective of transcription or Cas9 loading.

2.3 Nucleotide Composition

Nucleotide composition counts have revealed significant differences between functional and nonfunctional gRNAs. Adenine (A) content has the most significant mononucleotide composition bias, with functional gRNAs containing a greater number of As [6]. This is largely due to a regional preference in As toward the middle third of the gRNA, around positions 9–16 [5, 7, 10]. The higher A content in this region corresponds to greater sgRNA affinity to Cas9, as shown in Cas9 loading assays [7]. U and G content tended to be significantly less in functional gRNAs, although these differences were relatively less significant as compared to the overall preference for As. Dinucleotide and trinucleotide content can also be indicative of gRNA performance. The most significant dinucleotide and trinucleotides are GG and GGG, respectively, which are both depleted in functional gRNAs. Consistent with the mononucleotide biases, UU and GC dinucleotides are disfavored in functional gRNAs while the significantly favored dinucleotides all contain an A, namely, AG, CA, AC, and UA. Interestingly, there is no significant preference for AA, suggesting that the bias for A in the middle of the gRNA requires some degree of nucleotide heterogeneity.

2.4 Position-Specific Nucleotide Composition

Nucleotide usage at each individual position reveals conserved positions that are critical for efficient editing. Nucleotides proximal to the PAM sequence tend to be the most impactful, consistent with the observation that this region binds to the target sequence first during Cas9 target interrogation [16]. This feature-rich region is known as the "seed region" and is defined as positions 16–20 of the gRNA [21]. In addition to the importance of the seed region in target interrogation, nucleotides affecting Cas9 activity in the seed region are consistent with features affecting Cas9 loading [7]. Consistent with the TT motif detailed earlier, U nucleotides are disfavored at each of the four positions proximal to the PAM [6, 10, 12]. The most significant nucleotide is in position 20 at the 3′ end of the gRNA, where G is strongly favorable and C is strongly unfavorable [6, 10, 13]. A is also favorable, indicating an overall preference for purines in position 20. There is a similar preference for purines in position 19. Position 18 is the DNA cleavage site of Cas9 and there is a consistent enrichment of C in high efficiency gRNAs [5, 12]. In position 16, there is a similar trend in which C is preferred and G is disfavored. However, positional nucleotide preference in position 17 is in discordance, with G preferred in some datasets and disfavored in others. This may suggest that position 17 preferences are more susceptible to the variations in experimental conditions.

2.5 Thermodynamic Features

Thermodynamic features can affect gRNA secondary structure and Cas9 and target binding affinities, as well as properties of the target sequence such as those modulating the required DNA melting process. GC content can be used as a measure of thermodynamic stability due to the additional hydrogen bond in a GC pair compared to an AT pair. The overall consensus is that gRNAs with extreme GC content tend to be nonfunctional and a gRNA design should include between 35 and 80% GC content [6, 7, 10]. Regional GC content could also be important for Cas9 activity due to local interactions of the gRNA. When considering GC content for 10-nucleotide sequences, the strongest predictor of gRNA efficiency was in the first ten bases where better gRNAs had a higher GC content [9]. At a higher resolution of five nucleotides, higher GC content in positions 4–8 was most indicative of higher activity. In contrast, there is a preference for lower melting temperature in the middle of the gRNA, possibly related to the enrichment in A nucleotides noted earlier [22]. Finally, the six nucleotides proximal to the PAM favor higher GC content which is consistent with the strongest position-specific nucleotide preferences [23].

Another measure of thermodynamic stability can be observed by calculating free energy levels of the guide sequence. The self-folding free energy can be used to evaluate gRNA stability, which is indicative of the tendency to form secondary RNA structures through intramolecular interactions. Nonfunctional gRNAs tended to have a more negative self-folding free energy, corresponding to a greater propensity for self-folding compared to functional gRNAs [6] (Fig. 2a). This suggests that high gRNA stability is unfavorable for Cas9 activity. On the other hand, stability of the DNA/RNA hybrid in the R loop formation can also impact Cas9 activity. Strong DNA/RNA duplex binding is indicative of lower editing efficiency (Fig. 2b).

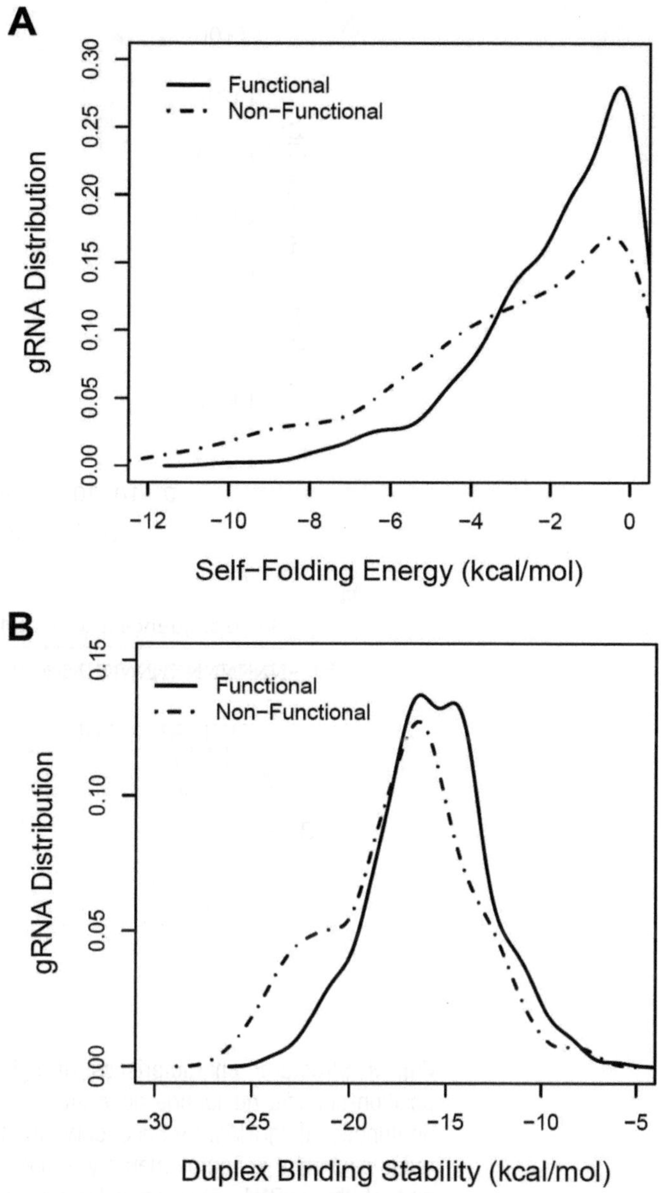

Fig. 2 Thermodynamic properties of the guide sequence (gRNA). Functional and nonfunctional gRNAs were compared in the analysis. (**a**) Structural stability of the gRNA as evaluated by self-folding free energy (ΔG). (**b**) Structural stability of the gRNA/target sequence duplex as evaluated by free energy calculation

2.6 Structural Features

RNA secondary structure is important in many biological processes, since the RNA structure can determine the nucleotide accessibility and resulting interactions at each locus. Conserved structural motifs within the sgRNA allow proper Cas9 loading as well as endonuclease activation [24]. Changes in the variable gRNA sequence can alter the sgRNA secondary structure, potentially

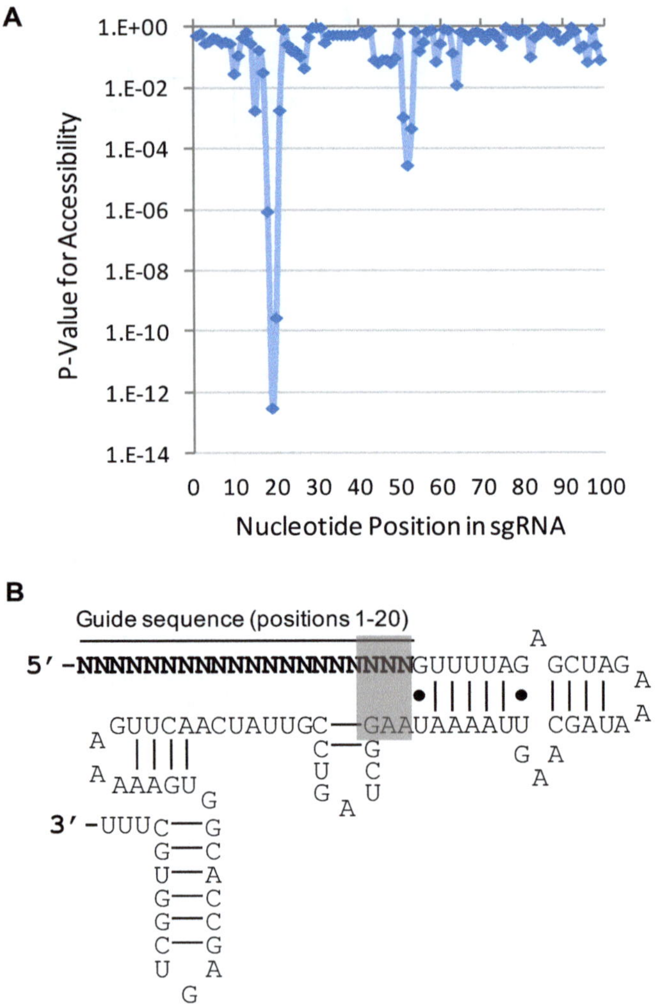

Fig. 3 Structural characteristics of sgRNAs. (**a**) Statistical significance of position-specific nucleotide accessibility of functional sgRNAs compared with nonfunctional sgRNAs. (**b**) Secondary structure of the sgRNA. The 20-nucleotide guide sequence is complementary to the target sequence and resides at the 5′ end of the sgRNA. The highlighted nucleotides could potentially base-pair, leading to an extended stem-loop structure

impacting Cas9 activity. In silico RNA folding has shown that functional gRNAs had significantly greater accessibility at positions 18–20 as well as positions 51–53 of the sgRNA [6] (Fig. 3a). These positions correspond to the 3′ end of the gRNA seed region and a region of the scaffold RNA respectively. Interestingly, one of the conserved motifs in the sgRNA is a hairpin loop between positions 21 and 50, aligning positions 18–20 with 51–53 in an antiparallel configuration (Fig. 3b). Base-pairing can occur between bases 19–20 and 52–53 if the sequences are complementary, resulting in decreased accessibility of the gRNA. This alteration in secondary structure could impede gRNA seed interactions with target or scaffold interactions with Cas9, resulting in the observed decrease in efficiency. Since the scaffold sequence in positions 52–53 is AA, the T nucleotides at the 3′ end of the gRNA could lead to additional base-pairing, resulting in decreased accessibility. This structure-related feature may explain the observed depletion in T at the PAM proximal end.

2.7 Genomic Context

In addition to the properties of the gRNA and target sequence, the molecular environment in which gene editing occurs could also affect Cas9 activity. Since editing success is ultimately measured in alteration of gene expression, it is also important to consider whether cleavage at a specific genomic locus is likely to functionally disrupt the gene. These concepts can be investigated by considering genomic and epigenomic biases in Cas9 assays.

One basic consideration is whether there are preferred or disfavored loci within a gene to be edited. Targeting the last coding exon of a protein coding gene is unfavorable compared to targeting earlier exons, due to the fact that a frameshift mutation originating from the 5′ end is more likely to disrupt protein function compared to alteration of a single exon [7, 10]. In addition, it is highly ineffective to target 5′- or 3′-untranslated regions (UTRs). Guides targeting exon-intron boundaries to disrupt splicing can be advantageous, but generally intronic regions should be avoided since targets 6 nt or more away from the boundary are ineffective. Since these protein-coding exon considerations do not apply to noncoding RNAs, an effective strategy is to focus on targeting functional RNA motifs such as the seed sequence or stem-loop structure of miRNAs [25, 26].

Epigenetic features of target sequences are independent of any sgRNA features. These additional features may account for differences in gRNA efficiency in distinct cell populations. Based on CTCF binding, histone modification, and chromatin-opening information, open chromatin is preferred for efficient editing [12]. This is consistent with the finding that target interrogation of Cas9 is slower in heterochromatin compared to euchromatin [27, 28]. In addition, recent studies have observed that DNA sequences with CpG methylation tend to be less efficient Cas9 targets [12, 21]. Cas9 activity was previously thought to be unaffected by methylation and the mechanism governing the observed bias against methylation is unknown [14, 23]. Further studies are thus needed to reconcile these differences and provide mechanistic explanations.

3 Cas9 Specificity

Off-target activity has been found to be associated with the CRISPR/Cas9 system, but the extent of nonspecific editing and degree of functional consequences vary across experimental settings [29]. Although Cas9 efficiency is low in off-target editing, they can produce false hits in genetic screens or add confounding variables to experimental studies. As gene therapies are increasingly gaining attention, there is also a safety concern regarding unintended gene editing effects. For these reasons, it is important to consider potential off-targets of each candidate gRNA.

Cas9 off-target activity is found in sequences which are similar to the intended target. Off-target profiles can thus be constructed by surveying for sequence similarity within the genome. An NGG PAM is considered a prerequisite for Cas9 editing; therefore, all 20-nt sequences adjacent to an NGG PAM should be considered. In addition, some alternative PAM sequences may be tolerated and should be considered, especially NAG, NCG, and NGA [11]. Cas9 may tolerate up to six mismatches in the 20-nt target sequence, depending on the position as well as the mismatched nucleotide identity [15]. The propensity for off-target activity directly corresponds to the number of mismatches, with two mismatches substantially reducing activity and three interspaced mismatches or five consecutive mismatches completely eliminating activity in most cases [14]. Overall, single-base rG:dT mismatches are tolerated across the entire gRNA sequence, while rC:dC mismatches result in the least activity. There is a greater tolerance for mismatches with the PAM distal end of the gRNA, as expected with the notion of a seed region proximal to the PAM. Nevertheless, 41% of purine-purine mismatches and 48% of T-T mismatches at position 19 resulted in off-target editing, highlighting the exceptions to the seed region model. In addition to symmetric mismatches, asymmetric mismatches which create RNA or DNA bulges must be considered separately. Bulges in the gRNA sequence can be tolerated only at positions 2 and 3. In contrast, DNA bulges can be tolerated in positions 19 and 20 in addition to positions 2 and 3. An exception to this is that in positions 19 and 20, a guanine DNA bulge rarely preserves activity.

As off-target profiling requires feature analysis of a massive number of sequences on a genome-wide level, it can be useful to restrict the search space to regions which are susceptible to gene editing. One approach is to only consider potential off-targets in coding regions, because Cas9 activity in noncoding regions is less likely to produce functional consequences [6, 10]. The greatly reduced search space allows for more robust computations as well as more stringent gRNA selection criteria.

4 Computational Tools

With the large number of sgRNA features affecting both efficiency and specificity of Cas9, coupled with the wide search space for off-targets in the genome, it is useful to employ a computational tool to maximize the probability of selecting viable gRNAs. Accurate predictions save time and money which would be otherwise spent on repeatedly testing each gRNA. The choice of tool to be used may depend on experimental design restrictions as well as desired prediction accuracy level. Tools using machine learning models which are trained on data similar to the testing environment are expected to produce better designs. Notably, there is a substantial difference in performance between models trained on gRNAs transcribed using the U6 promoter compared to models trained on gRNAs transcribed using the T7 promoter, even though many of the gRNA features are consistent [30]. In order to accommodate researchers using T7 promotor constructs, we include a tool trained on a T7 dataset in addition to the U6-based tools presented. The selected sgRNA design tools are described below.

4.1 WU-CRISPR

WU-CRISPR provides a precompiled gRNA library with both on-target efficiency and off-target profiles for up to five target sequences in all human and mouse genes [6]. Users can also enter custom sequences for additional on-target activity predictions. The underlying SVM learning model for gRNA selection was trained using data from a large pool of gRNAs whose activities were experimentally determined. Sequence features, thermodynamic features, and structural accessibility features were combined and used to train the model. In addition, WU-CRISPR employs a prefiltering process which excludes gRNAs with features that are highly associated with nonfunctional gRNAs. For off-target profiling, WU-CRISPR scans functional genomic regions for 20-nt sequence preceding an NGG or NAG PAM and labels sequences with a seed match or (>85%) similarity to the gRNA as potential off-targets. By utilizing stringent criteria for gRNA selection, WU-CRISPR prediction is highly precise while sacrificing the recall rate for all potent gRNAs. This is practical for usage as a design tool, as in practice the focus is on selecting a handful of functional gRNAs from a large number of candidates. WU-CRISPR is available at http://crispr.wustl.edu/.

4.2 CRISPR ML

CRISPR ML utilizes machine learning models from two separate studies to provide both efficiency and specificity predictions. The on-target model "Azimuth" uses a gradient-boosted regression tree model trained using two combined gRNA datasets [11]. Features considered in the model include sequence features, thermodynamic features, and location of the target within a gene. The off-target model "Elevation" uses a two-layer regression model to score each potential off-target including sequences containing NAG, NCG, NGA, NGC, NGG, NGT, or NTG PAM [15]. Training data for this model included a large pool of sequences with single-base mismatch from on-target gRNAs. Selection of gRNA to minimize off-target activity is then performed using a model based on gradient-boosted regression trees. Features considered by the first layer model include position and identity of each mismatch and whether the mismatch was a transition or a transversion. CRISPR ML is available at https://crispr.ml/.

4.3 CRISPRscan

CRISPRscan provides gRNA designs for multiple species, including human, mouse, zebrafish, and yeast [8]. On-target scoring is performed by the CRISPRscan algorithm which uses a linear regression model trained using sequence features from a large gRNA dataset. These gRNAs were transcribed in zebrafish embryos using a T7 promoter, making CRISPRscan particularly powerful for gRNA design in experiments using the T7 expression system. The off-target profile is computed based on a combination of heuristic rules and a separate machine learning model. Sequences with perfect gRNA seed match or up to two mismatches overall are considered off-targets. In addition, a CFD score is computed based on observed off-target frequencies for each type of single-base mismatch in each position [11]. CRISPRscan is available at http://www.crisprscan.org/.

5 Conclusions

Many features modulating either on-target or off-target activity have been identified, but the mechanisms driving many of these features remain unclear, making it difficult to confidently predict the precise activity profile of a gRNA. A successful way to overcome this challenge has been to utilize machine learning models, which can capture higher-order relationships within a large set of training data and generate predictions with high performance. However, gRNA design tools are not without limitations. Despite various strategies to reduce overtraining, prediction tools tend to perform worse under experimental conditions that are different from the training set. Target-specific features not captured in data used for training can also influence activity. At the current stage in Cas9 research, design tools cannot guarantee with certainty the functionality of predicted designs. Computational tools should thus be used in conjunction with experimental validation of gRNA activity in order to maximize gene editing success.

References

1. Doudna JA, Charpentier E (2014) Genome editing. The new frontier of genome engineering with CRISPR-Cas9. Science 346 (6213):1258096. https://doi.org/10.1126/science.1258096
2. Barrangou R, Fremaux C, Deveau H, Richards M, Boyaval P, Moineau S, Romero DA, Horvath P (2007) CRISPR provides acquired resistance against viruses in prokaryotes. Science 315(5819):1709–1712. https://doi.org/10.1126/science.1138140
3. Deltcheva E, Chylinski K, Sharma CM, Gonzales K, Chao Y, Pirzada ZA, Eckert MR, Vogel J, Charpentier E (2011) CRISPR RNA maturation by trans-encoded small RNA and host factor RNase III. Nature 471 (7340):602–607. https://doi.org/10.1038/nature09886
4. Jinek M, Chylinski K, Fonfara I, Hauer M, Doudna JA, Charpentier E (2012) A programmable dual-RNA-guided DNA endonuclease in adaptive bacterial immunity. Science 337 (6096):816–821. https://doi.org/10.1126/science.1225829
5. Xu H, Xiao T, Chen CH, Li W, Meyer CA, Wu Q, Wu D, Cong L, Zhang F, Liu JS, Brown M, Liu XS (2015) Sequence determinants of improved CRISPR sgRNA design. Genome Res 25(8):1147–1157. https://doi.org/10.1101/gr.191452.115
6. Wong N, Liu W, Wang X (2015) WU-CRISPR: characteristics of functional guide RNAs for the CRISPR/Cas9 system. Genome Biol 16:218. https://doi.org/10.1186/s13059-015-0784-0
7. Wang T, Wei JJ, Sabatini DM, Lander ES (2014) Genetic screens in human cells using the CRISPR-Cas9 system. Science 343 (6166):80–84. https://doi.org/10.1126/science.1246981
8. Moreno-Mateos MA, Vejnar CE, Beaudoin JD, Fernandez JP, Mis EK, Khokha MK, Giraldez AJ (2015) CRISPRscan: designing highly efficient sgRNAs for CRISPR-Cas9 targeting in vivo. Nat Methods 12(10):982–988. https://doi.org/10.1038/nmeth.3543
9. Labuhn M, Adams FF, Ng M, Knoess S, Schambach A, Charpentier EM, Schwarzer A, Mateo JL, Klusmann JH, Heckl D (2018) Refined sgRNA efficacy prediction improves large- and small-scale CRISPR-Cas9 applications. Nucleic Acids Res 46(3):1375–1385. https://doi.org/10.1093/nar/gkx1268
10. Doench JG, Hartenian E, Graham DB, Tothova Z, Hegde M, Smith I, Sullender M, Ebert BL, Xavier RJ, Root DE (2014) Rational design of highly active sgRNAs for CRISPR-Cas9-mediated gene inactivation. Nat Biotechnol 32(12):1262–1267. https://doi.org/10.1038/nbt.3026
11. Doench JG, Fusi N, Sullender M, Hegde M, Vaimberg EW, Donovan KF, Smith I, Tothova Z, Wilen C, Orchard R, Virgin HW, Listgarten J, Root DE (2016) Optimized sgRNA design to maximize activity and minimize off-target effects of CRISPR-Cas9. Nat Biotechnol 34(2):184–191. https://doi.org/10.1038/nbt.3437
12. Chuai G, Ma H, Yan J, Chen M, Hong N, Xue D, Zhou C, Zhu C, Chen K, Duan B, Gu F, Qu S, Huang D, Wei J, Liu Q (2018) DeepCRISPR: optimized CRISPR guide RNA design by deep learning. Genome Biol 19 (1):80. https://doi.org/10.1186/s13059-018-1459-4
13. Chari R, Mali P, Moosburner M, Church GM (2015) Unraveling CRISPR-Cas9 genome engineering parameters via a library-on-library approach. Nat Methods 12(9):823–826. https://doi.org/10.1038/nmeth.3473
14. Hsu PD, Scott DA, Weinstein JA, Ran FA, Konermann S, Agarwala V, Li Y, Fine EJ, Wu X, Shalem O, Cradick TJ, Marraffini LA, Bao G, Zhang F (2013) DNA targeting specificity of RNA-guided Cas9 nucleases. Nat Biotechnol 31(9):827–832. https://doi.org/10.1038/nbt.2647
15. Listgarten J, Weinstein M, Kleinstiver BP, Sousa AA, Joung JK, Crawford J, Gao K, Hoang L, Elibol M, Doench JG, Fusi N (2018) Prediction of off-target activities for the end-to-end design of CRISPR guide RNAs. Nat Biomed Eng 2(1):38–47. https://doi.org/10.1038/s41551-017-0178-6
16. Sternberg SH, Redding S, Jinek M, Greene EC, Doudna JA (2014) DNA interrogation by the CRISPR RNA-guided endonuclease Cas9. Nature 507(7490):62–67. https://doi.org/10.1038/nature13011
17. Anders C, Niewoehner O, Duerst A, Jinek M (2014) Structural basis of PAM-dependent target DNA recognition by the Cas9 endonuclease. Nature 513(7519):569–573. https://doi.org/10.1038/nature13579
18. Sen D, Gilbert W (1988) Formation of parallel four-stranded complexes by guanine-rich motifs in DNA and its implications for meiosis. Nature 334(6180):364–366. https://doi.org/10.1038/334364a0

19. Nielsen S, Yuzenkova Y, Zenkin N (2013) Mechanism of eukaryotic RNA polymerase III transcription termination. Science 340 (6140):1577–1580. https://doi.org/10.1126/science.1237934
20. Graf R, Li X, Chu VT, Rajewsky K (2019) sgRNA sequence motifs blocking efficient CRISPR/Cas9-mediated gene editing. Cell Rep 26(5):e1093–e1103. https://doi.org/10.1016/j.celrep.2019.01.024
21. Wu X, Scott DA, Kriz AJ, Chiu AC, Hsu PD, Dadon DB, Cheng AW, Trevino AE, Konermann S, Chen S, Jaenisch R, Zhang F, Sharp PA (2014) Genome-wide binding of the CRISPR endonuclease Cas9 in mammalian cells. Nat Biotechnol 32(7):670–676. https://doi.org/10.1038/nbt.2889
22. Peng H, Zheng Y, Blumenstein M, Tao D, Li J (2018) CRISPR/Cas9 cleavage efficiency regression through boosting algorithms and Markov sequence profiling. Bioinformatics 34 (18):3069–3077. https://doi.org/10.1093/bioinformatics/bty298
23. Ren X, Yang Z, Xu J, Sun J, Mao D, Hu Y, Yang SJ, Qiao HH, Wang X, Hu Q, Deng P, Liu LP, Ji JY, Li JB, Ni JQ (2014) Enhanced specificity and efficiency of the CRISPR/Cas9 system with optimized sgRNA parameters in Drosophila. Cell Rep 9(3):1151–1162. https://doi.org/10.1016/j.celrep.2014.09.044
24. Jiang F, Doudna JA (2017) CRISPR-Cas9 structures and mechanisms. Annu Rev Biophys 46:505–529. https://doi.org/10.1146/annurev-biophys-062215-010822
25. Zhao Y, Dai Z, Liang Y, Yin M, Ma K, He M, Ouyang H, Teng CB (2014) Sequence-specific inhibition of microRNA via CRISPR/CRISPRi system. Sci Rep 4:3943. https://doi.org/10.1038/srep03943
26. Jiang Q, Meng X, Meng L, Chang N, Xiong J, Cao H, Liang Z (2014) Small indels induced by CRISPR/Cas9 in the 5′ region of microRNA lead to its depletion and Drosha processing retardance. RNA Biol 11(10):1243–1249. https://doi.org/10.1080/15476286.2014.996067
27. Knight SC, Xie L, Deng W, Guglielmi B, Witkowsky LB, Bosanac L, Zhang ET, El Beheiry M, Masson JB, Dahan M, Liu Z, Doudna JA, Tjian R (2015) Dynamics of CRISPR-Cas9 genome interrogation in living cells. Science 350 (6262):823–826. https://doi.org/10.1126/science.aac6572
28. Horlbeck MA, Witkowsky LB, Guglielmi B, Replogle JM, Gilbert LA, Villalta JE, Torigoe SE, Tjian R, Weissman JS (2016) Nucleosomes impede Cas9 access to DNA in vivo and in vitro. elife 5. https://doi.org/10.7554/eLife.12677
29. Hartenian E, Doench JG (2015) Genetic screens and functional genomics using CRISPR/Cas9 technology. FEBS J 282 (8):1383–1393. https://doi.org/10.1111/febs.13248
30. Haeussler M, Schonig K, Eckert H, Eschstruth A, Mianne J, Renaud JB, Schneider-Maunoury S, Shkumatava A, Teboul L, Kent J, Joly JS, Concordet JP (2016) Evaluation of off-target and on-target scoring algorithms and integration into the guide RNA selection tool CRISPOR. Genome Biol 17(1):148. https://doi.org/10.1186/s13059-016-1012-2

Chapter 20

CRISPR-Cas9 Genome Editing in Human Cell Lines with Donor Vector Made by Gibson Assembly

Nirakar Sahoo, Victoria Cuello, Shreya Udawant, Carl Litif, Julie A. Mustard, and Megan Keniry

Abstract

CRISPR Cas9 genome editing allows researchers to modify genes in a multitude of ways including to obtain deletions, epitope-tagged loci, and knock-in mutations. Within 6 years of its initial application, CRISPR-Cas9 genome editing has been widely employed, but disadvantages to this method, such as low modification efficiencies and off-target effects, need careful consideration. Obtaining custom donor vectors can also be expensive and time-consuming. This chapter details strategies to overcome barriers to CRISPR-Cas9 genome editing as well as recent developments in employing this technique.

Key words CRISPR, Gibson assembly, Gene editing, FOXO3

1 Introduction

CRISPR Cas9 and related mechanisms are endogenous adaptive immune responses found widely in bacteria and archaea to fend off invading DNAs, such as those from viruses [1–8]. Short segments of invading DNA are inserted into a specialized bacterial chromosomal locus containing repetitive DNA elements, the clustered regularly interspaced short palindromic repeats (CRISPR) [9–15]. The resulting transcribed RNA is processed into several short RNAs (crRNAs) that are incorporated into a riboprotein complex [11–15]. The riboprotein complex then targets the foreign nucleic acids in a sequence-specific manner due to the presence of the crRNAs. CRISPR-associated proteins (Cas proteins) include those involved in transcription, RNA processing, and formation of the riboprotein complex. Researchers ingeniously co-opted this bacterial mechanism to devise heterologous systems to edit the genome of nearly any organism at will [16–22].

Mouldy Sioud (ed.), *RNA Interference and CRISPR Technologies: Technical Advances and New Therapeutic Opportunities*, Methods in Molecular Biology, vol. 2115, https://doi.org/10.1007/978-1-0716-0290-4_20,

1.1 Endogenous CRISPR Cas9 System from Bacteria

Approximately 50% of bacteria and 90% of archaea harbor an adaptive CRISPR system to mount an immunological response against phage (viruses) and other mobile genetic elements [23–25]. So far, at least six subtypes of CRISPR systems have been identified. First discovered by Ishino and colleagues in 1987, many variations of these systems have been identified and characterized [26]. At the forefront of this field are exciting, emerging issues. Chief among these is the ability of some bacterial phage to counter the CRISPR immune response by expressing anti-CRISPR proteins that prevent DNA binding and Cas3 activity [27–29]. Additional functional roles for CRISPR in bacteria include promoting genome evolution, bacterial virulence, and DNA repair [30–33].

1.2 Different Types of CRISPR

Numerous distinctive CRISPR/Cas systems have been discovered that show incredible diversity regarding genomic locus architecture and gene composition. This diversity in the CRISPR/Cas system helps prokaryotes accurately identify and neutralize invading bacteriophages. The most recent approach to classification of CRISPR/Cas systems includes two main subclassifications based on their effector modules (Fig. 1) [34, 35].

Class 1 systems include approximately 90% of the CRISPR-Cas loci of prokaryotes and employs multi-subunit effector complexes: types I, III, and the rare type IV. Type I systems contain the Cas-3 nuclease helicase; type III systems possess the signature gene Cas10. Type IV systems include an uncharacterized large subunit Csfl (Fig. 1) [34, 35]. Class 2 systems comprise ~10% of the CRISPR-Cas loci and contain single-subunit effector complexes, including types II, V, and VI. The type II system has a signature Cas9 nuclease. Type V systems consist of either Cas12a, Cas12b, or

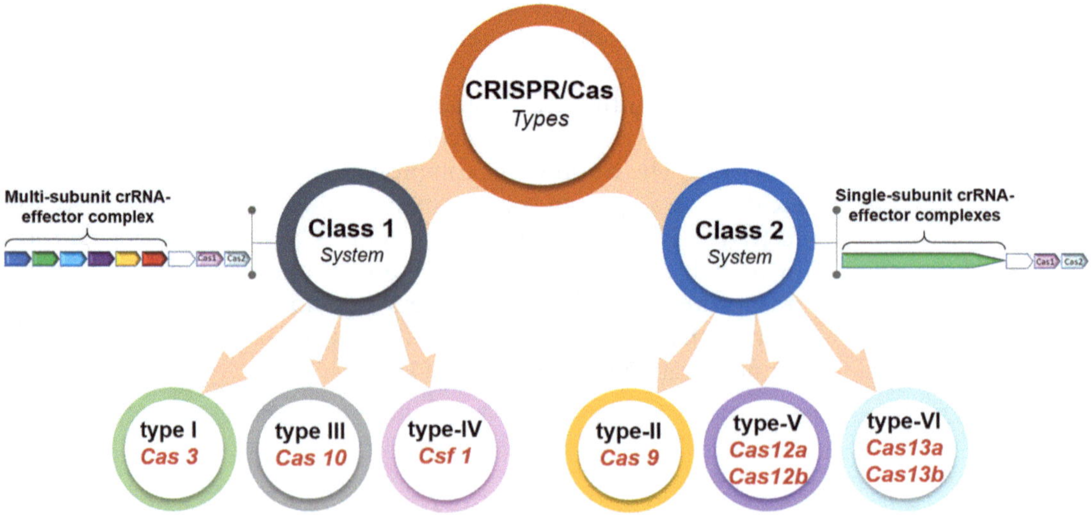

Fig. 1 Classification of CRISPR systems. CRISPR Class 1 and 2 systems are depicted

Cas12C, which are like Cas9. Type VI systems possess either Cas13a, Cas13b, or Cas13c (Fig. 1). Apparently, the class 2 systems convey more potential in gene editing and enhancing genetic assessment by using Cas9, Cas12a, Cas13a, and Cas13b [36].

The CRISPR Cas9 system remains the foremost and most extensively used for genetic engineering. In this system, the Cas9 protein is involved in processing the transcript into the mature crRNA [36]. Processing requires binding of a short trans-activating RNA (tracrRNA), which is complementary to the repeat sequence found in type II CRISPR systems. After processing the transcript, the Cas9 protein remains bound to the crRNA and the tracrRNA, ultimately leading the complex to cleave a target sequence.

1.3 Heterologous CRISPR Cas9 System Basics

CRISPR Cas9 genome editing using heterologous systems was pioneered by researchers including Dr. Doudna (UC Berkeley), Dr. Charpentier (currently at Max Planck), Dr. Church (Harvard), and Dr. Zhang (Broad Institute) [16, 19–22, 37]. In a breakthrough study, Drs. Doudna and Charpentier were the first scientists to show that a chimeric dual-tracrRNA:crRNA containing the sequence of a desired target could be engineered to direct the Cas9 nuclease to produce double-strand breaks in a sequence-specific manner [16]. This ground-breaking study ended with the statement *...could offer considerable potential for gene targeting and genome editing applications* [37]. In 2013, the employment of CRISPR Cas9 technology to genome edit in human cells (by Dr. Zhang and colleagues) using heterologous systems became a reality [19]. These systems utilized chimeric guide RNAs to target a desired sequence and an engineered human codon-optimized *S. pyogenes Cas9* (hSpCas9) in HEK 293FT cells [19]. Nuclear localization sequences were fused to hSpCas9 to ensure DNA access [19]. CRISPR heterologous systems are continually being refined to increase efficiency and reduce off-target effects in a vast array of biological systems from *C. elegans* to humans [9].

2 Materials

2.1 PCR to Amplify Gibson Assembly Components for Donor Vector Construction

1. Phusion DNA polymerase kit for PCR.
   - Phusion DNA polymerase (2 U/μL).
   - 5× Phusion HF buffer.
   - DMSO.
   - 50 mM $MgCl_2$ solution.
2. 25 mM stock dNTP.
3. UltraPure DNase/RNase-Free Distilled Water.
4. Primers obtained from Sigma (Table 1).
5. Template: CMV5-*FOXO3* prepared by miniprep kit.

Table 1
PCR primers utilized to amplify two separate *FOXO3* gene fragments (Arm 1 and Arm 2) for donor vector Gibson assembly reactions

Primer	Sequence*
FOXO3 ARM 1 F	5'- GTGCTTTACGGCACCTCGACCCCCCGGCACAACCTGTCACTGC-3'
FOXO3 ARM 1 R	5'- CCGTCTATCAGGGCGATGGCCGCTGTAGAGCATGGGCGAGAG-3'
FOXO3 ARM 2 F	5'- CAAATAAAGCATTTTTTTCACTCGGTGGAACTGCCACGGCTG-3'
FOXO3 ARM 2 R	5'- GAGTTTGGACAAACCACAACTAGGTCCAAGTCGCTGGGGAAC-3'

[a]Nucleotides in underlined blue correspond to sequences in pcDNA3 vector . Sequences in black were to amplify the indicated *FOXO3* arm [9]

2.2 Cloning Enzymes and E. coli Cells

1. *Dpn*I, *Dra*III, *Bsm*I.
2. Calf intestine phosphatase.
3. Gibson Assembly kit (NEB).
4. 5-Alpha competent *E. coli*.

2.3 Column Purification of DNA

1. DNA purification kit.
2. DNA Maxi Kit.

2.4 Propagation of Human Cell Lines

1. MEM (Minimal Essential Media).
2. DMEM (Dulbecco's Modified Eagle Medium).
3. RPMI (Roswell Park Memorial Institute 1640).
4. Penicillin/streptomycin mixture.
5. FBS (fetal bovine serum).
6. Trypsin (0.25%).

2.5 Transfection of Human Cell Lines

1. Scepter cell counter; can also use hemocytometer.
2. Nucleofector V solution (Lonza, Cologne, Germany).
3. *FOXO3* guide RNA plasmid (Sigma: HSL0001339461).
4. *FOXO3* guide RNA plasmid (Sigma: HSR0001339464).
5. CRISPR Cas9D10A-GFP Nickase vector (Sigma).

2.6 Selection and Isolation of Neomycin-Resistant Clones

1. G418.
2. Vaseline.
3. Cloning cylinders.
4. 1× PBS buffer.

2.7 Western Blot Analysis

1. Cell scraper.
2. 2× sample buffer: 125 mM Tris–HCL at pH 6.8, 2% sodium dodecyl sulfate (SDS), 10% 2-mercaptoethanol, 20% glycerol, 0.05% bromophenol blue, 8 M urea, all prepared with Ultra-Pure Distilled Water.
3. Polyacrylamide gel Mini-PROTEAN TGX 4-20, 10 W (Bio-Rad).
4. Protein ladder.
5. Polyacrylamide gel running buffer: 25 mM Tris-base, 190 mM glycine, 0.1% SDS.
6. Transfer buffer: 25 mM Tris-base, 190 mM glycine, 10% methanol.
7. PVDF (polyvinylidene fluoride) membrane.
8. TBS-T: 20 mM Tris pH 7.5, 150 mM NaCl, 0.1% Tween 20.
9. Carnation nonfat dry milk.
10. FOXO3 antibody (cell signaling).
11. GAPDH antibody (Santa Cruz Biotechnology).
12. Goat-anti-rabbit secondary antibody for FOXO3 blots (Fisher/Pierce).
13. Goat-anti mouse secondary antibody for GAPDH blots (Fisher/Pierce).
14. SuperSignal West Dura Extended Duration Substrate luminol (Fisher/Pierce).

3 Methods

3.1 Protocol to Disrupt FOXO3 Using CRISPR Cas9 with a Custom Donor Vector

The protocol highlighted in this chapter was published (in less detail) in Open Access BMC Molecular Biology; parts of this protocol are reproduced with permission from the initial journal [9]. This method employs a modified Cas9 nickase (Cas9D10A) purchased from Sigma (CAS9D10AGFPP). This mutant nuclease is from *Streptococcus pyogenes* and will only nick one strand of DNA. Cas9D10A retains the HNH domain activity (to cleave the target strand) and lacks the RuvC activity (which would normally cleave the nontarget strand).

3.2 Two Guide RNAs Were Employed to Target FOXO3

Guide RNAs contain a sequence resembling the repeat sequence in the CRISPR-Cas9 system plus a sequence to target the gene of interest [37]. The chimera guide RNA then takes the place of the endogenous crRNAs in binding Cas9 and targeting the nuclease to the specified sequence [37]. Guide RNAs were purchased from Sigma to nick the *FOXO3* genomic locus (on chromosome 6) 41 bases apart on opposite strands.

Sequences: We used two *FOXO3*-targetting gRNAs from Sigma: HSL0001339461, 5′-CTTACTGAAGGTGACAGGCTGG-3′ (chromosomal position: 108,663,888–108,663,909) and HSR0001339464, 5′-CACGGCTGACTGATATGGCAGG-3′ (chromosomal position 108,663,933–108,663,954). Researchers can also readily devise custom-guide RNAs using recommendations by Zhang et al. (among others, *see* Chapter 19) to target virtually any gene [38, 39].

3.3 Construction of a Donor Vector to Select for FOXO3 Disruption Mutants

To select for *FOXO3* disruption mutants, we prepared a donor vector containing the neomycin resistance gene via Gibson assembly using the commonly employed mammalian expression vector pcDNA3 as the backbone [9]. First, PCR was used to construct two "arms" containing *FOXO3* chromosomal sequences. Then, a two-step Gibson assembly cloning procedure was performed to insert the *FOXO3* arms on either side of the *NPTII* gene (encoding neomycin resistance) found in the pcDNA3 vector as illustrated in Fig. 2. The donor vector had *FOXO3* sequences of enough length to allow for homology directed repair in order to disrupt the chromosomal *FOXO3* gene. One *FOXO3* arm (termed **Arm 1**) in the donor vector included 418 bases (chromosomal position: 108663463-108,663,880) and the other *FOXO3* arm (termed **Arm 2**) had 750 bases (chromosomal position 108,663,921-108,664,670) (Fig. 2). Homology directed repair of the CRISPR Cas9D10A-induced nicks produced a truncated *FOXO3* gene in several human cell lines examined [9].

3.4 Gibson Assembly Protocol to Obtain Custom Donor Vector

Gibson assembly was developed by Daniel G. Gibson in 2009 [40]. The method joins together fragments containing identical sequences of DNA (typically ranging from 20 to 40 nucleotides long) on their ends. The first step of Gibson assembly involved 5′–3′ exonuclease activity that produced 3′ overhangs, which could then anneal to another fragment with a compatible overhang. By using PCR primers containing vector sequences, the pcDNA3 vector and *FOXO3* fragment had 3′ overhangs that could anneal. After annealing, a DNA polymerase extended from the 3′ ends to fill in missing bases. DNA ligase sealed the phosphodiester backbone [40]. In this manner two fragments with "overlapping" sequences could be joined. The amount of fragment overlap depends (in part) on the insert and vector sizes. In this protocol, an overlap of 20–30 bases was enough for efficient cloning. For larger inserts, 30–40 bases might be needed for efficient cloning. A guide to Gibson assembly can be found at: https://www.biocat.com/bc/files/Gibson_Guide_V2_101417_web_version_8.5_x_11_FINAL.pdf. We used a basic Gibson assembly reaction with three steps described above (NEB Gibson Assembly Kit E2611S); all took place at 50 °C in the span of 1 h. Specifics of the reactions including preparing the PCR products and vector are detailed in Subheading 3.5.

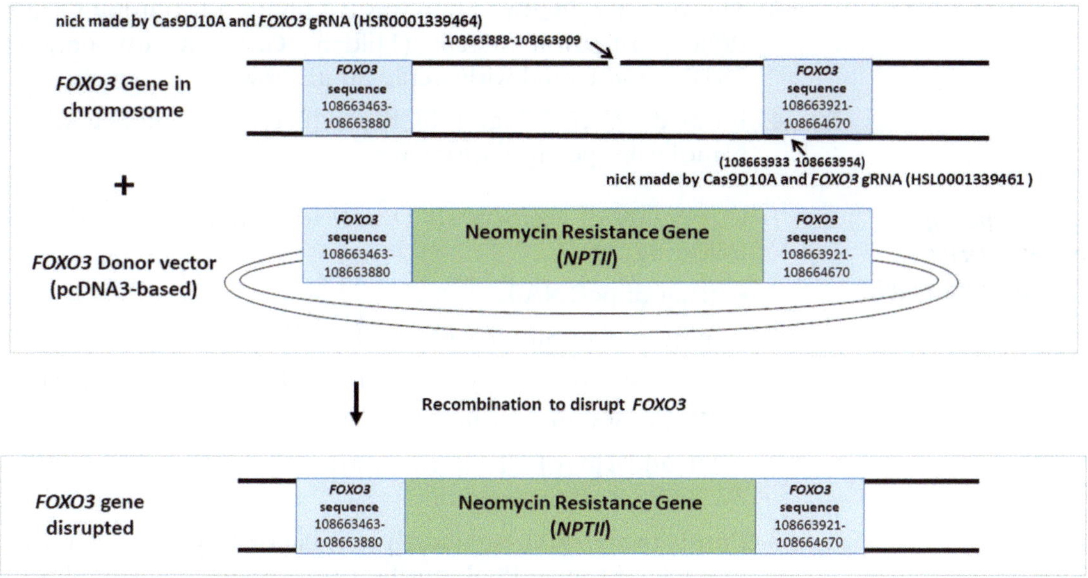

Fig. 2 Schematic of *FOXO3* gene disruption with neomycin resistance cassette (*NPTII*). Guide RNAs were employed to nick *FOXO3* gene in mammalian cell lines. The lesions were resolved by homology directed repair with a donor vector that contained a neomycin resistance gene (*NPTII*) flanked by *FOXO3* sequences (which were proximal to the CRISPR Cas9-generated nicked strands of *FOXO3* in the chromosome). This figure is adapted from Vazquez et al. [9]

3.5 PCR Reaction to Amplify FOXO3 Arm 1 to Be Inserted into pcDNA3 Vector at DraIII Site

1. Mix the following in a PCR tube.
   - 20 μL of 5× Phusion HF Buffer (Thermo-Fisher cat: F530S).
   - 8 μL 25 mM dNTP.
   - 1 μL of 50 μM forward primer (***FOXO3* ARM 1F**).
   - 1 μL of 50 μM reverse primer (***FOXO3* ARM 1 R**).
   - 1 μL of plasmid DNA template (200 ng/μL concentration contained *FOXO3* gene).
   - 3 μL of DMSO.
   - 1 μL of Phusion polymerase (Thermo-Fisher cat: F530S).
   - 59.5 μL of water.
   - 5.5 μL of 50 mM $MgCl_2$.

   Cycling parameters: 94 °C for 10 min (initial heat step done only once); then 94 °C for 1 min, 56 °C for 1 min, and 72 °C 3 min (three steps for forty cycles). One last extension of 72 °C for 20 min was followed by a hold at 4 °C until the reaction was collected.
2. After the PCR reactions were complete, add 2 μL of *Dpn*I (NEB R0176S, Ipswich, MA) to *FOXO3* Arm1 PCR reaction and digested at 37 °C for 2 h to remove the plasmid DNA template.

3. This digested PCR reaction was column purified using Qiagen PCR purification system (Hilden, Germany, cat no./ID: 28104) and eluted with sterile 30 μL of water.
4. Purified *FOXO3* Arm 1 PCR product was quantified with a NanoDrop spectrophotometer.

3.6 Preparing pcDNA3 Vector for Gibson Assembly

1. Cut pcDNA3 vector with *DraIII* for 2 h at 37 °C. Mix the following:
   - 1 μg of pcDNA3.
   - 4 μL of Cut Smart buffer (10×, NEB Ipswich, MA).
   - 3 μL of *DraIII* enzyme (NEB cat: R3510S, Ipswich, MA).
   - 27 μL of water for final volume of 40 μL.
2. After, add 1μL of CIP to the restriction digest and incubate at 37 °C for 1 additional hour.
3. Purify the *Dra*III-cut (and phosphatased) vector using, for example, Qiagen PCR purification system and eluted with sterile 30 μL of water.
4. Quantify with a NanoDrop spectrophotometer.

3.7 Gibson Assembly Reaction

This reaction consists at adding *FOXO3* Arm1 into the pcDNA3 vector at *Dra*III site.

1. Mix the following:
   - 10 μL of NEB Gibson Assembly Master Mix (E2611S NEB, Ipswich, MA).
   - 2 μL (10 ng) of prepared *Dra*III-cut pcDNA3 vector (*see* above for preparation details).
   - 8 μL of *FOXO3* Arm1 insert prepared as described above (80 ng).
2. Incubate the Gibson assembly reactions for 1 h at 50 °C.
3. Add 5 μL of the Gibson assembly reaction to 50 μL of 5-alpha competent *E. coli* and incubate for 30 min on ice.
4. Heat shock the sample at 42 °C for 30 s and place on ice for 5 min.
5. Add 250 μL of SOC recovery media to the sample and incubate for 1 h at 37 °C.
6. Prepare various dilutions of cells (1:10, 1:100, 1:1000) and plate onto LB plates with ampicillin (100 μg/mL) and incubate overnight at 37 °C. Using this protocol, hundreds of colonies were obtained and confirmed by restriction digest analysis and Sanger sequencing.

3.8 Adding Second FOXO3 Arm to CRISPR Cas9 Donor Vector

1. PCR was performed to amplify *FOXO3* Arm 2 exactly as described above (for *FOXO3* Arm 1, Subheading 3.5), except using Arm 2 primers (Table 1). The PCR product was cut with *Dpn*I to remove the template and column purified. The pcDNA3 vector that contained *FOXO3* Arm1 was prepared for Gibson assembly in order to add *FOXO3* Arm 2.
2. Digest the vector pcDNA3 *FOXO1* Arm 1 prepared in Subheading 3.7 with *BsmI* for 2 h at 65 °C.
   - 1 μg of pcDNA3 *FOXO3* Arm 1 vector.
   - 4 μL of Cut Smart buffer (10× NEB, Ipswich, MA).
   - 3 μL of *Bsm*I enzyme (NEB R0134S, Ipswich, MA).
   - 27 μL of water for final volume of 40 μL.
3. Add 1 μL CIP to the restriction digest and incubate the sample at 37 °C for 1 additional hour.
4. Purify the *Bsm*I-cut (and phosphatased) vector using, for example, Qiagen PCR purification system and elute with 30 μL of sterile water.
5. Quantify the purified vector with a NanoDrop spectrophotometer.
6. Transform *E. coli* 5-alpha cells with the Gibson assembly reaction to add *FOXO3* Arm 2 (to obtain final donor vector) as described in Subheading 3.7 for Arm1.
7. After confirmation of the desired construct by restriction digest analysis and Sanger sequencing plasmids were prepared for transfection of human cell lines by maxiprep (Qiagen Maxi Kit: 12163) according to manufacturer's instructions.

3.9 Transient Transfection of Human Cell Lines to Disrupt the FOXO3 Gene with a Neomycin Resistance Cassette

We disrupted the *FOXO3* gene in several human cell lines and compared mutation efficiencies. We chose to express Cas9D10A using transient transfections; *see* **Note 1**. Glioblastoma cell line U87MG was grown in MEM. Human embryonic kidney cells (HEK 293) were grown in DMEM. BT549 basal breast cancer cells were grown in RPMI 1640. All culture media were supplemented with 10% FBS and 5% penicillin/streptomycin (100 μg/mL each).

1. Cells were plated three days before the transfection in media that lacked antibiotic.
2. On transfection day, 70% confluent cells were treated with 0.25% trypsin solution for 3 min at 37 °C.
3. Cells were then mixed 1:10 with fresh media (lacking antibiotic) and were counted using a Scepter cell counter.
4. One million cells were put into 15-mL screw cap tubes and centrifuged at 100 × *g* for 10 min (Eppendorf 5810R

benchtop centrifuge) to pellet the cells, and then they were prepared for electroporation as indicated below.

5. Resuspend one million cells in 100 μL of Nucleofector V solution.
6. Add 250 ng of first guide RNA plasmid to nick *FOXO3* chromosome: HSL0001339461 (5′-CTTACTGAAGGTGACAGGCTGG-3′ chromosomal position: 108663888–108,663,909).
7. Add 250 ng of second guide RNA plasmid to nick *FOXO3* chromosome: HSR0001339464 (5′-CACGGCTGACTGATATGGCAGG-3′ chromosomal position 108,663,933–108,663,954).
8. Add 250 ng of the vector enabling Cas9D10A expression (CAS9D10AGFPP).
9. Add 1.0 μg of *FOXO3* donor vector (prepared as described above).
10. During electroporation with Lonza Nucleofector machine, use program P-20 for U87MG and BT549 cells and program X-001 for HEK 293 cells. Pre-warm the media for recovery ahead of time. Another useful practice is to have the recovery media ready in the plastic dropper included in the kit before electroporation in order to add it to the cells right away for recovery. Recovery media lacked antibiotic.
11. Cells recovered for 2 days in a 10-cm dish at 37 °C and 5% CO_2.

3.10 Selecting FOXO3 Disruption Clones Using Neomycin

Selecting clones takes weeks (and possibly longer if the mutation is deleterious to cell growth).

1. After recovery for 2 days in media lacking antibiotic, treat transformed cells with 0.25% trypsin for 3 min at 37 °C.
2. Add to the cells 10 mL of fresh media (containing 5% penicillin/streptomycin solution) and pass the cells through a cell strainer (BIOLOGIX 100-μm diameter cell strainer 1100) to eliminate clumps to better isolate single clones.
3. Ten plates were prepared with ~100,000 cells from strained cells.
4. Add neomycin the next day to select for single clones: G418 (0.5 mg/mL final concentration for U87MG and BT549 and 1.5 mg/mL for HEK 293 cells).
5. Incubate the cells for 1 month at 37 °C and 5% CO_2.

Using these conditions, approximately 2–10 clones were isolated from each 10-cm dish plate and single clones were purified using cloning cylinders.

3.11 Isolation of Single Clones from 10-cm Dishes

1. Wash 10-cm plate (with clones on it) twice with 2-mL sterile 1× PBS buffer. Important: make sure that all PBS is carefully removed from the plate. Residual PBS can make cloning cylinders move around.
2. Place the cloning cylinders onto the 10-cm plate (directly over individual colonies) using sterile forceps and autoclaved Vaseline (to make the cylinder stick to the plate).
3. Add 200 μL of 0.25% trypsin into the cloning cylinder and incubate at room temperature for 5 min.
4. After this, gently pipette up and down ten times and remove the 200-μL cell suspension. Place cells into fresh media in a well of a 6-well plate. Allow cells to grow for roughly a week.

3.12 Confirming Mutants by Western Blot Analysis

After cloned cells recovered for about a week, proceed to western blot analysis.

1. Treat the cells with 0.25% trypsin and replated at a 1:2 dilution to propagate the cells.
2. Prepare protein lysates for western blot analysis by adding 2× sample buffer to each well followed by scraping.
3. Collect the lysate from each well into a 1.5-mL microcentrifuge tube and heated for 10 min at 90 °C in a dry-bath heat block.
4. Separate protein lysates by sodium dodecyl sulfate–polyacrylamide gel (SDS-PAGE) electrophoresis at 100 V for 1 h.
5. Transfer onto a polyvinylidene fluoride (PVDF) membrane for an hour and 30 min and then block in a 5% milk solution prepared in 1× Tris-buffered saline with Tween 20 (TBST).
6. Incubate the membranes with FOXO3 or GAPDH antibodies overnight at 4 °C.
7. After incubation, wash the membranes for 20 min with TBST in 5-min intervals.
8. Incubate the membranes with secondary antibody (goat-anti-rabbit for FOXO3 blots or goat-anti-mouse for GAPDH blots) for 1 h at room temperature with gentle rocking.
9. After incubation, wash the membranes for 20 min in 5-min intervals.
10. Blots were developed with Super Signal West Dura Extended Duration Substrate luminol solution for 5 min. A Bio-Rad ChemDoc XRS+ molecular imager was used to detect light emitted from protein-containing complexes. Western blot SCN files from the (Bio-Rad ChemDoc XRS+) were analyzed using NIH Image J. Disruption of the *FOXO3* gene as described in this protocol led to a much smaller protein (349 amino acids versus 673). Therefore, western blot analysis allowed us to quickly screen hundreds of putative mutants

Table 2
CRISPR Cas9 mutation frequencies in three mammalian cell lines

Cell line	G418-resistant isolates screened by western blot analysis	Number of homozygous truncation mutant	Homozygous mutation frequency, %
U87MG Trial 1	77	4[a]	5%
U87MG Trial 2	50	3[a]	6%
BT549	56	5	9%
HEK293	77	1	1.3%

[a]Table 3 originally published in Vazquez et al. and reproduced with permission [9]

(which were cloned as G418 positive) for production of a truncated FOXO3 protein.

The results of the current gene editing protocols in several cell lines are summarized in Table 2 (*see* **Note 2**). One can see that mutation frequencies varied depending on the cell line. HEK 293 cells had the lowest mutation frequency of approximately 1.0% [9]. Of note, U87MG and BT549 cells had five-fold higher mutation frequencies than HEK 293 cells [9]. One possible reason for this is that HEK 293 cells are immortalized and have fewer mutations than the cancer cell lines U87MG and BT549. DNA repair mechanisms such as nonhomologous end joining may be more efficient in HEK 293 cells leading to fewer disruption mutants obtained via homology-directed repair of chromosomal lesions. Refinements to CRISPR technology including RNA editing to introduce novel mutations and inactivating Cas9 activity using small-molecule inhibitors have been recently developed and should be considered when devising new genome editing applications (*see* **Notes 3** and **4**).

3.13 Confirmation of FOXO3 Mutants by PCR and Sanger Sequencing

Homozygous disruption mutants of the *FOXO3* gene were also confirmed by PCR amplification of the disrupted region and Sanger sequencing. In this protocol, one million cells were removed from plates and DNA was prepared from the samples using basic techniques [9]. The DNA pellets were washed with 70% ethanol, dried, and re-suspended in sterile water. PCR was performed to identify clones that had *NPTII* integration into the *FOXO3* locus using the primers in Table 3. The forward primer for PCR was approximately 60 bases upstream of the *FOXO3* fragment found in Arm 1 of the *FOXO3* donor vector [9]. Therefore, primer sequences employed for mutant detection by PCR were absent from the donor vector and were only found in chromosomal *FOXO3.* The reverse primer

Table 3
Primers utilized to detect and sequence *FOXO3* gene disruption

Primer name	Sequence
FOXO3 F (for detection of disruption)	5′-GTGCTTCAGGATCGCTTCA-3′
Neo cassette R (for detection of disruption)	5′-TGCATGCTTTGCATACTTCTG-3′
FOXO3 seq. (for sequencing disruption mutants)	5′-CTCGGTTTTGGACCATTCTG-3′

[a]Table 4 originally published in Vazquez et al. and reproduced with permission [9]

was part of the *NPTII* cassette used to disrupt *FOXO3* [9]. PCR products (employed for detecting *FOXO3* disruption) were column purified, quantified with a NanoDrop spectrophotometer, and then analyzed by Sanger sequencing using primer in Table 3.

4 Notes

1. One major issue with CRISPR Cas9 genome engineering is off-target effects [41]. Our protocol took numerous approaches to decrease off-target effects from CRISPR Cas9 mutagenesis.
   - First, we employed transient expression of the Cas9D10A nuclease from a mammalian expression vector. This decreased the time that the nuclease was expressed in the cells, reducing the likelihood of multiple mutation events and greatly limiting the possibility of off-target mutations. Others have addressed this issue by employing an inducible form of Cas9 nuclease [42–44].
   - We utilized two guide RNAs to direct the Cas9D10A nickase in order to introduce a recombinogenic (staggered) double-strand break in the chromosomal locus of *FOXO3.* In this case, both strands needed to be targeted separately by the Cas9D10A nickase. This approach has been shown to increase specificity by 100-fold or more in other systems [45].
2. We initially tried to examine pools of neomycin-resistant cells directly after CRISPR Cas9 mutagenesis but found little *FOXO3* disruption. We concluded that the efficiency of the mutagenesis was not high enough to examine pools of mutagenized cells. Therefore, isolation of clones was required to obtain a sizeable pool of mutants to investigate. To accomplish this, we selected for disruption of the *FOXO3* gene using neomycin resistance followed by western blot analysis and sequencing. Using three tiers of screening helped to quickly isolate the desired mutants. Upon clone isolation, one can *see*

from Table 1 that our mutation frequencies were only 1–5% when seeking disruption alleles for *FOXO3*.

3. Efforts are continually being made to make CRISPR Cas9 genome editing more precise and efficient. CRISPR/Cas-based DNA and RNA editing tools have revolutionized biological research and extended the capability to manipulate genomes. CRISPR Cas9, which belongs to the class 2 type II CRISPR, is the most extensively used high-throughput tool for genome editing. Initially, CRISPR-Cas9 was a valuable tool for targeted DNA cleavage, but with the development of programmable RCas9, sequence-specific sites on RNA can now be recognized and cleaved for studying RNA regulation [46]. In RCas9, specific PAM-presenting oligonucleotide (PAMmers) target sites on RNA and avoids sites on genomic DNA, which stimulates the cleavage of RNA in a site-specific manner. CRISPR Cas13 is another powerful tool for precise RNA editing. Cas13a (previously known as C2c2) belongs to class 2 type VI CRISPR and is a single-effector RNA-guided RNase, which can be programmed for cleavage of specific sites of mRNA transcripts, mediated by the higher eukaryotes and prokaryotes nucleotide-binding (HEPN) domains [47]. The CRISPR Cas13 system is more efficient than RNA interference and can be used in mammalian and plant cells to knock down specific mRNA transcripts [48].

   REPAIR (RNA editing for programmable A to I replacement) technology further advances RNA editing. This novel technique employs Cas13 orthologs, dCas13b, which are catalytically inactive and can be used to convert a single A base to I (functionally equivalent to G) with the help of ADAR2 (adenosine deaminase acting on RNA type 2). With REPAIR, it is possible to target coding and noncoding RNA, and it has higher specificity. However, REPAIR can allow only single base A to I editing, which is a major drawback [49, 50]. To expand the editing capability to other nucleotide bases, Abudayyeh and his group proposed a programmable RESCUE system (RNA editing for specific C to U exchange), which is capable of a highly specific and precise C to U conversion in RNA. As an advantage, RESCUE can target amino acids encoding signals for posttranscriptional modifications including phosphorylation, glycosylation, and methylation [50]. REPAIR and RESCUE RNA editing tools have revolutionized the capability of CRISPR and expanded the ability to treat potential genetic diseases.

   Another advance has come from Sternberg and his group who developed a new fully programmable CRISPR/Cas model based on the bacterial Tn7-like transposons. In this system, highly specific DNA insertion in the genome is guided by the

RNA complex known as Cascade (type I CRISPR) without any double-strand breaks [51]. Transposon-based CRISPR/Cas models provide a new approach for site-specific insertion of DNA, which could be leveraged in the production of biopharmaceuticals, gene therapies, improvement of crops, and future development of knock-in methods [51].

4. As indicated above, off-target effects remain the major drawback for utilizing CRISPR-Cas9 as a genome editing tool, especially in the area of precision medicine. However, Wienert and colleagues have created a method, the DISCOVER-Seq, for better identification of off-target sites [52]. This technology is built on the idea that after a cut is made on a DNA strand, the cell recruits DNA repair machinery as part of the body's natural repair process. Finding those DNA repair factors would mean finding the locations of the CRISPR off-target effects [52]. This testing has been completed on induced pluripotent stem cells and patient cells and has shown to be more accurate than the initial CRISPR Cas9 system.

   After Cas9 completes its job by cleaving the target DNA strand, continued activity of the protein can become a potential danger by producing nontarget breaks in DNA strands. As a solution to this problem, current research is examining ways to make "off switches" to regulate the activity of Cas9 or similar proteins [53]. Much of the research on solving the problem of off-target effects began in 2012 at the University of Toronto where researchers attempted to find different ways to block the Cas9 enzyme and break its interaction with the guiding RNA strand [54–64]. A challenge was that Cas9 cuts the DNA with a specific pair of nuclease domains that must be disabled until needed. Amit Choudhary and colleagues focused on the interaction between Cas9 and a short DNA segment that Cas9 uses to begin cutting, the protospacer adjacent motif (PAM) sequence [65]. Luckily, Dr. Choudhary noted that since this is a short sequence, "the recognition [does not] have to be very strong." His team conducted different assays to discover which compounds could effectively disrupt the Cas9-PAM binding interaction. The result was a pair of small molecules that could penetrate cells and also inhibit this binding interaction without exhibiting off-target toxic effects [65].

   In 2019, Kocak and colleagues introduced a new way of increasing the precision and decreasing off-target effects of bacterial CRISPR Cas9 nucleases [66]. Researchers added hairpin secondary structures to the spacer region of single guide RNAs (hp-sgRNAs). It was discovered that these hp-sgRNAs added specificity when utilizing five different Cas12a or Cas9 protein variants [66]. The RNA secondary structure was created by elongating the hairpin on the 5′ end of the sgRNA to function as a steric barrier to R-loop formation. R-loop formation is critical to helping a trans-activator based on Cas9 from *Streptococcus pyogenes* (SpCas9) to function alongside the active nuclease [66]. Thus, by altering the strength of the RNA secondary structure, the R-loop could form at target sites, but not at off-target sites due to mispairing. This allowed for an increase in specificity by an average of 55-fold with five of the most commonly used Cas proteins in CRISPR [66].

Ultimately, much research remains to be done to increase the precision of CRISPR technologies. New future projects, such as introducing regulatory molecules along with the CRISPR technology, using the body's natural repair capabilities, or engineering new secondary structures, show promise in the field of cell biology and biomedicine [67–69].

References

1. Qi LS, Larson MH, Gilbert LA, Doudna JA, Weissman JS, Arkin AP, Lim WA (2013) Repurposing CRISPR as an RNA-guided platform for sequence-specific control of gene expression. Cell 152:1173–1183
2. van Duijn E, Barbu IM, Barendregt A, Jore MM, Wiedenheft B, Lundgren M, Westra ER, Brouns SJ, Doudna JA, van der Oost J, Heck AJ (2012) Native tandem and ion mobility mass spectrometry highlight structural and modular similarities in clustered-regularly-interspaced shot-palindromic-repeats (CRISPR)-associated protein complexes from *Escherichia coli* and *Pseudomonas aeruginosa*. Mol Cell Proteomics 11:1430–1441
3. Haurwitz RE, Sternberg SH, Doudna JA (2012) Csy4 relies on an unusual catalytic dyad to position and cleave CRISPR RNA. EMBO J 31:2824–2832
4. Sternberg SH, Haurwitz RE, Doudna JA (2012) Mechanism of substrate selection by a highly specific CRISPR endoribonuclease. RNA 18:661–672
5. Sashital DG, Jinek M, Doudna JA (2011) An RNA-induced conformational change required for CRISPR RNA cleavage by the endoribonuclease Cse3. Nat Struct Mol Biol 18:680–687
6. Jore MM, Lundgren M, van Duijn E, Bultema JB, Westra ER, Waghmare SP, Wiedenheft B, Pul U, Wurm R, Wagner R, Beijer MR, Barendregt A, Zhou K, Snijders AP, Dickman MJ, Doudna JA, Boekema EJ, Heck AJ, van der Oost J, Brouns SJ (2011) Structural basis for CRISPR RNA-guided DNA recognition by cascade. Nat Struct Mol Biol 18:529–536
7. Haurwitz RE, Jinek M, Wiedenheft B, Zhou K, Doudna JA (2010) Sequence- and structure-specific RNA processing by a CRISPR endonuclease. Science 329:1355–1358
8. Wiedenheft B, Zhou K, Jinek M, Coyle SM, Ma W, Doudna JA (2009) Structural basis for DNase activity of a conserved protein implicated in CRISPR-mediated genome defense. Structure 17:904–912
9. Vazquez N, Sanchez L, Marks R, Martinez E, Fanniel V, Lopez A, Salinas A, Flores I, Hirschmann J, Gilkerson R, Schuenzel E, Dearth R, Halaby R, Innis-Whitehouse W, Keniry M (2018) A protocol for custom CRISPR Cas9 donor vector construction to truncate genes in mammalian cells using pcDNA3 backbone. BMC Mol Biol 19:3
10. Lee JS, Kallehauge TB, Pedersen LE, Kildegaard HF (2015) Site-specific integration in CHO cells mediated by CRISPR/Cas9 and homology-directed DNA repair pathway. Sci Rep 5:8572
11. van Erp PB, Jackson RN, Carter J, Golden SM, Bailey S, Wiedenheft B (2015) Mechanism of CRISPR-RNA guided recognition of DNA targets in *Escherichia coli*. Nucleic Acids Res 43:8381–8391
12. Charpentier E (2015) CRISPR-Cas9: how research on a bacterial RNA-guided mechanism opened new perspectives in biotechnology and biomedicine. EMBO Mol Med 7:363–365
13. Rollins MF, Schuman JT, Paulus K, Bukhari HS, Wiedenheft B (2015) Mechanism of foreign DNA recognition by a CRISPR RNA-guided surveillance complex from *Pseudomonas aeruginosa*. Nucleic Acids Res 43:2216–2222

14. Dyda F, Hickman AB (2015) Mechanism of spacer integration links the CRISPR/Cas system to transposition as a form of mobile DNA. Mob DNA 6:9
15. Moore R, Spinhirne A, Lai MJ, Preisser S, Li Y, Kang T, Bleris L (2015) CRISPR-based self-cleaving mechanism for controllable gene delivery in human cells. Nucleic Acids Res 43:1297–1303
16. Wang H, Yang H, Shivalila CS, Dawlaty MM, Cheng AW, Zhang F, Jaenisch R (2013) One-step generation of mice carrying mutations in multiple genes by CRISPR/Cas-mediated genome engineering. Cell 153:910–918
17. Zhang Q, Rho M, Tang H, Doak TG, Ye Y (2013) CRISPR-Cas systems target a diverse collection of invasive mobile genetic elements in human microbiomes. Genome Biol 14:R40
18. Jiang W, Bikard D, Cox D, Zhang F, Marraffini LA (2013) RNA-guided editing of bacterial genomes using CRISPR-Cas systems. Nat Biotechnol 31:233–239
19. Cong L, Ran FA, Cox D, Lin S, Barretto R, Habib N, Hsu PD, Wu X, Jiang W, Marraffini LA, Zhang F (2013) Multiplex genome engineering using CRISPR/Cas systems. Science 339:819–823
20. Tzur YB, Friedland AE, Nadarajan S, Church GM, Calarco JA, Colaiacovo MP (2013) Heritable custom genomic modifications in *Caenorhabditis elegans* via a CRISPR-Cas9 system. Genetics 195:1181–1185
21. Friedland AE, Tzur YB, Esvelt KM, Colaiacovo MP, Church GM, Calarco JA (2013) Heritable genome editing in *C. elegans* via a CRISPR-Cas9 system. Nat Methods 10:741–743
22. DiCarlo JE, Norville JE, Mali P, Rios X, Aach J, Church GM (2013) Genome engineering in *Saccharomyces cerevisiae* using CRISPR-Cas systems. Nucleic Acids Res 41:4336–4343
23. Horvath P, Barrangou R (2010) CRISPR/Cas, the immune system of bacteria and archaea. Science 327:167–170
24. Horvath P, Coute-Monvoisin AC, Romero DA, Boyaval P, Fremaux C, Barrangou R (2009) Comparative analysis of CRISPR loci in lactic acid bacteria genomes. Int J Food Microbiol 131:62–70
25. Sorek R, Kunin V, Hugenholtz P (2008) CRISPR—a widespread system that provides acquired resistance against phages in bacteria and archaea. Nat Rev Microbiol 6:181–186
26. Ishino Y, Shinagawa H, Makino K, Amemura M, Nakata A (1987) Nucleotide sequence of the iap gene, responsible for alkaline phosphatase isozyme conversion in *Escherichia coli*, and identification of the gene product. J Bacteriol 169:5429–5433
27. Zhang H, Li Z, Daczkowski CM, Gabel C, Mesecar AD, Chang L (2019) Structural basis for the inhibition of CRISPR-Cas12a by Anti-CRISPR proteins. Cell Host Microbe 25: e814–e826
28. Wandera KG, Collins SP, Wimmer F, Marshall R, Noireaux V, Beisel CL (2019) An enhanced assay to characterize anti-CRISPR proteins using a cell-free transcription-translation system. Methods. https://doi.org/10.1016/j.ymeth.2019.05.014
29. Hoffmann MD, Aschenbrenner S, Grosse S, Rapti K, Domenger C, Fakhiri J, Mastel M, Borner K, Eils R, Grimm D, Niopek D (2019) Cell-specific CRISPR-Cas9 activation by microRNA-dependent expression of anti-CRISPR proteins. Nucleic Acids Res 47:e75
30. Babu M, Beloglazova N, Flick R, Graham C, Skarina T, Nocek B, Gagarinova A, Pogoutse O, Brown G, Binkowski A, Phanse S, Joachimiak A, Koonin EV, Savchenko A, Emili A, Greenblatt J, Edwards AM, Yakunin AF (2011) A dual function of the CRISPR-Cas system in bacterial antivirus immunity and DNA repair. Mol Microbiol 79:484–502
31. Bikard D, Hatoum-Aslan A, Mucida D, Marraffini LA (2012) CRISPR interference can prevent natural transformation and virulence acquisition during in vivo bacterial infection. Cell Host Microbe 12:177–186
32. Hatoum-Aslan A, Marraffini LA (2014) Impact of CRISPR immunity on the emergence and virulence of bacterial pathogens. Curr Opin Microbiol 17:82–90
33. Sampson TR, Saroj SD, Llewellyn AC, Tzeng YL, Weiss DS (2013) A CRISPR/Cas system mediates bacterial innate immune evasion and virulence. Nature 497:254–257
34. Makarova KS, Wolf YI, Alkhnbashi OS, Costa F, Shah SA, Saunders SJ, Barrangou R, Brouns SJ, Charpentier E, Haft DH, Horvath P, Moineau S, Mojica FJ, Terns RM, Terns MP, White MF, Yakunin AF, Garrett RA, van der Oost J, Backofen R, Koonin EV (2015) An updated evolutionary classification of CRISPR-Cas systems. Nat Rev Microbiol 13:722–736
35. Koonin EV, Makarova KS, Zhang F (2017) Diversity, classification and evolution of CRISPR-Cas systems. Curr Opin Microbiol 37:67–78
36. Tang Y, Fu Y (2018) Class 2 CRISPR/Cas: an expanding biotechnology toolbox for and beyond genome editing. Cell Biosci 8:59

37. Jinek M, Chylinski K, Fonfara I, Hauer M, Doudna JA, Charpentier E (2012) A programmable dual-RNA-guided DNA endonuclease in adaptive bacterial immunity. Science 337:816–821
38. Xie K, Zhang J, Yang Y (2014) Genome-wide prediction of highly specific guide RNA spacers for CRISPR-Cas9-mediated genome editing in model plants and major crops. Mol Plant 7:923–926
39. Li JF, Norville JE, Aach J, McCormack M, Zhang D, Bush J, Church GM, Sheen J (2013) Multiplex and homologous recombination-mediated genome editing in Arabidopsis and Nicotiana benthamiana using guide RNA and Cas9. Nat Biotechnol 31:688–691
40. Gibson DG, Young L, Chuang RY, Venter JC, Hutchison CA 3rd, Smith HO (2009) Enzymatic assembly of DNA molecules up to several hundred kilobases. Nat Methods 6:343–345
41. Fu Y, Foden JA, Khayter C, Maeder ML, Reyon D, Joung JK, Sander JD (2013) High-frequency off-target mutagenesis induced by CRISPR-Cas nucleases in human cells. Nat Biotechnol 31:822–826
42. Dow LE, Fisher J, O'Rourke KP, Muley A, Kastenhuber ER, Livshits G, Tschaharganeh DF, Socci ND, Lowe SW (2015) Inducible in vivo genome editing with CRISPR-Cas9. Nat Biotechnol 33:390–394
43. Polstein LR, Gersbach CA (2015) A light-inducible CRISPR-Cas9 system for control of endogenous gene activation. Nat Chem Biol 11:198–200
44. Zetsche B, Volz SE, Zhang F (2015) A split-Cas9 architecture for inducible genome editing and transcription modulation. Nat Biotechnol 33:139–142
45. Chiang TW, le Sage C, Larrieu D, Demir M, Jackson SP (2016) CRISPR-Cas9(D10A) nickase-based genotypic and phenotypic screening to enhance genome editing. Sci Rep 6:24356
46. O'Connell MR, Oakes BL, Sternberg SH, East-Seletsky A, Kaplan M, Doudna JA (2014) Programmable RNA recognition and cleavage by CRISPR/Cas9. Nature 516:263–266
47. Abudayyeh OO, Gootenberg JS, Konermann S, Joung J, Slaymaker IM, Cox DB, Shmakov S, Makarova KS, Semenova E, Minakhin L, Severinov K, Regev A, Lander ES, Koonin EV, Zhang F (2016) C2c2 is a single-component programmable RNA-guided RNA-targeting CRISPR effector. Science 353: aaf5573
48. Abudayyeh OO, Gootenberg JS, Essletzbichler P, Han S, Joung J, Belanto JJ, Verdine V, Cox DBT, Kellner MJ, Regev A, Lander ES, Voytas DF, Ting AY, Zhang F (2017) RNA targeting with CRISPR-Cas13. Nature 550:280–284
49. Cox DBT, Gootenberg JS, Abudayyeh OO, Franklin B, Kellner MJ, Joung J, Zhang F (2017) RNA editing with CRISPR-Cas13. Science 358:1019–1027
50. Abudayyeh OO, Gootenberg JS, Franklin B, Koob J, Kellner MJ, Ladha A, Joung J, Kirchgatterer P, Cox DBT, Zhang F (2019) A cytosine deaminase for programmable single-base RNA editing. Science 365:382–386
51. Klompe SE, Vo PLH, Halpin-Healy TS, Sternberg SH (2019) Transposon-encoded CRISPR-Cas systems direct RNA-guided DNA integration. Nature 571:219–225
52. Wienert B, Wyman SK, Richardson CD, Yeh CD, Akcakaya P, Porritt MJ, Morlock M, Vu JT, Kazane KR, Watry HL, Judge LM, Conklin BR, Maresca M, Corn JE (2019) Unbiased detection of CRISPR off-targets in vivo using DISCOVER-Seq. Science 364:286–289
53. Trasanidou D, Geros AS, Mohanraju P, Nieuwenweg AC, Nobrega FL, Staals RHJ (2019) Keeping crispr in check: diverse mechanisms of phage-encoded anti-crisprs. FEMS Microbiol Lett 366:fnz098
54. Bondy-Denomy J, Davidson AR, Doudna JA, Fineran PC, Maxwell KL, Moineau S, Peng X, Sontheimer EJ, Wiedenheft B (2018) A unified resource for tracking anti-CRISPR names. CRISPR J 1:304–305
55. Bondy-Denomy J, Garcia B, Strum S, Du M, Rollins MF, Hidalgo-Reyes Y, Wiedenheft B, Maxwell KL, Davidson AR (2015) Multiple mechanisms for CRISPR-Cas inhibition by anti-CRISPR proteins. Nature 526:136–139
56. Lee J, Mir A, Edraki A, Garcia B, Amrani N, Lou HE, Gainetdinov I, Pawluk A, Ibraheim R, Gao XD, Liu P, Davidson AR, Maxwell KL, Sontheimer EJ (2018) Potent Cas9 inhibition in bacterial and human cells by AcrIIC4 and AcrIIC5 anti-CRISPR proteins. MBio 9: e02321-18
57. Maxwell KL (2016) Phages fight back: inactivation of the CRISPR-Cas bacterial immune system by anti-CRISPR proteins. PLoS Pathog 12:e1005282
58. Maxwell KL (2017) The anti-CRISPR story: a battle for survival. Mol Cell 68:8–14
59. Maxwell KL, Garcia B, Bondy-Denomy J, Bona D, Hidalgo-Reyes Y, Davidson AR (2016) The solution structure of an anti-CRISPR protein. Nat Commun 7:13134

60. Pawluk A, Bondy-Denomy J, Cheung VH, Maxwell KL, Davidson AR (2014) A new group of phage anti-CRISPR genes inhibits the type I-E CRISPR-Cas system of *Pseudomonas aeruginosa*. MBio 5:e00896
61. Pawluk A, Davidson AR, Maxwell KL (2018) Anti-CRISPR: discovery, mechanism and function. Nat Rev Microbiol 16:12–17
62. Pawluk A, Shah M, Mejdani M, Calmettes C, Moraes TF, Davidson AR, Maxwell KL (2017) Disabling a type I-E CRISPR-Cas nuclease with a bacteriophage-encoded anti-CRISPR protein. MBio 8:e01751-17
63. Pawluk A, Staals RH, Taylor C, Watson BN, Saha S, Fineran PC, Maxwell KL, Davidson AR (2016) Inactivation of CRISPR-Cas systems by anti-CRISPR proteins in diverse bacterial species. Nat Microbiol 1:16085
64. Stanley SY, Maxwell KL (2018) Phage-encoded anti-CRISPR defenses. Annu Rev Genet 52:445–464
65. Gangopadhyay SA, Cox KJ, Manna D, Lim D, Maji B, Zhou Q, Choudhary A (2019) Precision control of CRISPR-Cas9 using small molecules and light. Biochemistry 58:234–244
66. Kocak DD, Josephs EA, Bhandarkar V, Adkar SS, Kwon JB, Gersbach CA (2019) Increasing the specificity of CRISPR systems with engineered RNA secondary structures. Nat Biotechnol 37:657–666
67. Raikwar SP, Kikkeri NS, Sakuru R, Saeed D, Zahoor H, Premkumar K, Mentor S, Thangavel R, Dubova I, Ahmed ME, Selvakumar GP, Kempuraj D, Zaheer S, Iyer SS, Zaheer A (2019) Next generation precision medicine: CRISPR-mediated genome editing for the treatment of neurodegenerative disorders. J Neuroimmune Pharmacol. https://doi.org/10.1007/s11481-019-09849-y
68. Schermer B, Benzing T (2019) Genome editing with CRISPR/Cas9: first steps towards a new era in medicine? Dtsch Med Wochenschr 144:276–281
69. Vaughan A, Yang IA (2019) CRISPR-Cas9 technology: a new direction for personalized medicine in respiratory disease? Respirology 24:614–615

Chapter 21

Genome Editing in Zebrafish Using High-Fidelity Cas9 Nucleases: Choosing the Right Nuclease for the Task

Sergey V. Prykhozhij, Anna Cordeiro-Santanach, Lucia Caceres, and Jason N. Berman

Abstract

Shortly after the development of the CRISPR/Cas9 system, it was recognized that it is prone to induce off-target mutations at significant frequencies. Therefore, there is a strong motivation to develop Cas9 enzymes with reduced off-target activity. Multiple rational design or selection approaches have been applied to develop several Cas9 versions with reduced off-target activities (high fidelity). To make these high-fidelity Cas9s available for model systems other than human cells and bacterial strains, as, for example, in zebrafish, new specialized expression vectors need to be developed. In this chapter, we focused on the HypaCas9 and HiFi Cas9 high-fidelity enzymes and incorporated the mutations of these Cas9 versions into a codon-optimized zebrafish Cas9 vector. This optimized vector was further improved by introducing an artificial polyadenine insert (A71) since polyadenylation is known to enhance mRNA translational efficiency. The Hypa-nCas9n and HiFi-nCas9n vectors were produced by single-site mutagenesis from pT3TS-nCas9n-A71 vector. We then tested the polyadenylated mRNAs for nCas9n, Hypa-nCas9n, HiFi-nCas9n, and HiFi-Cas9 protein for editing efficiency in five genome editing strategies and found that these high-fidelity Cas9 versions had different performances ranging from activity at 2–4 sites, where the wild-type nCas9n is active, indicating that these Cas9 versions have different sgRNA preferences. In summary, the developed new high-fidelity Cas9 vectors will enable researchers to perform much more accurate genome editing.

Key words CRISPR, Cas9, Zebrafish, HiFi Cas9, HypaCas9, Knock-in, sgRNA, Mutation, Genome editing, Off-target

1 Introduction

One of the latest revolutions in molecular biology has been the adaptation of the clustered regularly interspaced palindromic repeats (CRISPR)/Cas9 bacterial system to perform gene editing [1–5]. Briefly, this system is based on the ability of Cas9 to cut specific regions of double-stranded DNA (dsDNA) with the specificity provided by 18–20 bp single-guide RNAs (sgRNA) that can be designed (*see* Chapter 19). Upon a Cas9-induced cut, the cell

Mouldy Sioud (ed.), *RNA Interference and CRISPR Technologies: Technical Advances and New Therapeutic Opportunities*, Methods in Molecular Biology, vol. 2115, https://doi.org/10.1007/978-1-0716-0290-4_21,

uses its DNA repair mechanisms to bring together the cut ends. If the repair happens via error-prone nonhomologous end joining (NHEJ), it is likely to create insertions/deletions (indels) [6]. Upon addition of a template DNA molecule, the homologous repair (HR) pathway may produce more specific modifications, from point mutations to large DNA cassette insertions [7]. These two mechanisms, combined in various ways, account for all known CRIPR-mediated genome editing strategies from small indels to deletions of several 10s of kilobases (kb) as well as new gene segment knock-ins. Gene editing is of great importance in the era of personalized medicine to generate models for specific diseases/conditions or for its potential use in gene-corrective therapy.

Despite its great promise, CRISPR/Cas9 has been recognized to have a significant off-target activity at similar target sites shortly after its development [8–10]. In the last few years, there have been several attempts to create improved (high-fidelity) versions of the "wild-type" *Streptococcus pyogenes* Cas9 (SpCas9). Most of the improvements are based on two different strategies: modifications of the WT Cas9 sequence and delivering Cas9 and guide RNA as ribonucleoproteins (RNPs). Cas9-mediated cleavage is dependent on the interaction between the sgRNA and the target DNA, which requires strand separation to perform the cut. The authors of the enhanced SpCas9 (eSpCas9) study [11] hypothesized that Cas9 enzyme versions requiring more energy to bind the mismatching sequences would produce fewer off-target effects (*see* Chapters 1 and 2). The authors replaced specific amino acid residues with alanine to weaken off-target interactions and identified an eSpCas9 version containing 3 mutations (K848A, K1003A, and R1060A) that resulted in high on-target efficiency and reduced off-target activity at sites with 1 or 2 mismatches with the sgRNA. The SpCas9-HF1 study [12] took advantage of the findings that the contact between the Cas9 and the target DNA backbone (including hydrogen bonds) utilizes specific amino acids, N497, R661, Q695, and Q926 [13, 14]. The authors hypothesized that by reducing the contact between the enzyme and the target DNA region, the interaction can be made more specific. All combinations of alanine substitutions at these four codons (1, 2, 3, or all 4 at a time) were performed and tested. The more promising quadruple substitution variant was named SpCas9-HF1 and showed similar efficacy on-target as WT Cas9 but with reduced activity at off-target sites. Both SpCas9-HF1 and eSpCas9 variants contain substitutions reducing the binding energy to the sgRNA targets thus increasing the specificity but also decreasing activity at low Cas9 expression levels. In order to achieve better performance regarding off-target effects, Chen et al. [15] engineered a HypaCas9 variant with four point mutations (N692A/M694A/Q695A/H698A) which affect the conformation of the Cas9 protein. More specifically, these mutations affect the catalytic HNH domain of Cas9, requiring more energy to perform the cut. HypaCas9 showed similar on-target effects but less off-target effects than the WT Cas9 or SpCas9-HF1.

An innovation in the high-fidelity Cas9 screening efforts is the use of unbiased screening platforms. Casini et al. [16] isolated a series of Cas9 mutants via a yeast-based screening relying on targeting the selectable metabolic marker and a color marker for off-target sites. K526E mutation in the K526 residue contacting the protospacer adjacent motif (PAM) sequence was the most effective since it did not affect on-target nuclease activity but reduced the off-target activity. The authors then combined the most promising mutations (M495V/Y515N/K526E/R661L or VNEL), and after further substitutions of codon 661, the optimal VNEQ variant combination was generated, now known as evolved Cas9 (evoCas9). evoCas9 showed similar on-target efficacy but lower off-target effects than WT, eSpCas9, and SpCas9-HF1. The team at Integrated DNA Technologies used a bacterial screening system and designed the target site to be within the toxin gene (to test for on-target efficacy); they also engineered the sgRNA to have an off-target site within the antibiotic resistance sequence to test for target specificity [17]. Among the resulting mutants, this study identified a point mutation R691A in the Cas9 gene and named it high-fidelity (HiFi) Cas9. The authors of the HiFi Cas9 study also went beyond the plasmid delivery approach for the Cas9 enzyme and guide RNA(s). Instead, they suggest using ribonucleoprotein (RNP) delivery system consisting of three components: one or several crRNAs (CRISPR RNA), a tracrRNA (the two RNAs combine to form the complete guide RNA structure), and Cas9 protein. Delivery of the resulting RNPs offers several advantages over viral or plasmid delivery: the Cas9 RNP is functional immediately and the fact that the guide RNA and Cas9 are physically linked enhances the on-target efficiency and reduces the off-target effects. Comparison of HiFi Cas9 with the previously described high-fidelity Cas9 (eSpCas9(1.1) [11], SpCas9-HF1 [12], HypaCas9 [15], or WT Cas9, when all were delivered as RNPs, resulted in higher on-target efficiencies at multiple sites than the other high-fidelity Cas9s. Expectedly, HiFi Cas9 showed drastically lower off-target site efficiencies than the WT Cas9. Finally, the authors used HiFi Cas9 to perform *HBB* E6V knock-in and showed that it functions at the same efficiency as WT Cas9 but with lower off-target activity in human sickle cell disease-CD34^{+} hematopoietic stem and progenitor cells.

The discovery of new and improved versions of Cas9 enzymes promises a great leap towards precision medicine where CRIPR/Cas9 technology can be more precisely controlled. Thus, it is imperative that all these new high-fidelity Cas9 be tested in whole animal models; so far, they have mainly been explored in human cells/cell lines. In the last 30 years, zebrafish (*Danio rerio*) has emerged as an excellent model to study genetic diseases [18]. Their capacity to produce hundreds of transparent eggs per breeding facilitates experimental reproducibility and strong statistical power, as well as gene editing at the one-cell stage via microinjection. Even though zebrafish and humans are relatively distant in evolutionary terms, they both belong to the vertebrata subphylum and their genomes have high degree of conservation; 71% of human genes have at least one ortholog in zebrafish and 69% of zebrafish genes have at least one ortholog in humans [19]. Despite a lot of advancements in genome editing in zebrafish, vectors for high-fidelity Cas9 versions and their usage have not yet been published in zebrafish, where they can be useful for developing more precise mutants without the concealed challenge of off-target mutations. We now address this situation by describing the protocols for developing and testing vectors expressing zebrafish-optimized versions of HypaCas9 and HiFi Cas9, and we also test the commercially available HiFi Cas9 protein.

2 Materials

2.1 General Molecular Biological Reagents

1. Agarose.
2. 10× TBE (Tris/boric acid/EDTA) buffer, 5 L.
3. Gel loading dye, purple (6×), no SDS (NEB, B7025S).
4. Nuclease-free water.
5. Phase Lock Gel tubes (5PRIME, 2302800).
6. UltraPure phenol/chloroform/isoamyl alcohol (25:24:1) (Thermo Fisher Scientific, 15593031).
7. Glycogen.
8. SYBR® Green I nucleic acid gel stain (10,000× in DMSO).
9. Tris–EDTA (TE) buffer solution.
10. Taq DNA polymerase.
11. Q5® High-Fidelity 2× Master Mix (NEB, M0492).
12. Restriction enzymes: XhoI, DpnI, and SacII.
13. Calf intestinal phosphatase (CIP).
14. T4 polynucleotide kinase (PNK).
15. T4 DNA ligase.
16. MEGAshortscript T7 kit.
17. mMessage mMachine T3.
18. MAX Efficiency™ Stbl2™ Competent Cells.
19. QIAquick Gel Extraction kit.
20. QIAprep Spin Miniprep Kit.
21. LidLocks™ Microcentrifuge Tube Locks.

2.2 Plasmids and Oligonucleotides Used in Experiments

We cloned and deposited the following plasmids to Addgene: pT3TS-nCas9n-A71 (Addgene plasmid ID 125644), pT3TS-Hypa-nCas9n-A69 (Addgene plasmid ID 125645), and pT3TS-HiFi-nCas9n-A72 (Addgene plasmid ID 125646). The poly-A inserts in these plasmids can be unstable and we recommend using Stbl2 or other similar competent bacteria for propagating them. pT3TS-nCas9n (Addgene plasmid ID 46757) was previously published [20]. pCAG-T3-hCasD10A-pA (Addgene plasmid ID 51638) [21] was used for generating artificial polyA inserts.

2.3 Reagents for Growing and Handling Zebrafish

E3 medium: 5 mM NaCl, 0.17 mM KCl, 0.33 mM $CaCl_2$, 0.33 mM $MgSO_4$.

0.4% water solution of tricaine buffered to pH 7.2 with Tris–HCl pH 9.0.

3 Methods

3.1 Synthesis of sgRNAs and Cas9 mRNAs

Since the focus of the current chapter is on generating and using the high-fidelity Cas9 enzyme versions in zebrafish, we are going to describe the basic procedures of sgRNA and Cas9 mRNA synthesis. For further details on the application of CRISPR/Cas9 system in zebrafish, please refer to several excellent protocol and methods papers [22–26].

3.1.1 Synthesis and Purification of sgRNAs

1. Design your single-guide RNAs (sgRNA) using your favorite software. We and others have previously reviewed the design procedures [27, 28].
2. Order the oligos containing the spacer sequences of sgRNAs of the following structure: <u>GTAATACGACTCACTATAGG</u>-**N(20)**-gttttagagctagaaatagc (<u>T7 PROMOTER</u>, N(20)—spacer, overlap) and the rev_sgRNA_scaffold oligo. The sgRNA forward oligos we used in this protocol are listed in Table 1.
3. Dilute the forward sgRNA oligo (or their mix) and rev_sgRNA_scaffold oligo in water to a concentration of 25 μM.
4. Mix the following in a PCR tube for a 100-μL reaction:

10× buffer	10 μL
25 mM $MgSO_4$	6 μL
dNTP	2 μL
25 μM oligos	5 of each forward sgRNA and rev_sgRNA_scaffold oligos
Taq	1.5 μL
Water	71 μL

Table 1
Oligonucleotides used in the study

Oligo name	Sequence	Application
rev_sgRNA_scaffold	GGATCCGCACCGACTCGGTGCCAC TTTTTCAAGTTGATAACGGACTAGCCTTA TTTTAACTTGCTATTTCTAGCTCTAAAAC	sgRNA synthesis
SacII-pA_for	GATCCCGCGGGCACCAAAGAACCTG TAAACG	A_n addition
SacII-XbaI-pA_rev	GATACCGCGGTCTAGAGCATGCTACC TCGAGGGTTT	A_n addition
capn3b_sgR_ex4-1-sense	GTAATACGACTCACTATAGGAATGATGTCA TCCTGAAGGTTTAGAGCTAGAAATAGC	sgRNA synthesis
capn3b_sgR_ex4-2-sense	GTAATACGACTCACTATAGGTTCAGGG TCAGGCAAGCTAGTTTTAGAGCTAGAAA TAGC	sgRNA synthesis
capn3b_sgR_ex4-3-sense	GTAATACGACTCACTATAGGATGACATCA TTCAGGGTCGTTTTAGAGCTAGAAATAGC	sgRNA synthesis
capn3b_sgR_ex6-1_sense	GTAATACGACTCACTATAGGCTA TGGAGGACTTCACTGGGTTTTAGAGC TAGAAATAGC	sgRNA synthesis
capn3b_sgR_ex6-2_sense	GTAATACGACTCACTATAGGTTGTACAACA TCATGAGGAGTTTTAGAGCTAGAAATAGC	sgRNA synthesis
capn3b_sgR_ex6-3_sense	GTAATACGACTCACTATAGGTGAGGAAGGC TCTGAAGAGGTTTTAGAGCTAGAAATAGC	sgRNA synthesis
capn3b_GT-del_for	CCAATTGCTCCCTCTTTTCA	Genotyping
capn3b_GT-del_rev	GCACGAAAGCATCAAAGACA	Genotyping
p53-R242_sgRNA-3_sense	GTAATACGACTCACTATAGGTGCGTGTGTG TGCATGTCCGTTTTAGAGCTAGAAATAGC	sgRNA synthesis
p53-R242_SA_rev	TTTTCAGCAGTGTAGCAGCAAA	Genotyping
p53-R242_SA_for	ATACCAATCAATTGGACTCATCTCG	Genotyping
R144_sgRNA-3_sense	GTAATACGACTCACTATAGGATCCGGGG TTCGCTCATGATGTTTTAGAGCTAGAAA TAGC	sgRNA synthesis
R144_SA_for	TTGCAGTATTCACCGGACCT	Genotyping
R144_SA_rev	GCCTCCCTCGCCATAGTAAT	Genotyping
R144-WT_rev	CTCATGATGGGGGCATCT	Genotyping
R144H_KI_rev	CTCATGATGCGGGCAGTG	Genotyping
R144H-new_AS-Asymm-90	GAATGGATATGGAAAAAAAAATGTCTGTAC TATCTCCATCCGGGGTTCGCTCATGA TGCGGGCAGTGCCGGACCAC TTCAGCCACATGCT	Knock-in generation

(continued)

Table 1
(continued)

Oligo name	Sequence	Application
lmna_R471_sgRNA-2_for	GTAATACGACTCACTATAGGCC TGGAAGGAAAATTTGTGGTTTTAGAGC TAGAAATAGC	sgRNA synthesis
lmna_R471-site_for	CAGCACCAGTCGCACATC	Genotyping
lmna_R471-site_rev	AGCTGCCATAAGCAAAATCC	Genotyping
lmna_R471_WT_for	GACCTGGAAGGAAAATTTGTGC	Genotyping
lmna_R471_KI_for	GACCTGGAAGGAAAATTCGTTT	Genotyping
lmna_R471W_oligo	CAGCAAGCCTCAGCCAGCGGCCGCG TCACCGTTGACGAAGTCGACC TGGAAGGAAAATTCGTTTGGCTTAA TAACAAGTCTGACCAGGTG	Knock-in generation
dysf_e36_sgRNA_sense	GTAATACGACTCACTATAGGTGGAATG TGGCGGATCTGGTTTTAGAGCTAGAAA TAGC	sgRNA synthesis
Dysf-stop-KI_for	CACCTCAGAGGGAGCCAAA	Genotyping
Dysf-stop-KI_rev	AATTAAGGTTTAAACGCCATGA	Genotyping
dysf_e36_STOP-oligo	TCAGGTGGAATGTGGCGGATGTCATGGCG TTTAAACCTTAATTAAGCTGTTGTAGCTG TGGTTCAGTCGTGCGTC	Knock-in generation
hifi_ncas9_sdm_for	gcaAATTTTATGCAGCTGATTCACG	SDM[a]
hifi_ncas9_sdm_rev	GTTGGCGAAGCCATCACTTT	SDM
Hypa-nCas9_SDM_for	GCACTGATTGCCGATGACAGTCTGACC TTCAAAGAGG	SDM
Hypa-nCas9_SDM_rev	CGCAAAAGCGCGGTTGGCGAAGCCATCAC	SDM

[a]SDM—site-directed mutagenesis

5. Anneal and extend the two oligos together by placing the mixture in a PCR machine and run the following protocol: 94 °C for 5 min, 5 cycles of 94 °C for 30 s, 55 °C for 30 s, and 72 °C for 30 s followed once by extension at 72 °C for 2 min.
6. Prepare the samples by adding 10 μL of 100× SYBR Green I solution and 10 μL of 6× gel loading dye to each of the PCR reactions and run them on 2% agarose gel prepared with and immersed into 0.5× TBE buffer.
7. Purify the main PCR product by gel extraction using Qiagen Gel Extraction kit since the amplification in **step 5** produces some additional bands. Elute the DNA from the column with 50 μL of nuclease-free water.

8. Set up a 10-μL in vitro transcription reaction using MEGAshortscript T7 kit with the following components: 1 μL of each ATP, GTP, CTP, and UTP, 1 μL of 10× buffer, 4 μL of the sgRNA template from the previous step, and 1 μL of T7 enzyme. Incubate the reaction overnight at 37 °C (bacterial incubator or a PCR machine).
9. Precipitate the sgRNA by adding 125 μL of water, 15 μL of ammonium acetate solution (MEGAshortscript T7 kit), and 300 μL of 100% ethanol. Mix the tube contents by inversion several times and incubate at −20 °C for at least 15 min.
10. Centrifuge the precipitating sgRNA solution at the maximum speed at 4 °C for 10 min and remove most of the supernatant. Repeat centrifugation for 1 min and supernatant removal 1–2 times until no ethanol solution is present.
11. Resuspend the RNA pellet in 50 μL of water, measure the concentration on a NanoDrop or other spectrophotometer and adjust the concentration of the sgRNA solution to 1000 ng/μL or other desired standard concentration.
12. Analyze the integrity and size of sgRNAs by mixing 2.5 μL of their solution with 2.5 μL of 2× loading dye (mMessage mMachine T3), incubating at 65 °C for 3 min (in a PCR machine), cooling on ice, mixing with 1 μL of 100× SYBR Green I solution (by diluting the 10,000× SYBR Green stock with TE buffer), and running the resulting samples on 2% agarose gel (*see* **Note 1**).

3.1.2 Synthesis and Purification of Cas9 mRNAs

1. Use XhoI to digest the A_n-containing Cas9 expression plasmids. Set up a 100-μL digestion reaction containing 10 μg of the plasmid, 10 μL of 10× CutSmart buffer, 2 μL of the enzyme, and water to the 100-μL volume. Incubate the reaction at 37 °C overnight.
2. Add 200 μL of water to the digestion reaction, transfer to the Phase Lock Gel tube, and add 300 μL of phenol/chloroform/isoamyl alcohol. Shake the tube for 15 s and then centrifuge at the maximum speed for 5 min.
3. Collect the supernatant and transfer into a new 1.5-mL tube. Add 30 μL of 3 M sodium acetate pH 5.5, 1 μL of 20 mg/mL glycogen, and 600 μL of 100% ethanol.
4. Incubate the precipitation reactions at −20 °C for 30 min and then centrifuge the tubes for 10 min in the 4 °C centrifuge at the maximum speed.
5. Remove the supernatant completely and resuspend the pellet in 40 μL of water.
6. Set up the in vitro transcription reaction using the mMessage mMachine T3 with the following components: 10 μL of 2× NTP/CAP mix, 2 μL of 10× buffer, 6 μL of linearized DNA, and 2 μL of T3 enzyme.

7. Incubate the reaction for 4 h at 37 °C.
8. Add 30 μL of 7.5 M LiCl to the transcription reaction, mix, and incubate at −20 °C for 30 min. Precipitate the transcript by centrifugation for 10 min in the 4 °C centrifuge at the maximum speed.
9. Remove the supernatant and add 1 mL of 70% ethanol and centrifuge again for 5 min.
10. Remove the supernatant completely (by centrifuging twice for 1 min each—the second time is to remove residual ethanol) and resuspend the RNA pellet in 50 μL.
11. Measure the concentration of Cas9 mRNA and add water to adjust the concentration to 625 ng/μL or another desired concentration.
12. Analyze the integrity and size of Cas9 mRNA in the same way as in **step 11** of Subheading 3.1.1 but analyze the RNA samples on 1.5% agarose gel.

3.2 Addition of Artificial Encoded Poly-Adenine Tails to Cas9 Expression Vectors to Improve Translation Efficiency

When Cas9 is provided as mRNA, it is crucial that it becomes translated efficiently. Although nCas9n-SV40 mRNA from pT3TS-nCas9n vector works well, we considered the possibility that the translational efficiency of the standard vector might not be sufficient for the high-fidelity Cas9 versions. Yoshimi et al. [29] generated a Cas9 mRNA expression vector containing a poly-adenine stretch (81 nucleotides) that was able to cut DNA much earlier during mouse embryo development and improved the eventual overall frequency of knock-in events by about twofold compared to the Cas9 mRNA containing the standard polyA signal. We have decided to incorporate this feature in all of our Cas9 mRNA expression vectors to improve the activity of our Cas9 mRNAs. The source of the polyA insert was the pCAG-T3-hCasD10A-pA plasmid. The following protocol describes the cloning procedure to generate these Cas9 vectors.

1. Amplify the polyA insert from pCAG-T3-hCasD10A-pA using SacII-pA_for and SacII-XbaI-pA_rev (Table 1). The PCR included 10 ng of the vector, 5 μL of each 5 μM primer stock, and 25 μL of Q5® High-Fidelity 2× Master Mix in a final volume of 50 μL using the following program: 2 min at 98 °C followed by 35 cycles (30 s at 98 °C, 1 min at 72 °C) and 2 min at 72 °C for the final extension.
2. Run the PCR product on 2% agarose gel and extract using QIAquick Gel Extraction kit.
3. Digest the PCR product and the pT3TS-nCas9n vector with 2 μL of SacII enzyme in a 100-μL reaction for 3 h at 37 °C. The vector digestion reaction is treated with 2 μL of CIP phosphatase for 1 h at 37 °C to dephosphorylate the ends. After, the

vector and the polyA insert PCR product are run on 2% and 1% agarose gels, respectively, and purified using QIAquick Gel Extraction kit.

4. Ligate 100 ng of pT3TS-nCas9n SacII-digested vector with the polyA insert at a 10:1 ratio using T4 DNA ligase at 16 °C overnight followed by 10 min of inactivation at 65 °C.
5. Transform the ligations into MAX Efficiency™ Stbl2™ Competent Cells. From each reaction, take 4 μL and add to 40 μL of competent bacteria, and incubate for 20 min on ice, 30 s at 42 °C, and 2 min on ice. Then add 250 μL of SOC medium and incubate for 1 h at 37 °C with shaking. Plate bacteria on agar plates containing ampicillin.
6. Grow miniprep cultures from several colonies in LB medium with 100 μg/mL ampicillin at 37 °C overnight and prepare plasmid minipreps using QIAprep Spin Miniprep Kit.
7. Verify plasmids by sequencing the polyA insert with nCas9-n_end_for primer to ensure the correct orientation and the length of poly-adenine stretch (*see* **Note 2**). For our vectors (pT3TS-nCas9n-An, pT3TS-Hypa-nCas9n-An, pT3TS-HiFi-nCas9n-An), the lengths of poly-adenine stretches were 71, 69, and 72, respectively.

3.3 Site-Directed Mutagenesis of nCas9n to Create High-Fidelity Cas9 Vectors

As discussed in the Introduction, the new high-fidelity vectors are typically demonstrated to work comparably to their wild-type counterparts in specific systems where they were developed or tested. However, successful application of the original high-fidelity Cas9 constructs in other species is not always possible since they are not codon-optimized for another species and the high-fidelity mutations may even be generated in the background of another version of Cas9. Therefore, it is better to test the new high-fidelity mutations in the context of established Cas9 expression constructs. In the following, we describe the steps necessary to generate high-fidelity versions (HypaCas9 and HiFi Cas9) of the optimized zebrafish Cas9 expression vector (pT3TS-nCas9n).

1. Align Cas9 protein sequences of the wild-type (nCas9n), HypaCas9, and HiFi Cas9 proteins using Align tool from Vector NTI tool or a free online tool such as Clustal Omega (https://www.ebi.ac.uk/Tools/msa/clustalo/). The resulting alignment of wild-type nCas9n, HiFi-nCas9n, and Hypa-nCas9n in the region of interest is shown in Fig. 1a.
2. Design the primers for the site-directed mutagenesis (SDM) at a specific site in the plasmid by amplifying around the whole plasmid with the resulting PCR product containing the introduced mutations (Fig. 1b). The SDM primers we designed for HiFi-nCas9n and Hypa-nCas9n plasmids are listed in Table 1. In other cases when more than one site needs to be mutated, it

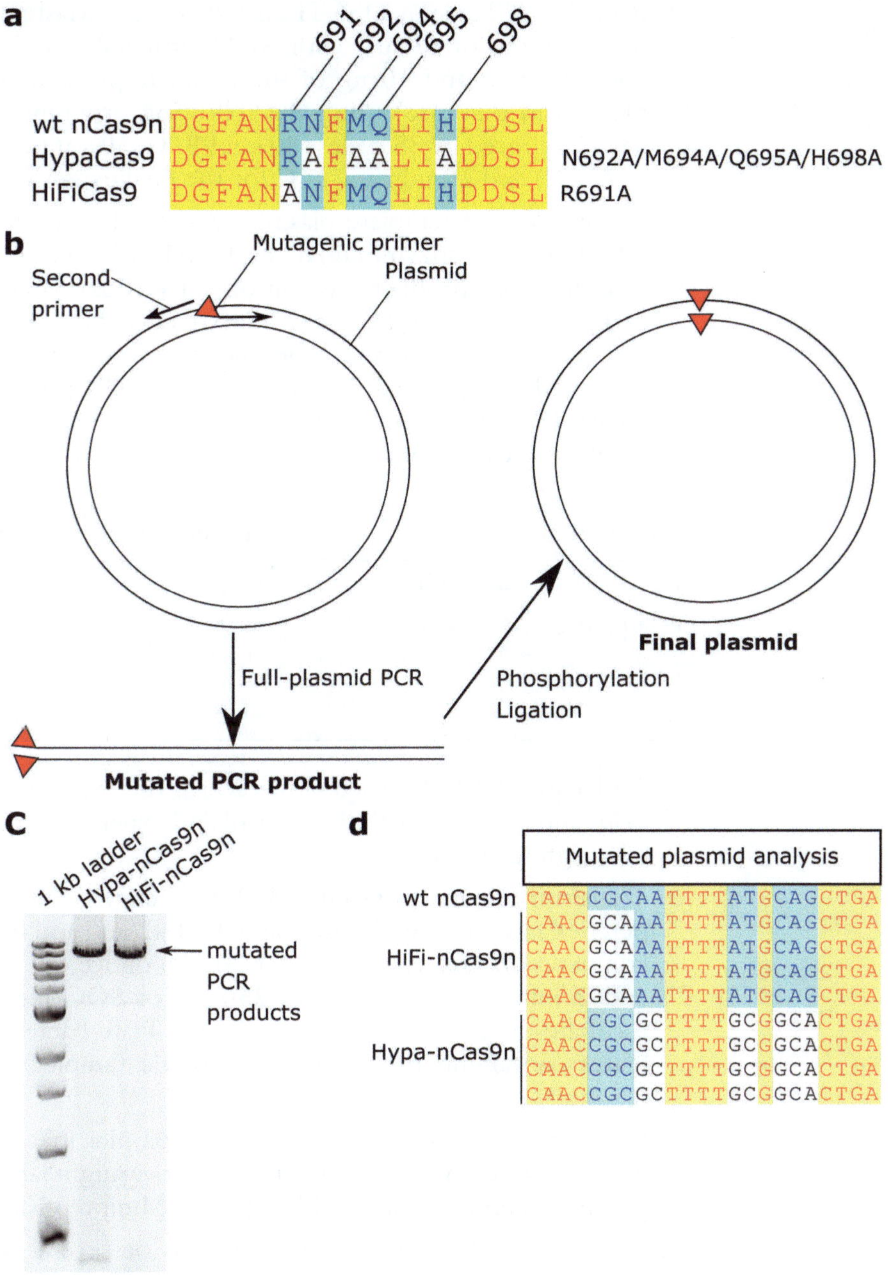

Fig. 1 Site-directed mutagenesis to create zebrafish codon-optimized high-fidelity Cas9 vectors for mRNA expression. (**a**) Alignment of wild-type nCas9n, HypaCas9, and HiFiCas9. (**b**) Primer design principle for site-directed mutagenesis (SDM), a single site. (**c**) Results of amplification of the SDM PCR products for pT3TS-Hypa-nCas9n-An and pT3TS-HiFi-nCas9n-An vectors. (**d**) Sequencing results of plasmids produced by SDM

is more efficient to use commercial kits such as QuikChange Multi Site-Directed Mutagenesis Kit and follow their oligo design guidelines and procedures.

3. Run the PCR using Q5® High-Fidelity 2× Master Mix in a final volume of 50 μL with SDM primers at 0.5-μM final concentration and 10 ng of the template plasmid (wild-type nCas9n in our case) using the following program: 2 min at 98 °C followed by 35 cycles (30 s at 98 °C, 30 s at 63 °C, 4 min at 72 °C) and 10 min at 72 °C for the final extension.

4. To remove the template plasmid, add to the SDM reactions 0.5 μL of DpnI enzyme and digest for 3 h at 37 °C (*see* **Note 3**). The digestion product was run on a 1% agarose gel (Fig. 1c) and purified using QIAquick Gel Extraction kit.

   Set up a control and experimental phosphorylation reactions with the purified PCR product using T4 polynucleotide kinase (PNK) enzyme and 10× T4 DNA ligase buffer and incubate the reactions at 37 °C for 1 h and at 65 °C for 20 min.

Component	Control	Experiment
10× T4 DNA ligase buffer	1	1
PCR product (>50 ng/μL)	8	8
PNK	–	1
Water	1	–

5. Add 1 μL of T4 DNA ligase to each of the experimental reactions and incubate both the control and experimental reactions overnight at 16 °C.

6. Transform the reactions into MAX Efficiency™ Stbl2™ Competent Cells. From each reaction, take 4 μL and add to 40 μL of competent bacteria, incubate for 20 min on ice, 30 s at 42 °C, and 2 min on ice. Then add 250 μL of SOC medium and incubate for 1 h at 37 °C with shaking. Plate bacteria on agar plates containing the required antibiotic (ampicillin for our plasmids).

7. Grow miniprep cultures from several colonies in LB medium with the relevant antibiotic at 37 °C overnight and prepare plasmid minipreps using QIAprep Spin Miniprep Kit.

8. Verify plasmids by digestion with one or several restriction enzymes generating characteristic fragments.

9. Sequence the mutated region in the positive clones and then complete genes, which were mutated. When we performed SDM to generate HiFi-nCas9n and Hypa-nCas9n plasmids, the clones with the expected digestion patterns had correct sequences (Fig. 1d).

One of the purposes of this chapter is to directly compare high-fidelity Cas9 versions in all 7 of our CRISPR/Cas9 strategies. This involved injecting 4 different Cas9 mixes (WT nCas9n, Hypa-

3.4 Testing of High-Fidelity Cas9 Versions Versus the Wild-Type Cas9

nCas9n, HiFi mRNA, HiFi Cas9 protein) with specific sgRNAs and, if necessary, a template oligo to generate precise mutations. The resulting injected embryos were genotyped as described below to assess the activities of different Cas9 versions. The sgRNAs chosen for these demonstration experiments were previously shown to work well with the WT nCas9n.

3.4.1 Microinjection of Zebrafish Eggs

1. Set up adult zebrafish in breeding tanks with dividers the day before the injection.
2. The next day, breed the fish for 10–15 min and collect the eggs in egg water.
3. Array the one cell-stage fish eggs in troughs of an agarose injection plate and orient them in such that the cell of each egg faces the injection needle.
4. Prepare the mixture of Cas9 and sgRNAs. For Cas9 mRNA injections, mix the following:
   (a) 625 ng of Cas9 mRNA.
   (b) 500 ng of sgRNA for individual sgRNA strategies.
   (c) 900 ng for 6-sgRNA deletion strategies.
   (d) 0.5 μL of 10-μM template oligo (optional, necessary for knock-in strategies).
   (e) Complete with water to 5 μL and 0.55 μL of 0.5% phenol red.

   For Cas9 protein injections, mix the following:
   (a) 2.5 μL of 1 μg/μL Cas9 protein solution.
   (b) Same amounts of sgRNAs as above.
   (c) 0.7 μL of 2 M KCl and complete with water to 4.5 μL.

   Incubate the mix at 37 °C in the bacterial incubator for 5 min, add 0.5-μL water or 0.5-μL 10-μM template oligo (optional, necessary for knock-in strategies), and finally add 0.55 μL of 0.5% phenol red. Template oligos used in this study are listed in Table 1.
5. Inject zebrafish eggs with the Cas9/sgRNA mix. The bolus diameter should not exceed 20% of the egg diameter (*see* **Note 4**).
6. Grow the injected fish to at least 48 h postfertilization (hpf) before genotyping.

3.4.2 Sample Preparation for Genotyping Genome-Edited Zebrafish

1. Before starting with sample preparation, treat embryos (48–72 hpf) with 0.02% tricaine in fish medium for several minutes.
2. For knock-in injected embryo genotyping, transfer them into PCR tubes and replace the fish medium with 40 μL of 50 mM NaOH.

3. Prepare 8 single-embryo extracts from uninjected embryos (wild-type control). Prepare extracts from injected and uninjected embryo pools by combining 50 embryos into a single tube and replacing the fish medium with 1 mL of 50 mM NaOH. The volume of the pooled samples and neutralization solution can be adjusted based on the number of available embryos.
4. Heat the samples at 95 °C for 10 min. Place the single-embryo samples into the PCR machine. For the pooled embryo samples, close the tubes with cap locks and invert the tubes manually several times in the first 3 min.
5. Vortex the single-embryo samples for 10 s or invert the pooled sample tubes several times.
6. Cool samples on ice for 5 min and neutralize the single-embryo and pooled samples with 4 and 100 μL of 1 M Tris–HCl pH 8.0, respectively.
7. Centrifuge the pooled samples at full speed for 5 min before proceeding with PCR reactions.

3.4.3 Genotyping Zebrafish Genome-Edited with Wild-Type and High-Fidelity Cas9s

CRISPR/Cas9 is now being used for a wide variety of genome editing strategies. Therefore, we use the high-fidelity Cas9 versions to test whether these newer Cas9 versions work similarly to the wild-type Cas9. We applied all of the Cas9 versions to one deletion strategy (*capn3b* gene), 3 individual sgRNA injections (*tp53* R242, *tp53* R144 sgRNA, and *lmna* sgRNA), and 3 knock-in strategies (*dysf* exon36 STOP cassette knock-in, *tp53* R144H knock-in, and *lmna* R471W knock-in). This knock-in approaches were described in detail previously [30] and here the essential information will be provided.

1. Calculate the annealing temperature (T_{anneal}) of your PCR using the NEB T_{m} Calculator online tool (http://tmcalculator.neb.com/#!/) or determine the optimal temperature empirically using gradient PCR. The details of our assays are listed in Table 2.
2. Use the touchdown PCR method: 94 °C for 3 min; *10 cycles* (94 °C for 30 s, T_{anneal} + 10 [with 1 °C decrease every cycle] for 30 s, 72 °C for 30 s [for longer assays use 1 min/kb]), *25 cycles* (94 °C for 30 s, T_{anneal} for 30 s, 72 °C for 30 s [for longer assays use 1 min/kb]). Use this PCR protocol for pooled samples and 8–16 single-embryo samples for knock-in strategies.
3. For knock-in strategies, run the wild-type assay on 4–8 uninjected (wild-type) embryos (PCR size control) and the knock-in assay on both 4–8 uninjected and 8–16 injected embryo samples. The knock-in assays are expected to fail or produce

Table 2
PCR assays used in the study

Assay	Primers	T_{anneal}, °C	Product size, bp	Extension, seconds
capn3b deletion assay	capn3b_GT-del_for capn3b_GT-del_rev	51	912	60
tp53 R242 site assay	p53-R242_SA_for p53-R242_SA_rev	51	255	30
tp53 R144 site assay	R144_SA_for R144_SA_rev	54	350	30
tp53 R144 wild-type assay	R144_SA_for R144-WT_rev	54	173	30
tp53 R144H knock-in assay	R144_SA_for R144H-new_KI_rev	54	173	30
Lmna R471 site assay	lmna_R471-site_for lmna_R471-site_rev	51	357	30
Lmna wild-type assay	lmna_R471_WT_for lmna_R471-site_rev	51	206	30
Lmna knock-in assay	lmna_R471_KI_for lmna_R471-site_rev	51	206	30
Dysf e36 STOP knock-in assay	Dysf-stop-KI_for Dysf-stop-KI_rev	51	285	30

different nonspecific bands in uninjected samples and serve as controls for the knock-in PCR assay. Include the uninjected embryo pool extract in the pooled sample PCR as control.

4. Run the resulting PCR products on an agarose gel of appropriate concentration as described above to assess whether amplification occurred. This step also provides direct visualization of deletion and knock-in PCR efficiencies but typically does not allow effective assessment of individual sgRNA strategies.
5. To evaluate one-sgRNA CRISPR/Cas9 strategies, run the heteroduplex mobility assay (HMA).
6. Make 15 mL of gel solution for typical gel cassettes (we use the Bio-Rad apparatus for making SDS-PAGE gels) with 1.5-mm spacers by mixing the following components (*see* **Note 5**).

Component	Volume for 15 mL of 6% gel solution	Volume for 15 mL of *X* % gel solution
10× TBE	750 μL	750 μL
Water	11.13 mL	14.13—*X*/2
30% acrylamide/ bisacrylamide (29:1)	3 mL	*X*/2
10% ammonium persulfate	105 μL	105 μL
TEMED	15 μL	15 μL

7. Cast your gels by filling the gel cassettes, insert combs, and let stand for 30 min.
8. Pre-run your gels for 15–30 min at the same voltage as you will use for running the samples. We use 70 V for our gel system, but you may need to use higher or lower voltages. Too high a voltage when running your gels invariably leads to poor results likely due to gel overheating. Determine the appropriate voltage for your system experimentally.
9. Prepare the samples. Use up to 8 μL of your PCR product and add 1 μL of 6× loading dye and 1 μL of 100× SYBR Green I solution.
10. Load 5–10 μL for best results and run the samples on the gel. The exact timing will depend on your gel system, voltage, and gel percentage. We typically run the HMA gels for 1.5–2 h depending on percentage (*see* **Note 6**).

3.4.4 Evaluating the Results of the Genome Editing Experiments

To aid users in evaluating their experimental results using high-fidelity Cas9 versions, we describe how we compared them to the previously used WT nCas9n RNA. The *capn3b* deletion strategy involving 6 sgRNAs (Fig. 2a) is straightforward to evaluate since after sgRNA and Cas9 injections the PCR assay produces small PCR products providing evidence of deletions and showing that WT nCas9n and all high-fidelity versions had significant activity. The pattern of deletions was different for different Cas9 versions suggesting that not every sgRNA was equally active with different Cas9 versions. Individual sgRNA injection strategies were tested by HMA, which identifies mutational activity by the appearance of slowly migrating PCR product heteroduplexes and gives an appearance of an upward smear or ladder. This method revealed some differences among the WT nCas9n and high-fidelity Cas9 versions (Fig. 2b–d). The *tp53* R242 sgRNA was active with all Cas9 versions except Hypa-nCas9n (Fig. 2b). By contrast, *tp53* R144 sgRNA was only active with WT nCas9n RNA (Fig. 2c). Interestingly, even though the HiFi-nCas9n RNA should have the same

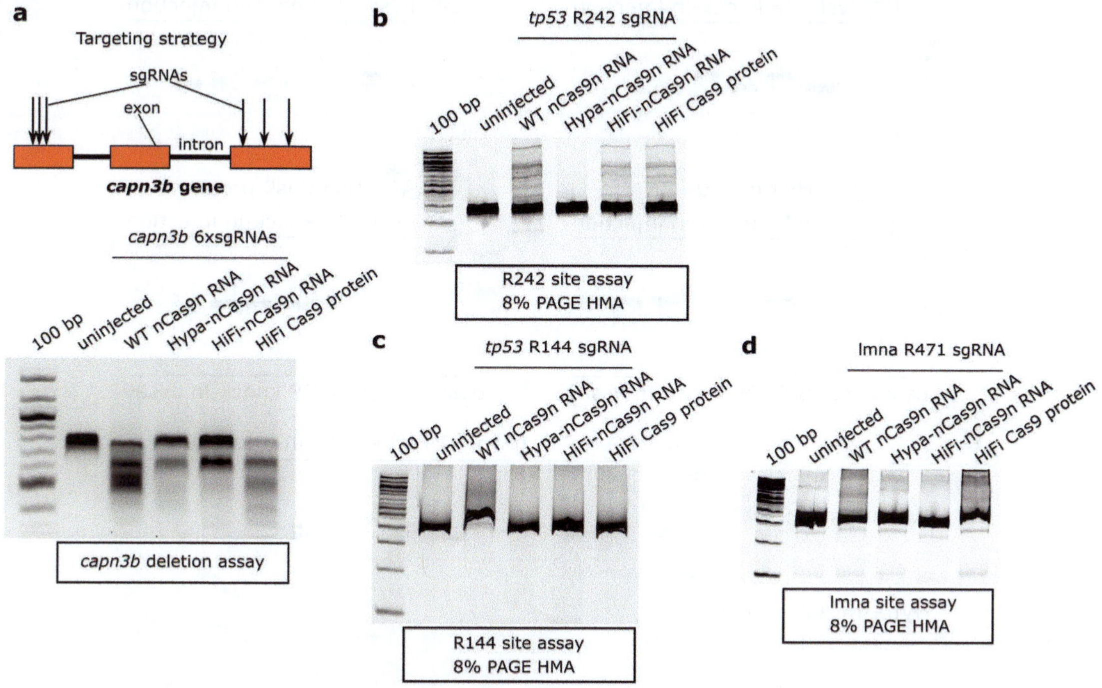

Fig. 2 Efficiency assessment of WT nCas9n and high-fidelity Cas9s on a *capn3b* deletion and 3 individual sgRNA CRISPR strategies illustrate their varying performance. (**a**) *capn3b* deletion strategy scheme shows sgRNAs targeting exons 4 and 6 (upper panel). The primers used for detection are not shown in the figure for simplicity. Analysis of embryos injected with the mix of 6 *capn3b* sgRNAs and different Cas9s shows that all the Cas9s have significant level of activity, the WT nCas9n and HiFi Cas9 protein being the most active (lower panel). (**b**) HMA analysis of Cas9 injections with *tp53* R242 sgRNA shows all except Hypa-nCas9n had significant activity at this site. (**c**) HMA analysis of Cas9 injections with *tp53* R144 sgRNA shows that only the WT nCas9n RNA was active. (**d**) HMA analysis of Cas9 injections with *lmna* R471 sgRNA shows activity of WT nCas9n RNA and HiFi Cas9 protein

protein sequence as HiFi Cas9 protein, only HiFi Cas9 protein and the WT nCas9n RNA performed well at the *lmna* R471 site (Fig. 2d).

For the knock-in CRISPR/Cas9 experiments we chose three strategies. The first one was previously published [31], and it involves introducing a stop codon cassette into exon 36 of *dysferlin* (*dysf*) using Cas9, an sgRNA, and a template oligo. All Cas9 versions performed comparably well to generate these knock-in events (Fig. 3a). By contrast, the *tp53* R144H knock-in only worked well with WT nCas9n RNA (Fig. 3b), which is fully consistent with the mutational analysis by HMA (Fig. 2c). The mutational results for *lmna* R471 sgRNA with different Cas9 versions showed that only HiFi Cas9 protein and WT nCas9n RNA worked well at that site. Therefore, we tested these versions with *lmna* R471W knock-in injections and they were both successful (Fig. 3c).

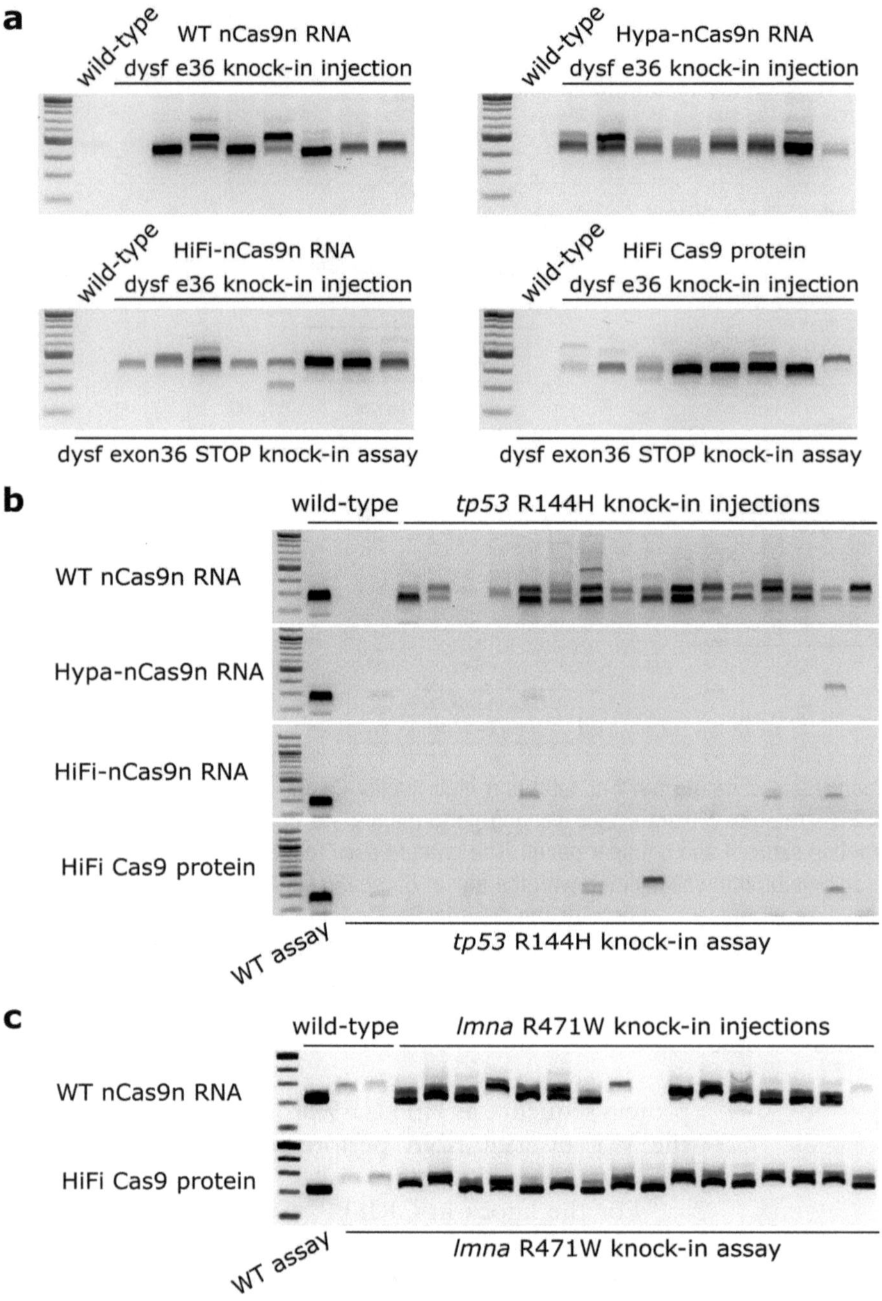

Fig. 3 Knock-in efficiency assessment using allele-specific PCR. (**a**) Knock-in of the stop codon containing cassette (STOP) into exon 36 of *dysferlin* (*dysf*) gene. Different panels show the results of the knock-in injections using dysf_e36_STOP-oligo, dysf e36 sgRNA, and the Cas9s: WT nCas9n RNA, Hypa-nCas9n RNA, HiFi-nCas9n RNA, or HiFi Cas9 protein. Wild-type genomic DNA samples were used as control and dysf e36 STOP knock-in assay was used for analysis. All Cas9 versions performed well in this knock-in experiment. (**b**) Knock-in of the R144H mutation into *tp53* gene using WT nCas9n RNA and 3 high-fidelity Cas9 versions. Different panels show the results of the knock-in injections with the R144H-new_AS-Asymm-90 oligo, R144 sgRNA, and each of the Cas9 versions. Only WT nCas9n RNA was successful at inserting this mutation into the genome. (**c**) Knock-in of the R471W mutation into *lmna* gene using WT nCas9n RNA and HiFi Cas9 protein. The panels show the results of the knock-in injections with the lmna_R471W_oligo, *lmna* R471 sgRNA, and each of the Cas9 versions. Both versions were successful at inserting this mutation into the genome

3.4.5 Conclusions

The WT nCas9n RNA worked in all previously tested strategies which was fully expected. This fact biases the comparison with the high-fidelity Cas9 versions. In a more complete set of experiments, the sgRNAs would need to be selected randomly since there may well be some sgRNA sites where only the high-fidelity Cas9s work well. Among the high-fidelity Cas9 versions, HiFi Cas9 protein was the most active (4 out of 5 unique strategies) followed by HiFi-nCas9n RNA (3 out of 5 unique strategies) and Hypa-nCas9n RNA (2 out of 5 unique strategies). Thus, we recommend that users of the present protocol seeking the most accurate genome editing procedures prioritize the HiFi Cas9 protein and HiFi-nCas9n RNA and, if needed, then expand their toolbox to incorporate other high-fidelity Cas9s.

4 Notes

1. When sgRNAs are analyzed on agarose gel without denaturation, one should typically *see* two bands. By contrast, this protocol ensures that sgRNAs are visualized as single bands on agarose gel of 100 nucleotides.
2. PCR products do not contain methylation and are therefore resistant to DpnI. By contrast, plasmids contain adenine methylation and can be digested by DpnI and are thereby inactivated.
3. The polyA insert is unstable in most bacterial strains. Even in the strains engineered for generating the unstable constructs such as Stbl2 cells, there is some instability of polyadenine stretches.
4. Many protocols give precise injection volume such as "1 nL." It is quite cumbersome to estimate such small volumes. Therefore, we prefer to inject different concentrations of active molecular ingredients such as Cas9/sgRNA mixes or morpholino oligonucleotides and use visual judgement to make sure that we do not inject too much. The viability and deformities after injections should also be used to guide the highest non-toxic dosages of the injected material.
5. The gel percentage for HMA is very important for optimal resolution. The following table provides a guide to choosing the gel percentage for your experiment:

Size range, bp	Polyacrylamide gel % for DNA analysis
100–1000	3.5
75–500	6
50–400	8
35–250	10
20–150	12

6. The progress of an HMA gel run can be monitored at several time points to choose the optimal one if your gel cassettes are transparent since the whole setup can be disassembled and reassembled easily.

References

1. Jinek M, Chylinski K, Fonfara I et al (2012) A programmable dual RNA-guided DNA endonuclease in adaptive bacterial immunity. Science 337:816–821
2. Gasiunas G, Barrangou R, Horvath P, Siksnys V (2012) Cas9—crRNA ribonucleoprotein complex mediates specific DNA cleavage for adaptive immunity in bacteria. Proc Natl Acad Sci U S A 109:2579–2586
3. Cho SW, Kim S, Kim JM, Kim J-S (2013) Targeted genome engineering in human cells with the Cas9 RNA-guided endonuclease. Nat Biotechnol 31:230–232
4. Cong L, Ran FA, Cox D et al (2013) Multiplex genome engineering using CRISPR/Cas systems. Science 339:819–823
5. Mali P, Yang L, Esvelt KM et al (2013) RNA-guided human genome engineering via Cas9. Science 339:823–826
6. Kass EM, Jasin M (2010) Collaboration and competition between DNA double-strand break repair pathways. FEBS Lett 584:3703–3708
7. Voit RA, McMahon MA, Sawyer SL, Porteus MH (2013) Generation of an HIV resistant T-cell line by targeted "stacking" of restriction factors. Mol Ther 21:786–795
8. Fu Y, Foden JA, Khayter C et al (2013) High-frequency off-target mutagenesis induced by CRISPR-Cas nucleases in human cells. Nat Biotechnol 31:822–826
9. Hsu PD, Scott DA, Weinstein JA et al (2013) DNA targeting specificity of RNA-guided Cas9 nucleases. Nat Biotechnol 31:827–832
10. Pattanayak V, Lin S, Guilinger JP et al (2013) High-throughput profiling of off-target DNA cleavage reveals RNA-programmed Cas9 nuclease specificity. Nat Biotechnol 31:839–843
11. Slaymaker IM, Gao L, Zetsche B et al (2016) Rationally engineered Cas9 nucleases with improved specificity. Science 351:84–89
12. Kleinstiver BP, Pattanayak V, Prew MS et al (2016) High-fidelity CRISPR-Cas9 nucleases with no detectable genome-wide off-target effects. Nature 529:490–495
13. Nishimasu H, Ran FA, Hsu PD et al (2014) Crystal structure of Cas9 in complex with guide RNA and target DNA. Cell 156:935–949
14. Anders C, Niewoehner O, Duerst A, Jinek M (2014) Structural basis of PAM-dependent target DNA recognition by the Cas9 endonuclease. Nature 513:569–573
15. Chen JS, Dagdas YS, Kleinstiver BP et al (2017) Enhanced proofreading governs CRISPR-Cas9 targeting accuracy. Nature 550:407–410
16. Casini A, Olivieri M, Petris G et al (2018) A highly specific SpCas9 variant is identified by in vivo screening in yeast. Nat Biotechnol 36:265–271
17. Vakulskas CA, Dever DP, Rettig GR et al (2018) A high-fidelity Cas9 mutant delivered as a ribonucleoprotein complex enables efficient gene editing in human hematopoietic stem and progenitor cells. Nat Med 24:1216–1224
18. Davis EE, Frangakis S, Katsanis N (2014) Interpreting human genetic variation with in vivo zebrafish assays. Biochim Biophys Acta 1842:1960–1970
19. Howe K, Clark MD, Torroja CF et al (2013) The zebrafish reference genome sequence and its relationship to the human genome. Nature 496:498–503
20. Jao L-E, Wente SR, Chen W (2013) Efficient multiplex biallelic zebrafish genome editing using a CRISPR nuclease system. Proc Natl Acad Sci U S A 110:13904–13909
21. Fujii W, Onuma A, Sugiura K, Naito K (2014) Efficient generation of genome-modified mice via offset-nicking by CRISPR/Cas system. Biochem Biophys Res Commun 445:791–794
22. Yin L, Jao L-E, Chen W (2015) Generation of targeted mutations in zebrafish using the CRISPR/Cas system. Methods Mol Biol 1332:205–217
23. Burger A, Lindsay H, Felker A et al (2016) Maximizing mutagenesis with solubilized CRISPR-Cas9 ribonucleoprotein complexes. Development 143:2025–2037
24. Xiao A, Zhang B (2016) Generation of targeted genomic deletions through CRISPR/Cas system in zebrafish. Methods Mol Biol 1451:65–79

25. Prykhozhij SV, Steele SL, Razaghi B, Berman JN (2017) A rapid and effective method for screening, sequencing and reporter verification of engineered frameshift mutations in zebrafish. Dis Model Mech 10:811–822
26. Gagnon JA, Valen E, Thyme SB et al (2014) Efficient mutagenesis by Cas9 protein-mediated oligonucleotide insertion and large-scale assessment of single-guide RNAs. PLoS One 9:e98186
27. Prykhozhij SV, Rajan V, Berman JN (2015) A guide to computational tools and design strategies for genome editing experiments in zebrafish using CRISPR/Cas9. Zebrafish 13:70–73
28. Cui Y, Xu J, Cheng M et al (2018) Review of CRISPR/Cas9 sgRNA design tools. Interdiscip Sci 10:455–465
29. Yoshimi K, Kunihiro Y, Kaneko T et al (2016) SsODN-mediated knock-in with CRISPR-Cas for large genomic regions in zygotes. Nat Commun 7:10431
30. Prykhozhij SV, Fuller C, Steele SL et al (2018) Optimized knock-in of point mutations in zebrafish using CRISPR/Cas9. Nucleic Acids Res 46:9252
31. Middel V, Zhou L, Takamiya M et al (2016) Dysferlin-mediated phosphatidylserine sorting engages macrophages in sarcolemma repair. Nat Commun 7:12875

Chapter 22

Next Generation of Adoptive T Cell Therapy Using CRISPR/Cas9 Technology: Universal or Boosted?

Sébastien Wälchli and Mouldy Sioud

Abstract

Adoptive T cell therapy (ACT) using either chimeric antigen receptor (CAR)- or T cell receptor (TCR)-engineered lymphocytes has emerged as a promising strategy to treat cancer. However, this therapy is still facing enormous challenges such as poor quality of autologous T cells, T cell exhaustion, and the immune suppressive tumor microenvironments. Additionally, graft-versus-host disease is an issue that must be addressed to allow the use of allogeneic T cells. Strategies to overcome these therapeutic challenges using gene editing technology are now being developed. One strategy is to disrupt TCR and/or MHC expression in healthy donor T cells to generate T cells for universal use. Another strategy is to improve the quality of patient's T cells by eliminating either the expression of selected immune checkpoint receptors or negative regulators of TCR signaling and/or T-cell homeostasis. Here, we review the use of CRISPR-Cas9 platform in T cell engineering with a focus on the development of universal T cells and boosted autologous cells for next-generation ACT.

Key words Adoptive T cell therapy, Chimeric antigen receptor, T cell receptor, CRISPR-Cas9, Tumor infiltrating lymphocytes, Graft-versus-host disease (GvHD), Gene editing

1 Introduction

The transfer of immune cells to obtain an antitumor effect was first shown in allogeneic hematopoietic cell transplantation (allo-HCT) and donor-leukocyte infusion (DLI), where the original donor's leukocytes were reinfused into the recipient at the time of relapse [1]. Unfortunately, both treatments are often associated with toxicity in the form of graft-versus-host disease (GvHD), a significant cause of morbidity and mortality [2]. Hence, safe and more specific treatments are required to harness the use of the immune system. In the 1980s, Rosenberg and colleagues developed a form of adoptive T cell therapy that uses tumor-infiltrating lymphocytes (TILs) [3]. This form of immunotherapy relies on the isolation of T cells with potent antitumor immune response from the immunosuppressive tumor microenvironments to a setting where they can

Mouldy Sioud (ed.), *RNA Interference and CRISPR Technologies: Technical Advances and New Therapeutic Opportunities*, Methods in Molecular Biology, vol. 2115, https://doi.org/10.1007/978-1-0716-0290-4_22,

expend, ex vivo, and then be injected back into the patient, who has in the meantime undergone treatment to deplete remaining lymphocytes, to favor the survival TILs. Injected tumor-specific T cells are expected to traffic to the tumor sites and kill tumor cells. This therapy has shown some efficiency in patients with advanced melanoma who had little or no benefit from conventional therapies [4]. However, TILs can only be generated from a subset of patients, and their preparation is time-consuming, technically challenging, and expensive [5, 6]. Moreover, several studies have demonstrated that TILs do sometimes not exert antitumor effects due to T cell exhaustion and/or reduced number of reactive clones [7, 8]. Despite these limitations, TIL therapy is now being enhanced by the use of new technologies such as deep sequencing to improve the quality of the injected T cells [9] or the use of genome-edited T cells (*see* Subheading 4 and Chapter 23).

2 Redirection of Cytotoxic T Cells to Kill Tumor Cells

Early studies by Dembic and colleagues showed that the transfer of the TCR genes into recipient T cells resulted in redirected antigen specificity of these T cells [10]. The identification of the coding sequence of tumor antigen-specific TCRs led to the development of so-called passive immunotherapy, where patient T cells were genetically modified to express selected receptors [11, 12]. Based on these early studies, several groups have demonstrated that TCR gene transfer could be used to circumvent cancer immune evasion [12]. The idea was to isolate normal T cells from blood, which is an easier process than isolating TILs, and then the cells are programmed to express a TCR specific for tumor antigen before injecting them back to the patient. Thus, all T cells were retargeted against the same tumor antigen in the context of a major histocompatibility complex (MHC) molecule. Therefore, a therapeutic TCR can be used in different patients if they share the same MHC class I molecules and the same mutation, if the TCR is directed against a neo-epitope. Mutations in cancer are the source of neoantigens that can be recognized by the immune system as foreign-like peptides called neoepitopes [13].

In addition to TCR, the genetic retargeting of T cells to tumor cells can be achieved by transduction with a chimeric antigen receptor (CAR) [14]. The prototype CAR was published 30 years ago, and even if the construct has evolved, the design usually contains the antigen binding region of a monoclonal antibody as a single-chain variable fragment (scFv) fused to a transmembrane domain and the intracellular signaling domain of other receptors such as the CD3ζ chain as illustrated in Fig. 1 [14]. CARs have further evolved into more complex and sensitive molecules by the incorporation of additional costimulatory domains into the CD3ζ-based

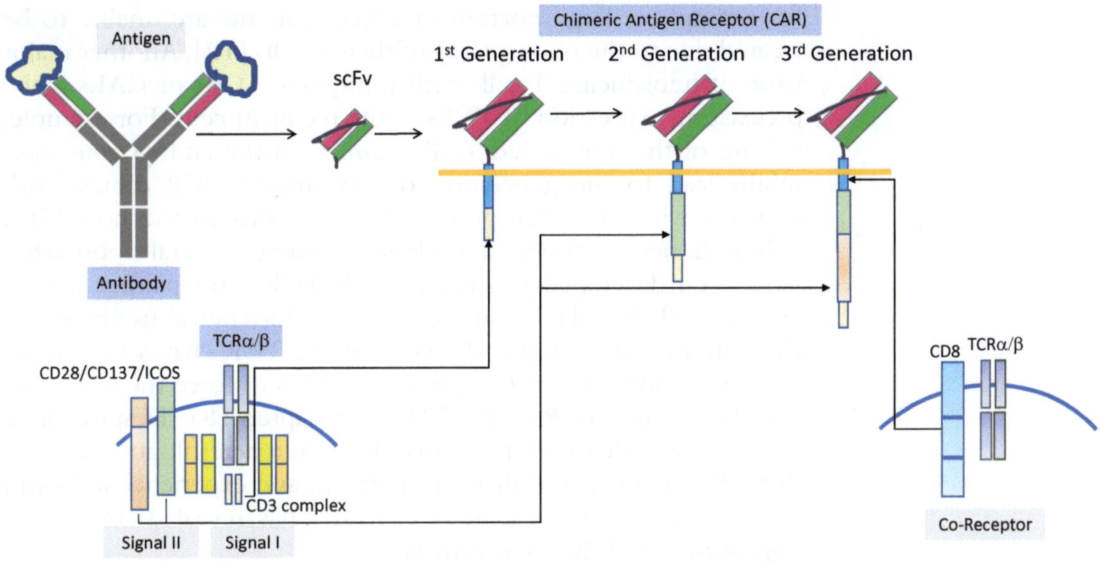

Fig. 1 Schematic representation of the three CAR generations. The single-chain variable fragment (scFv) derived from the original antibody forms the extracellular part. The transmembrane domain is normally derived from a T-cell resident protein such as CD8 or CD28. The intracellular signaling domain is composed of an arrangement of signaling motifs of receptors or adaptor proteins involved in TCR signaling. The different CAR generations follow the inclusion of more complex features. Signal I corresponds to the direct TCR signaling which is recapitulated by the CD3ξ ITAM, whereas signal II corresponds to coreceptor signaling domains

cytoplasmic domain, resulting in second- or third-generation CARs [15]. These modifications led to a significant increase in CAR T cell effector function, persistence, and survival.

In contrast to TCRs, CARs generally are confined to surface antigens and recognize tumor antigens in a MHC-independent manner, facilitating their universal use. Due to the high affinity for target recognition, CARs can generate extremely strong T-cell responses. Additionally, CAR expression is not restricted to T cells, but other effector cells such as NK cells can also express CARs [16]. CAR-mediated activation of NK cells bypasses inhibitory signals and overcomes NK resistance of tumor cells. We have recently shown that NK cells could express modified TCRs or normal TCRs when genetically reprogrammed [17, 18].

3 Weaknesses of the TCR/CAR Therapies

Most clinical ACT trials to date used autologous T cells collected from patients at the time of enrollment. While the use of these T cells has produced successful results particularly in the CD19 CAR treatment of B-cell malignancies, limitations and constraints exist;

for example, a large portion of cancer patients are unable to be treated due to the poor quality of their T cells [19]. Another major issue of transducing T cells with transgenic TCRs or CARs is the preexisting expression of TCRs in the recipient cells. For example, pairing of the introduced TCR chains with the endogenous ones might lead to the generation of new mixed TCR dimers with neo-reactivities that are potentially dangerous to the host [20], although never shown in a clinical setting. Several approaches have been developed to prevent such TCR cross-pairing and to generate efficient TCR transgenic T cells for clinical use by replacing the human constant domains of the TCR chains by murine ones or by adding of extra cysteine residues to force the pairing of the therapeutic TCR ([21, 22], *see* Chapter 18). Despite these efforts, the evaluation of the mispairing in a heterologous population of T cells seems difficult to perform and experience in "clean system" such as homogenous cell lines has revealed only slight improvement of the TCR activity.

Another important effort was to increase the TCR affinity or functional avidity by mutating selected position within the TCR-MHC contact surface, but at the cost of unpredicted specificity. Finally, repetitive tumor encountering can dampen effector functions by reprogramming the engineered T cells toward an exhausted state. CAR recognition of surface antigens or enhanced TCR binding to a target antigen can generate an extremely strong stimulation signal which might lead to T-cell exhaustion [23]. Accordingly, the strong promoters used to express the therapeutic constructs might saturate the protein production machinery. Additionally, the excess receptor presence on the plasma membrane might modify the basal signaling, and the combination of all these effects might also generate exhausted T cells.

To overcome the problem of poor T cell collections and long manufacture delay, investigators have developed "off-the-shelf" systems, mainly as CAR- T or NK cell solutions. The former ones are normally derived from partial or complete MHC matched allogeneic donor T cells [21]. However, the greatest barrier of implementing this therapy is the induction GvHD, which is primarily mediated through endogenous TCR activation of donor T cells against healthy tissues of the patient [24]. This effect was already reported before the use of redirected cells and is still responsible of the death of an important number of patients in clinical studies [24]. In addition to "off-the-shelf" systems, methods to overcome exhaustion and immunosuppressive signals such as the expression of checkpoint inhibitors like PD-1, CTLA-4, or TIGIT have been developed. Moreover, the potential presence of the target protein in the therapeutic T cells, which might lead to fratricide, is an issue that should be investigated. Fortunately, gene editing, particularly with the clustered regularly interspaced short palindromic repeats (CRISPR)-Cas9 system (*see* Chapters 19, 25 and 26), has provided solutions for the therapeutic challenges mentioned above.

4 Generation of Universal T Cells

Modification of T cells to make them universal emerged more than a decade ago. The strategy to make a cell universal is to remove the target and its receptor. From a manufacturing perspective, the ability to define the starting material for an allogeneic T cell product should allow greater control of the process and the generation of a more consistent and homogeneous product. Moreover, it should be possible to generate hundreds of therapeutic doses from a single GMP manufacturing run (Fig. 2). To achieve this goal, GvHD must be prevented by eliminating the expression of endogenous TCRs and/or MHC molecules on donor T cells. The first strategy consisted to target endogenous TCR by small interfering RNA (siRNA) to avoid GvHD and mixed dimers [25]. In order to obtain a stable expression of the product, the siRNA's coding sequence was introduced in the viral vectors delivering the TCR (*see* Chapter 18). It should be noted that siRNA strategy might saturate endogenous RNAi pathways such as miRNA, leading to toxic effects on the expressing T cells.

With the development of gene editing technologies, it is now possible to make various genetic modifications in primary human T cells to create safer and more effective T cell therapies (*see* Chapter 1). Among the methods, zinc-finger nucleases (ZFNs), transcription activator-like effector nucleases (TALENs), and the CRISPR-Cas 9 system are the most commonly used gene editing tools [26–30]. They can be designed to bind to specific genomic loci and create DNA double-strand breaks, repair of which leads to various genetic modifications. In contrast to ZFNs and TALENs, the CRISPR system relies upon easily engineered RNA molecules for sequence specificity versus DNA binding proteins that must be reconfigured for each targeted region [28, 30]. CRISPR-Cas 9 was first described as a gene editing tool in 2012 [28]. In just a few years, the technology has amplified in popularity due to its promise of making genome editing much faster, cheaper, and easier than ever before (*see* Chapter 19). In addition, CRISPR-Cas9 strategy has the advantage to generate permanent genetic modifications in a transient manner. Recently some protocols were released to improve the efficiency of transfecting primary T cells, thus increasing the number of modified cells in the sample before adding the therapeutic TCR or CAR genes. This was accomplished by electroporating target cells with CRISPR-Cas9 ribonucleoprotein (RNP) complexes, which will both be degraded after a few days leaving the transfected T cells stably modified without interfering with other cellular processes (*see* Chapter 23). Notably, rapid and efficient nonviral editing of endogenous genes in primary human cells will facilitate therapeutic manufacturing engineered primary human immune cells (*see* Chapters 25 and 26). The whole process and

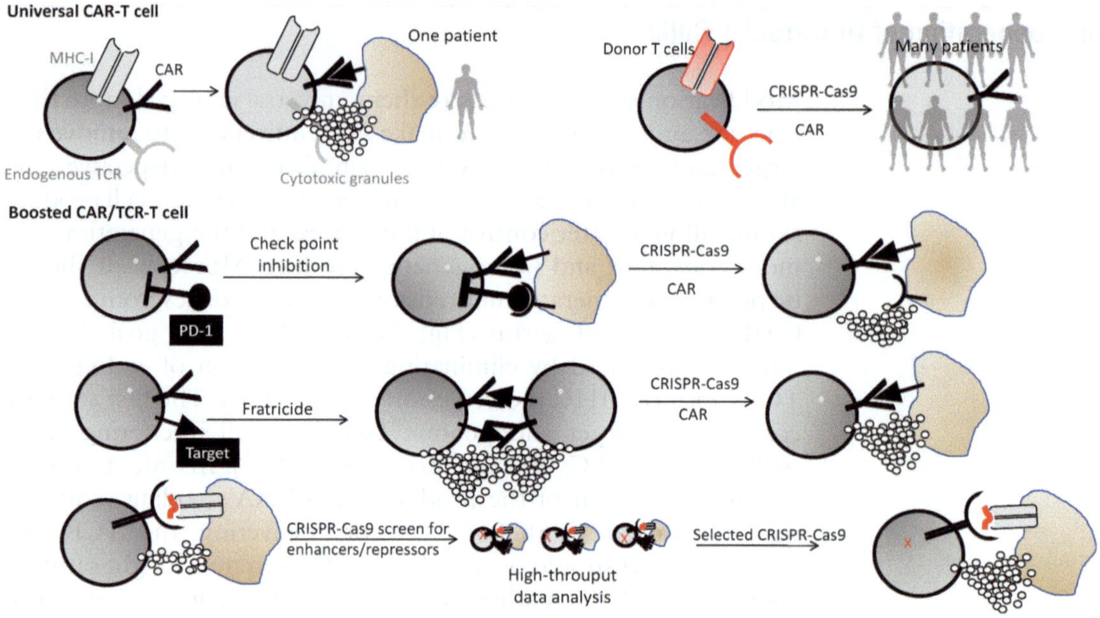

Fig. 2 Routes to the CRISPR-Cas9 integration in next generation immunotherapy. As discussed in the text, we foresee two major developments of the implementation of CRISPR-Cas9 in TCR/CAR ACT. The first one is the generation of universal T cells by removing the molecules which can lead to GvHD, namely MHC-I and TCR. Such modified T cells can be used in any cancer patient. The second route is the boosting of the T-cell product by blocking certain inhibitory molecules. The first group of molecules is represented by the checkpoint inhibitors, while the second group represents molecules expressed by T cells that might lead to fratricide killing upon CAR expression. The third group contains candidate molecules identified through whole genome screenings

required materials can be easily adapted to good manufacturing practices for clinical use.

The therapeutic success of CD19 CAR T cell therapy has been followed by the development of solutions to enable the use of this therapy in a large number of patients. In this respect, TALEN gene-edited TCR-deficient CAR T cells are being used in the clinic [31]. By contrast to TALEN, CRISPR-Cas9 system enables the simultaneous targeting of multiple loci (multiplexing). In 2017, Liu et al. reported the multiplex gene editing of CAR T cells [32]. The authors used a transient system to express both the sgRNA and the Cas9 protein and reported no toxicity. The engineered T cells were homogenous and showed antileukemia activity in a mouse xenograft model. Similarly, Ren and colleagues exploited a multiplex genome engineering to target β-2 microglobulin (B2M), TCR, and some additional checkpoint receptors [33]. Transient CRISPR/Cas9 system was also used. Because HLA class I molecules serve as major ligand inhibitors to natural killer (NK) cells, targeting β-2M may render engineered T cells sensitive to host NK cell recognition and destruction.

Although technically impressive, increasing the number of targets might also increase the number of unspecific modifications through wrong insertion, or simply due to unknown additional activities of the knockout genes. The group of Qasim presented an advanced system where the construct coding for the CAR was coupled with the anti-TCR sgRNA cassette [34]. The editing was subsequently triggered by electroporating Cas9 mRNA, thus avoiding the long-term presence of Cas9 and additional off-target modifications. The comparison of their construct with the original CD19 CAR showed that endogenous TCR elimination had no negative effect, but rather enhanced the quality and potency of the injected T cells [34]. Importantly the technology showed potential as a scalable platform for clinical use although further development is required for general use. Moreover, an isPSC cell line was generated from healthy peripheral blood T cells, which also contains a CD19-targeted CAR and biallelic TCR locus disruption. The engineered CAR T cells were fully functional in vitro [14]. An elegant approach was published by Sadelain and coworkers where the CRISPR-Cas9 system was exploited to simultaneously disrupt the *TRCA* exon, which encodes for the constant part of the TCR-α and replaced it with the CD19 CAR coding sequence. This method combines the endogenous TCR knockout with the advantage of maintaining the level of the therapeutic CAR at a "physiological" level, related to the natural TCR levels [35]. This strategy should avoid T cell exhaustion when compared with other CAR expression strategies. Indeed, the data were spectacular and the quality of the therapeutic T cells was enhanced. With respect to therapy, 5 clinical trials with CRISPR-Cas9 gene-edited universal allogeneic are to date (April 2019) reported in clinicaltrial.com, one being CAR T cells where both B2M and endogenous TCR-α have been deleted (NCT03166878).

5 Targeting the Un-targetable

Although several advances have been made in cancer treatments, some malignancies are difficult to target using the current T-cell-based ACT. Therefore, there is a need for additional targets for immunotherapy. Among the candidates, CD3, CD5, and CD7 were proposed as target for CAR therapy. Although the use of NK cells has been investigated, the use of the CRISPR-Cas9 system to knockout the endogenous protein in the redirected cells was found to be feasible for CD7 [36]. By deleting the endogenous TCR and CD7, Cooper and coworkers have created an "off-the-shelf" T cell product (UCART7) [37]. Two of the myeloid leukemia surface antigens CD33 and CD123 are also targeted by the CRISPR-Cas9 system. Currently, there are three trials on CAR T cell targeting CD33 and two trials targeting CD123 antigen in

China (e.g., NCT01864902, NCT02937103). In addition to leukemia cells, CD33 is also expressed in stem cells, rendering its use difficult. Interestingly, CRISPR-Cas9 deletion of CD33 in stem cells enabled the use of CAR T cell therapy in acute myeloid leukemia [38, 39].

6 Boosting Autologous T Cells

CRISPR-Cas9 technology has been used in screens to study the gene involved in positive outcome of immunotherapy (RESTIFIO) or immune functions [40]. These studies have generated valuable information concerning the genes that could potentially be deleted to improve the immune response. Such information could help to develop enhanced or boosted T cells. As indicate above, subsequent to sustained activation and/or extended rounds of expansion, T cells express inhibitory molecules and checkpoint receptors that block their effector function [23]. Immune checkpoints are normal "immunological brakes" that tumor cells exploit to escape from antigen-directed T cells. Several studies have shown that the use of monoclonal antibodies against certain immune checkpoints can rescue T cell exhaustion and restore T cell function. Unfortunately, only a fraction of patients responded to checkpoint inhibitors [41]. With the development of CRISPR-Cas 9 technology, it is possible to knockout genes involved in immunosuppression. Interestingly, PD1 knockout T cells showed enhanced function [42]. Similarly, PD-1 knockout in combination with different CARs was reported to enhance T cell responses against some cancers [43–45]. Another checkpoint inhibitor, LAG3, was also knocked out in combination with CAR and shown to provide enhanced T-cell activity [46]. Removing the break directly on the redirected cell has the advantage to avoid the off-target effects reminiscent to systemic injection of antibodies, and even if the PD1 deletion seems to be tolerated by T cells, the long-term effect is unknown. In addition, removing a checkpoint may upregulate the expression of other immunosuppressive signals. CRISPR clinical trials are mainly running in China. Only a few examples of TCR redirected T cells have been tested for CRISPR-Cas9 enhancement. Finally, multiplex CRISPR-Cas 9 genomic editing, allogeneic universal CAR T cells that are deficient in the TCR β chain, β2M, PD-1, and CTLA-4 have been generated [47]. These CAR T cells maintain function both in vitro and in vivo without causing GvHD. Additional disruption of PD-1 in these CAR T cells significantly improved their in vivo antitumor activities. It should be noted that a new generation of CAR T cells using our in house developed scFv antibodies is under investigation in our laboratory.

7 Concluding Remarks

With its simplicity of design and construction, the CRISPR-Cas9 system represents an appealing and facile approach for human genome engineering with broad applications in basic and applied medical research. TCR replacement by CRISPR-Cas9, or other means, will generate clinically useful T cells that do not encompass the dangers of TCR mispairing and that can be orders of magnitude more sensitive than the products currently being used. The production of allogeneic gene-edited T cells or CAR T cells for the generation of universal donor cells unable for inducing GvHD is encouraging. Engineered allogeneic cells would be available "off the shelf" eliminating the need for patient apheresis and reducing time to treatment. Although the frequency of gene editing in primary cells needs significant improvements, off-the-shelf solutions will most probably take over autologous strategies in the near future, mainly due to the standardization of the therapeutic products and cost reduction.

Acknowledgments

This work was supported by the Norwegian Cancer Society and South-Eastern Norway Regional Health Authority.

References

1. Singh AK, McGuirk JP (2016) Allogeneic stem cell transplantation: a historical and scientific overview. Cancer Res 76:6445–6451
2. Norkin M, Wingard JR (2017) Recent advances in hematopoietic stem cell transplantation. F1000Res 6:870
3. Rosenberg SA, Packard BS, Aebersold PM, Solomon D, Topalian SL, Toy ST, Simon P, Lotze MT, Yang JC, Seipp CA et al (1988) Use of tumor-infiltrating lymphocytes and interleukin-2 in the immunotherapy of patients with metastatic melanoma. A preliminary report. N Engl J Med 319:1676–1680
4. Geukes Foppen MH, Donia M, Svane IM, Haanen JBAG (2015) Tumor-infiltrating lymphocytes for the treatment of metastatic cancer. Mol Oncol 9:1918–1935
5. Lindenberg MA, Retèl VP, van den Berg JH, Geukes Foppen MH, Haanen JB, van Harten WH (2018) Treatment with tumor-infiltrating lymphocytes in advanced melanoma: evaluation of early clinical implementation of an advanced therapy medicinal product. J Immunother 41:413–425
6. vane IM, Verdegaal EM (2014) Achievements and challenges of adoptive T cell therapy with tumor-infiltrating or blood-derived lymphocytes for metastatic melanoma: what is needed to achieve standard of care? Cancer Immunol Immunother 63:1081–1091
7. Karagiannis P, Iriguchi S, Kaneko S (2016) Reprogramming away from the exhausted T cell state. Semin Immunol 28:35–44
8. Zikich D, Schachter J, Besser MJ (2016) Predictors of tumor-infiltrating lymphocyte efficacy in melanoma. Immunotherapy 8:35–43
9. Robbins PF, Lu Y-C, El-Gamil M, Li YF, Gross C, Gartner J, Lin JC, Teer JK, Cliften P, Tycksen E et al (2013) Mining exomic sequencing data to identify mutated antigens recognized by adoptively transferred tumor-reactive T cells. Nat Med 19:747–752
10. Dembic Z, Haas W, Weiss S, McCubrey J, Kiefer H, von Boehmer H, Steinmetz M (1986) Transfer of specificity by murine alpha

and beta T-cell receptor genes. Nature 320:232–238

11. Johnson LA, Heemskerk B, Powell DJ Jr, Cohen CJ, Morgan RA, Dudley ME, Robbins PF, Rosenberg SA (2006) Gene transfer of tumor-reactive TCR confers both high avidity and tumor reactivity to nonreactive peripheral blood mononuclear cells and tumor-infiltrating lymphocytes. J Immunol 177:6548–6559
12. Morgan RA, Dudley ME, Wunderlich JR, Hughes MS, Yang JC, Sherry RM, Royal RE, Topalian SL, Kammula US, Restifo NP et al (2006) Cancer regression in patients after transfer of genetically engineered lymphocytes. Science 314:126–129
13. Efremova M, Finotello F, Rieder D, Trajanoski Z (2017) Neoantigens generated by individual mutations and their role in cancer immunity and immunotherapy. Front Immunol 8:1679. eCollection 2017
14. Zhao J, Lin Q, Song Y, Liu D (2018) Universal CARs, universal T cells, and universal CAR T cells. J Hematol Oncol 11:132
15. Zhang C, Liu J, Zhong JF, Zhang X (2017) Engineering CAR-T cells. Biomark Res 5:22
16. Fesnak AD, June CH, Levine BL (2016) Engineered T cells: the promise and challenges of cancer immunotherapy. Nat Rev Cancer 16:566–581
17. Walseng E, Köksal H, Sektioglu IM, Fåne A, Skorstad G, Kvalheim G, Gaudernack G, Inderberg EM, Wälchli S (2017) A TCR-based chimeric antigen receptor. Sci Rep 7:10713
18. Mensali N, Dillard P, Hebeisen M, Lorenz S, Theodossiou T, Myhre MR, Fåne A, Gaudernack G, Kvalheim G, Myklebust JH et al (2019) NK cells specifically TCR-dressed to kill cancer cells. EBioMedicine 40:106–117
19. Sing N, Perazzelli J, Grupp SA, Barrett DM (2016) Early memory phenotypes drive T cells proliferation in patients with pediatric malignancies. Sci Transl Med 8:320ra3
20. van Loenen MM et al (2010) Mixed T cell receptor dimers harbor potentially harmful neoreactivity. Proc Natl Acad Sci U S A 107:10972–10977
21. Stauss HJ et al (2007) Monoclonal T-cell receptors: new reagents for cancer therapy. Mol Ther 15:1744–1750
22. Kuball J et al (2007) Facilitating matched pairing and expression of TCR chains introducd into human T cells. Blood 109:2331–2338
23. Wherry EJ, Kurachi M (2015) Molecular and cellualer insigjhts into T cell exhaustion. Nat Rev Immunol 15:486–499
24. Nassereddine S, Rafei H, Elbahesh E, Tabbara I (2017) Acute graft *Versus* host disease: a comprehensive review. Anticancer Res 37:1547–1555
25. Okamoto S et al (2009) Improved expression and reactivity of transduced tumor-specific TCRs in human lymphocytes by specific silencing of endogenous TCR. Cancer Res 69:9003–9011
26. Urnov FD et al (2010) Genome editing with engineered zinc finger nucleases. Nat Rev Genet 11:636–646
27. Bogdanove AJ, Voytas DF (2011) TAL effectors: customizable proteins for DNA targeting. Science 333:1843–1846
28. Doudna JA, Charpentier E (2014) Genome editing. The new frontier of genome engineering with CRISPR-Cas 9. Science 346:1258096
29. Salsman J, Dellaire G (2017) Precision genome editing in the CRISPR era. Biochem Cell Biol 95:187–201
30. Wiedenheft B, Sternberg SH, Doudna JA (2012) RNA-guided genetic silencing slencing systems in bacteria and archaea. Nature 482:331–338
31. Qasim W et al (2017) Molecular remission of infant B-ALL after infusion of universal TALEN gene-edited CAR T cells. Sci. Transl med 9:eaaj2013
32. Liu XX, Zhang Y, Cheng C, Cheng AW, Zhang X, Li N, Xia C, Wei X, Liu XX, Wang H (2017) CRISPR-Cas9-mediated multiplex gene editing in CAR-T cells. Cell Res 27:154–157
33. Ren J, Liu X, Fang C, Jiang S, June CH, Zhao Y (2017) Multiplex genome editing to generate universal CAR T cells resistant to PD1 inhibition. Clin Cancer Res 23:2255–2266
34. Georgiadis C, Preece R, Nickolay L, Etuk A, Petrova A, Ladon D, Danyi A, Humphryes-Kirilov N, Ajetunmobi A, Kim D et al (2018) Long terminal repeat CRISPR-CAR-coupled "universal" T cells mediate potent anti-leukemic effects. Mol Ther 26:1215–1227
35. Eyquem J, Mansilla-Soto J, Giavridis T, van der Stegen SJC, Hamieh M, Cunanan KM, Odak A, Gönen M, Sadelain M (2017) Targeting a CAR to the TRAC locus with CRISPR/Cas9 enhances tumour rejection. Nature 543:113–117
36. Gomes-Silva D, Srinivasan M, Sharma S, Lee CM, Wagner DL, Davis TH, Rouce RH, Bao G, Brenner MK, Mamonkin M (2017) CD7-edited T cells expressing a CD7-specific CAR for the therapy of T-cell malignancies. Blood 130:285–296

37. Cooper ML, Choi J, Staser K, Ritchey JK, Devenport JM, Eckardt K, Rettig MP et al (2018) An "off-the-shelf" fratricide-resistant CAR-T for the treatment of T cell hematologic malignancies. Leukemia 32:1970–1983
38. Humbert O, Laszlo GS, Sichel S, Ironside C, Haworth KG, Bates OM, Beddoe ME, Carrillo RR, Kiem H-P, Walter RB (2019) Engineering resistance to CD33-targeted immunotherapy in normal hematopoiesis by CRISPR/Cas9-deletion of CD33 exon 2. Leukemia 33:762–808
39. Kim MY, Yu K-R, Kenderian SS, Ruella M, Chen S, Shin T-H, Aljanahi AA, Schreeder D, Klichinsky M, Shestova O et al (2018) Genetic inactivation of CD33 in hematopoietic stem cells to enable CAR T cell immunotherapy for acute myeloid Leukemia. Cell 173:1439–1453. e19
40. Gao Q, Dong X, Xu Q, Zhu L, Wang F, Hou Y, Chao CC (2019) Therapeutic potential of CRISPR/Cas9 gene editing in engineered T-cell therapy. Cancer Med 8(9):4254–4264. https://doi.org/10.1002/cam4.2257
41. Sioud M (2018) T-cell cross-reactivity may explain the large variation in how cancer patients respond to checkpoint inhibitors. Scand J Immunol 87. https://doi.org/10.1111/sji.12643.
42. Zhang C, Peng Y, Hublitz P, Zhang H, Dong T (2018) Genetic abrogation of immune checkpoints in antigen-specific cytotoxic T-lymphocyte as a potential alternative to blockade immunotherapy. Sci Rep 8:5549
43. Hu W, Zi Z, Jin Y, Li G, Shao K, Cai Q, Ma X, Wei F (2019) CRISPR/Cas9-mediated PD-1 disruption enhances human mesothelin-targeted CAR T cell effector functions. Cancer Immunol Immunother 68:365–377
44. Su S, Zou Z, Chen F, Ding N, Du J, Shao J, Li L, Fu Y, Hu B, Yang Y et al (2017) CRISPR-Cas9-mediated disruption of PD-1 on human T cells for adoptive cellular therapies of EBV positive gastric cancer. Oncoimmunology 6: e1249558
45. Rupp LJ, Schumann K, Roybal KT, Gate RE, Ye CJ, Lim WA, Marson A (2017) CRISPR/Cas9-mediated PD-1 disruption enhances anti-tumor efficacy of human chimeric antigen receptor T cells. Sci Rep 7:737
46. Zhang Y, Zhang X, Cheng C, Mu W, Liu X, Li N, Wei X, Liu X, Xia C, Wang H (2017) CRISPR-Cas9 mediated LAG-3 disruption in CAR-T cells. Front Med 11:554–562
47. Ren J, Zhang X, Liu X, Fang C, Jiang S, June CH, Zhao Y (2017) A versatile system for rapid multiplex genome-edited CAR T cell generation. Oncotarget 8:17002–17011

Chapter 23

Engineering T Cells Using CRISPR/Cas9 for Cancer Therapy

Xingying Zhang, Chen Cheng, Wen Sun, and Haoyi Wang

Abstract

Recent advances in the development of gene editing technologies, especially the CRISPR/Cas 9 system, have substantially enhanced our ability to make precise and efficient changes in the genomes of various cells. In particular, the genetic engineering of T cells holds huge potential to improve the efficacy and safety of T cells-based cancer therapy. Due to its ease of use and high efficiency, CRISPR/Cas9 enables efficient gene knockout, site-specific knock-in, and genome-wide screen in T cells. Here we review the current progress of applying gene editing to T-cell therapy, focusing on the technical aspects of the CRISPR/Cas9 platform. We also discuss the challenges and future prospects.

Key words CRISPR-Cas9, T cells, Gene editing, Adoptive cell therapy, Immunotherapy

1 Introduction

Genome-editing technologies enable the generation of targeted genomic modifications with outstanding efficiency and specificity. The rationale of gene editing is that programmable site-specific nuclease induces DNA double-strand breaks (DSBs) at specific genomic loci, employing endogenous DNA repair pathways, such as nonhomologous end joining (NHEJ) or homology-directed repair (HDR), to generate various genetic modifications [1].

The most widely used gene editing tools include zinc finger nucleases (ZFNs) [2–8], transcription activator-like effector nucleases (TALENs) [9–12], and Clustered Regularly Interspaced Short Palindromic Repeats (CRISPR) and the CRISPR-associated protein 9 (Cas9) (CRISPR/Cas9) [13–21]. ZFNs and TALENs are chimeric nucleases composed of a modular DNA-binding domain and a sequence-independent nuclease domain Folk I. Design of a pair of ZFN with high specificity and activity is technically challenging and time-consuming, whereas generation of good TALENs

---

Xingying Zhang and Chen Cheng contributed equally to this work.

Mouldy Sioud (ed.), *RNA Interference and CRISPR Technologies: Technical Advances and New Therapeutic Opportunities*, Methods in Molecular Biology, vol. 2115, https://doi.org/10.1007/978-1-0716-0290-4_23,

requires complex molecular cloning. In comparison, Cas9 nuclease is guided by a short RNA molecule (single-guide RNA, sgRNA) that base-pairs directly with target DNA; thus, the specificity can be easily programmed by alternating gRNA sequences (*see* Chapter 19). Another advantage of the CRISPR/Cas9 system is the ease of multiplexing, enabling the introduction of multiple DSBs within the same cell via expressing multiple sgRNAs. Nonetheless, all these nucleases have been shown to achieve efficient genome editing in mammalian cells.

To date, genetically modified immune cells, primarily T cells, have been applied in the clinic for the treatment of cancers. The main strategy is to empower T cells to target cancer by introducing either natural T-cell antigen receptors (TCRs) or artificially designed chimeric antigen receptors (CARs) that recognize tumor antigen. TCR-T cells recognize the tumor-specific cytoplasmic peptide epitopes presented by the major histocompatibility complex (MHC) [22]. In contrast, target recognition by CAR is determined by the single-chain Fv fragment (scFv) of an antibody, thus bypassing the MHC restriction [23]. CD19-directed CAR-T therapy has shown remarkable performance in the clinic and has been recently approved by the FDA for the treatment of refractory B-cell leukemias and lymphomas [24–27].

Despite the recent progress in T-cell therapies, there remain major obstacles to be tackled for treating both hematologic and solid tumors. First, the complex and long manufacture procedure of autologous CAR-T cells is not suitable for patients with minimal healthy T cells or with need of immediate treatment. Generation of off-the-shelf universal CAR-T cells is highly desirable because it will not only solve these problems but also significantly reduce the product cost. Second, T-cell exhaustion and the immune suppressive tumor microenvironment pose serious challenges to the efficacy of CAR-T cells treating solid tumors. Third, major delivery systems derived from retrovirus or lentivirus insert the transgene semi-randomly into the genome, leading to heterogeneity of transgene expression and posing risks of insertional mutagenesis. Genome-editing technologies, especially the CRISPR/Cas9 system, hold great promise to overcome these challenges (Fig. 1). Knocking out (KO) immune checkpoint receptors can invigorate exhausted CAR-T cells and boost the antitumor efficacy. KO the TCR complex to avoid graft-versus-host disease (GVHD) and host rejection-related gene would facilitate the generation of universal CAR-T cells. Precise knock-in (KI) of CAR or TCR into a specific and safe genomic locus in T cells leads to the enhancement of antitumor responses and reduced risk of insertional mutagenesis. Due to ease of high-throughput assembly, CRISPR/Cas9 was also utilized in genome-wide screen for discovering novel functional genes in primary T cells.

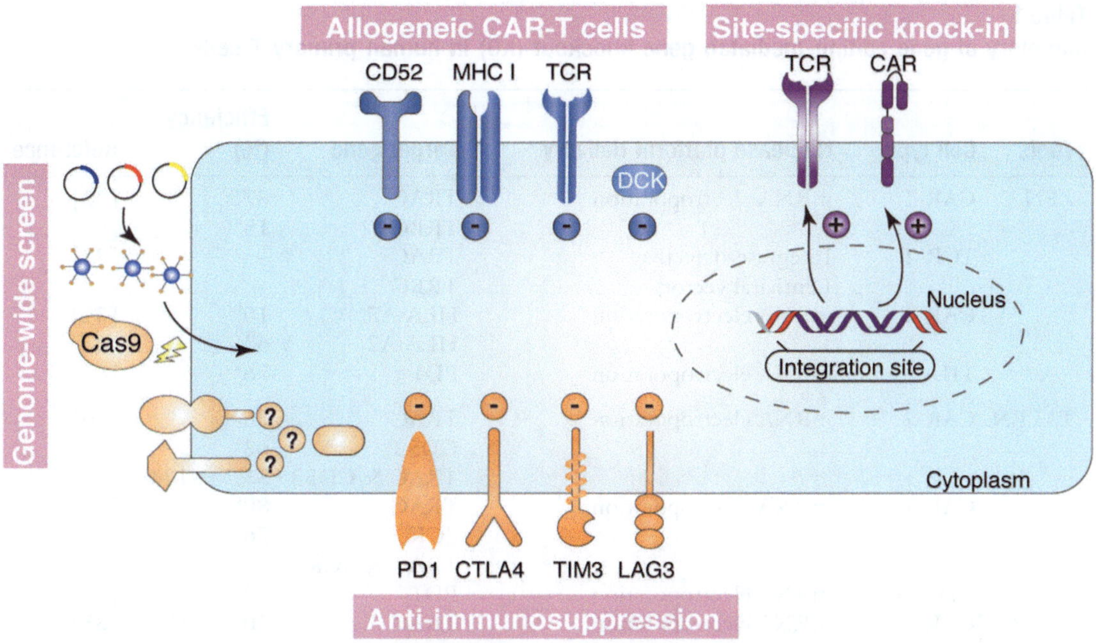

Fig. 1 Applications of CRISPR/Cas9 in primary human T cells for cancer therapy. CRISPR/Cas9 could be utilized for engineering primary human T cells to treat cancer in mainly three aspects. First, CRISPR/Cas9 platform enables multiple gene knockout to generate allogeneic CAR-T cells or to alleviate immunosuppression. Second, it allows site-specific knock-in of transgene in T cells. Lastly, Cas9 electroporation, coupled with pooled lentiviral sgRNAs transduction, makes possible the genome-wide screen to facilitate new targets discovery

Here we summarize the applications of genome-editing technologies in T-cell-based cancer therapy, mainly focusing on the technical aspect of CRISPR-Cas9 platform. Beyond cancer, gene editing in T cell also has therapeutic potential in treating infectious disease, autoimmune disease, and promoting transplantation tolerance. Due to the limited space of this review, only studies on engineering T cells for cancer therapy are reviewed. Readers interested in the other aspects are referred to the recent comprehensive or specialized reviews [28–30].

2 Gene KO in T Cells

The discovery by Dr. Maria Jasin in 1994 that DNA ends generated by a DSB in a mitotically dividing mammalian cell are recombinogenic, providing the foundation for the development of genome editing [1]. The first application of ZFNs to modify human primary T cells was to disrupt *CCR5* generating HIV-1 resistance CD4+ T cells [31]. In 2012, two independent groups used ZFNs to KO the endogenous TCRβ- and α-chain genes in a CD19-CAR-T or Wilms tumor 1 (WT1)-specific TCR-T cells, respectively [32, 33]. The

Table 1
Summary of gene editing-mediated gene knockout (KO) in human primary T cells

Tools	Cell type	Nuclease platform delivery	Target gene	Efficiency (%)	Reference
ZFN	CAR-T	mRNA electroporation	TRAC	37[a]	[32]
			TRBC	15[a]	
	TCR-T	Integrase-defective Lentiviral vectors	TRAC	–	[33]
			TRBC	–	
	CAR-T	mRNA electroporation	HLA-A3	10[b]	[34]
			HLA-A2	6[b]	
	TIL	mRNA electroporation	PD1	76[a]	[81]
TALEN	CAR-T	mRNA electroporation	TRAC	78[a]	[36]
			CD52	63[a]	
			TRAC & CD52	53.7 ± 12.9[a]	
	CAR-T	mRNA electroporation	TRAC	80[c]	[37]
			dCK	76[c]	
			TRAC & dCK	–	
	$CD8^+$ T cell	mRNA electroporation	PD1	53[c]	[82]
	CAR-T	mRNA electroporation	TRAC	70[c]	[38]
			CD52	70[c]	
			TRAC & CD52	64[a]	
CRISPR	$CD4^+$ T cell	Cas9 RNP electroporation	CXCR4	40[a]	[39]
				33[b]	
				55[c]	
			PD1	12[b]	
	T cell	Cas9 plasmid electroporation	PD1	66.67[d]	[40]
	CAR-T	Cas9 mRNA and sgRNA electroporation	TRAC	81.7[b]	[42]
			TRBC	49.3[b]	
			B2M	79.9[a]	
			PD1	–	
			TRAC & B2M	65.4[a]	
			TRAC & B2M & PD1	–	
	CAR-T	Cas9 RNP electroporation	TRAC	29.1[b]	[43]
				22.9[e]	
				31.1[a]	
			B2M	42.4[b]	
				54.1[e]	
				80.7[a]	
			PD1	29.2[b]	
				35.1[e]	
				64.7[d]	
			TRBC & B2M	48.9[a]	
			TRAC & B2M & PD1	–	
	$CD8^+$ T cell	Cas9 RNP electroporation	PD1	–	[46]
	T cell	Cas9 mRNA electroporation and transduction gRNAs-lentiviral vector	TRAC	90[a]	[51]
			TRAC & B2M	73[a]	
			TRAC & B2M & Fas	–	

(continued)

Table 1
(continued)

Tools	Cell type	Nuclease platform delivery	Target gene	Efficiency (%)	Reference
			TRAC & B2M & PD1 & CTLA4		
	CAR-T	Cas9 RNP electroporation	LAG3	50[a]	[45]
	CD8$^+$ T cell	Cas9 lentivirus transduction	CTLA4	–	[83]
	CAR-T	Cas9 RNP electroporation	DGKa	90[c]	[47]
	T cell	Cas9 mRNA and sgRNA electroporation	TRAC & CD7	72.8[a]	[48]
	CD8$^+$ T cell	Cas9 plasmid electroporation	CTLA4	–	[44]

[a]FACS (fluorescence-activated cell sorting analysis)
[b]Surveyor assay
[c]NGS, next-generation sequencing
[d]TA clone; RNP, ribonucleoprotein; –, KO efficiency not clearly indicated in the original study
[e]TIDE (Tracking of Indels by DEcomposition)

ZFNs were delivered either through integrase-defective lentiviral vectors or electrotransfer of messenger RNA. The gene disruption rates were 27–37% for *TRAC* and 4–15% for *TRBC* (Table 1). Both groups performed the ZFNs-mediate KO in collaboration with Sangamo BioSciences, a company with leading expertise in ZFN technology [4]. Their work laid the foundation for allogeneic T-cell-based immunotherapy. One year later, the group led by Laurence J. N. Cooper eliminated HLA-A in both human primary T cells and CD19-CAR-T cells using ZFNs [34].

After the DNA-binding code of TALEs was deciphered in 2009 [9–12], TALEN was developed as another powerful gene editing tool. Delivered by mRNA electroporation, TALENs were employed to KO TCR complex, resulting in ~60% *TRAC* and ~40% *TRBC* gene disruption rate [35–37]. TALENs also showed multiplex gene editing capability (*TRAC/CD52* double KO, dKO) in either primary T cells or CAR-T cells [36, 38]. Upon optimization of the mRNA format and cell handling, dKO efficiency could reach 53.7% (±12.9%). Clinical trials using TALEN-mediated TCR-knockout CAR-T cells are underway in patients with various malignancies (NCT03190278, NCT02808442, NCT02746952).

Although ZFNs and TALENs have shown promising KO efficacy [2, 9], the challenging design of ZFN and the complex molecular cloning of TALEN hinder their widespread adoption. With the advent of CRISPR/Cas9 and the demonstration of efficient gene editing in mammalian cells [13–21], many groups have applied this technology to editing various genes in human T cells. In 2015, *CXCR4*, a coreceptor for HIV entry, and *PDCD1* were disrupted using Cas9/single-guide RNA ribonucleoproteins (Cas9 RNPs) [39]. A group led by Xingxu Huang and Baorui Liu demonstrated that by optimizing the electroporation condition, delivery of CRISPR/Cas9 encoding plasmid could result in satisfactory transfection efficacy, cell viability, and KO efficiency [40]. However, some concerns arise with this deliver strategy. Plasmid DNA, albeit at a low frequency, may be incorporated into the genome, and the prolonged Cas9 expression may increase the off-target effect [41].

The first demonstration of CRISPR/Cas9-mediated multiplex gene editing in CAR-T cells was published by Carl June group and our group [42, 43]. In both studies, *TRAC*, *B2M*, and *PDCD1* were chosen as target genes. Electro-transfer of Cas9 mRNA and sgRNAs was utilized in Dr. Carl. H June's study. They systematically optimized the electroporation condition for delivering RNA into human T cells and achieved up to 80% of single-gene KO efficiency and 65% dKO efficiency. Using a different strategy, we delivered a ribonucleoprotein (RNP) consisting of the Cas9 protein in complex with sgRNAs and achieved efficient single and multiplex gene KO. In both studies, the gene-edited CAR-T cells showed potent antitumor efficacy in both in vitro assay and tumor xenograft mouse models. From then on, several groups reported CRISPR/Cas9-mediated KO of various genes in CAR-T cells or CD8+ T cells to eliminate various immune checkpoint genes (*PDCD1*, *LAG3*, *CTLA4*, *DGKα*) [44–47] or to generate allogenic CAR-T for the treatment of T-cell hematologic malignancies (*CD7*&*TRAC*) [48]. It is worth mentioning that the triphosphate at 5′ end of IVT (in vitro transcription)-sgRNA triggers the innate immunity of primary T cells, leading to cell death. Eliminating the triphosphate at 5′ end of IVT-sgRNA by CIP or using chemical synthesis sgRNA solved this problem [49, 50].

To further improve the multiplex gene-editing capacity of CRISPR/Cas9, Ren et al. incorporated multiple sgRNAs in the CAR encoding lentiviral vector. After viral transduction, Cas9 mRNA was electroplated into CAR-T cells to initiate the multiplex gene editing [51].

3 Gene KI in T Cells

In 2011, a group led by Luigi Naldini published the important proof of concept study using adenoviral vector-based ZFNs expression along with integrase-defective lentiviral vector-mediated delivery of targeting donor template for precise KI in primary human T cells [52]. They observed a 5% EGFP integration rate and showed that the integration of transgene cassette did not alter endogenous transcription at and around the AAVS1 insertion site, thus validating this site as a safe harbor for transgene integration in human T cells. Using megaTAL, a hybrid nuclease platform that combines a transcription activator-like effector (TALE) DNA-binding domain with an engineered, sequence-specific homing endonuclease,

Sather and colleagues integrated CAR gene into *CCR5* locus in primary human T cells. In this study, nuclease was delivered as mRNA along with adeno-associated virus (AAV)-mediated delivery of *CCR5*-targeting donor template [53].

In 2015, by electroporation of Cas9 RNP and 90 nt DNA oligo as HDR template, Schumann et al. achieved up to 20% efficiency of introducing a *HindIII* restriction site at CXCR4 locus in primary human T cells [39]. Compared to the 90 nt long ssDNA oligo donor, larger DNA donor encompassing a whole transgene cassette is more challenging to deliver. In 2017, Eyquem and colleagues delivered long donor DNA template via AAV6 transduction and Cas9 mRNA and sgRNA by electroporation, generating KI CAR-T cells [54]. They found that the KI efficacy was proportional to AAV dosage, and efficiency above 40% could be achieved at a multiplicity of infection of 10^6. Another group led by MacLeod used an engineered homing endonuclease for cleaving the *TRAC* locus and an AAV donor as repair template, achieving about 40% targeting efficiency [55]. All these studies used a two-step strategy, delivering nuclease by electroporation and HDR donor template by viral transduction.

To eliminate the necessity of viral vector, Roth et al. systematically optimized the electroporation conditions and demonstrated efficient KI and good cell viability could be obtained using double-strand DNA as HDR template along with Cas9 RNPs [56]. Because generation of clinical grade viral vectors is expensive and time consuming, this virus free strategy holds great potential. No matter which method was used, integration into endogenous TCR locus led to uniform CAR/TCR expression and enhanced T cell potency, compared to the random integrated CAR-T cells. Recently, a more elegant system leveraging CRISPR/Cpf1 with adeno-associated virus (AAV) was reported to build a stable CAR-T cell with HDR-mediated KI and immune-checkpoint KO (KIKO CAR-T cell) at high efficiency in one step [57]. This system enables double KI of two different CARs in the same T cells efficiently.

4 CRISPR/Cas9-Mediated Genome-Wide Screen in T Cells

Genome-wide loss-of-function screening is a powerful method to uncover novel insights into gene function. In T cells, this approach could be used to identify new targets for cancer immunotherapies. RNA interference (RNAi) is a powerful technology for loss-of-function screen in mammalian cells [58, 59]. However, it suffers from the incomplete gene repression and confounding off-target effects [60, 61]. Compared to RNAi, CRISPR/Cas9 in general has better specificity and less off-target interference, therefore becoming a preferred choice for genetic screening. The genome-scale loss-of-function screening in mammalian cells using CRISPR/Cas9 was first established in 2014 [62, 63]. Similar strategy was adopted using a pool of sgRNA-expressing lentivirus to generate a library of knockout cells that could be screened under both positive and negative selection. These pooled CRISPR screens in human cancer cell lines revealed gene targets that modulate responses to T cell-based therapies [64–66].

CRISPR screening in primary human T cells has been hindered by the infeasibility of creating clonal Cas9 expression primary T-cell line or the low transduction efficiency with Cas9-encoding lentiviral vectors [67]. Dr. Alexander Marson and colleagues combined pooled lentiviral sgRNAs transduction with Cas9 protein electroporation (SLICE) to enable loss-of-function pooled screening at genome-wide scale in primary human T cells [68]. SLICE screens identified regulators of T-cell proliferation, providing potential targets for immunotherapy. Ting et al. reported the similar design except that the Cas9 protein was preassembled with a nontargeting sgRNA [69]. They claimed that the association of nontargeting sgRNA in Cas9 RNP led to higher editing efficiency, which may be due to the enhanced Cas9 delivery when assembled with sgRNA.

5 Challenges and Prospects

The characteristics of primary human T cells, such as limited proliferative capacity, propensity to differentiate in culture, and low transfection efficiency, pose challenges to efficient gene editing. Through the great advancements of gene editing and delivery technologies reviewed above, we can now perform efficient KO, site-specific KI, and genome-wide screen in human primary T cells. However, several challenges remain.

5.1 Efficiency

Although easier than ZFNs and TALENs, highly efficient multiplex gene editing in primary human T cells using CRISPR/Cas9 is still challenging. In the previous studies, the multiplex editing efficiency is either not accurately quantified or suboptimal (Table 1). In addition, compared to the relatively high KO efficiency (Table 1), the general KI efficacy of large transgene fragment remains low (Table 2). Considering the complex tumor immune-suppressive microenvironment, blocking multiple pathways by knocking out several genes simultaneously or incorporating multiple transgene components might be necessary to achieve good efficacy. Therefore, further improving the efficiency of multiplex gene editing and transgene KI in T cells remains an important goal.

Table 2
Summary of gene editing-mediated gene knock-in (KI) in human primary T cells

Cell type	Nuclease delivery	HDR donor	Transgene	Target gene	KI efficiency (FACS)%	Reference
T cell	Adenoviral vector To deliver ZFNs	Integrase-defective Lentiviral vector	EGFP	CCR5	5	[52]
T cell	MegaTAL nuclease mRNA Electroporation	AAV vector	CD19 CAR	CCR5	8–30	[53]
CD4+ T cell	Cas9 RNP electroporation	Single-stranded DNA templates	HindIII	CXCR4	25[a]	[39]
T cell	Cas9 mRNA electroporation	AAV vector	CD19 CAR	TRAC	40	[54]
T cell	Engineered homing endonuclease electroporation	AAV vector	CD19 CAR	TRAC	40	[55]
T cell	Cas9 RNP electroporation	Double-stranded DNA Templates	NY-ESO-1 RAB11A	TRAC GFP	12.3 57.3	[56]
CD4+ T cell	LbCpfl mRNA electroporation	AAV6 for delivery of crRNA and HDR template	CD22 CAR	TRAC PDCD1	40	[57]

[a]NGS (next-generation sequencing)

5.2 Specificity

Whereas whole-genome sequencing of CRISPR/Cas9-edited cell lines has revealed a low incidence of off-target mutation [70, 71], other off-target analysis methods [72, 73] will likely identify more mutations imposed by gene editing. In addition, when multiple genes were edited, the risk of off-target mutation and chromosomal translocation will likely increase. Special safety concern should be taken into account as the gene edited T cells move toward clinical applications. However, currently there is no clear consensus as to the level of specificity that needs to be reached before these gene-edited T cells could be applied to treat a particular disease.

5.3 Better Delivery System

The major delivery systems manufacturing CAR-T cells for clinical therapy are retroviral and lentiviral-based vectors [74, 75]. Though no clear oncogenic consequences are reported so far, one study reported a lentiviral insertion in *TET2* gene led to dramatic clonal expansion of CAR-T cells in a patient, raising safety concerns [76]. Besides viral vectors, transposons such as *PiggyBac* [77] and *Sleeping Beauty* [78] *have* also been utilized for transgene expression in T cells. Clinical trials using SB-engineered CAR-T cells to treat patients with CD19+ hematologic cancers are ongoing [79]. A novel transposon system called CRISPR-associated transposase from cyanobacteria *Scytonema hofmanni* (ShCAST) that consists of Tn7-like transposase subunits and the type V-K CRISPR effector (Cas12k) [80] was recently discovered. The ShCAST system has the potential to serve as a novel site-specific gene integration strategy, leveraging the integration capacity of transposes and the targeting specificity of Cas proteins.

5.4 Truly "Universal" T Cells

Major efforts have been made to generate "universal," non-autologous T-cell products, but the cells generated using current strategies are less likely to have persistence as long as autologous T cells. Currently, the most advanced products were generated by knocking out the endogenous TCR and the pan-lymphocyte molecule *CD52* using TALENs, coupled with lentiviral vector transduction of a CD19-directed CAR. These allogeneic CAR-T cells have been used to treat patients with acute myeloid leukemia successfully (NCT03190278) [36]. To generate better universal off-the-shelf cell product, engineered T cells from a healthy donor should be able to expand substantially to support the treatment of hundreds even thousands of patients and optimally treat patients without lymphoid depletion treatment.

In conclusion, with the fast development of gene-editing technologies, genetically engineered T cells with better efficacy and safety will be developed to treat various cancers.

Acknowledgements

H. W is supported by National Key Research and Development Program of China (2018YFA0107703), Strategic Priority Research Program of the Chinese Academy of Sciences (No. XDA16010503), and National Natural Science Foundation of China (No. 31722036). We apologize if we unintentionally missed citing some of the work that deserved citation.

References

1. Rouet P, Smih F, Jasin M (1994) Introduction of double-strand breaks into the genome of mouse cells by expression of a rare-cutting endonuclease. Mol Cell Biol 14:8096–8106
2. Miller JC, Holmes MC, Wang J, Guschin DY, Lee YL, Rupniewski I, Beausejour CM, Waite AJ, Wang NS, Kim KA, Gregory PD, Pabo CO, Rebar EJ (2007) An improved zinc-finger nuclease architecture for highly specific genome editing. Nat Biotechnol 25:778–785
3. Wolfe SA, Nekludova L, Pabo CO (2000) DNA recognition by Cys2His2 zinc finger proteins. Annu Rev Biophys Biomol Struct 29:183–212
4. Urnov FD, Miller JC, Lee YL, Beausejour CM, Rock JM, Augustus S, Jamieson AC, Porteus MH, Gregory PD, Holmes MC (2005) Highly efficient endogenous human gene correction using designed zinc-finger nucleases. Nature 435:646–651
5. Porteus MH, Baltimore D (2003) Chimeric nucleases stimulate gene targeting in human cells. Science 300:763

6. Miller J, McLachlan AD, Klug A (1985) Repetitive zinc-binding domains in the protein transcription factor IIIA from Xenopus oocytes. EMBO J 4:1609–1614
7. Bibikova M, Golic M, Golic KG, Carroll D (2002) Targeted chromosomal cleavage and mutagenesis in drosophila using zinc-finger nucleases. Genetics 161:1169–1175
8. Bibikova M, Beumer K, Trautman JK, Carroll D (2003) Enhancing gene targeting with designed zinc finger nucleases. Science 300:764
9. Miller JC, Tan S, Qiao G, Barlow KA, Wang J, Xia DF, Meng X, Paschon DE, Leung E, Hinkley SJ, Dulay GP, Hua KL, Ankoudinova I, Cost GJ, Urnov FD, Zhang HS, Holmes MC, Zhang L, Gregory PD, Rebar EJ (2011) A TALE nuclease architecture for efficient genome editing. Nat Biotechnol 29:143–148
10. Li T, Huang S, Jiang WZ, Wright D, Spalding MH, Weeks DP, Yang B (2011) TAL nucleases (TALNs): hybrid proteins composed of TAL effectors and FokI DNA-cleavage domain. Nucleic Acids Res 39:359–372
11. Christian M, Cermak T, Doyle EL, Schmidt C, Zhang F, Hummel A, Bogdanove AJ, Voytas DF (2010) Targeting DNA double-strand breaks with TAL effector nucleases. Genetics 186:757–761
12. Boch J, Scholze H, Schornack S, Landgraf A, Hahn S, Kay S, Lahaye T, Nickstadt A, Bonas U (2009) Breaking the code of DNA binding specificity of TAL-type III effectors. Science 326:1509–1512
13. Mali P, Yang L, Esvelt KM, Aach J, Guell M, DiCarlo JE, Norville JE, Church GM (2013) RNA-guided human genome engineering via Cas9. Science 339:823–826
14. Cong L, Ran FA, Cox D, Lin S, Barretto R, Habib N, Hsu PD, Wu X, Jiang W, Marraffini LA, Zhang F (2013) Multiplex genome engineering using CRISPR/Cas systems. Science 339:819–823
15. Gasiunas G, Barrangou R, Horvath P, Siksnys V (2012) Cas9-crRNA ribonucleoprotein complex mediates specific DNA cleavage for adaptive immunity in bacteria. Proc Natl Acad Sci U S A 109:E2579–E2586
16. Jinek M, Chylinski K, Fonfara I, Hauer M, Doudna JA, Charpentier E (2012) A programmable dual-RNA-guided DNA endonuclease in adaptive bacterial immunity. Science 337:816–821
17. Sapranauskas R, Gasiunas G, Fremaux C, Barrangou R, Horvath P, Siksnys V (2011) The Streptococcus thermophilus CRISPR/Cas system provides immunity in *Escherichia coli*. Nucleic Acids Res 39:9275–9282
18. Deltcheva E, Chylinski K, Sharma CM, Gonzales K, Chao Y, Pirzada ZA, Eckert MR, Vogel J, Charpentier E (2011) CRISPR RNA maturation by trans-encoded small RNA and host factor RNase III. Nature 471:602–607
19. Garneau JE, Dupuis ME, Villion M, Romero DA, Barrangou R, Boyaval P, Fremaux C, Horvath P, Magadan AH, Moineau S (2010) The CRISPR/Cas bacterial immune system cleaves bacteriophage and plasmid DNA. Nature 468:67–71
20. Barrangou R, Fremaux C, Deveau H, Richards M, Boyaval P, Moineau S, Romero DA, Horvath P (2007) CRISPR provides acquired resistance against viruses in prokaryotes. Science 315:1709–1712
21. Bolotin A, Quinquis B, Sorokin A, Ehrlich SD (2005) Clustered regularly interspaced short palindrome repeats (CRISPRs) have spacers of extrachromosomal origin. Microbiology 151:2551–2561
22. Morgan RA, Dudley ME, Wunderlich JR, Hughes MS, Yang JC, Sherry RM, Royal RE, Topalian SL, Kammula US, Restifo NP, Zheng Z, Nahvi A, de Vries CR, Rogers-Freezer LJ, Mavroukakis SA, Rosenberg SA (2006) Cancer regression in patients after transfer of genetically engineered lymphocytes. Science 314:126–129
23. Wang M, Yin B, Wang HY, Wang RF (2014) Current advances in T-cell-based cancer immunotherapy. Immunotherapy 6:1265–1278
24. Turtle CJ, Hanafi LA, Berger C, Gooley TA, Cherian S, Hudecek M, Sommermeyer D, Melville K, Pender B, Budiarto TM, Robinson E, Steevens NN, Chaney C, Soma L, Chen X, Yeung C, Wood B, Li D, Cao J, Heimfeld S, Jensen MC, Riddell SR, Maloney DG (2016) CD19 CAR-T cells of defined CD4+:CD8+ composition in adult B cell ALL patients. J Clin Invest 126:2123–2138
25. Maude SL, Laetsch TW, Buechner J, Rives S, Boyer M, Bittencourt H, Bader P, Verneris MR, Stefanski HE, Myers GD, Qayed M, De Moerloose B, Hiramatsu H, Schlis K, Davis KL, Martin PL, Nemecek ER, Yanik GA, Peters C, Baruchel A, Boissel N, Mechinaud F, Balduzzi A, Krueger J, June CH, Levine BL, Wood P, Taran T, Leung M, Mueller KT, Zhang Y, Sen K, Lebwohl D, Pulsipher MA, Grupp SA (2018) Tisagenlecleucel in children and Young adults with B-cell lymphoblastic leukemia. N Engl J Med 378:439–448

26. Schuster SJ, Bishop MR, Tam CS, Waller EK, Borchmann P, McGuirk JP, Jager U, Jaglowski S, Andreadis C, Westin JR, Fleury I, Bachanova V, Foley SR, Ho PJ, Mielke S, Magenau JM, Holte H, Pantano S, Pacaud LB, Awasthi R, Chu J, Anak O, Salles G, Maziarz RT, Investigators J (2019) Tisagenlecleucel in adult relapsed or refractory diffuse large B-cell lymphoma. N Engl J Med 380:45–56
27. Neelapu SS, Locke FL, Bartlett NL, Lekakis LJ, Miklos DB, Jacobson CA, Braunschweig I, Oluwole OO, Siddiqi T, Lin Y, Timmerman JM, Stiff PJ, Friedberg JW, Flinn IW, Goy A, Hill BT, Smith MR, Deol A, Farooq U, McSweeney P, Munoz J, Avivi I, Castro JE, Westin JR, Chavez JC, Ghobadi A, Komanduri KV, Levy R, Jacobsen ED, Witzig TE, Reagan P, Bot A, Rossi J, Navale L, Jiang Y, Aycock J, Elias M, Chang D, Wiezorek J, Go WY (2017) Axicabtagene ciloleucel CAR T-cell therapy in refractory large B-cell lymphoma. N Engl J Med 377:2531–2544
28. Bailey SR, Maus MV (2019) Gene editing for immune cell therapies. Nat Biotechnol [Epub ahead of print]
29. Xiao Q, Guo D, Chen S (2019) Application of CRISPR/Cas9-based gene editing in HIV-1/AIDS therapy. Front Cell Infect Microbiol 9:69
30. Zhang Y, Mu W, Wang H (2017) Gene editing in T cell therapy. J Genet Genomics 44:415–422
31. Perez EE, Wang J, Miller JC, Jouvenot Y, Kim KA, Liu O, Wang N, Lee G, Bartsevich VV, Lee YL, Guschin DY, Rupniewski I, Waite AJ, Carpenito C, Carroll RG, Orange JS, Urnov FD, Rebar EJ, Ando D, Gregory PD, Riley JL, Holmes MC, June CH (2008) Establishment of HIV-1 resistance in CD4+ T cells by genome editing using zinc-finger nucleases. Nat Biotechnol 26:808–816
32. Torikai H, Reik A, Liu PQ, Zhou Y, Zhang L, Maiti S, Huls H, Miller JC, Kebriaei P, Rabinovich B, Lee DA, Champlin RE, Bonini C, Naldini L, Rebar EJ, Gregory PD, Holmes MC, Cooper LJ (2012) A foundation for universal T-cell based immunotherapy: T cells engineered to express a CD19-specific chimeric-antigen-receptor and eliminate expression of endogenous TCR. Blood 119:5697–5705
33. Provasi E, Genovese P, Lombardo A, Magnani Z, Liu PQ, Reik A, Chu V, Paschon DE, Zhang L, Kuball J, Camisa B, Bondanza A, Casorati G, Ponzoni M, Ciceri F, Bordignon C, Greenberg PD, Holmes MC, Gregory PD, Naldini L, Bonini C (2012) Editing T cell specificity towards leukemia by zinc finger nucleases and lentiviral gene transfer. Nat Med 18:807–815
34. Torikai H, Reik A, Soldner F, Warren EH, Yuen C, Zhou Y, Crossland DL, Huls H, Littman N, Zhang Z, Tykodi SS, Kebriaei P, Lee DA, Miller JC, Rebar EJ, Holmes MC, Jaenisch R, Champlin RE, Gregory PD, Cooper LJ (2013) Toward eliminating HLA class I expression to generate universal cells from allogeneic donors. Blood 122:1341–1349
35. Berdien B, Mock U, Atanackovic D, Fehse B (2014) TALEN-mediated editing of endogenous T-cell receptors facilitates efficient reprogramming of T lymphocytes by lentiviral gene transfer. Gene Ther 21:539–548
36. Poirot L, Philip B, Schiffer-Mannioui C, Le Clerre D, Chion-Sotinel I, Derniame S, Potrel P, Bas C, Lemaire L, Galetto R, Lebuhotel C, Eyquem J, Cheung GW, Duclert A, Gouble A, Arnould S, Peggs K, Pule M, Scharenberg AM, Smith J (2015) Multiplex genome-edited T-cell manufacturing platform for "off-the-shelf" adoptive T-cell immunotherapies. Cancer Res 75:3853–3864
37. Valton J, Guyot V, Marechal A, Filhol JM, Juillerat A, Duclert A, Duchateau P, Poirot L (2015) A multidrug-resistant engineered CAR T cell for allogeneic combination immunotherapy. Mol Ther 23:1507–1518
38. Qasim W, Zhan H, Samarasinghe S, Adams S, Amrolia P, Stafford S, Butler K, Rivat C, Wright G, Somana K, Ghorashian S, Pinner D, Ahsan G, Gilmour K, Lucchini G, Inglott S, Mifsud W, Chiesa R, Peggs KS, Chan L, Farzeneh F, Thrasher AJ, Vora A, Pule M, Veys P (2017) Molecular remission of infant B-ALL after infusion of universal TALEN gene-edited CAR T cells. Sci Transl Med 9:eaaj2013
39. Schumann K, Lin S, Boyer E, Simeonov DR, Subramaniam M, Gate RE, Haliburton GE, Ye CJ, Bluestone JA, Doudna JA, Marson A (2015) Generation of knock-in primary human T cells using Cas9 ribonucleoproteins. Proc Natl Acad Sci U S A 112:10437–10442
40. Su S, Hu B, Shao J, Shen B, Du J, Du Y, Zhou J, Yu L, Zhang L, Chen F, Sha H, Cheng L, Meng F, Zou Z, Huang X, Liu B (2016) CRISPR-Cas9 mediated efficient PD-1 disruption on human primary T cells from cancer patients. Sci Rep 6:20070
41. Kim S, Kim D, Cho SW, Kim J, Kim J-S (2014) Highly efficient RNA-guided genome editing in human cells via delivery of purified Cas9 ribonucleoproteins. Genome Research 24(6):1012–1019

42. Ren J, Liu X, Fang C, Jiang S, June CH, Zhao Y (2017) Multiplex genome editing to generate universal CAR T cells resistant to PD1 inhibition. Clin Cancer Res 23:2255–2266
43. Liu X, Zhang Y, Cheng C, Cheng AW, Zhang X, Li N, Xia C, Wei X, Liu X, Wang H (2017) CRISPR-Cas9-mediated multiplex gene editing in CAR-T cells. Cell Res 27:154–157
44. Zhang W, Shi L, Zhao Z, Du P, Ye X, Li D, Cai Z, Han J, Cai J (2019) Disruption of CTLA-4 expression on peripheral blood CD8 + T cell enhances anti-tumor efficacy in bladder cancer. Cancer Chemother Pharmacol 83:911–920
45. Zhang Y, Zhang X, Cheng C, Mu W, Liu X, Li N, Wei X, Liu X, Xia C, Wang H (2017) CRISPR-Cas9 mediated LAG-3 disruption in CAR-T cells. Front Med 11:554–562
46. Rupp LJ, Schumann K, Roybal KT, Gate RE, Ye CJ, Lim WA, Marson A (2017) CRISPR/Cas9-mediated PD-1 disruption enhances anti-tumor efficacy of human chimeric antigen receptor T cells. Sci Rep 7:737
47. Jung IY, Kim YY, Yu HS, Lee M, Kim S, Lee J (2018) CRISPR/Cas9-mediated knockout of DGK improves antitumor activities of human T cells. Cancer Res 78:4692–4703
48. Cooper ML, Choi J, Staser K, Ritchey JK, Devenport JM, Eckardt K, Rettig MP, Wang B, Eissenberg LG, Ghobadi A, Gehrs LN, Prior JL, Achilefu S, Miller CA, Fronick CC, O'Neal J, Gao F, Weinstock DM, Gutierrez A, Fulton RS, DiPersio JF (2018) An "off-the-shelf" fratricide-resistant CAR-T for the treatment of T cell hematologic malignancies. Leukemia 32:1970–1983
49. Kim S, Koo T, Jee HG, Cho HY, Lee G, Lim DG, Shin HS, Kim JS (2018) CRISPR RNAs trigger innate immune responses in human cells. Genome Res 28:367–373.
50. Mu W, Tang N, Cheng C, Sun W, Wei X, Wang H (2019) In vitro transcribed sgRNA causes cell death by inducing interferon release. Protein Cell 10:461–465
51. Ren J, Zhang X, Liu X, Fang C, Jiang S, June CH, Zhao Y (2017) A versatile system for rapid multiplex genome-edited CAR T cell generation. Oncotarget 8:17002–17011
52. Lombardo A, Cesana D, Genovese P, Di Stefano B, Provasi E, Colombo DF, Neri M, Magnani Z, Cantore A, Lo Riso P, Damo M, Pello OM, Holmes MC, Gregory PD, Gritti A, Broccoli V, Bonini C, Naldini L (2011) Site-specific integration and tailoring of cassette design for sustainable gene transfer. Nat Methods 8:861–869
53. Sather BD, Ibarra GSR, Sommer K, Curinga G, Hale M, Khan IF, Singh S, Song YM, Gwiazda K, Sahni J, Jarjour J, Astrakhan A, Wagner TA, Scharenberg AM, Rawlings DJ (2015) Efficient modification of CCR5 in primary human hematopoietic cells using a mega-TAL nuclease and AAV donor template. Sci Transl Med 7:307ra156
54. Eyquem J, Mansilla-Soto J, Giavridis T, van der Stegen SJ, Hamieh M, Cunanan KM, Odak A, Gonen M, Sadelain M (2017) Targeting a CAR to the TRAC locus with CRISPR/Cas9 enhances tumour rejection. Nature 543:113–117
55. MacLeod DT, Antony J, Martin AJ, Moser RJ, Hekele A, Wetzel KJ, Brown AE, Triggiano MA, Hux JA, Pham CD, Bartsevich VV, Turner CA, Lape J, Kirkland S, Beard CW, Smith J, Hirsch ML, Nicholson MG, Jantz D, McCreedy B (2017) Integration of a CD19 CAR into the TCR alpha chain locus streamlines production of allogeneic gene-edited CAR T cells. Mol Ther 25:949–961
56. Roth TL, Puig-Saus C, Yu R, Shifrut E, Carnevale J, Li PJ, Hiatt J, Saco J, Krystofinski P, Li H, Tobin V, Nguyen DN, Lee MR, Putnam AL, Ferris AL, Chen JW, Schickel JN, Pellerin L, Carmody D, Alkorta-Aranburu G, Del Gaudio D, Matsumoto H, Morell M, Mao Y, Cho M, Quadros RM, Gurumurthy CB, Smith B, Haugwitz M, Hughes SH, Weissman JS, Schumann K, Esensten JH, May AP, Ashworth A, Kupfer GM, Greeley SAW, Bacchetta R, Meffre E, Roncarolo MG, Romberg N, Herold KC, Ribas A, Leonetti MD, Marson A (2018) Reprogramming human T cell function and specificity with non-viral genome targeting. Nature 559:405–409
57. Dai X, Park JJ, Du Y, Kim HR, Wang G, Errami Y, Chen S (2019) One-step generation of modular CAR-T cells with AAV-Cpf1. Nat Methods 16:247–254
58. Berns K, Hijmans EM, Mullenders J, Brummelkamp TR, Velds A, Heimerikx M, Kerkhoven RM, Madiredjo M, Nijkamp W, Weigelt B, Agami R, Ge W, Cavet G, Linsley PS, Beijersbergen RL, Bernards R (2004) A large-scale RNAi screen in human cells identifies new components of the p53 pathway. Nature 428:431–437
59. Boutros M, Kiger AA, Armknecht S, Kerr K, Hild M, Koch B, Haas SA, Paro R, Perrimon N, Heidelberg Fly Array C (2004) Genome-wide RNAi analysis of growth and viability in drosophila cells. Science 303:832–835

60. Jackson AL, Burchard J, Schelter J, Chau BN, Cleary M, Lim L, Linsley PS (2006) Widespread siRNA "off-target" transcript silencing mediated by seed region sequence complementarity. RNA 12:1179–1187

61. Echeverri CJ, Beachy PA, Baum B, Boutros M, Buchholz F, Chanda SK, Downward J, Ellenberg J, Fraser AG, Hacohen N, Hahn WC, Jackson AL, Kiger A, Linsley PS, Lum L, Ma Y, Mathey-Prevot B, Root DE, Sabatini DM, Taipale J, Perrimon N, Bernards R (2006) Minimizing the risk of reporting false positives in large-scale RNAi screens. Nat Methods 3:777–779

62. Shalem O, Sanjana NE, Hartenian E, Shi X, Scott DA, Mikkelson T, Heckl D, Ebert BL, Root DE, Doench JG, Zhang F (2014) Genome-scale CRISPR-Cas9 knockout screening in human cells. Science 343:84–87

63. Wang T, Wei JJ, Sabatini DM, Lander ES (2014) Genetic screens in human cells using the CRISPR-Cas9 system. Science 343:80–84

64. Manguso RT, Pope HW, Zimmer MD, Brown FD, Yates KB, Miller BC, Collins NB, Bi K, LaFleur MW, Juneja VR, Weiss SA, Lo J, Fisher DE, Miao D, Van Allen E, Root DE, Sharpe AH, Doench JG, Haining WN (2017) In vivo CRISPR screening identifies Ptpn2 as a cancer immunotherapy target. Nature 547:413–418

65. Pan D, Kobayashi A, Jiang P, Ferrari de Andrade L, Tay RE, Luoma AM, Tsoucas D, Qiu X, Lim K, Rao P, Long HW, Yuan GC, Doench J, Brown M, Liu XS, Wucherpfennig KW (2018) A major chromatin regulator determines resistance of tumor cells to T cell-mediated killing. Science 359:770–775

66. Patel SJ, Sanjana NE, Kishton RJ, Eidizadeh A, Vodnala SK, Cam M, Gartner JJ, Jia L, Steinberg SM, Yamamoto TN, Merchant AS, Mehta GU, Chichura A, Shalem O, Tran E, Eil R, Sukumar M, Guijarro EP, Day CP, Robbins P, Feldman S, Merlino G, Zhang F, Restifo NP (2017) Identification of essential genes for cancer immunotherapy. Nature 548:537–542

67. Seki A, Rutz S (2018) Optimized RNP transfection for highly efficient CRISPR/Cas9-mediated gene knockout in primary T cells. J Exp Med 215:985–997

68. Shifrut E, Carnevale J, Tobin V, Roth TL, Woo JM, Bui CT, Li PJ, Diolaiti ME, Ashworth A, Marson A (2018) Genome-wide CRISPR screens in primary human T cells reveal key regulators of immune function. Cell 175 (1958–1971):e15

69. Ting PY, Parker AE, Lee JS, Trussell C, Sharif O, Luna F, Federe G, Barnes SW, Walker JR, Vance J, Gao MY, Klock HE, Clarkson S, Russ C, Miraglia LJ, Cooke MP, Boitano AE, McNamara P, Lamb J, Schmedt C, Snead JL (2018) Guide swap enables genome-scale pooled CRISPR-Cas9 screening in human primary cells. Nat Methods 15:941–946

70. Veres A, Gosis BS, Ding Q, Collins R, Ragavendran A, Brand H, Erdin S, Cowan CA, Talkowski ME, Musunuru K (2014) Low incidence of off-target mutations in individual CRISPR-Cas9 and TALEN targeted human stem cell clones detected by whole-genome sequencing. Cell Stem Cell 15:27–30

71. Tan EP, Li Y, Velasco-Herrera Mdel C, Yusa K, Bradley A (2015) Off-target assessment of CRISPR-Cas9 guiding RNAs in human iPS and mouse ES cells. Genesis 53:225–236

72. Crosetto N, Mitra A, Silva MJ, Bienko M, Dojer N, Wang Q, Karaca E, Chiarle R, Skrzypczak M, Ginalski K, Pasero P, Rowicka M, Dikic I (2013) Nucleotide-resolution DNA double-strand break mapping by next-generation sequencing. Nat Methods 10:361–365

73. Tsai SQ, Zheng Z, Nguyen NT, Liebers M, Topkar VV, Thapar V, Wyvekens N, Khayter C, Iafrate AJ, Le LP, Aryee MJ, Joung JK (2015) GUIDE-seq enables genome-wide profiling of off-target cleavage by CRISPR-Cas nucleases. Nat Biotechnol 33:187–197

74. Kochenderfer JN, Feldman SA, Zhao Y, Xu H, Black MA, Morgan RA, Wilson WH, Rosenberg SA (2009) Construction and preclinical evaluation of an anti-CD19 chimeric antigen receptor. J Immunother 32:689–702

75. Carpenter RO, Evbuomwan MO, Pittaluga S, Rose JJ, Raffeld M, Yang S, Gress RE, Hakim FT, Kochenderfer JN (2013) B-cell maturation antigen is a promising target for adoptive T-cell therapy of multiple myeloma. Clin Cancer Res 19:2048–2060

76. Fraietta JA, Nobles CL, Sammons MA, Lundh S, Carty SA, Reich TJ, Cogdill AP, Morrissette JJD, DeNizio JE, Reddy S, Hwang Y, Gohil M, Kulikovskaya I, Nazimuddin F, Gupta M, Chen F, Everett JK, Alexander KA, Lin-Shiao E, Gee MH, Liu X, Young RM, Ambrose D, Wang Y, Xu J, Jordan MS, Marcucci KT, Levine BL, Garcia KC, Zhao Y, Kalos M, Porter DL, Kohli RM, Lacey SF, Berger SL, Bushman FD, June CH, Melenhorst JJ (2018) Disruption of TET2 promotes the therapeutic efficacy of CD19-targeted T cells. Nature 558:307–312

77. Saha S, Nakazawa Y, Huye LE, Doherty JE, Galvan DL, Rooney CM, Wilson MH (2012) piggyBac transposon system modification of primary human T cells. J Vis Exp 69:e4235

78. Monjezi R, Miskey C, Gogishvili T, Schleef M, Schmeer M, Einsele H, Ivics Z, Hudecek M (2017) Enhanced CAR T-cell engineering using non-viral sleeping beauty transposition from minicircle vectors. Leukemia 31:186–194
79. Kebriaei P, Izsvak Z, Narayanavari SA, Singh H, Ivics Z (2017) Gene therapy with the sleeping beauty transposon system. Trends Genet 33:852–870
80. Strecker J, Ladha A, Gardner Z, Schmid-Burgk JL, Makarova KS, Koonin EV, Zhang F (2019) RNA-guided DNA insertion with CRISPR-associated transposases. Science 365:48–53
81. Beane JD, Lee G, Zheng Z, Mendel M, Abate-Daga D, Bharathan M, Black M, Gandhi N, Yu Z, Chandran S, Giedlin M, Ando D, Miller J, Paschon D, Guschin D, Rebar EJ, Reik A, Holmes MC, Gregory PD, Restifo NP, Rosenberg SA, Morgan RA, Feldman SA (2015) Clinical scale zinc finger nuclease-mediated gene editing of PD-1 in tumor infiltrating lymphocytes for the treatment of metastatic melanoma. Mol Ther 23:1380–1390
82. Menger L, Sledzinska A, Bergerhoff K, Vargas FA, Smith J, Poirot L, Pule M, Hererro J, Peggs KS, Quezada SA (2016) TALEN-mediated inactivation of PD-1 in tumor-reactive lymphocytes promotes Intratumoral T-cell persistence and rejection of established Tumors. Cancer Res 76:2087–2093
83. Shi L, Meng T, Zhao Z, Han J, Zhang W, Gao F, Cai J (2017) CRISPR knock out CTLA-4 enhances the anti-tumor activity of cytotoxic T lymphocytes. Gene 636:36–41

Chapter 24

CRISPR/Cas9-Mediated Genome Engineering of Primary Human B Cells

Kanut Laoharawee, Matthew J. Johnson, and Branden S. Moriarity

Abstract

The CRISPR/Cas9 system allows for site-specific gene editing and genome engineering of primary human cells. Here we describe methods for gene editing and genome engineering of B cells isolated from human peripheral blood mononuclear cells using CRISPR/Cas9. Editing frequencies of up to 90% and integration rates greater than 60% can be achieved with this method.

Key words B-cell engineering, CRISPR/Cas9, Genome engineering, Primary human B cells, rAAV

1 Introduction

The B-cell linage performs a critical role in the immune system, producing large amounts of antibody in response to immune challenge and providing long-term immunological memory through memory B cells and long-lived plasma cells [1]. Plasma cells in particular are unique in their ability among immune cells to produce a large amount of protein in the form of antibody while surviving for years or even decades [2]. These factors, combined with their ease of isolation from peripheral blood, make the B-cell lineage an excellent platform for novel cell therapies [3]. Previous methods of genome engineering of primary human B cells have used randomly integrating vectors, such as lentiviral vectors or transposons [4, 5]. These methods, along with engineering hematopoietic stem cells (HSCs) followed by differentiation into B cells, have allowed for the study of B cells, but lack the ability to target editing and thus carry an inherent risk of detrimental mutations.

Recently, multiple groups including ours have successfully developed methods for the precision engineering of primary human B cells [6, 7]. Here, we provide detailed methodology on primary human B-cell isolation from PBMCs via magnetic beads, expansion in cell culture using media containing CD40L, and

Mouldy Sioud (ed.), *RNA Interference and CRISPR Technologies: Technical Advances and New Therapeutic Opportunities*, Methods in Molecular Biology, vol. 2115, https://doi.org/10.1007/978-1-0716-0290-4_24,

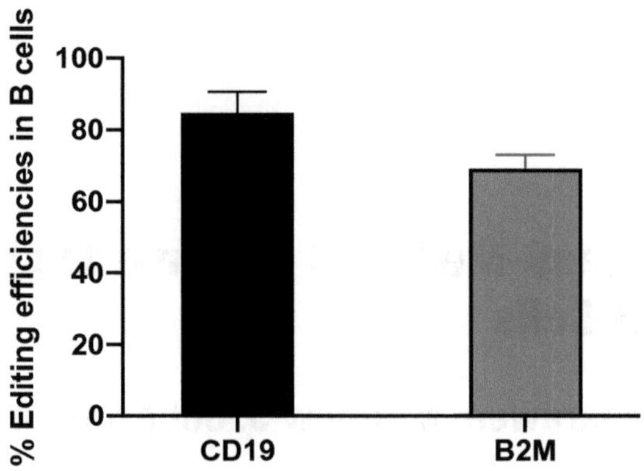

Fig. 1 CRISPR/Cas9-mediated gene editing of primary human B cells Electroporation of CRISPR/Cas9 into primary human B cells targeting different loci ($n = 2$ independent donors)

editing or engineering with CRISPR/Cas9. Using these methods, we routinely achieve editing frequencies up to 90% (Fig. 1) and site-specific integrations rates over 60% (Fig. 2a and b), while maintaining cell viability greater than 90% (Fig. 2c).

2 Materials

The reagents described below have been validated and successfully used by our lab, but alternative commercially equivalent reagents can also likely be optimized.

2.1 CRISPR/Cas9 System

2.1.1 Guide RNAs (sgRNAs or crRNA + tracrRNA)

1. There are a variety of online tools that can be used to design guide RNAs. We have used Benchling (www.benchling.com) and Synthego (www.synthego.com/products/bioinformatics/crispr-design-tool) to successfully design gRNAs. A series of 2–4 guide RNA sequences per target locus should be considered. On-target and off-target scores should be considered when selecting gRNAs as well. Higher on-target and lower off-target scores reflect guide RNAs predicted to be more successful.
2. gRNAs can be commercially ordered as chemically modified, (2′-*O*-methyl 3′ phosphorothioate modification) sgRNA (Syntego, CA or IDT, IA) without any further modification. However, crRNAs (IDT, IA) need to be complexed with tracrRNA (IDT, IA) to function. Thus, we prefer sgRNAs for this reason.
3. Nuclease-free molecular grade water, tris-EDTA buffer (TE buffer) or buffer T (Provided with Neon™ transfection system kit), can be used to reconstitute guide RNAs to a concentration of 1 mg/mL.

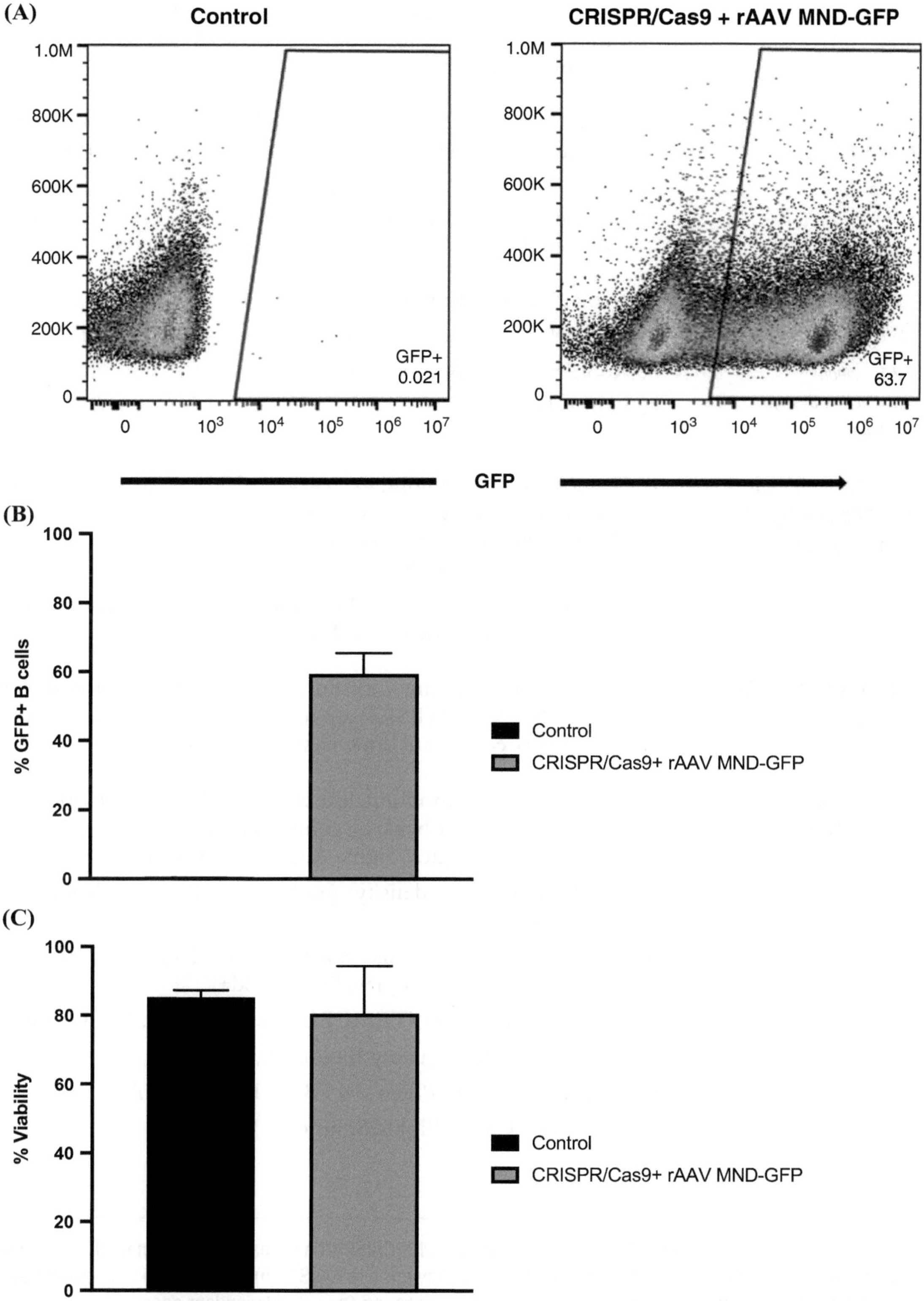

Fig. 2 Site-specific integration of an EGFP expression cassette in primary human B cells using CRISPR/Cas9 combined with rAAV6-mediated DNA donor delivery. (**a**) Example of flow cytometry scatter plots of GFP

4. Store reconstituted gRNAs at −20 °C to −80 °C. We often use buffer T to reconstitute and dilute guide RNAs to the desired concentration.

2.1.2 S.p.Cas9 (Protein or mRNA)

1. Alt-R® S.p.Cas9 V3 protein at 100 mg or 500 mg can be commercially purchased (IDT, IA) and stored at −20 °C.
2. Chemically modified (5-methoxyuridine modified) CleanCap® Cas9 mRNA at a concentration of 1 mg/mL can be commercially purchased (Trilink, CA) and stored at −80 °C.

2.2 Transfection System

2.2.1 Neon™ Transfection System (Invitrogen)

1. Neon™ transfection module (machine).
2. Neon™ transfection system kit (10 mL or 100 mL kit).
3. Neon™ transfection system pipette.
4. PCR tube strip.

2.3 Recombinant AAV6 Pseudotyped Vector (rAAV6)

1. Homology arms ranging from 400 to 1000 bp flank both sides of a cargo of interest (either with or without a promoter) on a rAAV backbone to be used as an expression cassette for packaging into rAAV6 vector.
2. We use a customized rAAV packaged through commercial source (Vigene Biosciences, MD).

2.4 B-Cell Isolation and Culture

IBC approval for working with human specimens and access to a proper biosafety level 2 (BSL2) are required for primary human cell culture. All procedures are done in sterile/aseptic manner.

2.4.1 Human Primary B-Cell Isolation

1. Peripheral blood mononuclear cells (PBMCs) can be obtained from a local blood bank or commercial sources (e.g., STEMCELLS Technologies, Sigma Aldrich, Lonza, etc.).
2. Ficoll-Paque plus density gradient media (GE healthcare Bio-Sciences, PA).
3. EasySep human B-cells isolation kit (Negative isolation, STEMCELL Technologies, BC, Canada).
4. "Big Easy" magnet (STEMCELL Technologies, BC, Canada).
5. ACK lysis buffer (Quality Biological, MD).
6. Separation buffer (PBS + 2% FBS + 1 mM EDTA).
7. CryoStor® CS10 (Biolife Solutions, WA).

Fig. 2 (continued) expression in B cells electroporated with CRISPR/Cas9 reagents and immediately transduced with rAAV6 encoding for MND-GFP-pA and homology arms to AAVS1. (**b**) GFP integration frequency and (**c**) cell viability of primary human B cells 5-day post-engineering ($n = 2$ independent donors)

2.4.2 B-Cell Expansion Medium and Culture Plates

1. Gibco™ IMDM 500 mL (ThermoFisher Scientific).
2. 2-Mercaptoethanol (55 mM).
3. MEGA CD40L (Enzo Life Sciences, NY).
4. CpG ODN 2006 (Invivogen, CA).
5. Animal free IL-2, IL-10, IL-15 lyophilized (Peprotech, NJ). Reconstitute and dilute to desired concentration using manufacturer's recommended reagents.
6. Fetal bovine serum (Invitrogen).
7. Penicillin–Streptomycin 100× (P/S, Sigma-Aldrich).
8. Base medium contains IMDM, 10% FBS, 1% P/S, and 55 μM 2-mercaptoethanol.
9. Complete B-cell medium contains base medium, 1 mg/mL CpG ODN 2006, 100 ng/mL MEGA CD40L, 50 ng/mL IL-2, 50 ng/mL IL-10, and 10 ng/mL IL-15. Prepare freshly before use.
10. Some 24-, 12-, or 6-well tissue culture plates or tissue culture flasks.

3 Methods

All of the procedures are done in the BSL-2 hood unless otherwise specified.

3.1 B-Cell Isolation

1. Transfer blood to a 50 mL conical tube.
2. Add 1× PBS to the blood in a 1:1 ratio and gently pipette up and down several times.
3. Add 15 mL Ficoll-Paque into a new 50 mL conical tube.
4. Layer the diluted blood on top of the tube containing Ficoll-Paque (tilt the tube and gently and slowly add blood on to the Ficoll-Paque reagent).
5. Centrifuge at 600 × *g* for 25 min without brake.
6. Gently move the tube back to the hood and discard the plasma layer (the top layer) by aspirating the plasma layer without disturbing the buffy coat layer.
7. Transfer buffy coat layer into a new 50 mL conical tube.
8. Centrifuge at 400 × *g* for 5 min and remove supernatant without disturbing the cell pellet.
9. Add 7 mL of ACK lysing buffer and incubate at room temperature for 3 min.
10. Bring up to 50 mL using 1× PBS.
11. Centrifuge at 400 × *g* for 5 min.

12. Gently discard the supernatant without disturbing the cell pellet (the pellet should appear white or slightly pinkish).
13. Resuspend the pellet with 25 mL separation buffer and count the cells.
14. Process with the EasySep human B-cell isolation kit (negative isolation kit) following the manufacturer's protocols.
15. The B cells are ready for immediate use or for storage for future use. To store cells for future use, resuspend cells at a density of 1×10^7 cells/mL in CryoStor® CS10 reagent, freeze in a "Mr. Frosty Freezing Container" (Thermo Fisher Scientific) or a controlled rate freezer, and store at −80 °C or liquid nitrogen for long-term storage.

3.2 B-Cell Expansion

Prepare and pre-warm the complete B-cell expansion media (Subheading 2.4.2) before thawing B cells. This expansion protocol allows mature naïve B cells to outgrow the other B-cell subpopulations (*see* **Note 1**).

3.2.1 Thaw and Expanding (Stimulating) B Cells

1. Pre-warm FBS in 37 °C water bath.
2. Thaw frozen B-cell aliquots in 37 °C water bath, immediately add 1 mL drop-wise of pre-warmed FBS into the thawed B-cell tube, and incubate at RT for 1 min.
3. Meanwhile add 2 mL of pre-warmed FBS into a 15 mL conical tube.
4. Transfer thawed B cells + FBS into the 15 mL conical tube in drop-wise manner and bring up to 15 mL with 1× PBS.
5. Centrifuge the tube at 400 × *g* for 5 min and discard supernatant without disturbing the pellet.
6. Resuspend the B-cell pellet with 1 mL of B-cell expansion medium (*see* **Note 2**).
7. Count B cells using hemocytometer or countess slide with trypan blue dye.
8. Culture B cells in T25 flask at 5×10^5 cells/mL.
9. B cells will be expanded for 2 days before engineering.

   You can skip **steps 1–6** if you are using freshly isolated B cells.

3.3 Transfection of B Cells with CRISPR/Cas9 Using Neon® Transfection System

B cells must be expanded for 2 days prior to electroporation of engineering reagents. All the preparations are done in the BSL2 hood unless otherwise stated. Below are two preparation methods for making CRISPR/Cas9 RNP complexes. Each method gives comparable editing efficiencies.

3.3.1 Preparing S.p. Cas9 Nuclease Protein/Chemically Modified sgRNA Ribonucleoprotein (RNP) Complex

1. Mix 5 μg (0.5 μL) of S.p.Cas9 nuclease∗∗ (IDT) with 1 μg of a selected sgRNA.
2. Incubate at room temperature for at least 20 min.
3. While waiting, go to Subheading 3.3.3 to prepare B cells for transfection.
4. The RNP complex is ready to use.

   This RNP complex concentration will be used for 300,000 B cells in 10 μL Neon tips. Scale up or down as need. ** If CleanCap® Cas9 mRNA is used with sgRNA, then simply mix 1 μg of the Cas9 mRNA to 1 μg sgRNA immediately prior to adding the cells without incubation (*see* **Note 3**).

3.3.2 Preparing S.p. Cas9 Nuclease Protein: crRNA/tracrRNA (IDT)

Preparing crRNA/tracrRNA Complex

1. Mix an equal molar amount of crRNA and tracrRNA in a PCR tube with cap.
2. Incubate at 95 °C for 5 min and allow to cool down on a bench top to room temperature.
3. Dilute the complex RNA to the desired concentration (we dilute to 1 mM/mL). The RNP complex can be stored at −20 °C for up to 6 months without losing activity [8].

Preparing Alt-R® RNP Complex

1. Dilute S.p.Cas9 nuclease V3 protein with buffer T (provided with Neon™ transfection system kit) to the desired concentration (usually to the same molar as the complex RNA in Subheading 3.3.2.1), mix gently with pipette, and avoid bubbles.
2. Mix equimolar amounts of the RNA complex and S.p.Cas9 nuclease V3 protein in a tube and avoid bubbles.
3. Incubate at room temperature for 5 min.
4. The Alt-R RNP complex is ready to use. Complex can be stored for up to 6 months at −80 °C [8].

   This RNP complex concentration will be used for 300,000 B cells in 10 μL Neon tips. Scale up or down as needed. If CleanCap® Cas9 mRNA is used with crRNA, then simply mix 1 mg of the Cas9 mRNA to 1 mM the complex RNA (from Subheading "Preparing crRNA/tracrRNA Complex") before adding the cells with no incubation step.

3.3.3 Prepare B-Cell Suspension for Transfection

1. Prepare the complete B-cell medium, transfer the desired volume to 24-well culture plate (use volume of media needed for a final cell density at 5 × 10^5 cells/mL), and place in a cell culture incubator at 37 °C, 5% CO_2 with humidity.
2. Transfer the 2-day stim (expanded) B cells from the flask to a clean conical tube.
3. Centrifuge at 400 × *g* for 5 min and carefully aspirate supernatant without disturbing the cell pellet.

4. Resuspend the cell pellet with 1 mL 1× PBS.
5. Count the cells using trypan blue dye.
6. Calculate the number of live B cells needed for the experiment.
7. Transfer the desired number of B cells to a clean 1.5 microcentrifuge tube and centrifuge at 400 × *g* for 5 min.
8. Carefully remove the supernatant without disturbing the pellet.
9. Resuspend the B cells in buffer T (provided with Neon™ transfection system kit) to final cell concentration of 30,000 cells/μL transfection reaction in a volume of 12 μL for use with a 10 μL Neon tip.
10. Optional: add 0.5 μL of chemically modified GFP mRNA (Trilink, CA) per 300,000 B cells to measure transfection efficiency (*see* **Note 4**).

Example

1. If the total volume of RNP complex from Subheading 3.3.1 is 1.5 μL.
2. Around 405,000 B cells should be resuspended in 11.5 μL buffer T, and then add 0.5 μL of GFP mRNA.
3. Total volume will be 405,000 B cells in 12 μL.

3.3.4 Setting Up a Neon Transfection System

1. For knock-in experiments, thaw rAAV6 vector and place on ice while preparing transfection reaction.
2. Place Neon transfection station in a BSL2 hood and turn it on.
3. Set the Neon transfection module to 1700 V, 20 ms, 1 pulse.
4. Place Neon electroporation chamber (provided with Neon™ transfection system kit) onto the Neon transfection station. Transfer 3 mL buffer E (provided with Neon™ transfection system kit) into the cuvette.

3.3.5 Preparing a Transfection Reaction

1. Prepare 24-well plate containing pre-warmed complete B-cell medium (*see* Subheading 3.3.1). This plate will be used for B-cell recovery and culture post-electroporation.
2. Transfer the desired volume of the RNP complex into a PCR tube.
3. Transfer a desired volume of B-cell suspension (with or without GFP mRNA) into the PCR tube. The total volume should be 13.5 μL. Gently mix the components and avoid bubbles (*see* **Note 5**).
4. With one single motion, slowly transfer 10 μL of the B cell/ RNP complex transfection reaction into the 10 μL Neon tip and Neon pipette for electroporation. (The transfection reaction in the Neon tip must be free of bubbles.)

5. Immediately transfer the electroporated cells into the 24-well plate containing B-cell expansion medium. Gently swirl the plate but DO NOT pipette to mix the cells. The cells should be cultured at a final density of 5×10^5 cells/mL.
6. For knock-in experiments, 500,000 MOI of rAAV6 vector can be added directly to electroporated B cells. Virus must be added within 30 min after electroporation. Swirl the plate gently to mix in the virus.
7. Culture the cells at 37 °C, 5% CO_2 with humidity.
8. After 24 h, observe the frequency of GFP-expressing cells under a fluorescent microscope to determine the estimated transfection efficiency.
9. Replenish the complete B-cell media every 3 days and keep the cells at 5×10^5 cells/mL for maximum cell expansion.
10. Cells are ready for downstream analysis on day 5 post-transfection.

4 Notes

1. Expanding B cells in culture will form large B-cell clusters. Break up these clusters to obtain accurate cell counts.
2. Aliquoting cytokines and MEGACD40 in small volumes can help to avoid repeated freeze–thaw cycles, which may reduce activity.
3. Cas9 mRNA can be substituted for Cas9 RNPs with only a slight reduction in post-editing cell expansion [6]. If Cas9 mRNA is used for knock-in experiments, then wait 6 h post-electroporation to add AAV6 vector.
4. Chemically modified GFP mRNA (Trilink, CA) can be used to track transfection efficiency.
5. Bubbles in the Neon™ tip will cause a spark to arc and will kill all cells in the electroporation. If a spark/arc is observed, then repeat the condition with a new sample. If bubbles prove to be a persistent problem, then increase the reaction volume or use a new Neon™ tip.

References

1. Radbruch A, Muehlinghaus G, Luger EO, Inamine A, Smith KGC, Dörner T et al (2006) Competence and competition: the challenge of becoming a long-lived plasma cell. Nat Rev Immunol 6(10):741–750
2. Slifka MK, Antia R, Whitmire JK, Ahmed R (1998) Humoral immunity due to long-lived plasma cells. Immunity 8(3):363–372
3. Spriggs MK, Armitage RJ, Strockbine L, Clifford KN, Macduff BM, Sato TA et al (1992) Recombinant human CD40 ligand stimulates B

cell proliferation and immunoglobulin E secretion. J Exp Med 176(6):1543–1550
4. Fusil F, Calattini S, Amirache F, Mancip J, Costa C, Robbins JB et al (2015) A lentiviral vector allowing physiologically regulated membrane-anchored and secreted antibody expression depending on B-cell maturation status. Mol Ther 23(11):1734–1747
5. Luo XM, Maarschalk E, O'Connell RM, Wang P, Yang L, Baltimore D (2008) Engineering human hematopoietic stem/progenitor cells to produce a broadly neutralizing anti-HIV antibody after in vitro maturation to human B lymphocytes. Blood 113(7):1422–1431
6. Johnson MJ, Laoharawee K, Lahr WS, Webber BR, Moriarity BS (2018) Engineering of primary human B cells with CRISPR/Cas9 targeted nuclease. Sci Rep 8(1):12144
7. Hung KL, Meitlis I, Hale M, Chen C-Y, Singh S, Jackson SW et al (2018) Engineering protein-secreting plasma cells by homology-directed repair in primary human B cells. Mol Ther 26 (2):456–467
8. Alt-R CRISPR-Cas9 System User Guide. https://sfvideo.blob.core.windows.net/sitefinity/docs/default-source/user-guide-manual/alt-r-crispr-cas9-user-guide-ribonucleoprotein-transfections-recommended.pdf?sfvrsn=1c43407_24. Accessed 11 Jun 2019

Chapter 25

Gene Editing in B-Lymphoma Cell Lines Using CRISPR/Cas9 Technology

Baoyan Bai, June Helen Myklebust, and Sébastien Wälchli

Abstract

Genome editing in eukaryotes has greatly improved through the application of targeted editing tools. The development of the CRISPR/Cas9 technology has facilitated genome editing in mammalian cells. However, efficient delivery of CRISPR components into cells growing in suspension remains a challenge. Here, we present a strategy for sequential delivery of the two essential components, Cas9 and sgRNA, into B-lymphoid cell lines. Stable Cas9 expression is obtained by retroviral transduction, before sgRNA is transiently delivered into the Cas9+ cells. This method improves the on-target efficiency of genome editing and, through the transient presence of sgRNA, reduces the potential off-target sites. The current method can be easily applied to other cell types that are difficult to edit with CRISPR/Cas9.

Key words B-lymphoma cells, CRISPR/Cas9, Genome editing, Cas9-expressing cells, Square wave electroporation

1 Introduction

The understanding of genetic effects underlying disease and malignancies is greatly facilitated by the ability to specifically alter endogenous genes with high accuracy. Genome engineering of cell lines or animal models has traditionally been accomplished through random mutagenesis. To improve genome editing, sequence-specific DNA nuclease technologies have enabled targeted modifications of endogenous genomic sequences with high efficiency [1–3]. By specifying a 20-nt targeting sequence as guide RNA, Cas9 nuclease from the microbial CRISPR adaptive immune system is an efficient tool for genome engineering as it generates DNA breaks at the binding site of sgRNA in eukaryotic cells [4–6]. As previously reviewed, the CRISPR/Cas9 technology relies on the delivery of two entities, namely, the Cas9 enzyme and the guiding sequence, sgRNA. It can be delivered into eukaryotic cells via plasmid DNA, RNA, ribonucleoprotein, or a combination of these (*see* Chapters

Mouldy Sioud (ed.), *RNA Interference and CRISPR Technologies: Technical Advances and New Therapeutic Opportunities*, Methods in Molecular Biology, vol. 2115, https://doi.org/10.1007/978-1-0716-0290-4_25,

19, 20, 25–27). However, a major pitfall remains the transfection efficiency of the CRISPR/Cas9 delivery. Indeed, most of the proof of principle studies were performed in adherent cell lines, such as Hek-293, which are known to be trivial to transfect [6, 7].

B-cell lines are notoriously resistant to most gene transfer techniques. Electroporation has therefore been widely used to study gene expression, but electroporation is poorly tolerated by most cell types. In addition, delivery of plasmid coding for CRISPR/Cas9 into the cells increases the potential for off-target effects [8, 9]. We have previously shown that retroviral transduction can be successfully used to establish ectopic expression of genes in B-lymphoma cell lines [10, 11]. However, constitutive co-expression of both Cas9 and sgRNAs may result in potential adverse effects. Although Cas9 expression alone has been shown to produce negligible DNA damage in cells, when combined with the presence of sgRNAs, the number of DNA damage foci increased significantly [12]. Thus, we assumed that a B-lymphoma cell line-based study could benefit from the creation of a Cas9-derived descendant which could then be transiently electroporated with a sgRNA of interest.

Here we describe a strategy for sequential delivery CRISPR reagents into B-lymphoma cells as illustrated in Fig. 1. First, a myc-tagged Cas9 was delivered into B-lymphoma cell lines using a retroviral vector carrying an antibiotic resistance gene. Positive cells are selected with puromycin, and Cas9 expression is further validated by detection of its myc tag through Western blotting (Fig. 2a). The Cas9-expressing cells are then cloned, expanded, and stocked for further use. Second, 3–4 sgRNAs are designed for each gene of interest (GOI) using the online tool CHOPCHOP (http://chopchop.cbu.uib.no/) [13, 14]. To reduce the amount of undesired effects, we search for potential off-target sites for each sgRNAs using the Cas-OFFinder (http://www.rgenome.net/cas-offinder/) [15]. Selected sgRNAs are then prepared by in vitro transcription, checked on agarose gel (Fig. 2b), and introduced into cells using an electroporation technique. The genome-editing efficiency of the GOIs was measured by direct Sanger sequencing (Fig. 2c, right panel) or by indirect readouts such as flow cytometry (Fig. 2c, left panel), immunofluorescence, and Western blotting. Using cells with stable Cas9 expression, we are able to determine the optimal amount of sgRNA for creating gene edited cells.

In this chapter, we illustrate our protocol by knocking out the CD79B gene in two different B-lymphoma cell lines. Our strategy allows the CRISPR system to work only for a limited time, as the sgRNA will be quickly degraded. To achieve the best balance between on-target efficiency and off-target effects, our system allows for titration of the amount of sgRNAs to avoid excessive delivery of sgRNA into the cells. For editing of primary B cells and stem cells, *see* Chapters 25–27.

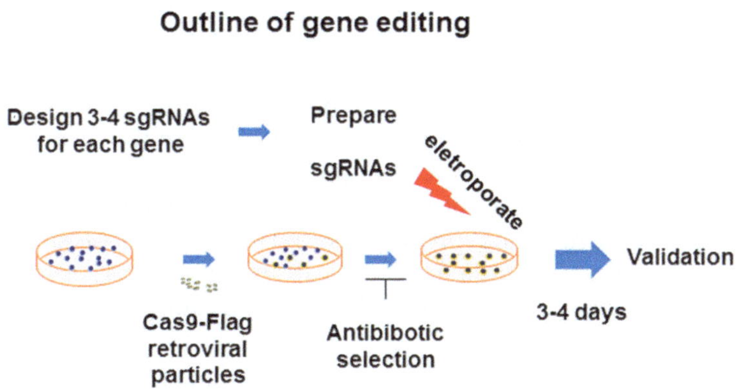

Fig. 1 Outline of gene editing. B-cell lines are first retrovirally transduced with Cas9-puromycin-containing particles and are further selected on antibiotic. Once a puromycin-resistant (Cas9 positive) population is selected, the cells are electroporated with sgRNAs in order to knock out the gene of interest

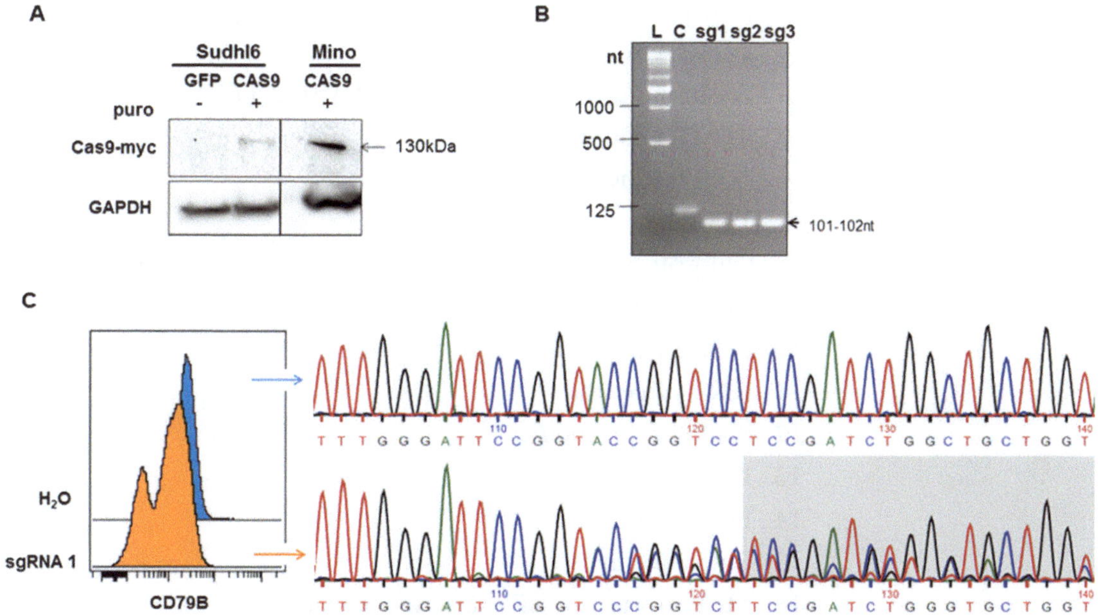

Fig. 2 Quality control. (**a**) Validation of Cas9 expression by Western blotting. From left to right: SU-DHL-6 cells transduced with a retroviral vector containing only the GFP gene serving as a negative control. SU-DHL-6 cells transduced with a retroviral vector containing the Myc-Cas9 gene and the puromycin resistance gene. Mino cells transduced with a retroviral vector containing the Myc-Cas9 gene and the puromycin resistance gene. (**b**) sgRNA size analysis before electroporation. Control RNA (**c**) 125 nt, and test sgRNAs were loaded on an agarose gel for size comparison. L: RNA ladder. (**c**) Validation of genome editing by CRISPR/Cas9. Surface staining of CD79B. Blue: Mino Cas9-expressing cells electroporated with H_2O. Orange: Mino Cas9-expressing cells electroporated with sgRNA 1. Sanger sequencing results from the same cells are shown

2 Materials

All solutions should be prepared in ultrapure deionized water, and all chemicals should be of analytical, and when possible, nuclease-free grade. Avoid contamination of nucleases from surfaces, equipment, or plastic ware. Use gloves throughout every step. Appropriate handling, procedures, and safety protocols should be adhered to when handling biological materials and potentially hazardous reagents. Institutional policies should be strictly adhered to in this regard.

2.1 Plasmids

1. MSCV_Cas9_puro (addgene #65655).
2. Retroviral package plasmids [16].

2.2 Reagents

1. Dulbecco's Modified Eagle's medium (DMEM).
2. 100× Penicillin–Streptomycin (Pen–Strep).
3. PBS buffer: 137 mM sodium chloride, 10 mM phosphate, 2.7 mM potassium chloride; pH 7.4.
4. Washing buffer: PBS supplement with 2% FCS.
5. RPMI medium.
6. Fetal bovine serum.
7. TRIzol.
8. Chloroform.
9. Isopropanol.
10. 80% ethanol in RNase free water.
11. RNase free water.
12. Anti-myc antibody.
13. Puromycin.
14. Fugene 6.
15. RetroNectin.
16. Megashortscript T7 transcription Kit.
17. DNeasy Blood and Tissue Kit.
18. Phusion High-Fidelity DNA Polymerase.

2.3 Cell Lines

The following cell cancer cell lines were used. Hek-P cell [16], Mino (http://www.lgcstandards-atcc.org/Products/All/CRL-3000.aspx?geo_country=no) and SU-DHL-6 (http://www.lgcstandards-atcc.org/products/all/CRL-2959.aspx?geo_country=no).

2.4 sgRNA Sequences Targeting the CD79B Gene

sgRNA1: 5′GACCTTTGGGATTCCGGTAC 3′.

sgRNA2: 5′GCTGGTACTGGCTCAGCTGC 3′.

sgRNA3: 5′GACCGGTACCGGAATCCCAA 3′.

2.5 PCR Primers for sgRNA1

Forward primer: 5′ATGGGAATCCTGGATGGGGA 3′.

Reverse primer: 5′TCAAAGCCACCATGAGGGTC 3′.

3 Methods

3.1 Establishing Cas9 Expressing Cells

Here we used the Mino and SU-DHL-6 cell lines as working models.

3.1.1 Determine the Puromycin Concentration Needed for a Given B-Cell Line

1. Day 1: Seed 3×10^5 Mino or SU-DHL-6 cells in 1 mL RPMI medium supplement with 10% FCS and 1× Pen–Strep into a 12-well plate.
2. Day 2: Replace the medium with selection medium supplemented with a range of puromycin concentrations (*see* **Note 1**).
3. Monitor the viability of the cells daily.
4. Replace the selection medium every 2–3 days.
5. Identify the minimum antibiotic concentration required to kill 100% of the cells (*see* **Note 2**). This concentration will be used to select Cas9-positive cells.

3.1.2 Prepare Cas9-Expressing Retroviral Particles

1. Day 1: seed 1.2×10^6 Hek-P cells in 3 mL DMEM supplemented with 10% FCS and 1× Pen–Strep in a 6 cm tissue culture dish at 37 °C.
2. Day 2: transfect Hek-P cells with retroviral constructs (package plasmids and Cas9 construct) using Fugene 6 (*see* **Note 3**).
3. Day 3. Replace with 3 mL fresh DMEM supplemented with 1% FCS and 1× Pen–Strep, and incubate at 32 °C.
4. Day 4: Transfer the viral supernatant containing the retroviral particles to a fresh tube and add 3 mL fresh DMEM supplemented with 1% FCS and 1× Pen–Strep to the Hek-P cells.
5. Day 5: Collect the viral supernatant into the same tube as used on Day 4.
6. Centrifuge the tube at 2000 × *g* for 5 min to remove the cell debris. The retroviruses can be kept at 4 °C for up to 1 week and should be kept at −80 °C for long-term storage.

3.1.3 Transduction of the Mino and SU-DHL-6 Cell Lines

1. Coat a non-treated 12-well plate with 20 μg/mL RetroNectin for at least 2 h at room temperature or overnight at 4 °C.
2. Remove the RetroNectin and wash each well with 2 mL 1× PBS with 2% FCS for 30 min at room temperature.

3. Remove the washing buffer.
4. Seed 3×10^5 cells (Mino cells and SU-DHL-6 cells, separately) in 1 mL RPMI medium supplemented with 10% FCS and 1× Pen–Strep in a 12-well plate.
5. Add 1 mL of the retroviral particles prepared in Subheading 3.1.2 to each well containing the Mino or SU-DHL-6 cells (*see* **Note 4**).
6. Centrifuge the plate at 900 × *g* for 1 h at 32 °C.
7. Incubate the plate at 37 °C in the presence of puromycin (concentration determined in Subheading 3.1.1) to select the Cas9-expressing cells 48 h after transduction.
8. Culture for 1 week to obtain stable cell lines; replace the medium containing puromycin every 2 days.
9. If possible, validate Cas9 expression using Western blotting (Fig. 2a).

3.2 Design of sgRNAs

Here we used the human CD79B gene as an example. The CD79B gene has 4 transcript isoforms. When designing sgRNAs, we usually choose sgRNAs targeting the exons that are common to all four transcripts. For sgRNAs with similar scores, we prioritize sgRNAs targeting exons located toward the 5′ end of the transcripts.

1. Design sgRNAs using the CHOPCHOP online tool (http://chopchop.cbu.uib.no/).
2. Select the 3–4 sgRNAs with the highest score.
3. Identify off-target sites for the selected sgRNAs using the Cas-OFFinder online tool (http://www.rgenome.net/cas-offinder/).
4. Proceed with sgRNAs with off-target scores above 80.

3.3 In Vitro Transcription of sgRNAs

sgRNAs can be chemically synthesized by commercial providers or prepared by in vitro transcription. For in vitro transcription, a DNA template containing the sgRNA1 sequence, T7 promoter, and the scaffold sequence (tracrRNA) is required.

1. Synthesize DNA templates for in vitro transcription of the sgRNAs.
2. Dissolve DNA template and adjust concentration to 50 ng/μL.
3. Prepare the in vitro transcription mixture: 10 μL NTP buffer mix, 2 μL T7 RNA polymerase, 5 μL DNA template from the previous step, and 13 μL dH_2O (*see* **Note 4**).
4. Incubate the mixture at 37 °C for at least 6 h.
5. Add 1 μL DNase I to the mixture and incubate at 37 °C for 15 min.

6. Add 400 μL TRIzol reagent to the mixture to stop transcription.
7. Incubate the mixture for 5 min at room temperature.
8. Add 80 μL cold chloroform and mix by vortexing.
9. Incubate the mixture at room temperature for 3 min.
10. Centrifuge the mixture for 15 min at 12,000 × *g* at 4 °C.
11. Carefully transfer the aqueous phase, which contains the RNA component, to a new tube (around 250 μL).
12. To the aqueous phase, add 250 μL 100% isopropanol to precipitate the RNA.
13. Mix the sample by inverting the tube several times and incubate at room temperature for 10 min.
14. Centrifuge for 10 min at 12,000 × *g* at 4 °C.
15. Carefully remove the supernatant from the tube. Do not disturb the RNA pellet.
16. Wash the RNA pellet with 1 mL cold 80% ethanol.
17. Centrifuge the tube for 5 min at 12,000 × *g* at 4 °C.
18. Remove the supernatant and air dry the RNA pellet for 10 min.
19. Dissolve the RNA pellet in 50 μL sterile RNase-free dH_2O.
20. Quantify the RNA sample using NanoDrop Spectrophotometer (*see* **Note 5**).
21. Visualize RNA integrity by agarose gel electrophoresis (Fig. 2b).

3.4 Electroporation of sgRNA into CAS9-Expressing Cells

1. Seed 4 × 10^6 Cas9-expressing cells in 4 mL culture medium the day before the electroporation.
2. Harvest the cells by centrifugation for 5 min at 300 × *g*.
3. Resuspend the cells in cold 800 μL RPMI medium, divide into two aliquots of 400 μL each, and incubate the cells on ice (*see* **Note 6**).
4. Set up the program for delivery of sgRNA in the square wave electroporator (ECM 830 from BTX): Voltage 500 V, pulse length 2 ms.
5. Add the desired amount of sgRNA to one of the 400 μL aliquots of cells prepared in **step 3** (*see* **Note 7**).
6. Mix well and transfer the cell–RNA mixture to a 4 mm cuvette, and close the cap.
7. Set the cuvette in the electroporator and press start.
8. After electroporation, immediately transfer the cells to 3 mL pre-warmed fresh culture medium.

9. Repeat **steps 5–8**, using the remaining 400 μL aliquots of cells, and replacing the sgRNA solution with sterile RNase-free dH_2O.
10. Return the cells to the cell incubator.

3.5 Extraction of Genomic DNAs from the Cells Electroporated with dH_2O and sgRNA Using the DNeasy Blood and Tissue Kit

1. Collect 5×10^5 cells from each well 3 days after electroporation.
2. Resuspend the cell pellets in 200 μL PBS.
3. Add 20 μL proteinase K buffer to the cell suspension.
4. Add 200 μL AL buffer to the cell suspension and mix by vortexing.
5. Incubate the lysate at 56 °C for 10 min.
6. Add 200 μL 100% ethanol to the lysate and mix well.
7. Load the mixture onto the DNeasy mini spin column.
8. Centrifuge at 6000 × *g* for 1 min, and discard the flow-through.
9. Add 500 μL AW1 buffer to the column and centrifuge the column at 6000 × *g* for 1 min, and then discard the flow-through.
10. Add 500 μL AW2 buffer to the column and centrifuge the column at maximum speed for 3 min.
11. Discard the flow-through and place the column in a clean collection tube.
12. Add 50 μL TE buffer to the column and incubate the column for 1 min at room temperature.
13. Centrifuge the column at 6000 × *g* for 1 min.
14. Collect the sample and measure the genomic DNA concentration using NanoDrop Spectrophotometer.

3.6 Amplify the Targeted DNA Regions

1. Design PCR primers for the targeted locus (sgRNA1-targeted locus) using the software Primer 3 (*see* **Note 8**).
2. Prepare the PCR master mix (100 μL per site): mix 20 μL 5× PCR buffer, 2 μL dNTP (10 mM each), 1 μL Phusion High-Fidelity DNA Polymerase, 5 μL primer mix (2 μM each), 5 μL genomic DNA for each reaction (200 ng DNA) and 67 μL dH_2O.
3. Set up the PCR program in a thermal cycler: stage 1, 98 °C for 3 min, stage 2, 34 cycles using the following conditions: 98 °C for 10 s, 65 °C for 20 s (*see* **Note 9**) and 72 °C for 30 s; stage 3 72 °C for 7 min; hold 4 °C.
4. Visualize the PCR product on an agarose gel.
5. Verify the gene editing by DNA sequencing.

3.7 Analysis of CRISPR Editing by Sanger Sequencing

There is continually being developed software to decompose sanger sequencing results, for example, TIDE https://tide.deskgen.com/ [17], or CRISPR-ID http://crispid.gbiomed.kuleuven.be/ [18]. Here, we use TIDE as an example.

1. Upload the sequencing results (abl format) to the TIDE website.
2. Paste the sgRNA sequence into the software, but do not include the PAM sequence.
3. Adjust parameters based on the length of the PCR product and click to update the view.
4. Analyze results.

4 Notes

1. We have tested concentrations from 200 ng/mL to 10 μg/mL.
2. Different antibiotics have different time kinetics. Puromycin selection takes about 1 week.
3. All reagents must be RNase free.
4. We usually transduce the cells with the same volume of retroviral supernatant as the cell culture volume. If you want to transduce more cells, then increase both volumes in a 1:1 ratio.
5. High-quality RNA is important for genome-editing efficiency and cell viability. The 260:280 ratio should be around 2.0 and the 260:230 ratio around 2.1.
6. The RPMI medium should not contain any antibiotics or serum.
7. The volume of the RNA solution should not be more than 10% of the cell suspension volume.
8. The targeted site should be in the middle of the PCR product, and the length of the PCR product should be at least 500 bp.
9. Calculate the annealing temperate using the online tool: http://tmcalculator.neb.com/#!/main.

References

1. Carlson DF, Tan W, Lillico SG, Stverakova D, Proudfoot C, Christian M, Voytas DF, Long CR, Whitelaw CB, Fahrenkrug SC (2012) Efficient TALEN-mediated gene knockout in livestock. Proc Natl Acad Sci U S A 109 (43):17382–17387
2. Geurts AM, Cost GJ, Freyvert Y, Zeitler B, Miller JC, Choi VM, Jenkins SS, Wood A, Cui X, Meng X, Vincent A, Lam S, Michalkiewicz M, Schilling R, Foeckler J, Kalloway S, Weiler H, Ménoret S, Anegon I, Davis GD, Zhang L, Rebar EJ, Gregory PD, Urnov FD, Jacob HJ, Buelow R (2009) Knockout rats via embryo microinjection of zinc-finger nucleases. Science 325(5939):433
3. Takasu Y, Kobayashi I, Beumer K, Uchino K, Sezutsu H, Sajwan S, Carroll D, Tamura T, Zurovec M (2010) Targeted mutagenesis in the silkworm Bombyx mori using zinc finger

nuclease mRNA injection. Insect Biochem Mol Biol 40(10):759–765

4. Chang N, Sun C, Gao L, Zhu D, Xu X, Zhu X, Xiong JW, Xi JJ (2013) Genome editing with RNA-guided Cas9 nuclease in zebrafish embryos. Cell Res 23(4):465–472
5. Cho SW, Kim S, Kim JM, Kim JS (2013) Targeted genome engineering in human cells with the Cas9 RNA-guided endonuclease. Nat Biotechnol 31(3):230–232. https://doi.org/10.1038/nbt.2507
6. Mali P, Yang L, Esvelt KM, Aach J, Guell M, DiCarlo JE, Norville JE, Church GM (2013) RNA-guided human genome engineering via Cas9. Science 339(6121):823–826
7. Hsu PD, Scott DA, Weinstein JA, Ran FA, Konermann S, Agarwala V, Li Y, Fine EJ, Wu X, Shalem O, Cradick TJ, Marraffini LA, Bao G, Zhang F (2013) DNA targeting specificity of RNA-guided Cas9 nucleases. Nat Biotechnol 31(9):827–832
8. Kim S, Kim D, Cho SW, Kim J, Kim JS (2014) Highly efficient RNA-guided genome editing in human cells via delivery of purified Cas9 ribonucleoproteins. Genome Res 24 (6):1012–1019
9. Liang X, Potter J, Kumar S, Zou Y, Quintanilla R, Sridharan M, Carte J, Chen W, Roark N, Ranganathan S, Ravinder N, Chesnut JD (2015) Rapid and highly efficient mammalian cell engineering via Cas9 protein transfection. J Biotechnol 208:44–53
10. Huse K, Bakkebø M, Wälchli S, Oksvold MP, Hilden VI, Forfang L, Bredahl ML, Liestøl K, Alizadeh AA, Smeland EB, Myklebust JH (2012) Role of Smad proteins in resistance to BMP-induced growth inhibition in B-cell lymphoma. PLoS One 7(10):e46117
11. Bollum LK, Huse K, Oksvold MP, Bai B, Hilden VI, Forfang L, Yoon SO, Wälchli S, Smeland EB, Myklebust JH (2017) BMP-7 induces apoptosis in human germinal center B cells and is influenced by TGF-beta receptor type I ALK5. PLoS One 12(5):e0177188
12. Shen B, Zhang W, Zhang J, Zhou J, Wang J, Chen L, Wang L, Hodgkins A, Iyer V, Huang X, Skarnes WC (2014) Efficient genome modification by CRISPR-Cas9 nickase with minimal off-target effects. Nat Methods 11(4):399–402
13. Labun K, Montague TG, Gagnon JA, Thyme SB, Valen E (2016) CHOPCHOP v2: a web tool for the next generation of CRISPR genome engineering. Nucleic Acids Res 44 (W1):W272–W276
14. Montague TG, Cruz JM, Gagnon JA, Church GM, Valen E (2014) CHOPCHOP: a CRISPR/Cas9 and TALEN web tool for genome editing. Nucleic Acids Res 42(Web Server issue):W401–W407
15. Bae S, Park J, Kim JS (2014) Cas-OFFinder: a fast and versatile algorithm that searches for potential off-target sites of Cas9 RNA-guided endonucleases. Bioinformatics 30 (10):1473–1475
16. Wälchli S, Løset GÅ, Kumari S, Johansen JN, Yang W, Sandlie I, Olweus J (2011) A practical approach to T-cell receptor cloning and expression. PLoS One 6(11):e27930
17. Brinkman EK, Chen T, Amendola M, van Steensel B (2014) Easy quantitative assessment of genome editing by sequence trace decomposition. Nucleic Acids Res 42(22):e168
18. Dehairs J, Talebi A, Cherifi Y, Swinnen JV (2016) CRISP-ID: decoding CRISPR mediated indels by sanger sequencing. Sci Rep 6:28973

Chapter 26

Gene Knockout in Hematopoietic Stem and Progenitor Cells Followed by Granulocytic Differentiation

Perihan Mir, Malte Ritter, Karl Welte, Julia Skokowa, and Maksim Klimiankou

Abstract

In this chapter, we present an optimized CRISPR/Cas9 RNP nucleofection approach for gene knockout (KO) in hematopoietic stem and progenitor cells (HSPCs). With experimentally proved active locus-specific sgRNAs, we routinely reach over 80% gene KO in HSPCs, thus avoiding the need for cell sorting or enrichment of targeted cell population. Additionally, we provide a protocol for in vitro granulocytic differentiation of HSPCs after gene KO and detailed description of granulocyte function tests which can be applied to study the effects of a particular gene KO.

Key words Hematopoietic stem and progenitor cells, Gene knockout, CRISPR/Cas9, Single-guide RNA, Ribonucleoprotein, Nucleofection, Granulocytic differentiation

1 Introduction

CRISPR (clustered regularly interspaced short palindromic repeat) DNA editing technology is becoming a revolutionary tool in biotechnology, medical, and biological research. In less than a decade after publishing the critical discovery that CRISPR/Cas9 can be reprogrammed to target a desired DNA locus in bacteria [1], RNA-programmable gene editing has transformed biological research, thanks to its extreme flexibility, efficiency, and ease of use (*see* Chapters 1 and 19). Currently, the most popular choice for genome editing is CRISPR/Cas9 system from *Streptococcus pyogenes*. Briefly, the complex of Cas9 protein and a guide RNA bind to DNA through base complementarity. This binding induces DNA cleavage [2]. Gene KO can be achieved by error-prone DNA repair mechanism called nonhomologous end joining pathway which induces insertions or deletions (indels) at position of the DNA cut site. Precise nucleotide sequence integration in desired

Mouldy Sioud (ed.), *RNA Interference and CRISPR Technologies: Technical Advances and New Therapeutic Opportunities*, Methods in Molecular Biology, vol. 2115, https://doi.org/10.1007/978-1-0716-0290-4_26,

DNA locus can be achieved by exploiting the homology-directed repair pathway.

CRISPR/Cas9 system is widely used because of the short and simple NGG protospacer-adjacent motif (PAM) sequence, the best characterized on- and off-target profiles, availability of online tools to assist in performing CRISPR/Cas9 experiments, and its existence in different formats (plasmid, RNA, protein) from an expanding number of companies. Over the past few years, many engineered Cas9 variants have been developed. Different Cas9 variants have been generated (e.g., high-fidelity Cas9 variants including eSpCas9 [3], SpCas9-HF1 [4], and HypaCas9 [5], catalytically dead Cas9 (dCas9) [6], Cas9 nickase [7], and Cas9 variant with expanded PAM recognition site [8]). This discovery enables targeted genome modification with unprecedented flexibility and efficiency (*see* Chapter 21).

Another critical part of CRISPR/Cas9 system is a guide RNA. Guide RNA can be used as single-guide RNA (sgRNA) or as a two-component system which includes CRISPR RNA (crRNA) and trans-activating CRISPR RNA (tracrRNA). Additionally, chemical modifications of sgRNA can substantially increase genome editing efficiency in human primary cells ([9], *see* Chapter 2). Availability of highly purified Cas9 protein and chemically modified guide RNA with enhanced stability and low immune-stimulatory properties from commercial vendors makes nucleofection of CRISPR/Cas9 RNP complex the cost-effective and efficient way of generating targeted gene KOs (*see* Chapter 26).

The advantages of CRISPR/Cas9 RNP complex nucleofection over transfection with Cas9 and sgRNA DNA (plasmid) or RNA or virus-assisted delivery are as follows: it is a nontoxic, well-tolerable genome editing approach with short but high activity window, short persistence of RNP complex in the nuclei reduces probability of cuts at off-target DNA loci, CRISPR/Cas9 RNP complex does not require cellular transcription or translation machinery, no/low induction of inflammatory or cell response against foreign DNA and RNA after RNP complex nucleofection, and visualization of RNP by labeling of Cas9 protein [10].

CRISPR/Cas9 gene KO efficiency depends on multiple factors. Among the most important is the on-target score of sgRNA. Although, a broad range of prediction algorithms and methods for on-target sgRNA activity estimation are available online [11–13], the functional sgRNA validation after in silico on-target activity analysis remains the most comprehensive way to identify a highly active locus-specific sgRNA. Usually, 3–5 sgRNAs are tested per gene or gene locus. In our hands, sgRNA efficiencies have varied from 10% to 99% of edited allele. The second most important factor for successful gene KO is the usage of a sufficient number of viable HSPCs (too less or too many cells per nucleofection reaction may lead to suboptimal results). Preliminary tests may help to determine the optimal amount of CRISPR/Cas9 RNP complex and the number of HSPCs used for nucleofection.

After successful induction of gene KO, HSPCs can be further differentiated in vitro to cells of the myeloid lineage using appropriate culture conditions. If myeloid differentiation is not disturbed by gene KO, then these cells are expected to demonstrate upregulation of myeloid specific surface markers, morphological maturation, and recapitulation of granulocyte functions. Thus, the coupling of targeted gene KO approach via CRISPR/Cas9 RNP nucleofection with a protocol of granulocytic differentiation of HSPCs and granulocyte function tests allows investigation of the role of certain genes in hematopoiesis in a fast and robust way. It should be noted that the entire protocol can be completed in 3 weeks (Fig. 1).

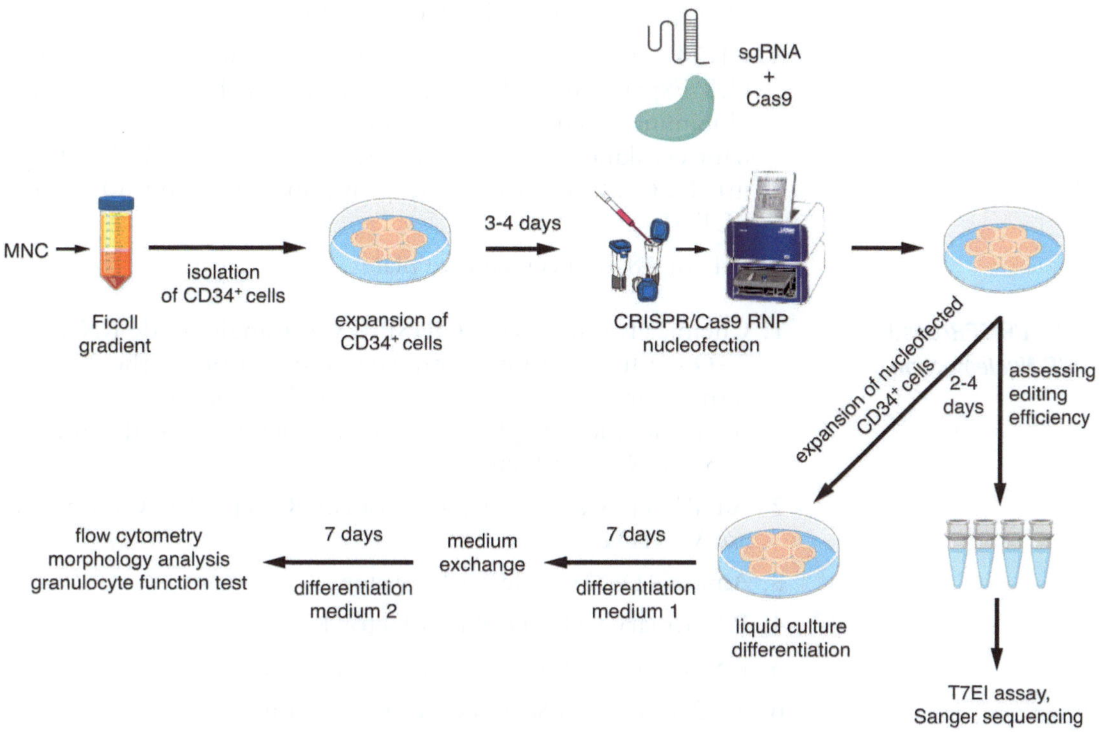

Fig. 1 Scheme of gene KO in HSPCs followed by in vitro granulocytic differentiation. $CD34^+$ cells are isolated from BM or mobilized blood MNCs and expanded in vitro for 3–4 days. After expansion $CD34^+$ cells are nucleofected with CRISPR/Cas9 RNP using the Lonza 4D nucleofection system. Knockout HSPCs are differentiated in vitro toward granulocyte lineage for 14 days and then analyzed by flow cytometry, morphologic analysis, and granulocyte function tests. Gene editing efficiency is assessed 2–4 days after nucleofection by T7EI assay and/or Sanger sequencing. Indel efficiency is determined by TIDE or ICE analysis web tools

2 Materials

All reagents, lab equipment, and web tools described in this section have been successfully tested by us, but other commercially available equivalents can be alternatively used after some optimization steps.

2.1 CD34⁺ Isolation and Expansion

1. Bone marrow (BM) or mobilized blood from healthy individuals or patients as source for $CD34^+$ cells.
2. Ficoll-Paque density gradient media (density: 1.077 g/mL) for isolation of BM or peripheral blood mononuclear cell (BMMNCs or PBMNCs, respectively).
3. Human CD34 MicroBead Kit for magnetic beads-based isolation of $CD34^+$ cells from BMMNCs or PBMNCs.
4. Cell freezing medium: PSC cryopreservation medium or 90% FCS supplemented with 10% DMSO.
5. $CD34^+$ expansion medium: Stemline II Hematopoietic Stem Cell Expansion medium supplemented with 10% FCS, 1% L-glutamine, 1% penicillin/streptomycin, and a human recombinant cytokine cocktail consisting of 20 ng/mL IL-3, 20 ng/mL IL-6, 20 ng/mL TPO, 50 ng/mL SCF, and 50 ng/mL FLT-3L.
6. 24- or 48-well cell culture plates.

2.2 CRISPR/Cas9 RNP Nucleofection

1. Chemically synthetized CRISPR/Cas9 single-guide RNA with 2′-*O*-methyl 3′phosphorothioate modifications at the 3′ and 5′ termini produced at small scale 15–20 nmol. For selection of the most efficient sgRNA, it is recommended to order and test 3–5 sgRNAs per locus.
2. Alt-R® S.p. Cas9 Nuclease V3 or Alt-R® S.p. HiFi Cas9 Nuclease V3 (IDT).
3. Amaxa 4D-Nucleofector™ System.
4. P3 Primary Cell 4D-Nucleofector Kit.
5. RNase-free aerosol-resistant pipette tips.
6. RNAse-free SafeSeal microcentrifuge tubes.
7. 1x TE Buffer pH 7.5.

2.3 Estimation of Gene-Editing Efficiency

1. DNA extraction kit.
2. Nanodrop or Qubit device and Qubit dsDNA High Sensitivity Assay Kit.
3. PCR: GoTaq Hot Start DNA Polymerase Kit, PCR primers for amplification of genome locus of interest.
4. PCR Nucleotide Mix.

5. T7 Endonuclease I Assay (T7EI) is a quick and cheap method of gene editing efficiency estimation which requires T7 Endonuclease I and NEB buffer 2.
6. PCR product purification kit or enzymatic cleanup using ExoSAP [14].
7. Sanger sequencing by commercial service provider or in-house Sanger sequencing facility.

2.4 Analysis of Granulocytic Differentiation

1. Colony-forming unit (CFU) assay: MethoCult™ Enriched (STEMCELL Technologies), IMDM with 2% FCS (STEMCELL Technologies), Antibiotic-Antimycotic Solution (Sigma Aldrich), 35 mm culture dishes, sterile 16 gauge needles, 1.5 in. long with a metal Luer Lock hub (STEMCELL Technologies).
2. Liquid culture differentiation: Differentiation medium 1 for day 0–7 consists of RPMI 1640 supplemented with 10% FCS, 1% penicillin/streptomycin, and human recombinant cytokines: 5 ng/mL SCF, 5 ng/mL IL-3, 5 ng/mL GM-CSF, and 10 ng/mL G-CSF. Differentiation medium 2 for day 7–14 consists of RPMI 1640 supplemented with 10% FCS, 1% penicillin/streptomycin, and 10 ng/mL G-CSF.
3. Morphology analysis: Shandon Cytospin centrifuge, Cytospin microscope slides, May–Grünwald, and Wright–Giemsa staining solutions.
4. FACS staining: antihuman CD45, antihuman CD34, antihuman CD33, antihuman CD11b, antihuman CD14, antihuman CD15, antihuman CD16 antibodies, PBS supplemented with 2% FCS and 0.02% sodium azide.

2.5 Analysis of Granulocyte Functions

1. Chemotaxis: 24-well Transwell reaction plates, inserts with 5 μm pore size, *N*-formyl-Met-Leu-Phe (fMLP), RPMI 1640 medium, bovine serum albumin (BSA), 5 mL FACS tubes, flow cytometer.
2. Phagocytosis: fluorescein-conjugated *Staphylococcus aureus* BioParticles, RPMI 1640 without Phenol red, 5 mL FACS tubes, flow cytometer capable of measuring FITC signal.
3. Netosis: IncuCyte® Live Cell Imaging system, Phorbol 12-myristate 13-acetate (PMA), RPMI 1640 without Phenol red, BSA, 96-well plate flat bottom, Sytox green dye, 0.001% Poly-L-Lysine Solution.
4. ROS production: ROS-Glo H_2O_2 Assay, bioluminescence reader, and 96-well plates.

3 Methods

3.1 sgRNA Design

1. Usually, the design of sgRNAs is performed with the assistance of online web tools. We routinely use the DESKGEN cloud platform (www.deskgen.com) or Benchling (www.benchling.com). Both online resources require a free registration.
2. For gene KO, make sure to target all possible isoforms of the gene of interest or design multiple sgRNAs (*see* **Note 1**).
3. Select all sgRNA target sites around the ideal position, and evaluate on-target and off-target prediction scores (*see* **Note 2**).
4. Choose from 3 to 5 sgRNAs per locus with the lowest off-target and highest on-target prediction score.
5. Test sgRNAs, as described in the following sections.

3.2 PCR Primers for Assessing CRISPR/Cas9 Gene-Editing Efficiency

1. Primer-BLAST (http://www.ncbi.nlm.nih.gov/tools/primer-blast) is a suitable online tool to design PCR primers [15] for assessing CRISPR/Cas9 gene editing efficiency by sequence trace decomposition.
2. Design PCR primers to ideally have a PCR product length of 700 bp.
3. To allow correct estimation of gene editing efficiency by sequence trace decomposition, the projected Cas9 cut site should be located around 200 bp away from reverse and forward primers used for sequencing.

3.3 CD34$^+$ Isolation and Expansion

1. Pipet 15 mL Ficoll-Paque medium into a 50 mL tube.
2. Dilute BM or mobilized PB sample 1:2 with PBS and carefully layer it onto the Ficoll-Paque medium (*see* **Note 3**).
3. Centrifuge at 500 × *g* for 25 min at room temperature (RT) with the centrifuge brake set off.
4. The mononuclear cell fraction is located in the interphase. Transfer the complete interphase layer to a sterile tube and wash the cells twice with 30 mL ice-cold PBS by centrifuging at 300 × *g* for 8 min at 8 °C.
5. Resuspend cells in ice-cold PBS and the determine cell count of BMMNCs or PBMNCs (*see* **Note 4**).
6. Resuspend cells in 300 μL MACS buffer. Use 200 μL FcR blocking and 100 μL CD34 microbeads for up to 10^8 cells (*see* **Note 5**).
7. Mix gently and incubate for 30 min at 4 °C in the refrigerator.
8. Wash cells with 5–10 mL MACS buffer by centrifugation at 300 × *g* for 10 min. Remove supernatant.

9. Resuspend cells in 500 μL MACS buffer.
10. Place the separation column in the magnetic field.
11. Rinse the column with 3 mL MACS buffer if you use a LS column, and if you use MS column, rinse with 500 mL prior to use (*see* **Note 6**).
12. Apply cell suspension onto the column (*see* **Note** 7). If desired, collect the CD34-negative cells in the flow-through fraction.
13. Wash LS column three times with 3 mL MACS buffer (if you use MS column, wash with 500 μL three times).
14. Remove column from magnetic field and put it onto a 15 mL tube.
15. Pipet 3 mL of MACS buffer on the column (1 mL for MS column) and flush out the cells by pushing the plunger into the column.
16. Determine number of $CD34^+$ HSPCs (*see* **Note 8**).
17. Either use cells freshly for gene editing experiment or freeze them for long-term preservation.
18. Perform cryopreservation in either 0.5 mL PSC cryopreservation medium or in 1 mL 90% FCS supplemented with 10% DMSO.
19. Defreeze $CD34^+$ HSPCs in a pre-warmed standard medium for suspension cells (e.g., RPMI 1640 with 10% FCS and 1% penicillin/streptomycin) (*see* **Note 9**). After centrifugation, resuspend the cells in hematopoietic stem cell expansion medium.
20. Culture $CD34^+$ HSPCs in $CD34^+$ expansion medium at a density of 2×10^5/mL (*see* **Note 10**) at 37 °C and 5% CO_2.
21. Determine cell count at day 3 or day 4 after culturing. If cells proliferate and cell viability is higher than 80% (e.g., by trypan blue), perform CRISPR/Cas9 RNP nucleofection.

3.4 CRISPR/Cas9 RNP Nucleofection

1. Assemble Cas9 protein with sgRNA for 30 min at RT. Use 136 pmol Cas9 protein and 330 pmol sgRNA for 1×10^6 cells.
2. For a single Nucleocuvette™ vessel, mix 18 μL of Supplement 1 and 82 μL P3 Primary Cell Nucleofector™ Solution from P3 Primary Cell Nucleofection Kit (Lonza) (*see* **Note 11**).
3. We recommend to use "mock" nucleofected cells to control the effect of nucleofection itself.
4. Add CRISPR/Cas9 RNP complex to the nucleofection buffer mix (*see* **Note 12**).
5. Wash cells once with PBS. Gently remove as much of the supernatant as possible.

6. Resuspend the cell pellet in the RNP-nucleofection buffer mix and transfer the mixture into the Nucleocuvette™ vessel, avoiding bubbles.
7. Place the Nucleocuvette™ vessel in the retainer of the 4D-Nucleofector X Unit.
8. Select and start the Nucleofector™ program CA-137 (*see* **Note 13**).
9. Carefully transfer the nucleofected cells into $CD34^+$ expansion medium, avoiding repeated aspiration of the sample (*see* **Note 14**). Culture $CD34^+$ cells, as described in Subheading 3.3.
10. After 2–4 days, harvest 1×10^5 cells for DNA isolation to determine the gene editing efficiency.

3.5 Estimation of CRISPR/Cas9 Gene-Editing Efficiency

1. Isolate DNA from harvested cells using the column-based QIAamp DNA Mini Kit according to the manufacturer's protocol and determine the DNA concentration using a Nanodrop or Qubit.
2. Alternatively, for quick estimation of KO efficiency, lyse $0.5–1 \times 10^5$ cells in 25–50 μL of QuickExtract DNA Extraction Solution (Lucigen) and then incubate at 65 °C for 6 min and at 98 °C for 2 min (*see* **Note 15**).
3. Run PCR for target genome locus of interest using the GoTaq Hot Start Polymerase Kit. Per reaction, pipet a master mix consisting of 4 μL of 5x Flexi buffer, 0.5 μL dNTP mix (10 mM each), 1 μL 25 mM $MgCl_2$, 1 μL forward and reverse primer (10 μM each), and 0.125 μL Go-Taq polymerase (5 U/μL). Add 100 ng DNA (or 2–4 μL of QuickExtract DNA Extraction Solution) to the PCR master mix and PCR-grade H_2O to a final volume of 20 μL. The PCR program is as follows: denaturation 1 min at 95 °C, annealing for 1 min at the temperature that is optimal for selected primer pair (usually, it is 5 °C below the Tm of the primers used), and extension 1 min at 72 °C. Repeat PCR cycles for 35 times. A final extension for 5 min at 72 °C is recommended.
4. To quickly estimate gene-editing efficiency, T7EI assay may be performed. T7EI recognizes and cleaves hetero-duplexed DNA (*see* **Note 16**). First, mix 11 μL of PCR product and 2 μL of NEB buffer 2. Anneal PCR sample in thermal cycler using the following hybridization conditions: denaturation at 95 °C for 10 min, annealing at 95–85 °C with ramp rate −2 °C/s followed by second annealing at 85–25 °C with ramp rate −0.3 °C/s. Add 2 μL of T7EI (20 U) to the sample

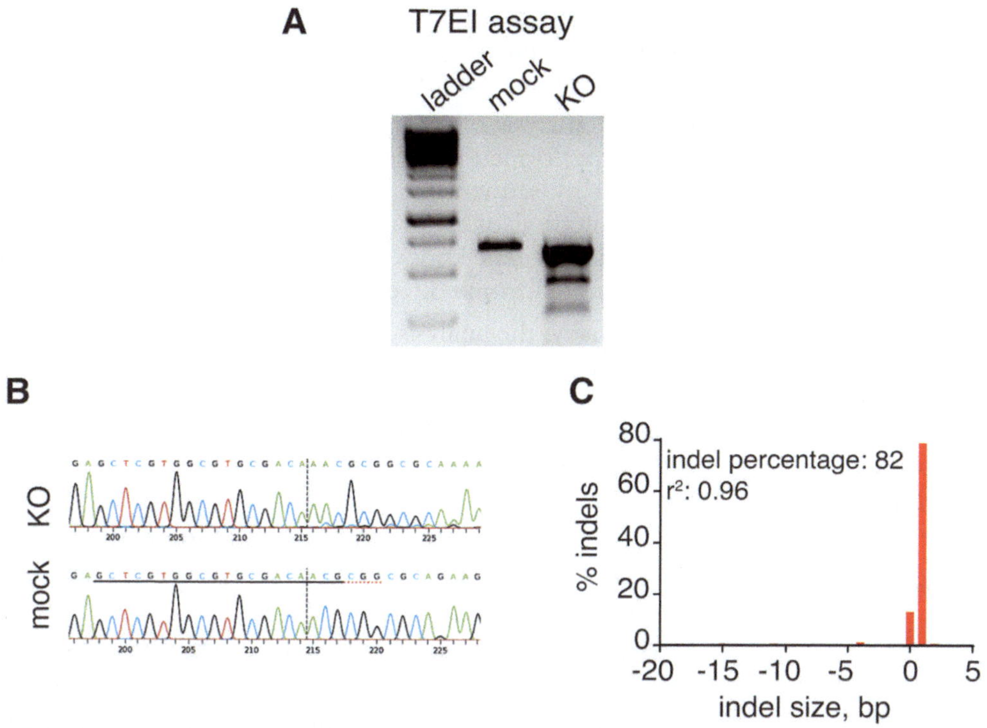

Fig. 2 (**a**) Agarose gel electrophoresis of PCR products after T7EI assay showing two extra bands in KO cells revealed the presence of heteroduplexed DNA resulting from gene editing. (**b**) Representative result of CRISPR/Cas9 RNP-mediated knockout in HSPCs showing Sanger sequencing of total population aligned to sequence track of mock cells. The position of DNA cut is depicted by dashed line. Target DNA sequence for guide RNA is underlined. (**c**) Representative ICE graph reconstructs the spectrum of indels in gene-edited HSPCs. Total editing efficiency is estimated approximately 82% with $r^2 = 0.96$ and one base insertion in the majority of cells. (**b** and **c**) were generated using ICE web tool from Synthego

and incubate at 37 °C for 1 h. Load sample into a 1.5% agarose gel and run it at 85 V for 1 h. Detect PCR products using a gel documentation system. In samples with successful gene editing, two low size extra bands are visible (Fig. 2a). The gene-editing efficiency can be calculated using density profile of the DNA bands by the following formula [16]:

$$\% \text{ gene modification} = 100 \times \left(1-(1-\text{fraction cleaved})^{1/2}\right)$$

5. For analysis of gene KO efficiency by sequence trace decomposition, PCR products from mock and CRISPR/Cas9 RNP nucleofected cells should be purified by column-based QIAquick PCR Purification Kit or enzymatic PCR clean-up using ExoSAP. Sanger sequencing traces have to be clean and have sufficient length to allow confident decomposition. Assess the gene modification efficiency using web tools such as TIDE assay [17] (https://tide.deskgen.com) or ICE analysis

(https://ice.synthego.com). For that, upload sgRNA sequence, sequencing traces (.abl file) of the mock sample, and the sample after gene editing (Fig. 2b, c).

6. Detailed analysis of mutation composition in the edited cells can be performed by cloning of PCR product into *E. coli* followed by sequencing of single bacteria clones. The most precise method of estimation of mutant allele frequency after gene editing/gene knockout is deep sequencing of PCR product generated from edited DNA locus.
7. Additionally, the analysis of off-target sites may be performed (*see* **Note 17**).

3.6 Characterization of Granulocytic Differentiation of HSPCs

3.6.1 Colony-Forming Unit Assay

1. Resuspend 3–10 × 10^3 cells (*see* **Note 18**) in 300 μL IMDM with 2% FCS, and then add 30 μL Antibiotic–Antimycotic Solution.
2. Add the cell suspension to 3 mL MethoCult™ Enriched medium. Vortex thoroughly, and wait until air bubbles disappear.
3. Seed the cell suspension into two 3.5 cm culture dishes (1.1 mL per dish) using sterile gauge needle.
4. Keep the plates in CO_2 cell culture incubator for 14 days and determine colony numbers and types at day 14 (*see* **Note 19**).

3.6.2 Liquid Culture Granulocytic Differentiation

1. After CRISPR/Cas9 RNP nucleofection, seed the cells at a density of 2 × 10^5/mL. During the first 7 days of differentiation, the cells are kept in differentiation medium 1 with medium exchange every second day (*see* **Note 20**).
2. At day 7, replace medium 1 with differentiation medium 2. The medium has to be changed every second day until day 14.
3. At day 14, perform flow cytometric analysis using antibodies specific for the following hematopoietic/myeloid markers: CD45 (leukocyte marker), CD34 (HSPC marker), CD33 (promyelocyte marker), CD11b (myeloid cell marker), CD14 (monocyte marker), and CD15 and CD16 (neutrophil markers). You may determine neutrophil percentage by gating on neutrophils as follows: $CD45^+CD11b^+CD15^+$, or $CD45^+CD11b^+CD16^+$, or $CD45^+CD15^+CD16^+$ cells.

3.6.3 Evaluation of Cell Morphology

1. Load the Cytoclip™ slide clips by fitting the filter card, the sample chamber, and the glass slide. Place an assemble Cytoclip™ slide clips in the slide clip support plate of the cytospin centrifuge ensuring that the centrifuge is balanced.
2. Pipette 2 × 10^4 cells from liquid culture differentiation (100 μL) into Cytofunnel™.
3. Centrifuge for 3 min at 200 × *g*.

4. Stain cytospin slides for 5 min in May–Grünwald stain, rinse shortly with ddH_2O, and then stain for 10 min in Wright-Giemsa stain. Afterwards rinse shortly with ddH_2O.
5. Determine cell morphology using a microscope.

3.7 Granulocyte Function Tests

3.7.1 Chemotaxis Assay

1. Resuspend the cells at 1×10^6 cells/mL in RPMI with 0.5% BSA and seed 200 μL of the cell suspension on top of the polycarbonate membrane with 5 μm pore size.
2. Fill the bottom well with RMPI 1640 medium supplemented with 0.5% BSA with or without 1 nM fMLP (*see* **Note 21**).
3. Incubate the plate for 2 h at 37 °C, 5% CO_2.
4. Add 70 μL of 50 mM EDTA to the lower chamber and transfer the plate to 4 °C for 30 min.
5. After incubation, remove the inserts, resuspend the cells vigorously, and transfer the cell suspension from the lower chamber into FACS tubes.
6. Count cells on flow cytometer.

3.7.2 Phagocytosis Assay

1. Resuspend 3–10 × 10^4 cells in 100 μL RPMI 1640 medium supplemented with 2% BSA.
2. Add fluorescent-labeled *Staphylococcus aureus* bioparticles in a ratio of 1:100 per cell and incubate at 37 °C for 2 h. Make sure to have untreated controls to determine background fluorescence.
3. After incubation, wash the cells twice with PBS supplemented with 2% BSA, resuspend cells in PBS with 2% BSA, and then measure the number of FITC-positive cells using f low cytometer.

3.7.3 Netosis Assay

1. Coat the wells of a flat bottom 96-well plate for 30 min at RT with 50 μL/well 0.0001% Poly-L-Lysin solution.
2. Remove the solution after coating and wash twice with PBS.
3. Seed 2 × 10^4 cells in 100 μL per well, add Sytox Green reagent to a final concentration of 250 nM, and let cells to adhere for 20 min at RT.
4. Carefully add stimulant of choice to induce net release. We used PMA at 100 nM.
5. Place the plates in the IncuCyte® live cell analysis system and perform phase-contrast and GFP imaging every 10 min at 20× magnifications overnight.
6. Analyzed the data on integrated software.

3.7.4 ROS Production

1. Seed the cells at a density of 1×10^5/mL, and then add fMLP at a final concentration of 10 nM to activate ROS production.
2. Incubate for 30 min at 37 °C, 5% CO_2.
3. Measure the level of hydrogen peroxide (H_2O_2) ROS by the ROS-Glo H_2O_2 Assay Kit according to manufacturer's protocol.

4 Notes

1. The preferred localization of guide RNA binding site for introduction of KO by CRISPR/Cas9 system is the first exon, which is present in all coding RNA transcripts. It is important to make sure that the selected DNA sequence is unique, thus reducing the probability of multiple binding of guide RNA to human genome.
2. Benchling and DESKGEN web sites provide on-target and off-target scores for every designed protospacer sequence. For comprehensive analysis of potential off-target sites, a bioinformatics-based tool, COSMID (CRISPR Off-target Sites with Mismatches, Insertions, and Deletions, https://crispr.bme.gatech.edu) [18], may be used. Additionally, calculation of microhomology-associated score can be helpful for the estimation of frequency of undesired in-frame deletion and selection of the gene locus with high out-of-frame score. Microhomology-Predictor tool [19] is available under http://www.rgenome.net/mich-calculator.
3. When layering the PB or BM sample onto Ficoll-Paque medium, pipet it gently to avoid mixing of medium and diluted BM/mobilized blood sample.
4. In case of cells clumps, use a nylon mesh (pre-separation filter with 30 μm, Miltenyi) to remove cell clumps prior to microbead labeling, to avoid a clogged column.
5. You may scale up or down the amount of beads and FcR blocking reagent, according to the cell number.
6. For 1×10^8 PBMNCs or BMMNCs, use either MS or LS columns.
7. For each step, wait until the column reservoir is empty before proceeding to the next step. At the same time, do not let the column to dry out.
8. We recommend to check the isolated $CD34^+$ cell fraction for its purity by staining the cells with an anti-CD34 antibody and flow cytometric analysis before proceeding with gene editing experiments.

9. Hematopoietic stem cell medium and cytokines are expensive cell culture reagents. RPMI 1640 cell medium or other equivalents may be used for the defreezing of $CD34^+$ cells with no adverse effects for the cells.
10. $CD34^+$ HSPCs proliferate better in a dense cell culture with close cell-to-cell contact. Do not seed them with a density lower than 2×10^5/mL.
11. For testing several sgRNAs, you may use the Lonza 16-well Nucleocuvette™ strip to nucleofect 1–10 $\times 10^5$ cells in a total volume of 20 μL. This way is cost-effective for saving CRISPR/Cas9 RNP and valuable primary cells.
12. The total volume of RNP-nucleofection buffer mix should not exceed 110 μL for nucleofection in the Nucleocuvette™ vessel.
13. Optional: we recommend to let the cells stay for 10 min after nucleofection at RT or in cell culture incubator before transferring them into the cell culture plate.
14. Lonza Nucleocuvette™ vessel may be reused for up to three times. Rinse the cuvettes three times with ddH_2O and once with EtOH. Then expose Nucleocuvette™ vessel with open lid to UV light for at least 30 min.
15. Cell lysis in the QuickExtract DNA Extraction Solution does not allow DNA concentration measurements. Usually, 2–5 μL of the lysate is sufficient for PCR. If PCR yields no product while the positive control shows specific PCR band, then dilute the quicklyzed DNA 1:3–1:5 with the QuickExtract DNA Extraction Solution and repeat PCR.
16. T7EI assay might not give an exact estimation of gene modification efficiency.
17. CRISPR/Cas9 RNP off-targets site analysis may be performed as follows:
    (a) In silico prediction of off-target DNA loci followed by Sanger sequencing of the top predicted off-target sites (low sensitivity, time consuming, low output).
    (b) Custom NGS panel designed for predicted off-target DNA regions (high sensitivity, long turnover time).
    (c) Whole genome or exome DNA sequencing with high depth (can be expensive, especially at the genome level, high quality, and amount of DNA is necessary).
18. $CD34^+$ cells lose stemness during in vitro culture. The number of cells plated for CFU experiment depends on how long cells were kept in culture. If $CD34^+$ cells are cultured more than 1 week, then use 10×10^3 cells per CFU experiment.

19. Place two 3.5 cm CFU dishes and an extra dish with 2 mL of PBS and without a lid in a 10 cm dish to avoid the 3.5 dishes from drying out.
20. It is not recommended to disturb the cells every second day by collecting the whole medium volume. Instead, without disturbing or soaking up the cells, gently remove as much medium as possible and add the same amount of fresh medium to the cells.
21. Each sample needs to be tested with and without chemoattractant (fMLP) to control for spontaneous migration. Additionally, each condition should be conducted at least in technical duplicates.

Acknowledgments

This work was supported by German Cancer Consortium (P.M. and J.S.), DFG (J.S., M.K.), the Fresenius Foundation (M.K.), and the Jose Carreras Leukemia Foundation (M. R.).

References

1. Jinek M, Chylinski K, Fonfara I, Hauer M, Doudna JA, Charpentier E (2012) A programmable dual-RNA-guided DNA endonuclease in adaptive bacterial immunity. Science 337 (6096):816–821
2. Jinek M, East A, Cheng A, Lin S, Ma E, Doudna J (2013) RNA-programmed genome editing in human cells. Elife 2:e00471
3. Slaymaker IM, Gao L, Zetsche B, Scott DA, Yan WX, Zhang F (2016) Rationally engineered Cas9 nucleases with improved specificity. Science 351(6268):84–88
4. Kleinstiver BP, Pattanayak V, Prew MS, Tsai SQ, Nguyen NT, Zheng Z, Joung JK (2016) High-fidelity CRISPR-Cas9 nucleases with no detectable genome-wide off-target effects. Nature 529(7587):490–495
5. Chen JS, Dagdas YS, Kleinstiver BP, Welch MM, Sousa AA, Harrington LB, Sternberg SH, Joung JK, Yildiz A, Doudna JA (2017) Enhanced proofreading governs CRISPR-Cas9 targeting accuracy. Nature 550 (7676):407–410
6. Qi LS, Larson MH, Gilbert LA, Doudna JA, Weissman JS, Arkin AP, Lim WA (2013) Repurposing CRISPR as an RNA-guided platform for sequence-specific control of gene expression. Cell 152(5):1173–1183
7. Mali P, Aach J, Stranges PB, Esvelt KM, Moosburner M, Kosuri S, Yang L, Church GM (2013) CAS9 transcriptional activators for target specificity screening and paired nickases for cooperative genome engineering. Nat Biotechnol 31(9):833–838
8. Hu JH, Miller SM, Geurts MH, Tang W, Chen L, Sun N, Zeina CM, Gao X, Rees HA, Lin Z, Liu DR (2018) Evolved Cas9 variants with broad PAM compatibility and high DNA specificity. Nature 556(7699):57–63
9. Hendel A, Bak RO, Clark JT, Kennedy AB, Ryan DE, Roy S, Steinfeld I, Lunstad BD, Kaiser RJ, Wilkens AB, Bacchetta R, Tsalenko A, Dellinger D, Bruhn L, Porteus MH (2015) Chemically modified guide RNAs enhance CRISPR-Cas genome editing in human primary cells. Nat Biotechnol 33 (9):985–989
10. Nasri M, Mir P, Dannenmann B, Amend D, Skroblyn T, Xu Y, Schulze-Osthoff K, Klimiankou M, Welte K, Skokowa J (2019) Fluorescent labeling of CRISPR/Cas9 RNP for gene knockout in HSPCs and iPSCs reveals an essential role for GADD45b in stress response. Blood Adv 3(1):63–71
11. Doench JG, Fusi N, Sullender M, Hegde M, Vaimberg EW, Donovan KF, Smith I, Tothova Z, Wilen C, Orchard R, Virgin HW, Listgarten J, Root DE (2016) Optimized sgRNA design to maximize activity and

minimize off-target effects of CRISPR-Cas9. Nat Biotechnol 34(2):184–191

12. Chari R, Mali P, Moosburner M, Church GM (2015) Unraveling CRISPR-Cas9 genome engineering parameters via a library-on-library approach. Nat Methods 12(9):823–826
13. Doench JG, Hartenian E, Graham DB, Tothova Z, Hegde M, Smith I, Sullender M, Ebert BL, Xavier RJ, Root DE (2014) Rational design of highly active sgRNAs for CRISPR-Cas9-mediated gene inactivation. Nat Biotechnol 32(12):1262–1267
14. Kim JB, Blackshaw S (2001) One-step enzymatic purification of PCR products for direct sequencing. Curr Protoc Hum Genet. Chapter 11:Unit 11 16. https://doi.org/10.1002/0471142905.hg1106s30
15. Ye J, Coulouris G, Zaretskaya I, Cutcutache I, Rozen S, Madden TL (2012) Primer-BLAST: a tool to design target-specific primers for polymerase chain reaction. BMC Bioinformatics 13:134
16. Guschin DY, Waite AJ, Katibah GE, Miller JC, Holmes MC, Rebar EJ (2010) A rapid and general assay for monitoring endogenous gene modification. Methods Mol Biol 649:247–256
17. Brinkman EK, Chen T, Amendola M, van Steensel B (2014) Easy quantitative assessment of genome editing by sequence trace decomposition. Nucleic Acids Res 42(22):e168
18. Cradick TJ, Qiu P, Lee CM, Fine EJ, Bao G (2014) COSMID: a web-based tool for identifying and validating CRISPR/Cas off-target sites. Mol Ther Nucleic Acids 3:e214
19. Bae S, Kweon J, Kim HS, Kim JS (2014) Microhomology-based choice of Cas9 nuclease target sites. Nat Methods 11(7):705–706

Chapter 27

CRISPR/Cas9 Genome Editing of Human-Induced Pluripotent Stem Cells Followed by Granulocytic Differentiation

Benjamin Dannenmann, Masoud Nasri, Karl Welte, and Julia Skokowa

Abstract

Research on patient-derived induced pluripotent stem cells (iPSCs) could immensely benefit from the implementation of CRISPR/Cas9 genome editing of iPSCs, creating unique opportunities such as the establishment of isogenic iPSC lines for disease modeling or personalized patient-specific drug screenings. Here we describe a stepwise protocol of safe, efficient, and selection-free CRISPR/Cas9-mediated gene correction or knockout in human iPSCs followed by 3D spin-embryoid body (EB)-based hematopoietic/neutrophilic iPSC-differentiation.

Key words CRISPR/Cas9, Human-induced pluripotent stem cells, Genome editing, Ribonucleoprotein, iPSC differentiation, Hematopoietic, and neutrophilic differentiation of iPSCs

1 Introduction

iPSC technology is a powerful tool for disease modeling, patient-specific drug screening, and paving the way for personalized medicine [1, 2]. Patient-derived iPSC cells can be used as an excellent tool to model the disease and reveal the key driving pathomechanisms behind it. An important challenge of iPSC technology is the high degree of variability between individual iPSC lines that are mainly due to differences in genetic background and reprogramming history [3]. Therefore, to have an optimal comparison in studies using iPSC, especially disease modeling, the role of the isogenic iPSC lines is indispensable. Combination of CRISPR/Cas9-based gene correction or knockout in human iPSCs (hiPSCs) derived from patients suffering from inherited or acquired hematopoietic disease with subsequent EB-based in vitro hematopoietic

Benjamin Dannenmann and Masoud Nasri contributed equally to this work.

Mouldy Sioud (ed.), *RNA Interference and CRISPR Technologies: Technical Advances and New Therapeutic Opportunities*, Methods in Molecular Biology, vol. 2115, https://doi.org/10.1007/978-1-0716-0290-4_27,

differentiation is a valuable tool to delineate the role of the gene of interest or gene mutations on hematopoietic differentiation in a fast, efficient, and reproducible manner [4–6].

In this chapter, we first explain in detail our CRISPR/Cas9 gene-editing strategy of iPSCs which is safe, as no viral vectors or plasmids are used, efficient, and selection free. We next describe our 3D spin-embryoid body (EB)-based iPSCs-differentiation protocol, which has many advantages over standard 2D iPSCs-differentiation protocols. The main advantage is the high amounts of HSPCs and mature granulocytes derived from iPSCs (more than 1×10^6 cells/10 EBs) and improved functionality of mature cells [4]. Another advantage is that the present protocol is based on the use of STEMdiff APEL medium, a fully defined, serum-free, and animal component free-differentiation medium which is highly versatile and allows the differentiation of cells to multiple hematopoietic lineages [7, 8]. By adapting variable cytokine combinations, our strategy allows efficient generation of not only a high number of $CD34^+/CD45^+$ cells and mature granulocytes but also monocytes or megakaryocytes. Furthermore, the current protocol displays high reproducibility between different experiments.

2 Materials

The reagents describing below have been successfully used by our group, but there are other commercially available alternatives that can be used and optimized.

2.1 Transfection and Electroporation of iPSCs with Cas9 Ribonucleoprotein (RNP)

1. Guide RNAs for target sequence ordered as lyophilized single-guide RNA with or without 2′-*O*-methyl 3′phosphorothioate (Integrated DNA Technologies).
2. Alt-R S.p Cas9 Nuclease V3 or Alt-R S.p HiFi Cas9 Nuclease V3 (Integrated DNA Technologies).
3. Ultramer DNA oligos, 4 nm synthesis scale, and standard desalting as purification method (Integrated DNA Technologies).
4. *Trans*IT-X2 transfection reagent (Mirus, Madison, USA).
5. Amaxa 4D-Nucleofector system and P3 primary Cell 4D-Nucleofector Kit (Lonza).

2.2 iPSCs Maintenance and EB-Based Myeloid Differentiation

1. StemFlex medium, RevitaCell supplement, and Geltrex LDEV-Free reduced growth factor basement membrane matrix (ThermoFisher Scientific).
2. STEMdiff APEL 2 Medium (Stem Cell Technologies).
3. ROCK inhibitor, FGF2, BMP4, VEGF, SCF, IL3, and G-CSF.

3 Methods

3.1 Design, Testing, and Validation of Guide RNA

1. Design guide RNAs using guide RNA designing web tools (e.g., CCTOP, Centre for Organismal Studies Heidelberg):
   (a) For gene-correction purpose, select 4–6 guide RNAs in a window of 40 nt distance from the desired mutation site with good off-target activity (*see* **Note 1**).
   The optimal window is 10 nt distance. Rank the guide RNAs based on the off-target activity profile.
   (b) For gene knockout purpose, select 4–6 guide RNAs in the coding sequence of the first or second exon of the gene of choice. Consider all possible transcripts and isoforms for a successful knockout.
2. Cross check the guide RNAs list in the second guide RNA designing website (e.g., DESKGEN).
3. Order the list of guide RNAs as crRNA or sgRNA via provider of choice (e.g., Integrated DNA Technologies, Coralville, USA).
4. Upon receiving the guide RNAs, spin them down briefly and resuspend to the final concentration of 100 μM.
5. For validation of guide RNAs, seed 10^5 HEK 293T cells 1 day before transfection in 12-well plate.
6. On the day of transfection, assemble 0.24 μL from each single-guide RNA (100 μM Stock, 24 nM final concentration) with 0.2 μL Cas9 protein (62 μM Stock, 12 nM final concentration) in a sterile tube.
7. Add 100 μL of OptiMEM I Reduced Serum Medium to each tube and mix gently.
8. Add 2 μL *Trans*IT-X2 transfection reagent (*see* **Note 2**) to each tube, mix gently, and incubate at room temperature for 20 min.
9. Add the *Trans*IT-X2: RNP complexes dropwise to the different areas of the well. Gently rock the plate forth and back and incubate at 37 °C for 48 h.
10. Harvest cells for isolation of genomic DNA and amplify the target genomic area by PCR. Upon successful PCR, send PCR products for Sanger sequencing.
11. Quantify the on-target efficiency of guide RNAs by Tideor ICE web tools [9, 10] and re-rank the guide RNA list based on highest on-target and lowest off-target activity.
12. For gene correction purpose, design the repair template as single-stranded oligonucleotide (ssODN) with homology arms of 50–90 nt surrounding the mutation site. Introduce a silent mutation at the PAM site to protect edited sequence from re-cutting by Cas9 after gene correction. Synthesize or order ssODN as 4 nmol Ultramer DNA oligonucleotide.

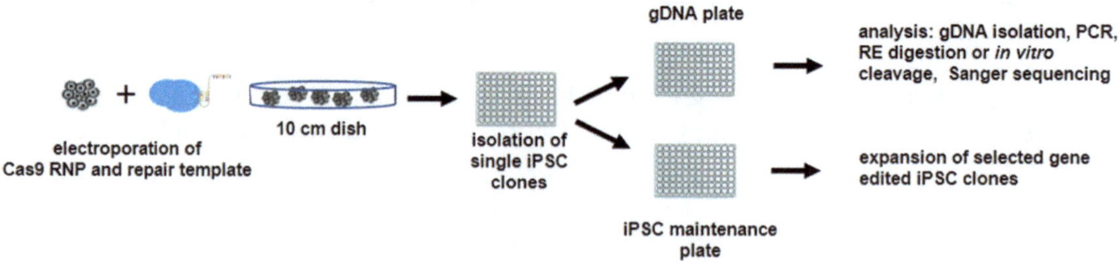

Fig. 1 Scheme of CRISPR/Cas9-based genome editing of human iPSCs

3.2 Electroporation of iPSCs with Cas9 Ribonucleoprotein (RNP)

Scheme of CRISPR/Cas9-based genome editing of human iPSCs is shown in Fig. 1.

1. Use the nucleofection protocol for the Amaxa 4D Nucleofection system with P3 Primary Cell 4D-Nucleofector X Kit.
2. For the electroporation of 10^6 cells, assemble 3.3 μL sgRNA and 15 μg Cas9 protein at room temperature for 30 min.
3. For detaching iPSCs, remove the medium and add Accutase, and then incubate until the cells start to detach (for approximately 3–5 min). Keep the number of pipetting to a minimum.
4. Wash cells two times with PBS prior to nucleofection to remove all traces of nucleases present in the culture medium. Centrifuge the cells at 300 × g for 5 min.
5. Remove the supernatant after centrifugation. For successful electroporation, it is important to completely remove the supernatant.
6. Prepare the nucleofection buffer according to the instructions of the P3 Primary Cell 4D-Nucleofector X Kit. Add ssODN and RNP mix to the nucleofection buffer.
7. Resuspend the cell pellet in the nucleofection buffer-RNP mix and transfer iPSCs suspension to the electroporation cuvette. Try to avoid any air bubble formation.
8. Place the cuvette in the Amaxa 4D Nucleofector and choose the CA-137 program.
9. After electroporation, incubate the cells in the cuvette for 10 min at room temperature.
10. Take up the cells gently in pre-warmed StemFlex medium supplemented with RevitaCell and transfer them to the Geltrex-coated 6-well plate.

3.3 Clonal Isolation and 96-Well Plate Culture of iPSCs

1. About 48 h post-electroporation, remove the medium, wash two times with PBS, and add 2 mL Accutase; incubate until the cells start to detach. It is important to have single cells at this step. Transfer the cells in 15 mL tube, centrifuge down at

300 × *g* for 5 min, discard the supernatant, and resuspend cells in Stemflex medium supplemented with RevitaCell.

2. Count cells.
3. Transfer 5000 iPSCs (*see* **Note 3**) to 10 mL Stemflex medium supplemented with RevitaCell, mix with 10 mL pipette few times, and transfer to the Geltrex-coated 10 cm dish. Rock forth and back. Put the dish back into the incubator.
4. Incubate the 10 cm dish for 7 days. Change the medium every 24 h with Stemflex medium without RevitaCell.
5. Coat the flat bottom 96-well plate with 100 μL Geltrex for at least 2 h at room temperature. Remove the Geltrex from the wells using a multichannel pump and add 100 μL Stemflex medium supplemented with RevitaCell to each well using a multichannel pipette.
6. After 7 days (*see* **Note 4**), wash the 10 cm dish gently with PBS. Add 10 mL PBS to the dish and incubate at room temperature for 5 min.
7. Put the 10 cm dish under the microscope (*see* **Note 5**) and using a 100 μL pipette that adjusted to 20 μL, gently scratch, pick, and transfer single iPS clone to a well of 96-well plate. Pipette few times to dissociate the clone. Pick 96 clones (*see* **Note 6**).
8. Put the 96-well plate containing 96 single iPSC clones to the incubator for three to four for days.
9. Coat at least 2 96-well plates (use flat bottom plates) with Geltrex (*see* **Note 7**) before use, remove the coating, and add 100 μL StemFlex Media supplemented with RevitaCell.
10. After 3–4 days, remove the media from 96-well plate using a multichannel pump (*see* **Note 8**), add 40 μL Accutase diluted 1:3 with PBS to each well, and incubate at 37 °C for 30 min (*see* **Note 9**) until the iPSCs start to dissociate.
11. Prepare a sterile reservoir filled with StemFlex Media supplemented with RevitaCell. Using a multichannel pipette, add 60 μL media to each well, pipette 4–5 times, and transfer 60 μL iPSCs containing media to the exact same position in a new Geltrex-coated 96-well plate. This plate will be used later for genomic DNA isolation (gDNA plate). Using the same tips, transfer the remaining 35 μL iPSCs containing media to the exact same position in the second Geltrex-coated 96-well plate. This plate will be stored in the incubator (maintenance plate).

3.4 Screening for Gene-Corrected iPSC Clones and Validation of Gene Correction

1. Using a multichannel pump, remove the media from all wells of the gDNA plate and wash each well two times with PBS.
2. Add 30 μL of QuickExtract DNA Extraction solution.
3. Heat the 96-well plate at 65 °C for 6 min and 98 °C for 2 min. The 96-well plate can be directly used as a template DNA in PCR reaction or stored at −20 °C.
4. To amplify the target region, prepare a PCR master mix, pipette it to PCR-specific 96-well plate, and add 2 μL of gDNA prepared in **step 11** of Subheading 3.3.
5. Perform the PCR reaction.
6. For gene correction purpose, use restriction endonucleases enzyme that produces a different pattern of digestion after gene correction (e.g., http://insilico.ehu.es) and perform restriction enzyme digestion of each clone.
7. For gene knockout purpose, instead of endonuclease restriction enzyme digestion, perform in vitro Cas9 cleavage using the same sgRNA that has been used for the knockout as following:
   (a) Prepare sgRNA at a concentration of 10 μM in a sterile microtube by diluting the sgRNA in nuclease-free buffer.
   (b) Prepare 1 μM RNP complex by adding 1.6 μL Cas9 enzyme (62 μM stock) and 88.4 μL PBS to the microtube containing 10 μM sgRNA.
   (c) Incubate at room temperature for 15 min.
   (d) Perform the in vitro digestion reaction by adding 1 μL of 1 μM Cas9 RNP, 1 μL IDT nuclease-free buffer, and 3 μL of unpurified PCR reaction from **step 5** of Subheading 3.4.
   (e) Incubate the reaction at 37 °C for 1 h.

   PCR product of knockout candidate iPSC clones should not be digested by Cas9 RNP as the target sequence is already modified.
8. Run the reaction from **steps 6** or **7** on 1% agarose gel and perform Sanger sequencing of the PCR products with a preferred pattern in order to validate gene correction or knock out in the selected iPSC clones.

3.5 Hematopoietic and Neutrophilic Differentiation of iPSCs Via Spin-EBs

The scheme of EB-based neutrophilic differentiation is shown in Fig. 2. A list of cytokines used for EB-differentiation is shown in Table 1.

1. Two days before EB formation, seed 4×10^5 mitotically inactivated SNL feeder cells on a 3.5 cm^2cell culture dish.

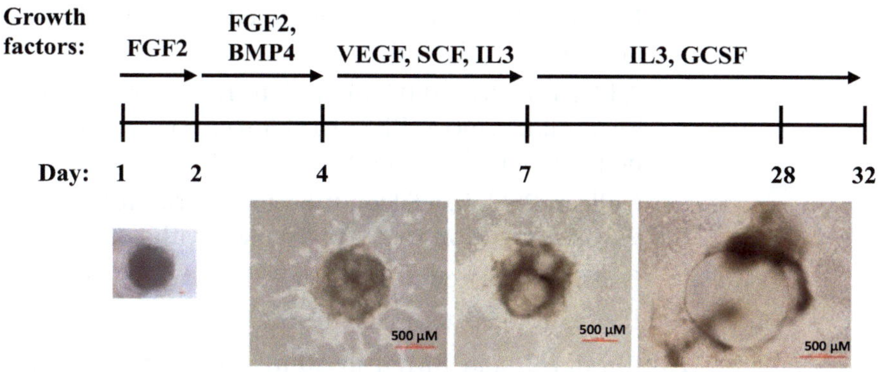

Fig. 2 Scheme of EB-based neutrophilic differentiation of human iPSCs

Table 1
List of cytokines used for EB-based hematopoietic iPSCs differentiation

Cytokine	Stock concentration	Final concentration	Function	Company
BMP4	20 μg/mL	20 ng/mL	Mesoderm induction	R&D
bFGF (FGF2)	100 μg/mL	20 ng/mL	Mesoderm differentiation	Peprotech
VEGF	40 μg/mL	40 ng/mL	Angiogenesis	R&D
SCF	100 μg/mL	50 ng/mL	Stem cell factor	Peprotech
IL-3	50 μg/mL	50 ng/mL	Myeloid differentiation	Peprotech
ROCK inhibitor (Y-27632dihydrochloride)	10 mM	10 nM	Inhibition of rho-kinase	Tocris
G-CSF	100 μg/mL	50 ng/mL	Granulocytic differentiation	Neupogen (Filgrastim)

2. One day before EB formation, seed approximately 50–70% confluent feeder-free iPSCs from one well of a 6-well plate to a 3.5 cm dish with SNL feeder cells in iPSCs maintenance medium supplemented with 20 ng/mL bFGF and 10 nM ROCK inhibitor (*see* **Note 10**).
3. On the day of EB formation (day 1), precool centrifuge to 4 °C, as it will increase efficiency of compact EB formation.
4. Prepare APEL medium for EB formation. From one 3.5 cm^2 dish, you can produce up to 30 EBs. For 30 EBs, 3 mL of APEL medium supplemented with ROCK (10 nM) and bFGF (20 ng/mL) are needed.

5. Remove iPSCs maintenance medium from iPSCs and wash once with 1 mL PBS.
6. Add 1 mL PBS/0.02% EDTA to the 3.5 cm^2 dish with iPSCs for 7 min, remove PBS/EDTA solution, and detach iPSCs in prepared APEL medium.
7. Collect iPSCs in a falcon tube and distribute100 μL per well of iPSCs suspension to a 96-well plate with conical bottom.
8. Centrifuge plate for 5 min at 435 × *g* and 4 °C.
9. Incubate the plate for 24 h at 37 °C.
10. On day 2, add BMP4 in 10 μL APEL medium per well for a final concentration of 20 ng/mL.
11. On day 3, prepare Geltrex-coated 6-well plates.
12. On day 4, plate EBs on Geltrex-coated wells (10 EBs/well) in 2 mL per well of APEL medium supplemented with VEGF (40 ng/mL), SCF (50 ng/mL), and IL-3 (50 ng/mL).
13. On day 7, change medium to APEL medium supplemented with IL-3 (50 ng/mL) and G-CSF (50 ng/mL). Hematopoietic cells start to appear on day 8.
14. On day 10, add 2 mL of APEL medium supplemented with IL-3 (50 ng/mL) and G-CSF (50 ng/mL).
15. On day 14, take floating cells for further analysis and exchange medium to the fresh APEL medium containing G-CSF (50 ng/mL) and IL-3 (50 ng/mL), 2 mL per well. At this stage, floating cells are mainly $CD34^+$ hematopoietic stem and progenitor cells that can be used for various downstream analyses such as colony-forming unit (CFU) assay, morphological analysis, or flow cytometry analysis of early hematopoietic markers of hematopoietic stem and progenitor cells (HSPCs) (*see* Subheading 3.6).
16. From day 14, flow cytometry analysis for hematopoietic stem cells (*see* Subheading 3.6) and mature myeloid cells (*see* Subheading 3.7) as well as cell counts of floating cells should be performed every 3–4 days to follow the myeloid maturation process. Floating cells mature stepwise from $CD34^+$ cells on day 14 to mature $CD15^+CD16^+$ neutrophils on day 28–32. The culture medium from day 14 to day 32 consists of APEL medium supplemented with IL-3 (50 ng/mL) and G-CSF (50 ng/mL), 2 mL medium per well.
17. On day 28 and day 32, mature neutrophils are analyzed using multicolor flow cytometry panel for mature myeloid cells (*see* Subheading 3.7 and **Note 11**).

Table 2
A list of antibodies used for the early panel FACS analysis

Antibody	Conjugation	Company	Volume in μL
CD33	BV421	BioLegend	2
CD45	BV510	BioLegend	2
KDR	AF647	BioLegend	2
CD41a	FITC	BD	0.5
CD235	FITC	BD	0.5
CD34	PE-Cy7	BD	5
CD43	PE	BD	5
7AAD		BD	2

3.6 Multicolor Flow Cytometry Analysis of the Early Stages of Hematopoietic Differentiation of iPSCs

Multicolor flow cytometry analysis for the evaluation of HSPCs in suspension cell fraction of EBs-based hematopoietic differentiation culture of iPSCs is performed on day 14 of culture.

1. Harvest 6×10^4 cells in 1 mL of ice-cold PBS and centrifuge for 5 min at $1200 \times g$. Usually, 3×10^4 cells are used for staining with specific antibody and 3×10^4 cells for unstained control.
2. Resuspend cells in 150 μL ice-cold FACS buffer.
3. Add 75 μL of cell suspension into each FACS tube on ice.
4. Add antibody as described in Table 2.
5. Incubate for 15 min on ice and protected from light.
6. Add 500 μL ice-cold FACS buffer to wash stained cells.
7. Centrifuge 5 min at $435 \times g$ and + 4 °C, and then discard supernatant.
8. Resuspend cell pellet in 100 μL ice-cold FACS buffer.
9. Evaluate stained cells on BD FACS Canto II. Representative images of flow cytometry are shown in Fig. 3.

3.7 Multicolor Flow Cytometry Analysis of the Late Stages of Myeloid Differentiation of iPSCs

Multicolor flow cytometry analysis of floating cells derived from EBs-based iPSCs culture in order to evaluate surface expression of myeloid markers is performed on day 28–32.

1. Harvest 6×10^4 cells in 1 mL of ice-cold PBS and centrifuge for 5 min at $12,000 \times g$. In addition, 3×10^4 cells are used for staining with specific antibody and 3×10^4 cells for unstained control.
2. Resuspend cells in 150 μL ice-cold FACS buffer.

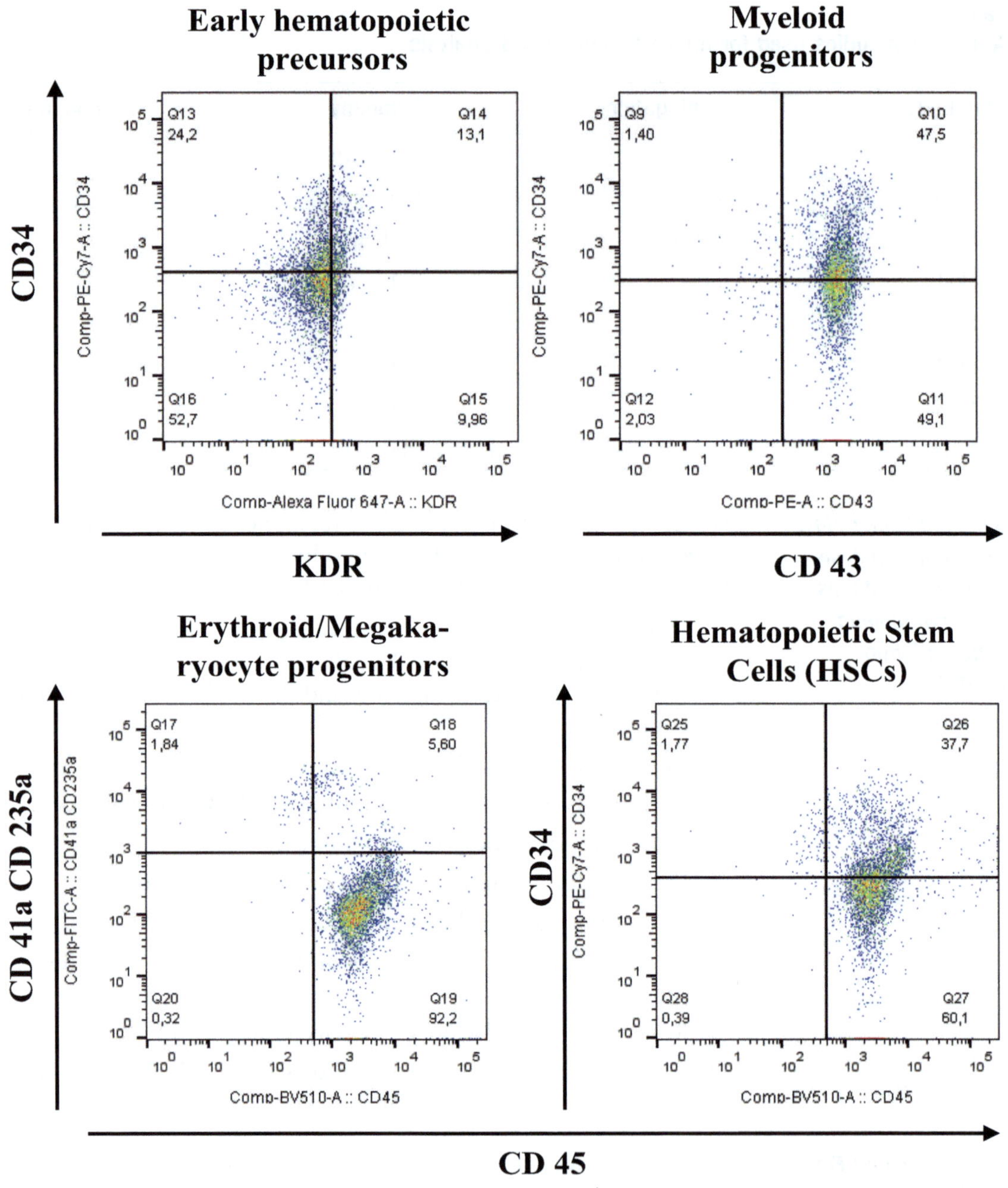

Fig. 3 Representative images of flow cytometry analysis of early stages of EB-based hematopoietic differentiation of healthy donor-derived iPSCs

3. Add 75 μL of cell suspension into each FACS tube to a final volume of 100 μL.
4. Add FACS antibodies as described in Table 3.
5. Incubate for 15 min on ice and protect from light.

Table 3
A list of antibodies used for the late panel FACS analysis

Antibody	Conjugation	Company	Volume in μL
CD33	BV421	BioLegend	2
CD45	BV510	BioLegend	2
CD15	PE	BD	5
CD16	FITC	BD	5
CD11b	PE-Cy7	BioLegend	2
CD14	APC-H7	BD	2
CD177	APC	BioLegend	0.5
7AAD		BD	2

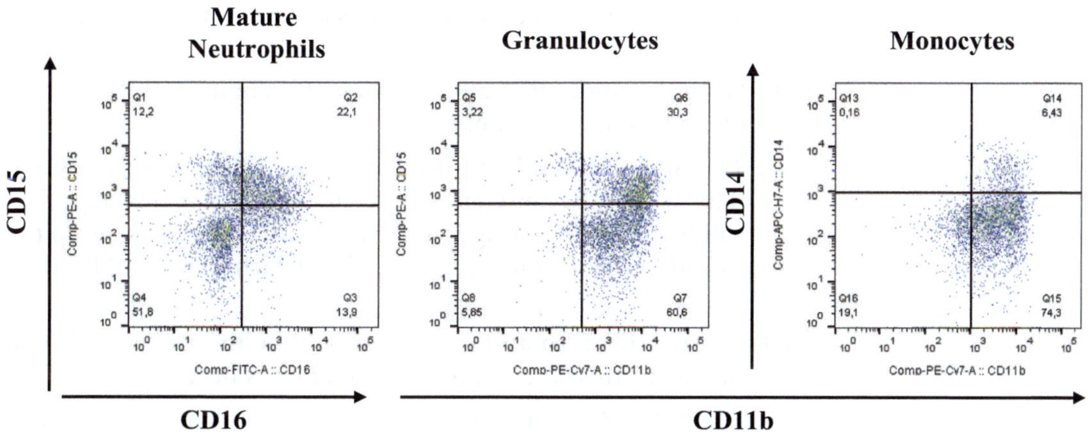

Fig. 4 Representative images of flow cytometry analysis of late stages of myeloid EB-based differentiation of healthy donor-derived iPSCs

6. Add 500 μL of ice-cold FACS buffer to wash stained cells.
7. Centrifuge 5 min at 435 × *g* and + 4 °C, and then discard supernatant.
8. Resuspend cell pellet in 100 μL of ice-cold FACS buffer.
9. Measure stained cells on BD FACS Canto II. Representative images of flow cytometry are shown in Fig. 4.

4 Notes

1. Most of the existing off-target prediction methods calculate scores based on the positions of the mismatches to the guide sequence [11]; a good off-target profile is a probability of

having no unintended cut with up to two mismatches (0:0, 1:0, 2:0). Sanger Institute webtool (https://www.sanger.ac.uk/htgt/wge/) can be used to assess the off-target profiles of selected sgRNAs.

2. Any RNP-compatible transfection reagent can be used at this step.
3. Cell number should be optimized between 5000 and 50,000 cells for each iPSC line.
4. The incubation time could be up to 14 days due to variations in different iPSC lines.
5. Keep the condition sterile and use a face mask. The probability of cell contamination at this step is high.
6. The number of colonies for picking can be adjusted based on the estimation of the gene modification efficiency in the total cell population.
7. The Geltrex-coated 96-well plates can be stored at 4 °C for up to 7 days.
8. Wash the tips in PBS after each use, to minimize the possibility of clonal cross-contamination.
9. This step could take up to 1 h in case of the high density of iPSC clones. Because diluted Accutase is used at this step, there will be no harm if longer incubation time is needed, but pipetting iPSC clones to accelerate the detachment could kill the cells.
10. To start EB formation, iPSCs should be more than 80% confluent. In this case, no feeder depletion is needed.
11. After day 32 of culture, the number of floating cells drops heavily.

Acknowledgments

This work was supported by the DFG (J.S.), the Fritz Thysen Foundation (B.D.), and the Jose Carreras Leukemia Foundation (B.D.).

References

1. Kim C (2015) iPSC technology-powerful hand for disease modeling and therapeutic screen. BMB Rep 48(5):256
2. Hockemeyer D, Jaenisch R (2016) Induced pluripotent stem cells meet genome editing. Cell Stem Cell 18(5):573–586
3. de Bruin O, Lisa M, Volpi S, Musunuru K (2015) Novel genome-editing tools to model and correct primary immunodeficiencies. Front Immunol 6:250
4. Dannenmann B et al (2019) Human iPSC-based model of severe congenital neutropenia reveals elevated UPR and DNA damage in CD34+ cells preceding leukemic transformation. Exp Hematol 71:51–60

5. Lachmann N, Ackermann M, Frenzel E et al (2015) Large-scale hematopoietic differentiation of human induced pluripotent stem cells provides granulocytes or macrophages for cell replacement therapies. Stem Cell Reports 4 (2):282–296
6. Nasri M et al (2019) Fluorescent labeling of CRISPR/Cas9 RNP for gene knockout in HSPCs and iPSCs reveals an essential role for GADD45b in stress response. Blood Adv 3 (1):63–71
7. Ng ES, Davis RP, Azzola L, Stanley EG, Elefanty AG et al (2005) Forced aggregation of defined numbers of human embryonic stem cells into embryoid bodies fosters robust, reproducible hematopoietic differentiation. Blood 106(5):1601–1603
8. Ng ES, Davis R, Stanley EG, Elefanty AG (2008) A protocol describing the use of a recombinant protein-based, animal product-free medium (APEL) for human embryonic stem cell differentiation as spin embryoid bodies. Nat Protoc 3(5):768–776
9. Brinkman EK et al (2014) Easy quantitative assessment of genome editing by sequence trace decomposition. Nucleic Acids Res 42 (22):e168
10. Hsiau T et al. (2018) Inference of CRISPR Edits from Sanger Trace Data. bioRxiv. 251082
11. Lin J, Wong K-C (2018) Off-target predictions in CRISPR-Cas9 gene editing using deep learning. Bioinformatics 34(17):i656–i663

Index

A

B

C

D

E

F

G

H

Mouldy Sioud (ed.), *RNA Interference and CRISPR Technologies: Technical Advances and New Therapeutic Opportunities*, Methods in Molecular Biology, vol. 2115, https://doi.org/10.1007/978-1-0716-0290-4,

MIX
Papier aus verantwortungsvollen Quellen
Paper from responsible sources
FSC® C105338

If you have any concerns about our products,
you can contact us on
ProductSafety@springernature.com

In case Publisher is established outside the EU,
the EU authorized representative is:
Springer Nature Customer Service Center GmbH
Europaplatz 3, 69115 Heidelberg, Germany

Printed by Libri Plureos GmbH
in Hamburg, Germany